Stochastic Partial Differential Equations with Lévy Noise

Recent years have seen an explosion of interest in stochastic partial differential equations where the driving noise is discontinuous. In this comprehensive monograph, two leading experts apply the evolution equation approach to the analysis of the solutions. Most of the results appear here for the first time in book form, and the volume is sure to stimulate further research in this important field.

The authors give a detailed analysis of Lévy processes in infinite dimensions and their reproducing kernel Hilbert spaces; cylindrical Lévy processes are constructed in terms of Poisson random measures; stochastic integrals are introduced. Stochastic parabolic and hyperbolic equations on domains of arbitrary dimensions are studied, and applications to statistical and fluid mechanics as well as to finance are also investigated.

Ideal for researchers and graduate students in stochastic processes and partial differential equations, this self-contained text will also be of interest to those working on stochastic modeling in finance, statistical physics and environmental science.

SZYMON PESZAT is an Associate Professor in the Institute of Mathematics at the Polish Academy of Sciences.

JERZY ZABCZYK is a Professor in the Institute of Mathematics at the Polish Academy of Sciences. He is an author (with G. Da Prato) of three earlier books for Cambridge University Press: *Stochastic Equations in Infinite Dimensions* (1992), *Ergodicity for Infinite Dimensional Systems* (1996) and *Second Order Partial Differential Equations in Hilbert Spaces* (2002), and several books for other publishers.

ENCYCLOPEDIA OF MATHEMATICS AND ITS APPLICATIONS

All the titles listed below can be obtained from good booksellers or from Cambridge University Press. For a complete series listing visit
http://www.cambridge.org/uk/series/sSeries.asp?code=EOM

50   F. Borceux   *Handbook of Categorical Algebra I*
51   F. Borceux   *Handbook of Categorical Algebra II*
52   F. Borceux   *Handbook of Categorical Algebra III*
53   V. F. Kolchin   *Random Graphs*
54   A. Katok and B. Hasselblatt   *Introduction to the Modern Theory of Dynamical Systems*
55   V. N. Sachkov   *Combinatorial Methods in Discrete Mathematics*
56   V. N. Sachkov   *Probablistic Methods in Discrete Mathematics*
57   P. M. Cohn   *Skew Fields*
58   R. Gardner   *Geometric Tomography*
59   G. A. Baker, Jr., and P. Graves-Morris   *Padé Approximants, 2nd edn*
60   J. Krajicek   *Bounded Arithmetic, Propositional Logic and Complexity Theory*
61   H. Groemer   *Geometric Applications of Fourier Series and Spherical Harmonics*
62   H. O. Fattorini   *Infinite Dimensional Optimization and Control Theory*
63   A. C. Thompson   *Minkowski Geometry*
64   R. B. Bapat and T. E. S. Raghavan   *Nonnegative Matrices with Applications*
65   K. Engel   *Sperner Theory*
66   D. Cvetkovic, P. Rowlinson and S. Simic   *Eigenspaces of Graphs*
67   F. Bergeron, G. Labelle and P. Leroux   *Combinatorial Species and Tree-Like Structures*
68   R. Goodman and N. Wallach   *Representations and Invariants of the Classical Groups*
69   T. Beth, D. Jungnickel and H. Lenz   *Design Theory I, 2nd edn*
70   A. Pietsch and J. Wenzel   *Orthonormal Systems for Banach Space Geometry*
71   G. E. Andrews, R. Askey and R. Roy   *Special Functions*
72   R. Ticciati   *Quantum Field Theory for Mathematicians*
73   M. Stern   *Semimodular Lattices*
74   I. Lasiecka and R. Triggiani   *Control Theory for Partial Differential Equations I*
75   I. Lasiecka and R. Triggiani   *Control Theory for Partial Differential Equations II*
76   A. A. Ivanov   *Geometry of Sporadic Groups I*
77   A. Schinzel   *Polynomials with Special Regard to Reducibility*
78   H. Lenz, T. Beth and D. Jungnickel   *Design Theory II, 2nd edn*
79   T. Palmer   *Banach Algebras and the General Theory of * -Algebras II*
80   O. Stormark   *Lie's Structural Approach to PDE Systems*
81   C. F. Dunkl and Y. Xu   *Orthogonal Polynomials of Several Variables*
82   J. P. Mayberry   *The Foundations of Mathematics in the Theory of Sets*
83   C. Foias *et al.*   *Navier–Stokes Equations and Turbulence*
84   B. Polster and G. Steinke   *Geometries on Surfaces*
85   R. B. Paris and D. Karminski   *Asymptotics and Mellin–Barnes Integrals*
86   R. McEliece   *The Theory of Information and Coding, 2nd edn*
87   B. Magurn   *Algebraic Introduction to K-Theory*
88   T. Mora   *Solving Polynomial Equation Systems I*
89   K. Bichteler   *Stochastic Integration with Jumps*
90   M. Lothaire   *Algebraic Combinatorics on Words*
91   A. A. Ivanov and S. V. Shpectorov   *Geometry of Sporadic Groups II*
92   P. McMullen and E. Schulte   *Abstract Regular Polytopes*
93   G. Gierz *et al.*   *Continuous Lattices and Domains*
94   S. Finch   *Mathematical Constants*
95   Y. Jabri   *The Mountain Pass Theorem*
96   G. Gasper and M. Rahman   *Basic Hypergeometric Series, 2nd edn*
97   M. C. Pedicchio and W. Tholen (eds.)   *Categorical Foundations*
98   M. E. H. Ismail   *Classical and Quantum Orthogonal Polynomials in One Variable*
99   T. Mora   *Solving Polynomial Equations Systems II*
100 E. Olivieri and M. Eulália Vares   *Large Deviations and Metastability*
102 L. W. Beineke *et al.* (eds.)   *Topics in Algebraic Graph Theory*
103 O. Staffans   *Well-Posed Linear Systems*
105 M. Lothaire   *Applied Combinatorics on Words*
106 A. Markoe   *Analytic Tomography*
107 P. A. Martin   *Multiple Scattering*
108 R. A. Brualdi   *Combinatorial Matrix Classes*
110 M.-J. Lai and L. L. Schumaker   *Spline Functions on Triangulations*
111 R. T. Curtis   *Symmetric Generation of Groups*
112 H. Salzmann, T. Grundhöfer, H. Hähl and R. Löwen   *The Classical Fields*

Stochastic Partial Differential Equations with Lévy Noise

An Evolution Equation Approach

S. PESZAT and J. ZABCZYK

Institute of Mathematics,
Polish Academy of Sciences

CAMBRIDGE
UNIVERSITY PRESS

CAMBRIDGE UNIVERSITY PRESS
Cambridge, New York, Melbourne, Madrid, Cape Town, Singapore, São Paulo

Cambridge University Press
The Edinburgh Building, Cambridge CB2 8RU, UK

Published in the United States of America by Cambridge University Press, New York

www.cambridge.org
Information on this title: www.cambridge.org/9780521879897

© Szymon Peszat and Jerzy Zabczyk 2007

First published 2007

Printed in the United Kingdom at the University Press, Cambridge

A catalogue record for this publication is available from the British Library

ISBN   978-0-521-87989-7   hardback

Contents

Preface   *page* ix

Part I   Foundations   1

1 Why equations with Lévy noise?   3
   1.1   Discrete-time dynamical systems   3
   1.2   Deterministic continuous-time systems   5
   1.3   Stochastic continuous-time systems   6
   1.4   Courrège's theorem   8
   1.5   Itô's approach   9
   1.6   Infinite-dimensional case   12
2 Analytic preliminaries   13
   2.1   Notation   13
   2.2   Sobolev and Hölder spaces   13
   2.3 L^p- and C_ρ-spaces   15
   2.4   Lipschitz functions and composition operators   16
   2.5   Differential operators   17
3 Probabilistic preliminaries   20
   3.1   Basic definitions   20
   3.2   Kolmogorov existence theorem   22
   3.3   Random elements in Banach spaces   23
   3.4   Stochastic processes in Banach spaces   25
   3.5   Gaussian measures on Hilbert spaces   28
   3.6   Gaussian measures on topological spaces   30
   3.7   Submartingales   31
   3.8   Semimartingales   36
   3.9   Burkholder–Davies–Gundy inequalities   37

4 Lévy processes   38
   4.1 Basic properties   38
   4.2 Two building blocks – Poisson and Wiener processes   40
   4.3 Compound Poisson processes in a Hilbert space   45
   4.4 Wiener processes in a Hilbert space   50
   4.5 Lévy–Khinchin decomposition   52
   4.6 Lévy–Khinchin formula   56
   4.7 Laplace transforms of convolution semigroups   57
   4.8 Expansion with respect to an orthonormal basis   62
   4.9 Square integrable Lévy processes   65
   4.10 Lévy processes on Banach spaces   72
5 Lévy semigroups   75
   5.1 Basic properties   75
   5.2 Generators   78
6 Poisson random measures   83
   6.1 Introduction   83
   6.2 Stochastic integral of deterministic fields   85
   6.3 Application to construction of Lévy processes   87
   6.4 Moment estimates in Banach spaces   90
7 Cylindrical processes and reproducing kernels   91
   7.1 Reproducing kernel Hilbert space   91
   7.2 Cylindrical Poisson processes   100
   7.3 Compensated Poisson measure as a martingale   105
8 Stochastic integration   107
   8.1 Operator-valued angle bracket process   107
   8.2 Construction of the stochastic integral 111
   8.3 Space of integrands   114
   8.4 Local properties of stochastic integrals 117
   8.5 Stochastic Fubini theorem   118
   8.6 Stochastic integral with respect to a Lévy process   121
   8.7 Integration with respect to a Poisson random measure   125
   8.8 L^p-theory for vector-valued integrands   130

Part II   Existence and Regularity   137

9 General existence and uniqueness results   139
   9.1 Deterministic linear equations   139
   9.2 Mild solutions   141
   9.3 Equivalence of weak and mild solutions   148
   9.4 Linear equations   155

9.5	Existence of weak solutions	164
9.6	Markov property	167
9.7	Equations with general Lévy processes	170
9.8	Generators and a martingale problem	174
10	**Equations with non-Lipschitz coefficients**	**179**
10.1	Dissipative mappings	179
10.2	Existence theorem	183
10.3	Reaction–diffusion equation	187
11	**Factorization and regularity**	**190**
11.1	Finite-dimensional case	190
11.2	Infinite-dimensional case	193
11.3	Applications to time continuity	197
11.4	The case of an arbitrary martingale	199
12	**Stochastic parabolic problems**	**201**
12.1	Introduction	201
12.2	Space–time continuity in the Wiener case	208
12.3	The jump case	214
12.4	Stochastic heat equation	219
12.5	Equations with fractional Laplacian and stable noise	223
13	**Wave and delay equations**	**225**
13.1	Stochastic wave equation on $[0, 1]$	225
13.2	Stochastic wave equation on $\mathbb{R}^d$ driven by impulsive noise	230
13.3	Stochastic delay equations	238
14	**Equations driven by a spatially homogeneous noise**	**240**
14.1	Tempered distributions	240
14.2	Lévy processes in $S'(\mathbb{R}^d)$	241
14.3	RKHS of a square integrable Lévy process in $S'(\mathbb{R}^d)$	242
14.4	Spatially homogeneous Lévy processes	246
14.5	Examples	248
14.6	RKHS of a homogeneous noise	253
14.7	Stochastic equations on $\mathbb{R}^d$	255
14.8	Stochastic heat equation	256
14.9	Space–time regularity in the Wiener case	261
14.10	Stochastic wave equation	267
15	**Equations with noise on the boundary**	**272**
15.1	Introduction	272
15.2	Weak and mild solutions	275
15.3	Analytical preliminaries	277
15.4	L^2 case	279
15.5	Poisson perturbation	282

Part III   Applications   285

16   Invariant measures   287
  16.1   Basic definitions   287
  16.2   Existence results   289
  16.3   Invariant measures for the reaction–diffusion equation   297

17   Lattice systems   299
  17.1   Introduction   299
  17.2   Global interactions   300
  17.3   Regular case   303
  17.4   Non-Lipschitz case   305
  17.5   Kolmogorov's formula   306
  17.6   Gibbs measures   307

18   Stochastic Burgers equation   312
  18.1   Burgers system   312
  18.2   Uniqueness and local existence of solutions   314
  18.3   Stochastic Burgers equation with additive noise   317

19   Environmental pollution model   322
  19.1   Model   322

20   Bond market models   324
  20.1   Forward curves and the HJM postulate   324
  20.2   HJM condition   327
  20.3   HJMM equation   332
  20.4   Linear volatility   342
  20.5   BGM equation   347
  20.6   Consistency problem   350

Appendix A   Operators on Hilbert spaces   355
Appendix B C_0-semigroups   365
Appendix C   Regularization of Markov processes   388
Appendix D   Itô formulae   391
Appendix E   Lévy–Khinchin formula on $[0, +\infty)$ 394
Appendix F   Proof of Lemma 4.24   396
List of symbols   399
References   403
Index   415

Preface

This book is an introduction to the theory of stochastic evolution equations with Lévy noise. The theory extends several results known for stochastic partial differential equations (SPDEs) driven by Wiener processes. We develop a general framework and discuss several classes of examples both with general Lévy noise and with Wiener noise. Our approach is functional analytic and, as in Da Prato and Zabczyk (1992a), SPDEs are treated as ordinary differential equations in infinite-dimensional spaces with irregular coefficients. In many respects the Lévy noise theory is similar to that for Wiener noise, especially when the driving Lévy process is a square integrable martingale. The general case reduces to this, owing to the Lévy–Khinchin decomposition. The functional analytic approach also allows us to treat equations with a so-called cylindrical Lévy noise and implies, almost automatically, that solutions to equations with Lévy noise are Markovian. In some important cases, however, a càdlàg version of the solution does not exist.

An important role in our approach is played by a generalization of the concept of the reproducing kernel Hilbert space to non-Gaussian random variables, and its independence of the space in which the random variable takes values. In some cases it proves useful to treat Poissonian random measures, with respect to which many SPDEs have been studied, as Lévy processes properly defined in appropriate state spaces.

The majority of the results appear here for the first time in book form, and the book presents several completely new results not published previously, in particular, for equations driven by homogeneous noise and for dissipative systems.

Several monographs have been devoted to stochastic ordinary differential equations driven by discontinuous noise: see Métivier (1982), Protter (2005), Applebaum (2005) and Cont and Tankov (2004), the last two of which are devoted entirely to the case of Lévy noise.

To the best of our knowledge the only monograph devoted to SPDEs with general noise is Kallianpur and Xiong (1995), which covers mainly linear equations.

The papers by Chojnowska-Michalik (1987) on Ornstein–Uhlenbeck processes and by Kallianpur and Pérez-Abreu (1988) were the first to discuss SPDEs with Lévy noise. Then, after a period of 10 years, articles on the subject started to appear again; see e.g. Albeverio, Wu and Zhang (1998), Applebaum and Wu (2000), Bo and Wang (2006), Fournier (2000, 2001), Fuhrman and Röckner (2000), Hausenblas (2005), Mueller (1998), Mytnik (2002), Knoche (2004), Saint Loubert Bié (1998), Stolze (2005) and Peszat and Zabczyk (2006).

Infinite-dimensional calculus, but not SPDEs, with Lévy noise in a disguised form appeared in the late 1990s, in papers on mathematical finance devoted to the bond market; see Björk *et al.* (1997), Björk, Kabanov and Runggaldier (1997) and Eberlein and Raible (1999). This list is certainly not exhaustive.

The book starts with an introductory chapter outlining the interplay between stochastic dynamical systems, Markov processes and stochastic equations with Lévy noise. It turns out that all discrete-time stochastic dynamical systems on arbitrary linear Polish spaces can be represented as solutions to stochastic difference equations, in which the noise, of random-walk type, enters the equation linearly. An analogous situation occurs for continuous-time stochastic dynamical systems on $\mathbb{R}^d$. Here the noise is a Lévy process and the stochastic difference equation is replaced by a stochastic equation of Itô type. To some extent this is also true in infinite dimensions. That is why stochastic evolution equations with Lévy noise are of particular interest.

Chapters 2 and 3 are devoted respectively to analytic and probabilistic preliminaries. Basic definitions related to differential operators and function spaces are recalled, together with fundamental concepts from the theory of stochastic processes in finite- and infinite-dimensional spaces.

Lévy processes in infinite-dimensional spaces are studied in Chapter 4. The chapter starts with explicit constructions of Wiener and Poisson processes. Then it deals with the Lévy–Khinchin decomposition and the Lévy–Khinchin formula. Integrability properties are also studied.

In Chapter 5 transition semigroups of Lévy processes are considered and, in particular, their generators.

The important concept of a Poisson random measure is discussed in Chapter 6. An application to the construction of Lévy processes is given as well. Some moment estimates are derived.

In Chapter 7 we introduce the concept of the reproducing kernel Hilbert space (RKHS) of a square integrable Lévy process. Then we study so-called cylindrical processes and calculate their reproducing kernels. It is also shown that Poisson random measures can be treated as Lévy processes with values in sufficiently large Hilbert spaces. This identification is behind the majority of the results in this book. It is also shown that cylindrical processes are distributional derivatives of Lévy sheets.

Chapter 8 concerns stochastic integration, first with respect to square integrable Hilbert-space-valued martingales and, as an application, with respect to general Lévy processes. The construction of the operator angle bracket is explained and a class of integrands is characterized. The final sections are devoted to integration with respect to a Poisson measure and integration in L^p-spaces.

Part II of the book deals with the existence of solutions and their regularity. Chapter 9 starts with a semigroup treatment of the Cauchy problem for deterministic evolution equations. Next, weak solutions and mild solutions to stochastic equations are introduced and their equivalence is established. The existence of weak solutions to linear equations is proven as well. If the noise evolves in the state space then càdlàg regularity of the solution is proved, using the Kotelenez maximal inequality and, in parallel, a dilation theorem, as in Hausenblas and Seidler (2006). We provide an example which shows that the solutions are not càdlàg in general. Finally, the existence of weak solutions is established and their Markov property is proved.

In Chapter 10 we show that in some cases the Lipschitz assumption can be relaxed.

Chapter 11 is devoted to the so-called factorization method, introduced in Da Prato, Kwapień and Zabczyk (1987). The method allows us, in particular, to prove the continuity of the stochastic convolution of an arbitrary semigroup with a Wiener process.

In Chapters 12 and 13, the general theory of the previous chapters is applied to stochastic parabolic problems and to stochastic wave, delay and transport equations. Lévy noise is then treated as a Lévy process in extended, Sobolev-type, spaces. Parabolic equations of a similar type were dealt with in, for example, Albeverio, Wu and Zhang (1998), Saint Loubert Bié (1998) and Applebaum and Wu (2000). They are discussed here in a unified way and for general partial differential operators. Sharp regularity results for the Wiener noise, using the factorization method, are obtained as well. Stochastic wave equations driven by Lévy processes have not been studied previously in the literature.

In Chapter 14 we develop a theory of stochastic equations with spatially homogenous Lévy noise of both parabolic and hyperbolic type on $\mathbb{R}^d$. Results known already for Wiener noise are extended to this more involved case.

The final chapter of the second part of the book, Chapter 15, is devoted to equations in which the noise enters through the boundary.

Part III is devoted to selected applications. Our aim is to show the applicability of the theory to specific models of physical and economic character. In particular, models in statistical mechanics, fluid dynamics and finance are studied in greater detail. In Chapter 16 we give a self-contained treatment of invariant measures for dissipative systems with Lévy noise and in Chapter 17 we consider lattice systems. The Burgers equation is studied in Chapter 18, and, after a brief discussion of a model for environmental pollution in Chapter 19, in Chapter 20 we present some applications of stochastic infinite-dimensional analysis to mathematical models of the bond market.

In Appendix A we give proofs of some results on linear operators often used in the text. In Appendix B we gather basic results on the theory of C_0-semigroups and provide important results on specific semigroups used in the book. Special attention is paid to semigroups with non-local generators. This allows us to extend the results proved for equations with differential operators. However, owing to space limitations, stochastic equations with non-local linear parts are not discussed in this book. In Appendix C the existence of càdlàg versions of Markov processes is proved. This leads to a simple proof of the existence of a càdlàg version of an arbitrary Lévy process. In Appendix D we recall the Itô formulae for semimartingales.

A list of symbols is given before the Index.

This book grew out of lectures and papers presented by the authors. We have used some material from Peszat and Zabczyk (2004) as well as many unpublished notes.

Acknowledgements   We thank J. Jakubowski, J. Noble, and J. Trzeciak for reading some parts of the manuscript and for indicating several mistakes and misprints. We also thank J. K. Kowalski for typing several sections of the text and the anonymous referees for useful remarks and comments. Finally, we would like to thank S. Parkinson and the Cambridge University Press staff for all their work on the typescript.

Financial support from the Polish Ministry of Education and Science via grant 1 PO3A 034 29 is gratefully acknowledged. The authors are also thankful for very good working conditions at their home institution, the Institute of Mathematics, in the Polish Academy of Sciences.

Part I

Foundations

1

Why equations with Lévy noise?

The book is devoted to stochastic evolution equations with Lévy noise. Such equations are important because, roughly speaking, stochastic dynamical systems, or equivalently Markov processes, can be represented as solutions to such equations. In this introductory chapter, it is shown how that is the case. To motivate better the construction of the associated stochastic equations, the chapter starts with discrete-time systems.

1.1 Discrete-time dynamical systems

A deterministic discrete-time dynamical system consists of a set E, usually equipped with a σ-field $\mathcal{E}$ of subsets of itself, and a mapping F, usually measurable, acting from E into E. If the position of the system at time $t = 0, 1, \ldots$, is denoted by $X(t)$ then by definition $X(t + 1) = F(X(t))$, $t = 0, 1, \ldots$ The sequences $(X(t), \ t = 0, 1, \ldots)$ are the so-called *trajectories* or *paths* of the dynamical system, and their asymptotic properties are of prime interest in the theory. The set E is called the *state space* and the transformation F determines the dynamics of the system.

If the present state x determines only the probability $P(x, \Gamma)$ that at the next moment the system will be in the set Γ then one says that the system is stochastic. Thus a *stochastic dynamical system* consists of the state space E, a σ-field $\mathcal{E}$ and a function $P = P(x, \Gamma)$, $x \in E$, $\Gamma \in \mathcal{E}$, such that, for each $\Gamma \in \mathcal{E}$, $P(\cdot, \Gamma)$ is a measurable function and, for each $x \in E$, $P(x, \cdot)$ is a probability measure. We call P the *transition function* or *transition probability*. A deterministic system is a particular case of a stochastic system with $P(x, \cdot) = \delta_{F(x)}$, where δ_r denotes the Dirac measure at r. We define, by induction, the *probability of visiting* sets

3

$\Gamma_1, \dots, \Gamma_k$ at times $1, \dots, k$, starting from x by

$$P(x, \Gamma_1, \dots, \Gamma_k) = \int_{\Gamma_1} P(x, dx_1) P(x_1, \Gamma_2, \dots, \Gamma_k).$$

The stochastic analogue of the trajectory of a deterministic dynamical system is called a Markov chain.

Definition 1.1   A *Markov chain* X with transition probability P starting from $x \in E$ is a sequence $(X(t), t = 0, 1, \dots)$ of E-valued random variables on a probability space $(\Omega, \mathcal{F}, \mathbb{P})$, such that

(i) $X(0) = x,$ $\mathbb{P}$-a.s.,
(ii) $\mathbb{P}\big(X(j) \in \Gamma_j, j = 1, \dots, k\big) = P(x, \Gamma_1, \dots, \Gamma_k), \quad \forall \Gamma_1, \dots, \Gamma_k \in \mathcal{E}.$

Let P be a transition probability on a Polish space[1] E. By the Kolmogorov existence theorem (see Theorem 3.7), there is a probability space and a Markov chain with transition probability P. It turns out that an arbitrary stochastic dynamical system on a Polish space can be regarded as a solution of the stochastic difference equation

$$X(0) = x, \qquad X(t+1) = F(X(t), Z(t+1)), \qquad t = 0, 1, \dots, \qquad (1.1)$$

where $Z(1), Z(2), \dots$ is a sequence of independent identically distributed random variables (i.i.d.s). In the engineering literature, a sequence of this type is called *discrete-time white noise*. We have the following representation result.

Theorem 1.2   *Let E be a Polish space, and let $\mathcal{E} = \mathcal{B}(E)$ be the family of its Borel sets. Then, for any transition probability P, there exists a measurable mapping $F: E \times [0, 1] \mapsto E$ such that, for any sequence of independent random variables $Z(1), Z(2), \dots$ with uniform distribution on $[0, 1]$ and for any $x \in E$, the process X given by (1.1) is a Markov chain with transition probability P.*

Proof   We follow Kifer (1986). First we construct F in the case where E is a countable set and $E = \mathbb{R}$. Let $E = \{1, 2, \dots\} = \mathbb{N}$ and let $p_{n|m} = P(n, \{1\}) + \dots + P(n, \{m\})$, $n, m \in \mathbb{N}$. Define $F(n, r) = m$ for $r \in [p_{n|m-1}, p_{n|m})$. Measurability is obvious. If Z has a uniform distribution on $[0, 1]$ then $\mathbb{P}(\{\omega : F(n, Z(\omega)) = m\}) = P(n, \{m\})$, as required.

If $E = \mathbb{R}$ then we first define $F_0(x, a) := P(x, (-\infty, a])$ for $x \in \mathbb{R}$, $a \in \mathbb{R}$, and set $F(x, r) := \inf\{a : r \le F_0(x, a)\}, r \in [0, 1]$. It is clear that if Z has a uniform distribution on $[0, 1]$ then $\mathbb{P}(\{\omega : F(x, Z(\omega)) \in \Gamma\}) = P(x, \Gamma)$ for $x \in \mathbb{R}$ and $\Gamma \in \mathcal{B}(\mathbb{R})$. By the so-called *Borel isomorphism theorem*, actually due to K. Kuratowski (see Kuratowski 1934, Dynkin and Yushkevich 1978 or Srivastava

---

[1] That is, a separable metric space that is complete with respect to some equivalent metric.

1998), any uncountable Polish space is measurably isomorphic to $\mathbb{R}$ and the result follows. □

Remark 1.3 The result is a version of the famous Skorokhod embedding theorem. A similar theorem, but for controlled stochastic dynamical systems, can be found in Zabczyk (1996), pp. 26–7.

Once the result is proved for complete separable metric spaces E, it can be generalized to all spaces E that are measurably isomorphic to such spaces, that is, to all Borel spaces.

If E is a linear space it is convenient to reformulate (1.1) slightly; namely, we write

$$dX(t) := X(t) - X(t-1), \qquad Y(t) := Z(1) + \cdots + Z(t),$$
$$dY(t) := Y(t) - Y(t-1)$$

and change the function F to $\widetilde{F}(x, r) := -x + F(x, r)$. Then

$$dX(t) = \widetilde{F}(X(t-1), dY(t)).$$

Considering an embedding $r \mapsto \delta_r$ of the interval $[0, 1]$ into the space of all finite measures on $[0, 1]$, we arrive at an equation in which the noise enters linearly. Namely, we set $G(x)\lambda := \int_{[0,1]} \tilde{F}(x, r)\lambda(dr)$ and $\widetilde{Z}(t) := \delta_{Z(t)}$, $L(t) := \widetilde{Z}(1) + \ldots + \widetilde{Z}(t)$, $dL(t) := \widetilde{Z}(t)$. Then

$$dX(t) = G(X(t-1))\, dL(t), \qquad t = 1, 2, \ldots, \qquad (1.2)$$

and the increments of L are independent. Thus the diffusion operator G acts on the increments of the noise in a linear way. An analogous result holds for continuous-time stochastic dynamical systems with values in $\mathbb{R}^d$; see Section 1.5. The main ideas come from Itô (1951).

1.2 Deterministic continuous-time systems

Deterministic continuous-time dynamical systems are families $(F_t, \ t \geq 0)$ of transformations from a given state space E into E satisfying the semigroup property $F_t F_s = F_{t+s}, t, s \geq 0$. The trajectory starting from x is the mapping $X(t) = F_t(x)$ of the parameter t. Are the dynamical systems always solutions of differential equations? The answer is obviously no! Differential equations are well defined only on rather special state spaces. Even if we assume that the state space is $E = \mathbb{R}^d$ and that the transformation $(t, x) \mapsto F_t(x)$ is continuous, there are still dynamical systems not defined by differential equations, as the following example shows.

Example 1.4   Consider a continuous but nowhere differentiable function $f : \mathbb{R} \mapsto \mathbb{R}$ and define

$$F_t \begin{pmatrix} x \\ y \end{pmatrix} = \begin{pmatrix} x + t \\ f(x+t) + y - f(x) \end{pmatrix}, \qquad t \geq 0, \ x, y \in \mathbb{R}.$$

Then the trajectories of F_t are nowhere differentiable and consequently (F_t) cannot be a solution to an equation of the form $\mathrm{d}F_t(z)/\mathrm{d}t = A(F_t(z))$, where $A : \mathbb{R}^2 \mapsto \mathbb{R}^2$.

However, if $E = \mathbb{R}^d$ and all trajectories of a given dynamical system are continuously differentiable then they are solutions of the ordinary differential equation $\mathrm{d}X(t)/\mathrm{d}t = A(X(t))$, where $A(x) := \lim_{t \downarrow 0}(1/t)(F_t(x) - x)$.

If E is an infinite-dimensional space then the answer can again be positive provided that the flow is not too pathological.

Example 1.5   Assume that E is a Banach space and that, for each $t \geq 0$, F_t is a continuous linear transformation on E and, for each $x \in E$, $t \mapsto F_t(x)$ is a continuous mapping. Then in a proper sense $\mathrm{d}F_t(x)/\mathrm{d}t = A(F_t(x))$, $t > 0$, where A is usually an unbounded linear operator on E. In fact (F_t) is a C_0-semigroup and A is its generator; see Chapter 9 and Appendix B. Often we write $F_t = \mathrm{e}^{At}$.

Example 1.6   Assume that E is a Hilbert space, that the transformations F_t are *contractions*, i.e. $|F_t(x) - F_t(y)| \leq |x - y|$ for $t \geq 0$ and $x, y \in E$, and that, for each x, $t \mapsto F_t(x)$ is a continuous function. Then $\mathrm{d}F_t(x)/\mathrm{d}t = A(F_t(x))$, $t > 0$, where A is a so-called dissipative, usually unbounded and non-linear, operator; see Chapter 10.

A differential equation does not always uniquely determine the flow of its solutions. There are many subtleties here and interesting results; see for instance Hartman (1964).

1.3 Stochastic continuous-time systems

In analogy to discrete-time dynamical systems, a *stochastic continuous-time dynamical system* is a family (P_t) of stochastic kernels $P_t(x, \Gamma), t \geq 0, x \in E, \Gamma \in \mathcal{E}$. We interpret $P_t(x, \Gamma)$ as the probability that the system will be in a set Γ at time t, provided that its initial position is x. More precisely, we have the following definition.

Definition 1.7   A family of probability measures $P_t(x, \cdot)$ on E is said to be a *transition probability* if:

(i) for each $x \in E$, $P_0(x, \cdot) = \delta_x$;
(ii) for all $\Gamma \in \mathcal{E}$ and $t \geq 0$, the function $E \ni x \mapsto P_t(x, \Gamma) \in \mathbb{R}$ is measurable;
(iii) the family satisfies the *Chapman–Kolmogorov equation*

$$P_{t+s}(x, \Gamma) = \int_E P_t(x, dy) P_s(y, \Gamma), \qquad \forall t, s \geq 0, \ \forall \Gamma \in \mathcal{E}.$$

Note that the transition function of a deterministic dynamical system is given by $P_t(x, \cdot) = \delta_{F_t(x)}$. The transition function defines the *Markov* or *transition semigroup of operators* acting on the space $B_b(E)$ of all bounded measurable functions on E by the formula

$$P_t \varphi(x) := \int_E P_t(x, dy) \varphi(y), \qquad x \in E, \ t \geq 0, \ \varphi \in B_b(E). \tag{1.3}$$

Often the transition function and transition semigroups will be denoted in the same way, by (P_t). Note that the semigroup property $P_t P_s = P_{t+s}$, $t, s \geq 0$, is a consequence of the Chapman–Kolmogorov equation.

A *stochastic realization of a transition function* is a stochastic process $(X(t), t \geq 0)$, with values in E and defined on a fixed probability space $(\Omega, \mathcal{F}, \mathbb{P})$, such that

$$\mathbb{P}(\{\omega \colon X(\omega, t_j) \in \Gamma_j, \ j = 1, \ldots, k\}) = P^{t_1, \ldots, t_k}(x, \Gamma_1, \ldots, \Gamma_k),$$

where, similarly to the discrete-time case,

$$P^{t_1, \ldots, t_k}(x, \Gamma_1, \ldots, \Gamma_k) = \int_{\Gamma_1} P_{t_1}(x, dx_1) P^{t_2 - t_1, \ldots, t_k - t_{k-1}}(x_1, \Gamma_2, \ldots, \Gamma_k).$$

The process X is *Markov with transition function* (P_t); see also Definition 3.22. As in the discrete-time case, if E is a Polish space then, by the Kolmogorov existence theorem, for any transition function (P_t) there is a stochastic realization.

Following the discrete-time case, one can ask whether a stochastic realization of a given transition probability can be constructed as a solution of the stochastic equation

$$dX(t) = G(X(t-)) \, dL(t), \qquad t \geq 0, \ X(0) = x, \tag{1.4}$$

understood as the integral equation

$$X(t) = x + \int_0^t G(X(t-)) \, dL(t), \qquad t \geq 0,$$

where L is a process with independent increments and G is a transformation from the *noise space* into the state space. The answer was given in Itô (1951) and is, basically, yes! Since the noise enters linearly we can expect that, as in the discrete-time case, the noise space will be rather large and, in general, will be an infinite-dimensional space of measures. The process L might have jumps and this is why it is necessary to consider the left limit $X(t-)$ in the equation rather than $X(t)$. This is also required for solvability of the equation; see subsection 9.2.1.

1.4 Courrège's theorem

Assume now that $E = \mathbb{R}^d$. Then we have the following theorem from Courrège (1965/66). In its formulation $C_0(\mathbb{R}^d)$ and $C_0^\infty(\mathbb{R}^d)$ denote, respectively, the space of continuous functions vanishing at infinity[2] and the space of infinitely differentiable functions vanishing at infinity with all their derivatives. We denote by $M_s^+(d \times d)$ the space of all symmetric non-negative-definite $d \times d$ matrices and by D and Tr the Fréchet derivative (i.e. the gradient) and the trace operators, respectively. We denote by χ_Γ the indicator function of a set Γ and by $\langle \cdot, \cdot \rangle$ the scalar product on $\mathbb{R}^d$, with corresponding norm $| \cdot |$. Finally, $a \wedge b$ denotes the minimum of two numbers a and b; see Section 2.1.

Theorem 1.8 (Courrège) *Let P be a transition semigroup such that, for $\varphi \in C_0(\mathbb{R}^d)$, $P_t\varphi \in C_0(\mathbb{R}^d)$ and $P_t\varphi \to \varphi$ uniformly as $t \to 0$. In addition, for all $\varphi \in C_0^\infty(\mathbb{R}^d)$ and $x \in \mathbb{R}^d$, let the function $t \mapsto P_t\varphi(x)$ be differentiable. Then there exist transformations $a \colon \mathbb{R}^d \mapsto \mathbb{R}^d$ and $Q \colon \mathbb{R}^d \mapsto M_s^+(d \times d)$ and a family $v(x, \cdot)$, $x \in \mathbb{R}^d$, of measures concentrated on $\mathbb{R}^d \setminus \{0\}$ and satisfying*

$$\int_{\mathbb{R}^d} \big(|y|^2 \wedge 1\big) v(x, dy) < \infty, \qquad \forall x \in \mathbb{R}^d,$$

such that for all $\varphi \in C_0^\infty(\mathbb{R}^d)$ and $x \in \mathbb{R}^d$

$$
\begin{aligned}
A\varphi(x) :&= \lim_{t \downarrow 0} \frac{P_t\varphi(x) - \varphi(x)}{t} \\
&= \langle a(x), D\varphi(x) \rangle + \tfrac{1}{2}\mathrm{Tr}\, Q(x)D^2\varphi(x) \\
&\quad + \int_{\mathbb{R}^d} \big(\varphi(x+y) - \varphi(x) - \chi_{[0,1]}(|y|)\langle y, D\varphi(x)\rangle\big) v(x, dy).
\end{aligned}
$$

In subsection 5.2.4 we will prove the Courrège theorem in the particular case of a translation-invariant semigroup (P_t), that is, for the so-called *Lévy semigroup*. Namely, we will show in Theorems 4.27 and 5.3 that if

$$P_t(x, \Gamma) = P_t(x + h, \Gamma + h) \qquad \forall x, h \in \mathbb{R}^d, \ \forall t \geq 0, \ \forall \Gamma \in \mathcal{B}(\mathbb{R}^d)$$

then the functions a and Q are constant and all the measures $v(x, \cdot)$ are identically equal to a certain measure v. The process corresponding to the triple (a, Q, v) is

---

[2] That is, having the limit 0 at infinity.

a Lévy process. It can be represented in the following way:

$$X(t) = ta + Q^{1/2}W(t) + Z(t), \qquad t \geq 0,$$

where W is a Wiener process with covariance I and Z is an independent discontinuous process. Moreover, if a set Γ has a positive distance from 0 then

$$\nu(\Gamma) = \mathbb{E} \sum_{t \leq 1} \chi_\Gamma(Z(t) - Z(t-)).$$

For a modern treatment of Courrège's work we refer the reader to Jacob (2001).

1.5 Itô's approach

Courrège gave in 1965 a precise analytical description of all sufficiently regular Markov semigroups on $\mathbb{R}^d$. It had been discovered by Itô in the 1950s, however, that a large class of Markov processes can be constructed as solutions of appropriately formulated stochastic equations.

Let us consider first a special case, where the operator A is given by

$$A\varphi(x) = \int_E (\varphi(x + y) - \varphi(x))\nu(x, \mathrm{d}y). \tag{1.5}$$

Here the $\nu(x, \cdot)$, $x \in E$, are finite measures on a linear space E. In this case, the construction of the corresponding Markov process is easy for those familiar with compound Poisson processes, discussed in detail in Section 4.3. Namely, the process with generator A is constructed as a solution of (1.4), with L a compound Poisson process on a properly chosen linear space U. The situation is described in the following propositions, 1.9 and 1.10.

Proposition 1.9 *Let E and U be linear spaces, and let L be a compound Poisson process on U with jump intensity ν. For every family $(G(x), x \in E)$ of measurable linear transformations between U and E, (1.4) has a unique solution $(X(t, x), t \geq 0)$. Let (P_t) be the transition function corresponding to X. Then, for every bounded measurable function φ,*

$$\lim_{t \downarrow 0} \frac{P_t\varphi(x) - \varphi(x)}{t} = \int_E (\varphi(x + y) - \varphi(x))\nu(x, \mathrm{d}y),$$

where

$$\nu(x, \cdot) = G(x) \circ \nu := \nu\left((G(x))^{-1}\right) \tag{1.6}$$

is the image of the measure ν under the linear transformation $G(x)$.

Proof Let $0 = \tau_0 < \tau_1 < \tau_2 < \cdots$ be the jump times of the process L and let $Z_1, Z_2, \ldots$ be the consecutive jumps. Then (1.4) can be solved separately on each interval $[\tau_j, \tau_{j+1})$, $j = 0, 1, \ldots$, and the existence and uniqueness of the solution follow. Moreover, $X(t) = x$ for $t \in [0, \tau_1)$ and $X(t) = x + G(x)Z_1$ for $t \in [\tau_1, \tau_2)$.

Let $(\Pi(t), t \geq 0)$ be a Poisson process (see subsection 4.2.1) with jump times $\tau_1, \tau_2, \ldots$ Then the intensity of Π is equal to $a := \nu(U)$ and the distribution of each Z_j is equal to ν/a. We have

$$\varphi(X(t)) = \varphi(x)\chi_{\{\Pi(t)=0\}} + \varphi(x + G(x)Z_1)\chi_{\{\Pi(t)=1\}} + \varphi(\zeta)\chi_{\{\Pi(t)>1\}},$$

where ζ is a random variable. Consequently,

$$P_t\varphi(x) = \varphi(x)\mathbb{P}(\Pi(t) = 0) + \mathbb{E}\,\varphi(x + G(x)Z_1)\mathbb{P}(\Pi(t) = 1)$$
$$+ \mathbb{E}\,\varphi(\zeta)\chi_{\{\Pi(t)>1\}}$$
$$= \varphi(x)e^{-at} + \frac{1}{a}\int_E \varphi(x + G(x)y)\nu(dy)\,e^{-at}at + \mathbb{E}\,\varphi(\zeta)\chi_{\{\Pi(t)>1\}}.$$

However,

$$\limsup_{t\downarrow 0} \frac{1}{t}\left|\mathbb{E}\left(\varphi(\zeta)\chi_{\{\Pi(t)>1\}}\right)\right| \leq \limsup_{t\downarrow 0} \|\varphi\|_\infty \frac{\mathbb{P}(\Pi(t) > 1)}{t} = 0,$$

where $\|\varphi\|_\infty$ is the supremum norm of φ; see Section 2.1. Therefore

$$\lim_{t\downarrow 0} \frac{P_t\varphi(x) - \varphi(x)}{t} = \varphi(x)\left(\lim_{t\downarrow 0} \frac{e^{-at} - 1}{t}\right) + \int_E \varphi(x + G(x)y)\nu(dy)$$
$$= -a\varphi(x) + \int_E \varphi(x + G(x)y)\nu(dy),$$

as required.   □

In order to finish the construction of the process whose generator A is given by (1.5) we need to prove that the family $\nu(x, \cdot)$, $x \in E$, can be represented as in (1.6). We will show this under the assumption that ν is a transition probability.

Proposition 1.10   *Assume that $\nu(x, \cdot)$, $x \in E$, is a transition probability on a Polish space E. Then there exist a linear space U and linear transformations $G(x) \colon U \mapsto E$ such that (1.6) holds true.*

Proof   Define U to be the linear space of finite measures on $[0, 1]$. Let ℓ_1 be Lebesgue measure on $[0, 1]$. By Theorem 1.2 there exists a transformation $F \colon E \times [0, 1] \mapsto E$ such that $\nu(x, \cdot) = F(x, \cdot) \circ \ell_1$, $x \in E$. Let $j \colon [0, 1] \mapsto U$ be the embedding given by $x \mapsto \delta_x$. Then one obtains the desired representation with $\nu := j \circ \ell_1$ and $G(x)\lambda := \int_{[0,1]} F(x, y)\lambda(dy)$ for $\lambda \in U$.   □

In general a Markov process X on $\mathbb{R}^d$ can be constructed starting from a uniform movement $L_1(t) = t$, $t \geq 0$, a standard[3] Wiener process $L_2(t) = W(t)$, $t \geq 0$, on

---

[3] The covariance of $W(1)$ is the identity matrix I.

$\mathbb{R}^d$ and a Lévy process L_3, the so-called *Cauchy process* (see also Example 4.37) with Lévy measure $\nu_3(dx) = |x|^{-d-1} dx$. More precisely, the infinitesimal generator A_3 of L_3 is given by

$$A_3 \varphi(x) = \int_{\mathbb{R}^d} (\varphi(x + y) - \varphi(x))\nu_3(dy).$$

The process L_3 determines a family $\pi(t, \Gamma)$, $t \geq 0$, $\Gamma \in \mathcal{B}(\mathbb{R}^d \setminus \{0\})$, of Poisson processes with intensity $\nu_3(\Gamma)$; see subsection 4.2.1. Namely, let

$$\pi(t, \Gamma) := \sum_{s \leq t} \chi_\Gamma(L_3(s) - L_3(s-)),$$

where $L_3(s-)$ denotes the left limit at s, and let $\widehat{\pi}(t, \Gamma) := \pi(t, \Gamma) - \nu_3(\Gamma)t$, $t \geq 0$. Under mild conditions on the family $\nu(x, \cdot)$, one can select functions $\tilde{a}$ and F in such way that the generator A appearing in Theorem 1.8 has the form

$$A\varphi(x) = \langle \tilde{a}(x), D\varphi(x) \rangle + \tfrac{1}{2}\mathrm{Tr}\, Q(x)D^2\varphi(x)$$
$$+ \int_{\mathbb{R}^d} (\varphi(x + F(x, y)) - \varphi(x)$$
$$- \chi_{[0,1]}(|y|)\langle F(x, y), D\varphi(x) \rangle)\nu_3(dy).$$

Essentially F should be a regular function such that, for each $x \in \mathbb{R}^d$, $\nu(x, \cdot)$ is the image of ν_3 under the transformation $F(x, \cdot)$. Under appropriate regularity conditions on $\tilde{a}$, Q and F there exists a solution X to the stochastic differential equation

$$dX(t) = \tilde{a}(X(t))\, dt + Q^{1/2}(X(t))\, dW(t) + \int_{\{|y| \geq 1\}} F(X(t-), y)\pi(dt, dy)$$
$$+ \int_{\{|y| < 1\}} F(X(t-), y)\widehat{\pi}(dt, dy),$$
$$X(0) = x.$$

Moreover, X is a Markov process with transition function (P_t). The above stochastic equation should be understood as the integral equation

$$X(t) = x + \int_0^t \tilde{a}(X(s))ds + \int_0^t Q^{1/2}(X(s))\, dW(s)$$
$$+ \int_0^t \int_{\{|y| \geq 1\}} F(X(s-), y)\pi(ds, dy)$$
$$+ \int_0^t \int_{\{|y| < 1\}} F(X(s-), y)\widehat{\pi}(ds, dy),$$

the integrals on the right-hand side being respectively Lebesgue, Itô and Poissonian. Stochastic integrals were introduced by K. Itô.

One can treat the random measures π, $\widehat{\pi}$ as measure-valued Lévy processes, setting $Z_0(t, \cdot) = \pi(t, \cdot)$, $Z_1(t, \cdot) = \widehat{\pi}(t, \cdot)$. If we define the noise process as the Lévy process $L = \big((t, W(t), Z_0(t), Z_1(t)), \ t \geq 0\big)$ and, for each $x \in \mathbb{R}^d$, the transformation $G(x, \cdot)$ acting on each component as, respectively, multiplication by $\tilde{a}(x)$, the linear mapping $Q^{1/2}(x)$ and integration of the function $F(x, \cdot)$, the above stochastic equation can be written, similarly to the discrete-time case, in the form of (1.4).

For a modern and detailed treatment of the Itô approach we refer the reader to Stroock (2003).

1.6 Infinite-dimensional case

One may expect that a similar result to the Courrège theorem holds true in the case when the state space for the equation is infinite-dimensional, but to the best of our knowledge the proof is still missing. The reason is that in the infinite-dimensional case the situation is much more complex. The difficulties grow to the same extent as when going from ordinary to partial differential equations. In addition, examples show that noise may enter through the boundary conditions. Such highly irregular equations will also be considered in the present book.

2

Analytic preliminaries

In this chapter Sobolev and Hölder spaces are introduced. Then Aronson estimates for Green functions are formulated. First, though, we present some notation.

2.1 Notation

Let $a, b \in \mathbb{R}$. We adopt the notation $a \wedge b := \min\{a, b\}, a \vee b := \max\{a, b\}, a^+ := a \vee 0, a^- := (-a) \vee 0$. Clearly $a = a^+ - a^-, a \in \mathbb{R}$. We set $\delta_{a,b} = 0$ if $a \neq b$ and $\delta_{a,a} = 1$.

Throughout the book $(B, |\cdot|_B)$ denotes a Banach space. On B we consider the σ-field $\mathcal{B}(B)$ of Borel sets. We denote by $C_b(B)$ the space of all bounded continuous functions ϕ on B equipped with the *supremum norm*

$$\|\varphi\|_\infty := \sup_{x \in B} |\varphi(x)|.$$

Usually U and H stand for Hilbert spaces. In this book, unless otherwise indicated, all Banach and Hilbert spaces are real and separable. Let $(E, \mathcal{E})$ be a measurable space. We denote by $B_b(E)$ the space of all bounded measurable functions on E.

2.2 Sobolev and Hölder spaces

Let $n \in \mathbb{N}$ and $p \geq 1$ and let $\mathcal{O} \subseteq \mathbb{R}^d$ be a possibly unbounded open set. We denote by $W^{n,p}(\mathcal{O})$ the *Sobolev space* of functions p-integrable with all derivatives up to order n. Recall that $W^{n,p}(\mathcal{O})$ consists of all locally summable functions $\psi : \mathcal{O} \mapsto \mathbb{R}$ such that, for each multi-index α with $|\alpha| \leq n$, the weak derivative[1] $\partial^{|\alpha|}\psi/\partial\xi^\alpha$

---

[1] That is, in the sense of the theory of distributions.

belongs to $L^p(\mathcal{O}) := L^p(\mathcal{O}, \mathcal{B}(\mathcal{O}), \mathrm{d}\xi)$. Then $W^{n,p}(\mathcal{O})$ equipped with the norm

$$\|\psi\|_{W^{n,p}(\mathcal{O})} := \left(\int_{\mathcal{O}} \sum_{|\alpha| \le n} \left| \frac{\partial^{|\alpha|}\psi}{\partial \xi^{\alpha}}(\xi) \right|^p \mathrm{d}\xi \right)^{1/p}$$

is a Banach space. We denote by $W_0^{n,p}(\mathcal{O})$ the subspace of $W^{n,p}(\mathcal{O})$ that is the completion of the space $C_c^n(\mathcal{O})$ of compactly supported functions having continuous derivatives up to order n with respect to the $W^{n,p}(\mathcal{O})$-norm. We denote by $C_b^\infty(\mathcal{O})$ the class of functions having bounded derivatives of all orders.

Assume now that $\mathcal{O}$ is a bounded domain. Given $\gamma \in (0, 1)$ we denote by $C^\gamma(\mathcal{O})$ the space of all Hölder continuous functions $\psi : \overline{\mathcal{O}} \mapsto \mathbb{R}$ with Hölder exponent γ. On $C^\gamma(\mathcal{O})$ we consider the norm

$$\|\psi\|_{C^\gamma(\mathcal{O})} := \sup_{\xi \in \mathcal{O}} |\psi(\xi)| + \sup_{\xi, \eta \in \mathcal{O}, \xi \ne \eta} \frac{|\psi(\xi) - \psi(\eta)|}{|\xi - \eta|^\gamma}.$$

Finally, let $C^{n+\gamma}(\mathcal{O})$ be the space of all functions ψ having all derivatives of order $\le n$ Hölder continuous with exponent γ. We set

$$\|\psi\|_{C^{n+\gamma}(\mathcal{O})} := \sum_{|\alpha| \le n} \sup_{\xi \in \mathcal{O}} \left| \frac{\partial^{|\alpha|}\psi}{\partial \xi^{\alpha}}(\xi) \right| + \sum_{|\alpha| = n} \left\| \frac{\partial^{|\alpha|}\psi}{\partial \xi^{\alpha}} \right\|_{C^\gamma(\mathcal{O})}.$$

Assume now that $\mathcal{O}$ is a bounded domain in $\mathbb{R}^d$ with boundary of class C^1. We recall the classical Sobolev embedding theorem. For its proof we refer the reader to Evans (1998) or Gilbarg and Trudinger (1998).

Theorem 2.1 *Let $p \ge 1$, $n \in \mathbb{N}$.*

 (i) *Assume that $n < d/p$. Let q be such that $1/q + 1/p = 1$. Then $W^{n,p}(\mathcal{O})$ is continuously embedded into $L^q(\mathcal{O})$.*
 (ii) *Assume that $n > d/p$. Let $\gamma = [d/p] + 1 - d/p$, if d/p is not an integer, and let γ be an arbitrary strictly positive number < 1 if d/p is an integer. Then $W^{n,p}(\mathcal{O})$ is continuously embedded into $C^{n-[d/p]-1+\gamma}(\mathcal{O})$.*

Later we will need the Poincaré inequality; see e.g. Gilbarg and Trudinger (1998). For brevity we denote by $|\cdot|_{L^p}$ the norm on $L^p(\mathcal{O})$ and by ω_d the volume of the unit ball in $\mathbb{R}^d$.

Theorem 2.2 (Poincaré) *For $u \in W_0^{1,p}(\mathcal{O})$ and $1 \le p < \infty$,*

$$|u|_{L^p} \le \left(\omega_d^{-1} \ell_d(\mathcal{O}) \right)^{1/d} |Du|_{L^p}.$$

2.3 L^p- and C_ρ-spaces

In this book we denote by ℓ_d or $d\xi$ Lebesgue measure on $\mathbb{R}^d$. Where it will not lead to confusion we will write L^p instead of $L^p(\mathcal{O})$. In the case of equations on $\mathcal{O} = \mathbb{R}^d$, it will be useful to consider weighted L^p-spaces. In fact we will consider two classes of weights.

Exponential weights   Given $\rho \in \mathbb{R}$, let $\vartheta_\rho \in C^\infty(\mathbb{R}^d)$ be such that

(i) $\forall\, \xi \in \mathbb{R}^d, \quad \vartheta_\rho(\xi) > 0$ and $\vartheta_\rho(\xi) = \vartheta_\rho(-\xi)$,
(ii) $\forall\, \xi : |\xi| \geq 1, \quad \vartheta_\rho(\xi) = e^{-\rho|\xi|}$.

We will assume that, for all ρ and ξ, $\vartheta_{-\rho}(\xi) = (\vartheta_\rho(\xi))^{-1}$. We also assume that $\vartheta_0(\xi) = 1$ for $\xi \in \mathbb{R}^d$.

Define $L^p_\rho := L^p(\mathbb{R}^d, \mathcal{B}(\mathbb{R}^d), \vartheta^p_\rho(\xi)\,d\xi)$, $p \in [1, \infty)$, $\rho \in \mathbb{R}$. In the definition of L^p_ρ we introduce the weight ϑ^p_ρ instead of the more natural ϑ_ρ, in order to ensure that the semigroups generated by elliptic operators transform L^p_ρ into L^q_ρ for all p, q and ρ; see appendix section B.2. This would not have been the case if we had defined L^p_ρ as $L^p(\mathbb{R}^d, \mathcal{B}(\mathbb{R}^d), \vartheta_\rho(\xi)\,d\xi)$.

Clearly, if $\rho < \rho'$ then L^p_ρ is continuously embedded into $L^p_{\rho'}$. Moreover, $L^p(\mathbb{R}^d) = L^p_0$ and constant functions belong to an arbitrary L^p_ρ-space with $\rho > 0$.

Let C_ρ be the space of all continuous functions ψ such that $|\psi(\xi)|\vartheta_\rho \times (\xi) \to 0$ as $|\xi| \to \infty$. We equip C_ρ with the weighted supremum norm

$$|\psi|_{C_\rho} := \|\psi\vartheta_\rho\|_\infty = \sup_{\xi \in \mathbb{R}^d} |\psi(\xi)|\vartheta_\rho(\xi).$$

Polynomial weights   For $\rho \in \mathbb{R}$ let $\theta_\rho(\xi) := \left(1 + |\xi|^2\right)^{-\rho}, \xi \in \mathbb{R}^d$, and let $\mathcal{L}^p_\rho := L^p(\mathbb{R}^d, \mathcal{B}(\mathbb{R}^d), \theta^p_\rho(\xi)\,d\xi)$. Obviously $L^p_0 = \mathcal{L}^2_0$, and if $\rho < \rho'$ then $\mathcal{L}^p_\rho$ is continuously embedded into $\mathcal{L}^p_{\rho'}$. Note that non-zero constant functions belong to $\mathcal{L}^p_\rho$ if and only if $p\rho > d/2$.

For polynomial weights, an analogue of C_ρ is the $\mathcal{C}_\rho$-space of all continuous functions ψ such that $|\psi(\xi)|\theta_\rho(\xi) \to 0$ as $|\xi| \to \infty$. The space $\mathcal{C}_\rho$ is equipped with the norm $|\psi|_{\mathcal{C}_\rho} := \|\psi\theta_\rho\|_\infty$.

Remark 2.3   For some problems (see e.g. Capiński and Peszat 2001), it is necessary to consider polynomial rather than exponential weights. Polynomial weights have the advantage that one can construct (for example using wavelets) an orthonormal basis of $\mathcal{L}^2_\rho$ which is simultaneously a Schauder basis of all $\mathcal{L}^p_\rho$-spaces, $p \geq 2$, and the corresponding Sobolev weighted spaces; see García-Cuerva and Kazarian (1985) and Capiński and Peszat (2001).

2.4 Lipschitz functions and composition operators

Most of our existence results will hold true under various Lipschitz and linear-growth conditions imposed on non-linear components. Given a normed space $(E, |\cdot|_E)$ we denote by $\mathrm{Lip}\,(E)$ the space of all Lipschitz continuous functions $\psi : E \mapsto \mathbb{R}$ equipped with the norm $\|\psi\|_\infty + \|\psi\|_{\mathrm{Lip}}$, where $\|\psi\|_\infty$ is the supremum norm (see Section 2.1) and

$$\|\psi\|_{\mathrm{Lip}} := \sup_{x \neq y} \frac{|\psi(x) - \psi(y)|}{|x - y|_E}$$

is the *smallest Lipschitz constant of* ψ.

For our further purposes it will be convenient to introduce the following definitions. Here we assume that λ is a non-negative measure on the measurable space $(S, \mathcal{S})$.

Definition 2.4   Let $p \in [1, \infty)$. We denote by $\mathrm{Lip}\,(p, S, \lambda)$ the class of all functions $f : S \times \mathbb{R} \mapsto \mathbb{R}$ such that

$$|f(\sigma, u)| \leq \phi(\sigma) + K|u|, \qquad |f(\sigma, u) - f(\sigma, v)| \leq K|u - v|$$

with a constant $K < \infty$ and a function $\phi \in L^p(S, \mathcal{S}, \lambda)$.

We will consider equations with non-linear mappings of the so-called *composition* or *Nemytskii* type. Namely, given $f : \mathcal{O} \times \mathbb{R} \mapsto \mathbb{R}$ we denote by N_f the operator that associates with a function $\psi : \mathcal{O} \mapsto \mathbb{R}$ the function

$$(N_f \psi)(\xi) := f(\xi, \psi(\xi)), \qquad \xi \in \mathcal{O}. \tag{2.1}$$

The following lemma deals with the Lipschitz property of composition operators. Its easy verification is left to the reader.

Lemma 2.5

(i) *Assume that* $\mathcal{O} \in \mathcal{B}(\mathbb{R}^d)$. *Then for every* $f \in \mathrm{Lip}\,(p, \mathcal{O}, \ell_d)$ *the composition operator* N_f *is a Lipschitz mapping from* $L^p(\mathcal{O})$ *to* $L^p(\mathcal{O})$.

(ii) *If* $f \in \mathrm{Lip}\,(p, \mathbb{R}^d, \vartheta_\rho^p(\xi)\,\mathrm{d}\xi)$ *then* N_f *is a Lipschitz mapping from* L_ρ^p *to* L_ρ^p.

(iii) *If* $f \in \mathrm{Lip}\,(p, \mathbb{R}^d, \theta_\rho^p(\xi)\,\mathrm{d}\xi)$ *then* N_f *is a Lipschitz mapping from* $\mathcal{L}_\rho^p$ *to* $\mathcal{L}_\rho^p$.

(iv) *Assume that* $f : \mathbb{R} \mapsto \mathbb{R}$ *is a Lipschitz-continuous function. Then, for all* $p \geq 1$ *and* $\rho > 0$ N_f *is a Lipschitz mapping on* L_ρ^p. *If* $\rho \leq 0$ *then* N_f *is a Lipschitz mapping on* L_ρ^p *if and only if* $f(0) = 0$.

(v) *Assume that* $f : \mathbb{R} \mapsto \mathbb{R}$ *is a Lipschitz continuous function. If* $p\rho > d/2$ *then* N_f *is Lipschitz on* $\mathcal{L}_\rho^p$. *If* $p\rho \leq d/2$ *then* N_f *is Lipschitz on* $\mathcal{L}_\rho^p$ *if and only if* $f(0) = 0$.

Composition operators form an important class of mappings arising in nonlinear analysis. In this book we shall be concerned mostly with their Lipschitz property. Conditions for their continuity or differentiability can be obtained but using less simple arguments. For a more complete presentation we refer the reader to Ambrosetti and Prodi (1995).

2.5 Differential operators

Let

$$\mathcal{A} = \mathcal{A}(\xi, D) = \sum_{|\alpha| \leq 2m} a_\alpha(\xi) \frac{\partial^{|\alpha|}}{\partial \xi^\alpha}, \qquad \xi \in \mathcal{O}, \tag{2.2}$$

be an elliptic differential operator of order $2m$. We will consider $\mathcal{A}$ either on $\mathcal{O} = \mathbb{R}^d$ or on a bounded domain $\mathcal{O}$.

If $\mathcal{O}$ is bounded then $\mathcal{A}$ is considered together with boundary operators

$$B_j = B_j(\xi, D) = \sum_{|\alpha| \leq r_j} b_{j,\alpha}(\xi) \frac{\partial^{|\alpha|}}{\partial \xi^\alpha}, \qquad \xi \in \partial\mathcal{O}, \ j = 1, \ldots, m,$$

of orders not greater than $2m - 1$. We assume that:

(i) $\partial\mathcal{O}$ is of class $C^{2m+\gamma}$ for some $\gamma > 0$;
(ii) $\{a_\alpha\} \subset C_b^\infty(\mathcal{O})$ and $\mathcal{A}$ is uniformly elliptic on $\overline{\mathcal{O}}$, that is, there exists a $\delta > 0$ such that

$$(-1)^m \sum_{|\alpha|=2m} a_\alpha(\xi) \prod_{i=1}^d v_i^{\alpha_i} \leq -\delta|v|^{2m}, \qquad \xi \in \overline{\mathcal{O}}, \ v \in \mathbb{R}^d;$$

(iii) $0 \leq r_j \leq 2m - 1$ and $\{b_{j,\alpha}\} \subseteq C^{2m-r_j+\gamma}(\partial\mathcal{O})$ for a $\gamma > 0$;
(iv) the system $\{B_j\}$ fulfils uniformly the complementarity condition on $\partial\mathcal{O}$ (see Solonnikov 1965, Eidelman 1969 or Eidelman and Zhitarashu 1998).

Owing to limitations of space we do not give the definition of the complementarity condition mentioned in (iv). The hypothesis of (iv) is satisfied by the so-called *system of Dirichlet conditions* $B_j = \partial^j/\partial\mathbf{n}^j$, $j = 0, 1, \ldots, m - 1$, where $\mathbf{n}$ is the normal interior to the boundary.

We will frequently employ some properties of the *Green function* $\mathcal{G}$ for the system $\{\mathcal{A}, B_1, \ldots, B_m\}$. Namely, there is a function $\mathcal{G} \colon (0, +\infty) \times \mathcal{O} \times \mathcal{O} \mapsto \mathbb{R}$, continuously differentiable with respect to the first variable and having continuous derivatives of orders less than or equal to $2m$ with respect to the second variable, such that for every $\eta \in \mathcal{O}$

$$\left(\frac{\partial}{\partial t} - \mathcal{A}\right) \mathcal{G}(\cdot, \cdot, \eta) = 0 \qquad \text{on } (0, +\infty) \times \mathcal{O},$$

$$B_j \mathcal{G}(\cdot, \cdot, \eta) = 0 \qquad \text{on } (0, +\infty) \times \partial \mathcal{O}$$

and

$$\lim_{t \downarrow 0} \int_{\mathcal{O}} \mathcal{G}(t, \xi, \eta) \psi(\eta) \, d\eta = \psi(\xi), \qquad \forall \xi \in \mathcal{O}, \ \forall \psi \in C_b(\overline{\mathcal{O}}).$$

In the case $\mathcal{O} = \mathbb{R}^d$ we assume, as for a bounded domain, that $\mathcal{A}$ is uniformly elliptic (on $\mathbb{R}^d$) and has coefficients of class $C_b^\infty(\mathbb{R}^d)$. Obviously in this case we do not need to impose any boundary conditions. In fact, the decay at infinity of a solution to the equation for $\mathcal{A}$ is implied by the integrability assumptions. As in the case of a bounded region there is a continuous function $\mathcal{G} \colon (0, +\infty) \times \mathbb{R}^d \times \mathbb{R}^d \mapsto \mathbb{R}$, continuously differentiable with respect to the first variable and having continuous derivatives of orders less than or equal to $2m$ with respect to the second variable, such that

$$\left(\frac{\partial}{\partial t} - \mathcal{A}\right) \mathcal{G}(\cdot, \cdot, \eta) = 0 \qquad \text{on } (0, +\infty) \times \mathbb{R}^d, \ \forall \eta \in \mathbb{R}^d,$$

$$\lim_{t \downarrow 0} \int_{\mathbb{R}^d} \mathcal{G}(t, \xi, \eta) \psi(\eta) \, d\eta = \psi(\xi), \qquad \forall \xi \in \mathbb{R}^d, \ \forall \psi \in C_b(\mathbb{R}^d).$$

As in the case of a bounded domain, we call $\mathcal{G}$ the Green function for the operator $\mathcal{A}$.

Assume that $\mathcal{G}$ is the Green function either for the system $(\mathcal{A}, \{B_j\})$ on a bounded region $\mathcal{O}$ or for an $\mathcal{A}$ considered on $\mathcal{O} = \mathbb{R}^d$. In our future considerations, an important role will be played by so-called *parabolic semigroups*, that is, by families of operators $S = (S(t), \ t \geq 0)$ defined by setting $S(0)$ equal to the identity operator and

$$S(t)\psi(\xi) = \int_{\mathcal{O}} \mathcal{G}(t, \xi, \eta) \psi(\eta) \, d\eta, \qquad t > 0. \tag{2.3}$$

In fact, S is the C_0-semigroup generated by a proper realization of $\mathcal{A}$; see Appendix B. In appendix section B.2 we derive some regularity properties of S from the following classical *Aronson estimate*.

Theorem 2.6 (Aronson) *For each $T > 0$ there exist constants $K_1, K_2 > 0$ such that*

$$\left| \frac{\partial^{|\alpha|} \mathcal{G}}{\partial \xi^\alpha} (t, \xi, \eta) \right| \leq K_1 t^{-|\alpha|/(2m)} \mathfrak{g}_m(K_2 t, |\xi - \eta|),$$

$$\forall t \in (0, T], \ \forall \xi, \eta \in \mathcal{O}, \ \forall \alpha \colon |\alpha| \leq 2m, \tag{2.4}$$

where

$$\mathfrak{g}_m(t,r) = t^{-d/(2m)} \exp\left\{ -\left(t^{-1}r^{2m}\right)^{1/(2m-1)} \right\}, \qquad t > 0, \ r > 0. \qquad (2.5)$$

It is worth noting that the above theorem provides the same upper estimates for a Green kernel considered on a bounded region and one considered on the whole space. For the proof we refer the reader to Eidelman (1969) or Solonnikov (1969).

3

Probabilistic preliminaries

In this chapter we set up the notation for, and present infinite-dimensional analogues of, classical probabilistic concepts. We recall, mostly without proof, basic results on real-valued martingales. We define the expectation and conditional expectation of a random variable with values in a Banach space and introduce the concept of Gaussian measure on an infinite-dimensional space. Then we deal with various classes of stochastic processes taking values in a Banach space.

3.1 Basic definitions

Throughout the book $(\Omega, \mathcal{F}, \mathbb{P})$ denotes a complete[1] probability space.

Many arguments in the theory of stochastic processes depend on Dynkin's π–λ theorem; see Dynkin (1965) and Theorem 3.2 below. This theorem is concerned with the smallest σ-field $\sigma(\mathcal{K})$ containing all elements of a family $\mathcal{K}$.

Definition 3.1 A collection $\mathcal{K}$ of subsets of Ω is said to be a π-*system* if it is closed under the formation of finite intersections. A collection $\mathcal{M}$ of subsets of Ω is a λ-*system* if it contains Ω and is closed under the formation of complements and of countable disjoint unions.

Theorem 3.2 (Dynkin) *If a λ-system $\mathcal{M}$ contains a π-system $\mathcal{K}$ then $\mathcal{M} \supset \sigma(\mathcal{K})$.*

Proof Denote by $\mathcal{K}_0$ the smallest λ-system containing $\mathcal{K}$. This is equal to the intersection of all λ-systems containing $\mathcal{K}$. Then $\mathcal{K}_0 \subset \sigma(\mathcal{K})$. To prove the opposite inclusion we first show that $\mathcal{K}_0$ is a π-system. Let $A \in \mathcal{K}_0$ and define $\mathcal{K}_A = \{B : B \in \mathcal{K}_0 \text{ and } A \cap B \in \mathcal{K}_0\}$. It is easy to check that $\mathcal{K}_A$ is closed under the

---

[1] That is, $A \in \mathcal{F}$, $B \subset A$ and $\mathbb{P}(A) = 0$ imply that $B \in \mathcal{F}$.

formation of complements and countable disjoint unions and that if $A \in \mathcal{K}$ then $\mathcal{K}_A \supset \mathcal{K}$. Thus, for $A \in \mathcal{K}, \mathcal{K}_A = \mathcal{K}_0$ and we have shown that if $A \in \mathcal{K}$ and $B \in \mathcal{K}_0$ then $A \cap B \in \mathcal{K}_0$. But this implies that $\mathcal{K}_B \supset \mathcal{K}$ and, consequently, that $\mathcal{K}_B = \mathcal{K}_0$ for any $B \in \mathcal{K}_0$. It is now an easy exercise to show that if a π-system is closed under the formation of complements and countable disjoint unions then it is a σ-field. This completes the proof. □

Let $(E, \mathcal{E})$ be a measurable space. Any measurable mapping $X \colon \Omega \mapsto E$ is called an *E-valued random variable*, a *random element in E* or a *random variable in E*. We denote by $\mathcal{L}(X)(\Gamma) := \mathbb{P}(\omega \in \Omega \colon X(\omega) \in \Gamma)$, $\Gamma \in \mathcal{E}$, the *law* of X, i.e. the distribution of the random element X in E.

Let I be a time interval. In fact, I is the set of non-negative real numbers $\mathbb{R}_+$, or a finite interval $[0, T]$ or, in the discrete-time case, a subset of non-negative integers $\mathbb{Z}_+ = \{0, 1, \ldots\}$. On I we consider the Borel σ-field $\mathcal{B}(I)$. Any family $X = (X(t), t \in I)$ of random elements in E is called a *stochastic process in E* or an *E-valued stochastic process*.

A *filtration* is any non-decreasing family of σ-fields $\mathcal{F}_t \subset \mathcal{F}, t \in I$. If

$$\mathcal{F}_t = \mathcal{F}_{t+} := \bigcap_{s > t} \mathcal{F}_s, \qquad \forall t \in I,$$

then the filtration $(\mathcal{F}_t)$ is called *right-continuous*. Note that for an arbitrary filtration $(\mathcal{F}_t)$, the filtration $(\mathcal{F}_{t+})$ is also right-continuous. The quadruple $(\Omega, \mathcal{F}, (\mathcal{F}_t), \mathbb{P})$ is called a *filtered probability space*. We say that an E-valued stochastic process X is *adapted to the filtration* $(\mathcal{F}_t)$ if, for every $t \in I$, $X(t)$ is $\mathcal{F}_t$-measurable.

Definition 3.3 An E-valued stochastic process X is *measurable* if it is a measurable mapping from $I \times \Omega$ to E, where the product σ-field $\mathcal{B}(I) \times \mathcal{F}$ is considered on $I \times \Omega$.

Let $\mathcal{P}_I$ denote the σ-field of *predictable sets*, that is, the smallest σ-field of subsets of $I \times \Omega$ containing all sets of the form $(s, t] \cap I \times A$, where $s, t \in I$, $s < t$ and $A \in \mathcal{F}_s$. For simplicity we write $\mathcal{P}$ instead of $\mathcal{P}_{[0,\infty)}$.

Definition 3.4 A stochastic process X taking values in a measurable space $(E, \mathcal{E})$ is called *predictable* if it is a measurable mapping from $I \times \Omega$ to E, where the σ-field $\mathcal{P}_I$ is considered on $I \times \Omega$.

Definition 3.5 Let $(X(t), t \in I)$ be an E-valued stochastic process defined on $(\Omega, \mathcal{F}, \mathbb{P})$. An E-valued process $(Y(t), t \in I)$ is said to be a *modification* of X if $\mathbb{P}(X(t) = Y(t)) = 1$ for each $t \in I$.

It is clear that a modification Y of X has the same finite-dimensional distributions as X. If there exists a modification Y of X which has, $\mathbb{P}$-a.s., continuous trajectories then we say that X *has a continuous modification*. If there exists a measurable or predictable modification Y of X then we say that X *has a measurable or predictable modification* Y.

Definition 3.6 Let $(\Omega, \mathcal{F}, (\mathcal{F}_t), \mathbb{P})$ be a filtered probability space. We say that $\tau \colon \Omega \mapsto [0, +\infty]$ is a *stopping time* (with respect to $(\mathcal{F}_t)$) if, for every $t \in I$, $\{\tau \leq t\} \in \mathcal{F}_t$.

Let τ be a stopping time. We denote by $\mathcal{F}_\tau$ the collection of events $A \in \mathcal{F}$ such that $A \cap \{\tau \leq t\} \in \mathcal{F}_t$ for all $t \in I$. It is easy to verify that $\mathcal{F}_\tau$ is a σ-field, the so-called σ-*field of events prior to* τ, and that τ is $\mathcal{F}_\tau$-measurable.

Given an E-valued stochastic process X we denote by $\left(\mathcal{F}_t^X\right)$ the *filtration generated by* X. That is, for every $t \in I$, $\mathcal{F}_t^X := \sigma\left(X(s) \colon s \in I, \, s \leq t\right)$ is the smallest σ-field under which all random elements $X(s)$, $s \in I$, $s \leq t$, are measurable. We denote by $\overline{\mathcal{F}}_t^X$ the smallest σ-field containing $\mathcal{F}_t^X$ and all sets from $\mathcal{F}$ of $\mathbb{P}$-measure zero. Note that the filtration $\left(\overline{\mathcal{F}}_{t+}^X\right)$ is right-continuous and that each $\overline{\mathcal{F}}_{t+}^X$ contains all sets of $\mathcal{F}$ of $\mathbb{P}$-measure 0. In fact, in this book we will usually take it for granted that $(\mathcal{F}_t)$ is a right-continuous filtration such that every $\mathcal{F}_t$ contains all sets of $\mathcal{F}$ of $\mathbb{P}$-measure 0. We will then say that $(\mathcal{F}_t)$ satisfies the *usual conditions*.

3.2 Kolmogorov existence theorem

Let $(E, \mathcal{E})$ be a measurable space. A stochastic process X is usually described in terms of the measures it induces on products of E; namely, if $(X(t), \, t \in I)$ is an E-valued process then, for each sequence $(t_1, t_2, \ldots, t_k)$ of distinct elements of I, $(X(t_1), \ldots, X(t_k))$ is a random vector with values in the product $E \times E \times \cdots \times E$ equipped with the product σ-field $\mathcal{E} \times \mathcal{E} \times \cdots \times \mathcal{E}$. The probability measures on E^k given by $\lambda_{t_1,\ldots,t_k} := \mathcal{L}(X(t_1), \ldots, X(t_k))$ are called *finite-dimensional distributions* of X. Note that the finite-dimensional distributions of a stochastic process $(X(t), \, t \in I)$ satisfy two *consistency conditions*.

(i) For arbitrary $A_i \in \mathcal{E}$, $i = 1, 2, \ldots, k$, and for any permutation π of $\{1, 2, \ldots, k\}$, $\lambda_{t_1,\ldots,t_k}(A_1 \times \cdots \times A_k) = \lambda_{t_{\pi 1},\ldots,t_{\pi k}}(A_{\pi 1} \times \cdots \times A_{\pi k})$.

(ii) For arbitrary $A_i \in \mathcal{E}$, $i = 1, 2, \ldots, k-1$, $\lambda_{t_1,\ldots,t_{k-1}}(A_1 \times \cdots \times A_{k-1}) = \lambda_{t_1,\ldots,t_k}(A_1 \times \cdots \times A_{k-1} \times E)$.

Theorem 3.7 (Kolmogorov) *Assume that E is a separable complete metric space and $\mu_{t_1,\ldots,t_k}$ is a family of distributions on E^k, $k \in \mathbb{N}$, satisfying (i) and (ii). Then on some probability space $(\Omega, \mathcal{F}, \mathbb{P})$ there exists a stochastic process $(X(t), \, t \in I)$ having the $\lambda_{t_1,\ldots,t_k}$ as its finite-dimensional distributions.*

Kolmogorov's theorem is an easy consequence of the following results, due to C. Carathéodory (1918) and S. Ulam (see Oxtoby and Ulam 1939), which are of independent interest.

Definition 3.8 A collection $\mathcal{E}_0$ of subsets of E is said to be a *field* or *algebra* if $E \in \mathcal{E}_0$, $A^c \in \mathcal{E}_0$ for every $A \in \mathcal{E}_0$ and $\bigcup_{k=1}^n A_k \in \mathcal{E}_0$ for arbitrary $A_1, \ldots, A_n \in \mathcal{E}_0$.

Theorem 3.9 (Carathéodory) *Assume that λ is a non-negative function on a field $\mathcal{E}_0$ such that, for any sequence $\{A_i\} \subset \mathcal{E}_0$ satisfying $A_i \cap A_j = \emptyset$ for $i \neq j$ and $\bigcup_i A_i \in \mathcal{E}_0$,*

$$\lambda\left(\bigcup_{i=1}^\infty A_i\right) = \sum_{i=1}^\infty \lambda(A_i).$$

Then there exists a unique extension of λ to a finite measure on the σ-field generated by $\mathcal{E}_0$.

Theorem 3.10 (Ulam) *Assume that λ is a probability measure on a separable complete metric space. Then for any $\varepsilon > 0$ there exists a compact set $K \subset E$ such that $\lambda(K) \geq 1 - \varepsilon$.*

3.3 Random elements in Banach spaces

We have the following basic inequality due to Lévy and Ottaviani, the proof of which can be found in Kwapień and Woyczyński (1992).

Proposition 3.11 (Lévy–Ottaviani) *If $X(1), \ldots, X(n)$ are independent random variables with values in a (not necessarily separable) Banach space B then, for any $r > 0$,*

$$\mathbb{P}\left(\max_{k \leq n} |X(1) + \cdots + X(k)|_B > r\right)$$
$$\leq 3 \max_{k \leq n} \mathbb{P}\left(|X(1) + \cdots + X(k)|_B > \frac{r}{3}\right).$$

The following corollary to the Lévy–Ottaviani inequality is part of the Itô–Nisio theorem; see Itô and Nisio (1968) or Kwapień and Woyczyński (1992).

Corollary 3.12 *If $(X(n), n \in \mathbb{N})$ are independent random variables in B such that $\sum_{n=1}^N X_n$ converges in probability as $N \uparrow \infty$ then $\sum_{n=1}^N X_n$ converges $\mathbb{P}$-a.s.*

A B-valued random variable X is *integrable* if

$$\mathbb{E}|X|_B := \int_\Omega |X(\omega)|_B \, \mathbb{P}(d\omega) < \infty.$$

If $\mathbb{E}\,|X|_B^2 < \infty$ then X is called *square integrable*. Since B is separable, for any B-valued random variable the real-valued function $\omega \mapsto |X(\omega)|_B$ is measurable. Given an integrable B-valued random variable, we define its *expectation*

$$\mathbb{E}\,X = \int_\Omega X(\omega)\,\mathbb{P}(d\omega)$$

as a Bochner integral. Let us recall that the *Bochner integral* is initially defined for *simple random variables*

$$X = \sum_{j=1}^n x_j \chi_{A_j}, \tag{3.1}$$

where $n \in \mathbb{N}$, $\{x_j\} \subset B$ and $\{A_j\} \subset \mathcal{F}$, by the formula

$$\mathbb{E}\,X = \sum_{j=1}^n x_j\,\mathbb{P}(A_j).$$

Clearly, for any simple random variable,

$$|\mathbb{E}\,X|_B \le \mathbb{E}\,|X|_B. \tag{3.2}$$

Now, let X be an arbitrary integrable random variable. Then there is a sequence (X_m) of simple random variables such that for any ω the sequence $(|X_m(\omega) - X(\omega)|_B)$ decreases to 0. Indeed, let $B_0 = \{u_1, u_2, \ldots\}$ be a countable dense subset of B. Write $X_m(\omega) := u_{k_m(\omega)}$, $\omega \in \Omega$, $m \in \mathbb{N}$, where

$$k_m(\omega) := \min\left\{k \le m\colon |X(\omega) - u_k|_B = \min_{l\le m}\{|X(\omega) - u_l|_B\}\right\}.$$

Then (X_m) has the desired property. We define $\mathbb{E}\,X$ as the limit of $\mathbb{E}\,X_m$. Note that $\mathbb{E}\,X$ does not depend on the choice of the approximating sequence and that (3.2) holds. An integrable B-valued random variable is said to be of *mean zero* if $\mathbb{E}\,X = 0$.

Proposition 3.13 *Let $\mathcal{V}$ be a sub-σ-field of $\mathcal{F}$ and let X be a B-valued integrable random variable. Then, up to a set of $\mathbb{P}$-measure 0, there is a unique integrable $\mathcal{V}$-measurable B-valued random variable $\mathbb{E}\,(X|\mathcal{V})$ such that*

$$\int_A X(\omega)\,\mathbb{P}\,(d\omega) = \int_A \mathbb{E}\,(X|\mathcal{V})\,(\omega)\,\mathbb{P}\,(d\omega), \qquad \forall\,A \in \mathcal{V}.$$

Proof Assume first that X is a simple random variable given by (3.1). Set

$$\mathbb{E}\,(X|\mathcal{V}) = \sum_{j=1}^n \alpha_j\,\mathbb{P}\,(A_j|\mathcal{V}).$$

Here $\mathbb{P}(A_j|\mathcal{V})$ denotes the classical conditional probability; see e.g. Kallenberg (2002). Next, for an integrable X define $\mathbb{E}(X|\mathcal{V})$ as the $L^1(\Omega, \mathcal{V}, \mathbb{P}; B)$-limit of $\mathbb{E}(X_m|\mathcal{V})$, where (X_m) is a sequence of simple random variables appearing in the construction of the Bochner integral. □

Definition 3.14   We call $\mathbb{E}(X|\mathcal{V})$ the *conditional expectation of X given $\mathcal{V}$.*

Recall that a sequence $X(n)$ of random variables in B is called *uniformly integrable* if

$$\lim_{r \to \infty} \sup_n \int_{\{|X(n)|_B \geq r\}} |X(n)|_B \, d\mathbb{P} = 0.$$

In the proposition below we gather some basic properties of the conditional expectation of B-valued random variables. These results are well known for real-valued random variables, so we leave their easy verification to the reader.

Proposition 3.15   *Let X, Y be integrable random elements in B, let $a, b \in \mathbb{R}$ and let $\mathcal{V}$ be a sub-σ-field $\mathcal{F}$. Then the following hold.*

  (i) $\mathbb{E}(aX + bY|\mathcal{V}) = a\,\mathbb{E}(X|\mathcal{V}) + b\,\mathbb{E}(Y|\mathcal{V})$, $\mathbb{P}$-*a.s.*

  (ii) *If T is a continuous linear operator from B to a real separable Banach space B_1 then $\mathbb{E}(TX|\mathcal{V}) = T\,\mathbb{E}(X|\mathcal{V})$, $\mathbb{P}$-a.s.*

 (iii) *If X is $\mathcal{V}$-measurable and ζ is a real-valued integrable random variable such that ζX is integrable then $\mathbb{E}(\zeta X|\mathcal{V}) = X\,\mathbb{E}(\zeta|\mathcal{V})$, $\mathbb{P}$-a.s.*

 (iv) *If $\mathcal{U}$ is a sub-σ-field of $\mathcal{V}$ then $\mathbb{E}(X|\mathcal{U}) = \mathbb{E}(\mathbb{E}(X|\mathcal{V})|\mathcal{U})$, $\mathbb{P}$-a.s.*

  (v) *If X is independent of $\mathcal{V}$ then $\mathbb{E}(X|\mathcal{V}) = \mathbb{E}\,X$, $\mathbb{P}$-a.s.*

 (vi) *If $f : \mathbb{R} \mapsto \mathbb{R}$ is a convex function such that $f(|X|_B)$ is integrable then $f(|\mathbb{E}(X|\mathcal{V})|_B) \leq \mathbb{E}(f(|X|_B)|\mathcal{V})$, $\mathbb{P}$-a.s.*

 (vii) *Let $(X(n))$ be a sequence of uniformly integrable random variables in B such that $X(n)$ converges $\mathbb{P}$-a.s. to X. Then $\mathbb{E}(X(n)|\mathcal{V}) \to \mathbb{E}(X|\mathcal{V})$, $\mathbb{P}$-a.s.*

(viii) *Assume that $(\mathcal{V}_n, \ n \in \mathbb{N})$ is an increasing family of σ-fields such that $\mathcal{V} = \sigma(\mathcal{V}_n : n \in \mathbb{N})$, that is, that $\mathcal{V}$ is the smallest σ-field containing each $\mathcal{V}_n$. Then $\mathbb{E}(X|\mathcal{V}_n) \to \mathbb{E}(X|\mathcal{V})$, $\mathbb{P}$-a.s.*

3.4 Stochastic processes in Banach spaces

In this section we present basic definitions concerning stochastic processes taking values in a separable Banach space B.

A B-valued stochastic process $X = (X(t), \ t \in I)$ is *integrable* (*square integrable*) if the random variables $|X(t)|_B, \ t \in I$, are integrable (square integrable). A process X has *mean zero* if the random elements $X(t), t \in I$, have mean zero. The definition below deals with different aspects of continuity.

Definition 3.16 A B-valued stochastic process X is *right-continuous* if

$$X(t+) := \lim_{s \downarrow t, s \in I} X(s) = X(t), \qquad \forall t \in I.$$

If, furthermore,

$$X(t) = \lim_{s \to t, s \in I} X(s), \qquad \forall t \in I,$$

then X is called *continuous*. We say that X is *càdlàg* (continu à droite et limites à gauche) if X is right-continuous and has left limits, that is, if for every $t \in I$ the limit $\lim_{s \in I, s \uparrow t} X(s) =: X(t-)$ exists.

The simple proof of the following result is left to the reader.

Proposition 3.17 *Assume that $(X(t),\ t \geq 0)$ and $(Y(t),\ t \geq 0)$ are càdlàg processes defined on the same probability space $(\Omega, \mathcal{F}, \mathbb{P})$ and that X is a modification of Y. Then there is a set $\Omega_0 \in \mathcal{F}$ of $\mathbb{P}$-measure 1 such that $X(t, \omega) = Y(t, \omega)$ for all $(t, \omega) \in [0, \infty) \times \Omega_0$.*

Definition 3.18 A B-valued stochastic process $X = (X(t),\ t \in I)$ is *stochastically continuous* or *continuous in probability* if

$$\lim_{s \to t, s \in I} \mathbb{P}(|X(t) - X(s)|_B > \varepsilon) = 0, \qquad \forall \varepsilon > 0,\ \forall t \in I.$$

Note that if $I \subset \mathbb{R}$ is compact then any stochastically continuous process X is uniformly stochastically continuous, that is,

$$\forall \varepsilon > 0,\ \exists \delta > 0: \forall t, s \in I: |t - s| \leq \delta, \qquad \mathbb{P}(|X(t) - X(s)|_B > \varepsilon) < \varepsilon.$$

Definition 3.19 A B-valued square integrable stochastic process X is *mean-square continuous* if

$$\lim_{s \to t,\ s \in I} \mathbb{E}\,|X(t) - X(s)|_B^2 = 0, \qquad \forall t \in I.$$

The theorem below, combining results of A. N. Kolmogorov, M. Loève and N. N. Chentsov, provides a simple moment condition for the existence of a Hölder-continuous version of a random field on $\mathbb{R}^d$. For its proof we refer the reader to Kallenberg (2002). A more general version of this result can be found in Walsh (1986).

Theorem 3.20 (Kolmogorov–Loève–Chentsov) *Let (E, ρ) be a complete metric space, and let $\bigl(X(v),\ v \in \mathbb{R}^d\bigr)$ be a family of E-valued random variables. Assume that there exist $a, b, c > 0$ such that*

$$\mathbb{E}\bigl(\rho(X(v), X(u))\bigr)^a \leq c|u - v|^{d+b}, \qquad u, v \in \mathbb{R}^d.$$

Then X has a modification that is locally Hölder continuous with exponent $\alpha \in (0, b/a)$.

Proposition 3.21 *Any measurable stochastically continuous $(\mathcal{F}_t)$-adapted B-valued process $(X(t),\ t \geq 0)$ has a predictable modification.*

Proof From the stochastic continuity there is a partition $0 = t_{m,0} < t_{m,1} < \cdots < t_{m,n(m)} = m$ such that, for $t \in (t_{m,k}, t_{m,k+1}]$,

$$\mathbb{P}\left(|X(t_{m,k}) - X(t)|_B > 2^{-m}\right) \leq 2^{-m}, \qquad k = 0, 1, \ldots, n(m) - 1.$$

Define

$$X_m(\omega, t) = \chi_{\{0\}}(t)X(\omega, 0) + \sum_{k=0}^{n(m)-1} \chi_{(t_{m,k}, t_{m,k+1}]}(t)X(\omega, t_{m,k}).$$

Then X_m is predictable. Denote by A the set of all $(\omega, t) \in \Omega \times [0, T]$ for which the sequence $(X_m(\omega, t))$ converges. Then A is a predictable set, and the process $\widetilde{X}$ defined by

$$\widetilde{X}(\omega, t) = \chi_A(\omega, t) \lim_{m \to \infty} X_m(\omega, t)$$

is predictable. We now show that $\widetilde{X}$ is a modification of X. By the Borel–Canteli lemma, for each $t \in [0, T]$ there is a set $\Omega_t \in \mathcal{F}$ of $\mathbb{P}$-measure 1 such that for $\omega \in \Omega_t$ there is an $m_0 \in \mathbb{N}$ such that

$$|X_m(\omega, t) - X(\omega, t)|_B \leq 2^{-m}, \qquad \forall m \geq m_0.$$

Therefore $\{t\} \times \Omega_t \subset A$, and $X(\omega, t) = \widetilde{X}(\omega, t)$ for $\omega \in \Omega_t$. $\qquad\square$

Let $P_t(x, \Gamma), t \geq 0, x \in E, \Gamma \in \mathcal{B}(E)$, be a transition function on a metric space E (see Definition 1.7), let (P_t) be the corresponding transition semigroup, given by (1.3), and let $(\mathcal{F}_t)$ be a filtration on $(\Omega, \mathcal{F}, \mathbb{P})$.

Definition 3.22 An E-valued $(\mathcal{F}_t)$-adapted process $X = (X(t),\ t \geq 0)$ is said to be *Markov* with respect to $(\mathcal{F}_t)$ and (P_t) if, for every $t, h \geq 0$ and $\varphi \in B_b(E)$,

$$\mathbb{E}\big(\varphi(X(t + h))\big|\mathcal{F}_t\big) = P_h\varphi(X(t)), \qquad \mathbb{P}\text{-a.s.}$$

Often $\mathcal{F}_t = \sigma(X(s), s \leq t)$. If X is a process such that, for $t, h \in [0, \infty)$,

$$\mathbb{E}\big(\varphi(X(t + h))\big|\sigma(X(s), s \leq t)\big) = \mathbb{E}\big(\varphi(X(t + h))\big|\sigma(X(t))\big)$$

then X is called *general Markov*.

The following result comes from Kinney (1953). For the convenience of the reader the proof is presented in Appendix C.

Theorem 3.23 (Kinney) *If X is Markov with respect to (P_t) and*

$$\limsup_{t \downarrow 0} \ _{x \in E} P_t(x, B^c(x, r)) = 0, \qquad \forall r > 0, \tag{3.3}$$

then X has a càdlàg modification.

The definition below extends the concept of a martingale to a class of processes taking values in a separable Banach space.

Definition 3.24 A family $(X(t),\ t \in I)$, of integrable B-valued random variables is said to be a *B-valued martingale with respect to a filtration* $(\mathcal{F}_t)$, or simply a *martingale*, if it is $(\mathcal{F}_t)$-adapted and satisfies

$$\mathbb{E}\left(X(t)|\mathcal{F}_s\right) = X(s), \qquad \forall\, t, s \in I,\ t \geq s.$$

As a direct consequence of Proposition 3.15(v) we have the following result.

Proposition 3.25 *Let X be a B-valued integrable process. Assume that, for all $t, s \in I$, $t > s$, the random variable $X(t) - X(s)$ is independent of $\mathcal{F}_s$. Then the process $Y(t) := X(t) - \mathbb{E}\, X(t)$, $t \in I$, is a martingale.*

Definition 3.26 A B-valued process $(X(t),\ t \geq 0)$ is said to be a *B-valued local martingale with respect to a filtration* $(\mathcal{F}_t)$, or simply a *local martingale*, if it is $(\mathcal{F}_t)$-adapted and there exists a sequence of $(\mathcal{F}_t)$ stopping times $\tau_n \uparrow \infty$ such that the processes $X^{\tau_n} := \left(X(t \wedge \tau_n)\, t \geq 0\right)$ are martingales.

Definition 3.27 An adapted process $(V(t),\ t \geq 0)$ is said to be of *finite variation* if there exists a random variable $\widetilde{V}$, whose values are B-valued measures on $[0, \infty)$, such that $V(t) = \widetilde{V}([0, t])$ for $t \geq 0$.

Definition 3.28 An adapted process $(V(t),\ t \geq 0)$ is said to be of *local finite variation* if there exists a sequence of $(\mathcal{F}_t)$ stopping times $\tau_n \uparrow \infty$ such that the processes $V^{\tau_n} := \left(V(t \wedge \tau_n),\ t \geq 0\right)$ are of finite variation.

Definition 3.29 A B-valued $(\mathcal{F}_t)$-adapted process $(X(t),\ t \geq 0)$ is said to be a *B-valued semimartingale with respect to the filtration* $(\mathcal{F}_t)$, or simply a *semimartingale*, if $X(t) = M(t) + V(t)$, $t \in [0, \infty)$, where M is a local martingale and V is an adapted process with locally finite variation.

3.5 Gaussian measures on Hilbert spaces

A measure λ on $\left(\mathbb{R}^d, \mathcal{B}(\mathbb{R}^d)\right)$ is Gaussian if and only if its characteristic functional is given by

$$\int_{\mathbb{R}^d} e^{i\langle \eta, \xi \rangle_U} \lambda(d\xi) = \exp\left\{ i\langle \eta, m \rangle_U - \tfrac{1}{2} \langle Q\eta, \eta \rangle_U \right\}, \qquad \eta \in \mathbb{R}^d,$$

where $m = (m_1, \ldots, m_d) \in \mathbb{R}^d$ and $Q = [q_{i,j}] \in M_s^+(d \times d)$. Such a measure is denoted by $\mathcal{N}(m, Q)$. Moreover, m and Q are respectively its mean and covariance:

$$m_i = \int_{\mathbb{R}^d} \xi_i \, \mathcal{N}(m, Q)(\mathrm{d}\xi), \qquad q_{i,j} = \int_{\mathbb{R}^d} (\xi_i - m_i)(\xi_j - m_j) \mathcal{N}(m, Q)(\mathrm{d}\xi).$$

If $m = 0$ then λ is said to be *centered*. Assume that Q is invertible. Then $\mathcal{N}(m, Q)$ has density

$$\frac{\sqrt{\det Q^{-1}}}{(2\pi)^{d/2}} \exp\left\{-\tfrac{1}{2} \left\langle Q^{-1}(\xi - m), \xi - m \right\rangle\right\}.$$

Definition 3.30 A random variable X with values in a Hilbert space U is *Gaussian (centered Gaussian)* if, for all $x \in U$, the real-valued random variable $\langle X, x \rangle_U$ has a Gaussian (centered Gaussian) distribution. A random process X taking values in U is *Gaussian* if, for all $t_1, \ldots, t_n, (X(t_1), \ldots, X(t_n))$ is a Gaussian random element in U^n.

The definitions of the mean vector and the covariance matrix can be extended to the infinite-dimensional case, thanks to the following theorem.

Theorem 3.31 *Assume that X is a centered Gaussian random variable with values in a Hilbert space U. Then $\mathbb{E}\,|X|_U^2 < \infty$. Moreover,*

$$\mathbb{E}\,e^{s|X|_U^2} \le \frac{1}{\sqrt{1 - 2s\,\mathbb{E}\,|X|_U^2}}, \qquad \forall s < \frac{1}{2\,\mathbb{E}\,|X|_U^2}. \tag{3.4}$$

Proof Assume first that X is a random variable in $(\mathbb{R}^n, \langle \cdot, \cdot \rangle)$ with the law $\mu = \mathcal{N}(0, Q)$. Take an orthonormal basis of $\mathbb{R}^n$ consisting of eigenvectors of Q. Let $\{\gamma_j\}$ be the corresponding eigenvalues, that is, $Qe_j = \gamma_j e_j$, $j = 1, \ldots, n$. Then $X = \sum_{j=1}^n \langle X, e_j \rangle e_j, \langle X, e_j \rangle, j = 1, \ldots, n$, are independent and the law of $\langle X, e_j \rangle$ is $\mathcal{N}(0, \gamma_j)$. Consequently,

$$\mathbb{E}\,e^{s|X|^2} = \mathbb{E}\exp\left\{s\sum_{j=1}^n \langle X, e_j \rangle^2\right\} = \prod_{j=1}^n \mathbb{E}\,e^{s\langle X, e_j \rangle^2} = \prod_{j=1}^n \frac{1}{\sqrt{1 - 2s\gamma_j}}$$

$$= \frac{1}{\sqrt{(1 - 2s\gamma_1)\cdots(1 - 2s\gamma_n)}} \le \frac{1}{\sqrt{1 - 2s(\gamma_1 + \cdots + \gamma_n)}}.$$

Since $\mathbb{E}\,|X|^2 = \gamma_1 + \cdots + \gamma_n$, we have (3.4).

Let us now choose an arbitrary orthonormal basis $\{e_n\}$ of U. Then $X = \sum_n \langle X, e_n \rangle_U e_n$ and $X = \lim_{n\to\infty} X_n$, $X_n = \sum_{j=1}^n \langle X, e_j \rangle_U e_n$. Since X_n is an n-dimensional Gaussian random variable, applying (3.4) for $s = -1$ we obtain

$$\mathbb{E}\,e^{-|X_n|_U^2} \le \frac{1}{\sqrt{1 + 2\,\mathbb{E}\,|X|_U^2}}.$$

Note that $|X_n|_U^2 \uparrow |X|_U^2$, so that $\mathbb{E}\,|X_n|_U^2 \uparrow \mathbb{E}\,|X|_U^2$ and therefore

$$\mathbb{E}\,e^{-|X|_U^2} \le \frac{1}{\sqrt{1 + 2\,\mathbb{E}\,|X|_U^2}}.$$

If $\mathbb{E}\,|X|_U^2$ were equal to ∞ then $|X|_U^2$ would equal ∞ with probability 1, which is not the case, so $\mathbb{E}\,|X|_U^2 < \infty$.

Now let $s < \left(2\,\mathbb{E}\,|X|_U^2\right)^{-1}$, and let $\{X_n\}$ be the finite-dimensional projections of X defined as above. Then $\mathbb{E}\,|X_n|_U^2 \le \mathbb{E}\,|X|_U^2$. Hence, for any n,

$$\mathbb{E}\,e^{s|X_n|_U^2} \le \frac{1}{\sqrt{1 - 2s\,\mathbb{E}\,|X_n|_U^2}} \le \frac{1}{\sqrt{1 - 2s\,\mathbb{E}\,|X|_U^2}}.$$

Letting $n \uparrow \infty$ gives

$$\mathbb{E}\,e^{s|X|_U^2} \le \frac{1}{\sqrt{1 - 2s\,\mathbb{E}\,|X|_U^2}}.$$

$\square$

A (non-trivial) generalization of Theorem 3.31 to Gaussian random variables taking values in a Banach space is due to M. X. Fernique; see Fernique (1970) or e.g. Kuo (1975).

It follows from the theorem that for every centered Gaussian random variable X there exists a trace-class non-negative operator $Q\colon U \mapsto U$ (see Appendix A), called the *covariance operator of* X, such that

$$\mathbb{E}\,\langle X, x \rangle_U \langle X, y \rangle_U = \langle Qx, y \rangle_U.$$

It is easy to see that $\operatorname{Tr} Q = \mathbb{E}\,|X|_U^2$. More generally, if $\mathbb{E}\,X = m$ then the covariance operator of X is the covariance operator of $X - m$.

It is easy to see that if $X = (X_1, \ldots, X_d)$ is Gaussian in $\mathbb{R}^d$ then the random variables $X_1, \ldots, X_d$ are independent if and only if the covariance matrix is diagonal.

3.6 Gaussian measures on topological spaces

Let E be a topological vector space. Denote by E^* the dual space and by $(\cdot, \cdot)$ the canonical bilinear form on $E \times E^*$. On E we consider the σ-field $\mathcal{B}(E)$ of Borel sets. For any finite subset $\{x_1, \ldots, x_n\}$ of E^* we denote by $\tau_{x_1,\ldots,x_n}$ the mapping $E \ni y \mapsto ((y, x_1), \ldots, (y, x_n)) \in \mathbb{R}^n$.

Definition 3.32 A measure λ on $(E, \mathcal{B}(E))$ is *Gaussian* if for any finite subset $\{x_1, \ldots, x_n\}$ of E^* the transport measure $\lambda \circ \tau^{-1}_{x_1,\ldots,x_n}$ is Gaussian on $\mathbb{R}^n$. The measure λ is *centered* or equivalently has *mean zero* or is *symmetric*, if, for any $\{x_1, \ldots, x_n\}$, $\lambda \circ \tau^{-1}_{x_1,\ldots,x_n}$ has mean zero.

Clearly any Gaussian measure on E is a probability measure. To check that λ is Gaussian it is sufficient to show that, for every $x \in E^*$, $\lambda \circ \tau^{-1}_x$ is Gaussian. Indeed, since finite-dimensional Gaussian measures are characterized by their Fourier transforms, a finite-dimensional measure $\lambda \circ \tau^{-1}_{x_1,\ldots,x_n}$ is Gaussian if and only if $\lambda \circ \tau^{-1}_{t_1 x_1 + \cdots + t_n x_n}$ is Gaussian for any vector $(t_1, \ldots, t_n) \in \mathbb{R}^n$.

We say that an E-valued random variable is Gaussian if its law $\mathcal{L}(X)$ is Gaussian.

Definition 3.33 Let X be a centered Gaussian random variable in E. Then its covariance form is the bilinear mapping $K: E^* \times E^* \mapsto \mathbb{R}$ given by

$$K(x, y) = \int_E (x, z)(y, z)\lambda(\mathrm{d}z), \qquad x, y \in E^*.$$

Given $m \in E$ we denote by T_m the *translation operator*; that is, $T_m(x) = m + x$, $x \in E$. Note that if there is an $m \in E$ such that $\lambda \circ T_m^{-1}$ is a centered Gaussian measure then

$$(m, x) = \int_E (y, x)\lambda(\mathrm{d}y), \qquad \forall x \in E^*.$$

Thus, if it exists, m is uniquely determined and is called the *mean* of λ. If λ is a Gaussian measure with mean m and if $\lambda \circ T_m^{-1}$ has covariance Q then we write $\lambda = \mathcal{N}(m, Q)$.

3.7 Submartingales

We now recall the classical concepts of the submartingale and supermartingale. Let $I \subset [0, \infty)$.

Definition 3.34 A real-valued integrable process $(X(t), t \in I)$ is a *submartingale with respect to* the filtration $(\mathcal{F}_t)$ if it is $(\mathcal{F}_t)$-adapted and

$$\mathbb{E}(X(t)|\mathcal{F}_s) \geq X(s), \qquad \mathbb{P}\text{-a.s.}, \qquad \forall s < t, \ s, t \in I.$$

X is called a *supermartingale* if $-X$ is a submartingale.

Theorem 3.35 *Let $p \geq 1$ and let $(X(t), t \in I)$ be a B-valued martingale. If $\mathbb{E}|X(t)|_B^p < \infty$ for $t \in I$ then the process $(|X(t)|_B^p, t \in I)$ is a submartingale.*

Proof The function $f(a) = |a|^p$, $a \in \mathbb{R}$, is convex and therefore by Proposition 3.15(vi), for $t > s$,

$$|X(s)|_B^p = \left|\mathbb{E}(X(t)|\mathcal{F}_s)\right|_B^p \leq \left|\mathbb{E}(|X(t)|_B|\mathcal{F}_s)\right|^p \leq \mathbb{E}(|X(t)|_B^p|\mathcal{F}_s).$$

$\square$

3.7.1 Doob optional-sampling theorem

The following Doob optional-sampling theorems in discrete and continuous time are of fundamental importance in the theory of submartingales. For their proofs we refer the reader to e.g. Kallenberg (2002).

Theorem 3.36 (Doob optional sampling) *Let $(X(n)$, $n = 1, \ldots, k)$ be a submartingale (supermartingale, martingale) relative to $(\mathcal{F}_n)$. Let $\tau_1, \ldots, \tau_m$ be an increasing sequence of $(\mathcal{F}_n)$-stopping times with values in the set $\{1, \ldots, k\}$. Then the sequence $(X(\tau_i)$, $i = 1, \ldots, m)$ is also a submartingale (supermartingale, martingale) with respect to $(\mathcal{F}_{\tau_i})$.*

From the discrete-time result Theorem 3.36 one can deduce the following continuous-time result.

Theorem 3.37 (Doob optional sampling) *Let $(X(t)$, $t \in [0, T])$ be a right-continuous submartingale (supermartingale, martingale) relative to $(\mathcal{F}_t)$. Let $\tau_1, \ldots, \tau_m$ be an increasing sequence of stopping times with values in $[0, T]$. Then the sequence $(X(\tau_i)$, $i = 1, \ldots, m)$ is also a submartingale (supermartingale, martingale) with respect to $(\mathcal{F}_{\tau_i})$.*

3.7.2 Doob submartingale inequality

The following fundamental result is called the Doob submartingale inequality. Recall that $X^+(T)$ is the positive part of $X(T)$ (see Section 2.1).

Theorem 3.38 (Doob inequality) *Assume that $(X(t)$, $t \geq 0)$ is a right-continuous submartingale. Then*

$$r\,\mathbb{P}\left(\sup_{t\in[0,T]} X(t) \geq r\right) \leq \mathbb{E}\,X^+(T), \qquad \forall r > 0, \ \forall T \geq 0.$$

The theorem is an easy consequence of its discrete-time version.

Proposition 3.39 *Let $(X(n)$, $n = 1, \ldots, k)$ be a submartingale. Then*

$$r\,\mathbb{P}\left(\max_n X(n) \geq r\right) \leq \mathbb{E}\,X^+(k), \qquad \forall r > 0.$$

Proof Note that $A = \{\max_n X(n) \geq r\} = \bigcup_{n=1}^k A_n$, $A_1 = \{X(1) \geq r\}$ and

$$A_n = \{X(1) < r\} \cap \cdots \cap \{X(n-1) < r\} \cap \{X(n) \geq r\}, \quad n = 2, \ldots, k.$$

We have $A_n \in \mathcal{F}_n$ and $X(n) \geq r$ on A_n. Therefore

$$\mathbb{E}\left(X(k)\chi_{A_n}\right) = \mathbb{E}\left(\mathbb{E}(X(k) \mid \mathcal{F}_n)\chi_{A_n}\right) \geq \mathbb{E}\left(X(n)\chi_{A_n}\right) \geq r\mathbb{P}(A_n).$$

Consequently,

$$r \, \mathbb{P}(A) \leq \mathbb{E}\left(X(k)\chi_{A_k}\right) \leq \mathbb{E} X^+(k).$$

$\square$

Proof of Theorem 3.38 To prove this theorem we choose an increasing sequence (Q_k) of finite subsets of $[0, T]$, containing T and such that $Q = \bigcup_k Q_k$ is dense in $[0, T]$. For every $\varepsilon \in (0, r)$,

$$\left\{ \sup_{t \in [0,T]} X(t) \geq r \right\} \subset \bigcup_k \left\{ \max_{t \in Q_k} X(t) \geq r - \varepsilon \right\}.$$

Consequently,

$$\mathbb{P}\left(\sup_{t \in [0,T]} X(t) \geq r \right) \leq \frac{1}{r - \varepsilon} \mathbb{E} X^+(T).$$

Taking $\varepsilon \downarrow 0$ we obtain the required result. $\square$

The following Doob regularity theorem states a condition under which a (real-valued) submartingale has a càdlàg modification. It holds for submartingales and thus also for supermartingales; it has been successfully applied also to wide classes of stochastic processes, implying the existence of their càdlàg versions (see e.g. Theorem 3.41 below). A proof of the Doob theorem can be found in e.g. Rogers and Williams (2000).

Theorem 3.40 (Doob regularity) *Any stochastically continuous submartingale $(X(t), t \in I)$ has a càdlàg modification.*

As a corollary to Theorems 3.38 and 3.40 we will derive the following result.

Theorem 3.41 *Let $(M(t), t \geq 0)$ be a stochastically continuous square integrable martingale taking values in a Hilbert space $(U, \langle \cdot, \cdot \rangle_U)$. Then M has a càdlàg modification (denoted also by M) satisfying*

$$\mathbb{P}\left(\sup_{t \in [0,T]} |M(t)|_U \geq r \right) \leq \frac{\mathbb{E} |M(T)|_U^2}{r^2}, \qquad \forall T \geq 0, \, \forall r > 0. \tag{3.5}$$

Moreover,

$$\mathbb{E} \sup_{t \in [0,T]} |M(t)|_U^\alpha \leq \frac{2}{2-\alpha} \left(\mathbb{E} |M(T)|_U^2 \right)^{\alpha/2}, \qquad \forall T \geq 0, \ \forall \alpha \in (0,2). \quad (3.6)$$

Proof Let $\{e_k\}$ be an orthonormal basis of U and let

$$M_n(t) = \sum_{k=1}^n \langle M(t), e_k \rangle_U e_k, \qquad n \in \mathbb{N}, \ t \geq 0.$$

Then each M_n is a martingale. Since each $\langle M, e_k \rangle_U$ is a stochastically continuous real-valued martingale, and hence a submartingale, it has a càdlàg version. Hence each M_n has a càdlàg version. Clearly $\sum_{k=n+1}^m \langle M(t), e_k \rangle_U^2, t \geq 0$, is a submartingale. Hence, by Doob's submartingale inequality,

$$I_{n,m} := \mathbb{P} \left(\sup_{t \in [0,T]} |M_n(t) - M_m(t)|_U \geq r \right)$$

$$= \mathbb{P} \left(\sup_{t \in [0,T]} \sum_{k=n+1}^m \langle M(t), e_k \rangle_U^2 \geq r^2 \right) \leq \frac{1}{r^2} \mathbb{E} \sum_{k=n+1}^m \langle M(t), e_k \rangle_U^2,$$

and hence $I_{n,m} \to 0$ as $n, m \to \infty$. By a standard application of the Borel–Canteli lemma, we can find a subsequence $M_{n_l}, l = 1, 2, \dots$, which $\mathbb{P}$-a.s. converges uniformly to a càdlàg process, a modification of M. Moreover, again by Doob's submartingale inequality, we have (3.5). To prove (3.6), note that, for $Y := \sup_{t \in [0,T]} |M(t)|_U^\alpha$ and $b := \mathbb{E} |M(T)|_U^2$,

$$\mathbb{E} Y = \mathbb{E} \int_0^\infty \chi_{[0,Y]}(r) \, dr = \int_0^\infty \mathbb{E} \chi_{[0,Y]}(r) \, dr = \int_0^\infty \mathbb{P}(Y \geq r) \, dr$$

$$\leq b^{\alpha/2} + \int_{b^{\alpha/2}}^\infty \mathbb{P}(Y \geq r) \, dr \leq b^{\alpha/2} + \mathbb{E} |M(T)|_U^2 \int_{b^{\alpha/2}}^\infty r^{-2/\alpha} \, dr$$

$$\leq b^{\alpha/2} + \mathbb{E} |M(T)|_U^2 \frac{b^{(\alpha/2)(-2/\alpha+1)}}{2/\alpha - 1} = \frac{2}{2-\alpha} \left(\mathbb{E} |M(T)|_U^2 \right)^{\alpha/2}.$$

$\square$

3.7.3 Doob–Meyer decomposition

Given a filtration $(\mathcal{F}_t)$ and a $T \in [0, \infty)$, we denote by $\Sigma_{[0,T]}$ the family of all stopping times τ satisfying $\mathbb{P}(\tau \leq T) = 1$.

Definition 3.42 A right-continuous submartingale $X = (X(t), \ t \geq 0)$ with respect to $(\mathcal{F}_t)$ belongs to the *class (DL)* if for any $T \in [0, \infty)$ the random variables $\left(X(\tau), \ \tau \in \Sigma_{[0,T]} \right)$ are uniformly integrable.

We present the following fundamental result. For its proof we refer the reader to Kallenberg (2002), Rogers and Williams (2000) or Jakubowski (2006).

Theorem 3.43 (Doob–Meyer) *A càdlàg submartingale X of class (DL) admits a unique decomposition $X(t) = N(t) + A(t)$, $t \geq 0$, where N is a martingale and A is a predictable process starting from 0 with increasing trajectories.*

We will apply the Doob–Meyer theorem to $\left(|M(t)|_B^2, \, t \geq 0\right)$, where M is a square integrable martingale taking values in B. To this end we need the following lemmas. The first is a consequence of Proposition 3.15(vi).

Lemma 3.44 *If M is a B-valued martingale with respect to $(\mathcal{F}_t)$ then the process $(|M(t)|_B, \, t \geq 0)$ is a real-valued submartingale. If $\psi : \mathbb{R} \mapsto \mathbb{R}$ is a convex function and $\mathbb{E} \, |\psi(|M(t)|_B)| < \infty$ for $t \geq 0$ then the process $(\psi(|M(t)|_B), \, t \geq 0)$ is a submartingale.*

Lemma 3.45 *Let M be a square integrable B-valued right-continuous martingale. Then the process $|M|_B^2 = \left(|M(t)|_B^2, t \geq 0\right)$ is a submartingale of class (DL).*

Proof By Lemma 3.44 (see also Theorem 3.35), $|M|_B^2$ is a submartingale. Let $T < \infty$ be fixed, and let $\tau \in \Sigma_{[0,T]}$. Then, applying Doob's optional sampling theorem (Theorem 3.37), we obtain $|M(\tau)|_B^2 \leq \mathbb{E}\left(|M(T)|_B^2 \big| \mathcal{F}_\tau\right)$. Hence

$$\lim_{r \to \infty} \sup_{\tau \in \Sigma_{[0,T]}} \int_{\{|M(\tau)|_B^2 \geq r\}} |M(\tau)|_B^2 \, d\mathbb{P} \leq \lim_{r \to \infty} \sup_{\tau \in \Sigma_{[0,T]}} \int_{\{|M(\tau)|_B^2 \geq r\}} |M(T)|_B^2 \, d\mathbb{P}.$$

Since

$$\sup_{\tau \in \Sigma_{[0,T]}} \mathbb{P}\left(|M(\tau)|_B^2 \geq r\right) \leq \frac{\mathbb{E} \, |M(T)|_B^2}{r^2},$$

the lemma follows. $\qquad\qquad\square$

We denote by $\mathcal{M}^2(B)$ the class of all square integrable B-valued martingales $M = (M(t), \, t \geq 0)$ with respect to $(\mathcal{F}_t)$ such that $(|M(t)|_B, \, t \geq 0)$ is càdlàg. Note that if M is stochastically continuous then by the Doob regularization theorem the submartingale $|M|_B^2$ has a càdlàg modification. If B is a Hilbert space then, by Theorem 3.41, M has a càdlàg modification. In fact, in the Hilbert case we will always assume that the elements of $\mathcal{M}^2(B)$ are càdlàg. Now, if $M \in \mathcal{M}^2(B)$ then by the Doob–Meyer decomposition theorem there is a unique increasing predictable process $(\langle M, M \rangle_t, \, t \geq 0)$, called the *angle bracket* or *predictable-variation process* of M, such that $\langle M, M \rangle_0 = 0$ and $|M(t)|_B^2 - \langle M, M \rangle_t, t \geq 0$, is a martingale. Given $M, N \in \mathcal{M}^2(B)$ we define

$$\langle M, N \rangle := \tfrac{1}{4}\big(\langle M + N, M + N \rangle - \langle M - N, M - N \rangle\big).$$

For simplicity we will write $\mathcal{M}^2$ instead of $\mathcal{M}^2(\mathbb{R})$.

Remark 3.46   Since, for $M, N \in \mathcal{M}^2$,

$$M(t)N(t) = \tfrac{1}{4}\left(|M(t) + N(t)|^2 - |M(t) - N(t)|^2\right),$$

the process $M(t)N(t) - \langle M, N \rangle_t$, $t \geq 0$, is a martingale. More generally, if $(U, \langle \cdot, \cdot \rangle_U)$ is a Hilbert space and $M, N \in \mathcal{M}^2(U)$ then $\langle M(t), N(t) \rangle_U - \langle M, N \rangle_t$, $t \geq 0$, is a martingale.

3.8 Semimartingales

Generally, if $(X(t), t \geq 0)$ is a process and τ is a stopping time, we denote by X^τ the process $(X(t \wedge \tau), t \geq 0)$. Given any class $\mathcal{X}$ of processes (e.g. martingales, submartingales or supermartingales), we denote by $\mathcal{X}_{\text{loc}}$ the class of processes such that there is a sequence of stopping times $\tau_n \uparrow \infty$ for which $X^{\tau_n} \in \mathcal{X}$ for every $n \in \mathbb{N}$.

We call the elements of $\mathcal{M}_{\text{loc}}(B)$ and $\mathcal{M}^2_{\text{loc}}(B)$ *local martingales* and *local square integrable martingales*, respectively.

Let $\mathcal{BV}$ be the class of all real-valued adapted càdlàg processes with trajectories of bounded variation on every finite time interval.

Definition 3.47   We say that a real-valued process X is a *semimartingale* if it is càdlàg, adapted and can be written in the form $X = M + A$, where $M \in \mathcal{M}^2$ and $A \in \mathcal{BV}$. A process X is a *local semimartingale* if $X = M + A$, where $M \in \mathcal{M}^2_{\text{loc}}$ and $A \in \mathcal{BV}_{\text{loc}}$.

We need the concept of the *quadratic variation process* $[M, M]$ of a martingale M. Its definition and properties are provided by the theorem below. For its proof we refer the reader to Métivier (1982), Theorem 18.6.

Theorem 3.48   *For every $M \in \mathcal{M}^2$ there exists an increasing adapted càdlàg process $[M, M]$, called the* quadratic variation *of M, having the following properties.*

(i) *For every sequence $\pi_n = \left(0 < t_0^n < t_1^n < \cdots\right)$ of partitions of $[0, \infty)$ such that $t_k^n \to \infty$ as $k \to \infty$ and $\lim_{n \to \infty} \sup_j \left(t_{j+1}^n - t_j^n\right) = 0$, one has*

$$[M, M]_t = \lim_{n \to \infty} \sum_j \left(M\left(t_{j+1}^n \wedge t\right) - M\left(t_j^n \wedge t\right)\right)^2,$$

where the limit is in $L^1(\Omega, \mathcal{F}, \mathbb{P})$.
(ii) *$M^2 - [M, M]$ is a martingale.*
(iii) *If M has continuous trajectories then $\langle M, M \rangle = [M, M]$.*

We define $[M, N]$, $M, N \in \mathcal{M}^2$, by polarization:

$$[M, N] := \tfrac{1}{4}\big([M + N, M + N] - [M - N, M - N]\big).$$

Note that

$$[M, N]_t = \lim_{n \to \infty} \sum_j \big(M\big(t^n_{j+1} \wedge t\big) - M\big(t^n_j \wedge t\big)\big)\big(N\big(t^n_{j+1} \wedge t\big) - N\big(t^n_j \wedge t\big)\big).$$

3.9 Burkholder–Davies–Gundy inequalities

Certain basic inequalities for semimartingales involve the quadratic variation and are known as the Burkholder–Davis–Gundy (BDG) inequalities. First we consider the case of martingales with continuous trajectories. Recall (see Theorem 3.48(iii)) that in this case $\langle M, M \rangle = [M, M]$.

Theorem 3.49 (Burkholder–Davis–Gundy) *For every $p > 0$ there is a constant $C_p \in (0, \infty)$ such that for any real-valued continuous martingale M with $M_0 = 0$, and for any $T \geq 0$,*

$$C_p^{-1}\, \mathbb{E}\, \langle M, M \rangle_T^{p/2} \leq \mathbb{E}\, \sup_{t \in [0,T]} |M_t|^p \leq C_p\, \mathbb{E}\, \langle M, M \rangle_T^{p/2}.$$

The next result is an extension of the BDG inequality to the class of discontinuous martingales.

Theorem 3.50 (Burkholder–Davis–Gundy) *For every $p \geq 1$ there is a constant $C_p \in (0, \infty)$ such that for any real-valued square integrable càdlàg martingale M with $M_0 = 0$, and for any $T \geq 0$,*

$$C_p^{-1}\, \mathbb{E}\,[M, M]_T^{p/2} \leq \mathbb{E}\, \sup_{t \in [0,T]} |M_t|^p \leq C_p\, \mathbb{E}\,[M, M]_T^{p/2}.$$

For the proofs of the above theorems we refer the reader to Kallenberg (2002), Theorems 17.7 and 26.12.

4

Lévy processes

The main theme of this book concerns stochastic equations driven by processes with stationary independent increments. This chapter is devoted to the properties of such processes. Their structure is determined by the Lévy–Khinchin formula. The chapter starts with general facts on the regularity of trajectories and exponential integrability. Then the building blocks, Poisson and Wiener processes, are introduced, the Lévy–Khinchin decomposition and formula are proved and the properties of square integrable Lévy processes are investigated.

4.1 Basic properties

Processes with independent increments can be defined in any linear space E equipped with a σ-field $\mathcal{E}$ such that the addition and subtraction operations are measurable. The most important cases are when E is the real line, or $\mathbb{R}^d$ or a Banach space. We start with the general case, however.

Definition 4.1   A stochastic process $L = (L(t),\ t \geq 0)$ taking values in E has *independent increments* if, for any $0 \leq t_0 < t_1 < \cdots < t_n$, the $(E, \mathcal{E})$-valued random variables $L(t_1) - L(t_0),\ L(t_2) - L(t_1),\ \ldots,\ L(t_n) - L(t_{n-1})$ are independent. If the law $\mathcal{L}(L(t) - L(s))$ of $L(t) - L(s)$ depends only on the difference $t - s$ then we say that L has *stationary*, or *time-homogeneous, independent increments*. If in addition E is a Banach space, $L(0) = 0$ and the process L is stochastically continuous then L is called a *Lévy process*.

Lévy processes with values in the space of tempered distributions are studied separately, in Chapter 14.

Let L be a Lévy process on a Banach space E and let μ_t be the law of the random variable $L(t)$. Then, denoting by $\mu * \nu$ the convolution of the measures μ and ν, we have

(i) $\mu_0 = \delta_0$ and $\mu_{t+s} = \mu_t * \mu_s$ for all $t, s \geq 0$,

(ii) $\mu_t(\{x : |x|_E < r\}) \to 1$ as $t \downarrow 0$ for every $r > 0$.

Note that (ii) is equivalent to the statement

(iii) μ_t converges weakly to δ_0 as $t \downarrow 0$.

Definition 4.2   The family (μ_t) of measures satisfying the above conditions is called a *convolution semigroup of measures* or, sometimes, an *infinitely divisible family*. Sometimes μ_1 is called an infinitely divisible measure.

Note that every Lévy process is also Markov with transition probability (see Definition 1.7) $P_t(x, \Gamma) = \mu_t(\Gamma - x), t \geq 0, \Gamma \in \mathcal{B}(E), x \in E$, and that the corresponding semigroup is given by

$$P_t \varphi(x) = \int_E \varphi(x + y) \, \mu_t(dy). \tag{4.1}$$

By the Kolmogorov theorem for each convolution semigroup of measures (μ_t) there exists a Lévy process having the distribution μ_t at time t. Explicit constructions of Lévy processes will be given in Section 4.5 and also in Section 6.3, where we discuss the concept of random measures. The following basic results will be used often.

Theorem 4.3   *Every Lévy process has a càdlàg modification.*

Proof   If $B(x, r)$ denotes the ball in E with its center at x and with radius r then $P_t(x, B^c(x, r)) = \mu_t(B^c(0, r))$ and

$$\limsup_{t \downarrow 0} \, _x P_t\left(x, B^c(x, r)\right) = \lim_{t \downarrow 0} \mu_t\left(B^c(0, r)\right) = 0.$$

By Theorem 3.23 the result follows.   □

Given a càdlàg process L we define the *process of jumps of* L by $\Delta L(t) := L(t) - L(t-), t \geq 0$. The following result of De Acosta (1980) is a special case of a more general theorem of Rosinski (1995).

Theorem 4.4 (De Acosta)   *Assume that $(L(t), t \geq 0)$ is a càdlàg Lévy process in a Banach space B with jumps bounded by a fixed number $c > 0$; that is, $|\Delta L(t)|_B \leq c$ for every $t \geq 0$. Then, for any $\beta > 0$ and $t \geq 0$,*

$$\mathbb{E} \, e^{\beta |L(t)|_B} < \infty. \tag{4.2}$$

Proof   We give here a short proof of a weaker result of Kruglov (1972), that (4.2) holds for some $\beta > 0$. This will be sufficient for our purposes. Our considerations are based on Protter (2005). Write $\tau_0 = 0$ and

$$\tau_{n+1} = \inf\{t \geq \tau_n : |L(t) - L(\tau_n)|_B \geq c\}, \qquad n = 0, 1, \ldots$$

Since L has independent and stationary increments, the random variables $(\tau_{n+1} - \tau_n, n = 0, 1, \ldots)$ are independent and have the same distribution. Consequently, for $n = 1, 2, \ldots$,

$$\mathbb{E}\,e^{-\tau_n} = \mathbb{E}\,e^{-(\tau_n - \tau_{n-1}) + \cdots + (\tau_1 - \tau_0)} = \prod_{j=1}^{n} \mathbb{E}\,e^{-(\tau_j - \tau_{j-1})} = \left(\mathbb{E}\,e^{-\tau_1}\right)^n =: (\alpha)^n.$$

By Chebyshev's inequality,

$$\mathbb{P}\big(|L(t)|_B > 2nc\big) \leq \mathbb{P}\big(\tau_n < t\big) \leq e^t (\alpha)^n.$$

Note that $\alpha \in (0, 1)$. Let $\gamma \in (0, \log 1/\alpha)$. Then

$$\mathbb{P}\big(\exp\left\{\frac{\gamma}{2c}|L(t)|_B\right\} > e^{\gamma n}\big) \leq e^t (\alpha)^n,$$

and hence

$$
\begin{aligned}
\mathbb{E}\,\exp\left\{\frac{\gamma}{2c}|L(t)|_B\right\} &= \int_0^\infty \mathbb{P}\left(\exp\left\{\frac{\gamma}{2c}|L(t)|_B\right\} > s\right) ds \\
&\leq \sum_{n=0}^\infty \mathbb{P}\left(\exp\left\{\frac{\gamma}{2c}|L(t)|_B\right\} > e^{\gamma n}\right) e^{\gamma(n+1)} \\
&\leq e^{t+\gamma} \sum_{n=0}^\infty \left(\alpha e^\gamma\right)^n < \infty.
\end{aligned}
$$

$\square$

In this book we are concerned mainly with the case when E is a Hilbert space $(U, \langle \cdot, \cdot \rangle_U)$ and $\mathcal{E}$ is the σ-field of Borel sets $\mathcal{B}(U)$.

4.2 Two building blocks – Poisson and Wiener processes

The real-valued *Poisson* and *Wiener* processes play a fundamental role in the theory of Lévy processes. In fact any Lévy process can be built from them in a constructive way. The following subsections are concerned with the direct constructions and basic properties of Poisson and Wiener processes.

4.2.1 Poisson processes

A Lévy process with values in $\mathbb{Z}_+ = \{0, 1, \ldots\}$, which is increasing and has a finite number of jumps on any finite interval, each jump equalling 1, is called a *Poisson process*; see Proposition 4.9 and Definition 4.8 below. We will construct a Poisson process using random variables with an exponential distribution.

Proposition 4.5 *Assume that Z is a positive random variable such that, for all $t, s \geq 0$, $\mathbb{P}(Z > t + s \mid Z > t) = \mathbb{P}(Z > s)$. Then Z has an exponential distribution with parameter a, that is, there exists a constant $a > 0$ such that $\mathbb{P}(Z > t) = e^{-at}$ for $t \geq 0$.*

Proof Let $G(s) := \mathbb{P}(Z > s), s \geq 0$. Then $G(t + s) = G(s)G(t)$ for all $t, s \geq 0$. The function G is right-continuous and positive, so the functional equation has a unique solution of the required form. □

Let $\zeta_1, \zeta_2, \ldots$ be a sequence of independent random variables such that $\mathbb{P}(\zeta_n = 1) = p$ and $\mathbb{P}(\zeta_n = 0) = 1 - p$, $n \in \mathbb{N}$. Let $Z = \inf\{n: \zeta_n = 1\}$. Then ζ has a *geometric distribution* with parameter p, that is, $\mathbb{P}(Z = k) = (1 - p)^{k-1}p$, $k = 1, 2, \ldots$

Lemma 4.6 *Let $\alpha \geq 0$, and let Z_n have a geometric distribution with parameter $p_n = \alpha/n$. Let λ_n be the distribution of Z_n/n. Then (λ_n) converges weakly to an exponential distribution with parameter α.*

Proof The characteristic functional of Z_n/n is

$$\widehat{\lambda}_n(z) := \mathbb{E} \exp\left\{\frac{izZ_n}{n}\right\} = \sum_{k=1}^{\infty} \exp\left\{\frac{izk}{n}\right\} \left(1 - \frac{\alpha}{n}\right)^{k-1} \frac{\alpha}{n}$$

$$= \frac{\alpha}{n} e^{iz/n} \left(1 - \left(1 - \frac{\alpha}{n}\right) e^{iz/n}\right)^{-1}, \qquad z \in \mathbb{R}.$$

Then $\widehat{\lambda}_n(z)$ converges to the characteristic functional of the exponential distribution

$$\alpha \int_0^{\infty} e^{izr} e^{-\alpha r} \, dr = \frac{\alpha}{\alpha - iz}, \qquad z \in \mathbb{R},$$

and the result follows. □

Given $a \in [0, +\infty]$ we denote by $\mathcal{P}(a)$ the *Poisson distribution with parameter* a; that is, $\mathcal{P}(+\infty)(\{+\infty\}) = 1$ and, for $a < \infty$, $\mathcal{P}(a)(\{k\}) = (a^k/k!)e^{-a}$, $k = 0, 1, \ldots$ We leave to the reader the proof of the following well-known lemma.

Lemma 4.7 *Assume that (X_n) is a sequence of independent random variables with Poisson distributions $\mathcal{P}(a_n)$. Then $X = \sum X_n$ has Poisson distribution $\mathcal{P}(a)$, with $a = \sum a_n$. Moreover, the Laplace transform of $\mathcal{P}(a)$ is equal to*

$$\sum_{k=0}^{\infty} e^{-rk} \mathcal{P}(a)(\{k\}) = \sum_{k=1}^{\infty} e^{-rk} \frac{a^k}{k!} e^{-a} = \exp\left\{a(e^{-r} - 1)\right\}, \qquad r > 0,$$

if $a < \infty$ and 0 if $a = \infty$.

Definition 4.8   A *Poisson process with intensity a* is a Lévy process $\Pi = (\Pi(t),$ $t \geq 0)$ such that, for every $t \geq 0$, $\Pi(t)$ has the Poisson distribution $\mathcal{P}(at)$.

The following proposition provides the construction and main properties of the Poisson process.

Proposition 4.9

(i) *Let (Z_n) be a sequence of independent exponentially distributed random variables with parameter a. Then the formula*

$$\Pi(t) = \begin{cases} 0 & \text{if } t < Z_1, \\ k & \text{if } t \in [Z_1 + \cdots + Z_k, Z_1 + \cdots + Z_{k+1}), \end{cases} \tag{4.3}$$

defines a Poisson process with intensity a.

(ii) *Conversely, given a Poisson process with intensity a defined on a probability space $(\Omega, \mathcal{F}, \mathbb{P})$, there exists a sequence (Z_n) of independent random variables defined on $(\Omega, \mathcal{F}, \mathbb{P})$ having an exponential distribution with parameter a such that formula (4.3) holds.*

(iii) *If Π is a Poisson process with intensity a then, for all $z \in \mathbb{C}$ and $t \geq 0$, $\mathbb{E}\, e^{z\Pi(t)} = \exp\{at(e^z - 1)\}$.*

(iv) *If Π is a Poisson process then it has only jumps of size 1, that is,*

$$\mathbb{P}\big(\Delta\Pi(t) := \Pi(t) - \Pi(t-) \in \{0, 1\}\big) = 1, \qquad t \geq 0. \tag{4.4}$$

Conversely, any $\mathbb{Z}_+$-valued Lévy process Π satisfying (4.4) is a Poisson process.

Proof of (i)   Let λ denote the exponential distribution with parameter a, and let (X_i) be a sequence of independent random variables having exponential distributions with parameter a. Then the distribution of $Z_1 + \cdots + Z_n$ is $\lambda^{*n} := \lambda * \lambda * \cdots * \lambda$, where $*$ is the convolution operator. For $n \geq 1$ the measure λ^{*n} has the density

$$g_n(r) = a\frac{(ar)^{n-1}}{(n-1)!}e^{-ar}, \qquad r > 0.$$

Note that, for $k = 1, 2, \ldots,$

$$\begin{aligned} \mathbb{P}\big(\Pi(t) = k\big) &= \mathbb{P}\big(Z_1 + \cdots + Z_k \leq t < Z_1 + \cdots + Z_{k+1}\big) \\ &= \int_0^t g_k(r)e^{-a(t-r)}\, dr = \frac{a^k}{(k-1)!}e^{-at}\int_0^t r^{k-1}\, dr \\ &= \frac{(at)^k}{k!}e^{-at}. \end{aligned}$$

This proves that the law of $\Pi(t)$ is $\mathcal{P}(at)$. We now prove that Π has stationary independent increments. Fix $0 \le t_1 < \cdots < t_k$. For each $n \in \mathbb{N}$, let $\zeta_1^n, \zeta_2^n, \ldots$ be a sequence of independent random variables such that $\mathbb{P}\left(\zeta_m^n = 1\right) = \alpha/n$ and $\mathbb{P}\left(\zeta_m^n = 0\right) = 1 - \alpha/n$ for $m = 1, 2 \ldots$ Let $\Pi^n(m)$ be the number of "successes" (occurrences of 1) in the sequence $\zeta_1^n, \zeta_2^n, \ldots, \zeta_m^n$. Define $m_l^n := [nt_l]$, where $[s]$ denotes the integer part of s. By definition, for each n, the random variables $\Pi^n\left(m_1^n\right), \Pi^n\left(m_2^n\right) - \Pi^n\left(m_1^n\right), \ldots, \Pi^n\left(m_k^n\right) - \Pi^n\left(m_{k-1}^n\right)$ are independent. By a straightforward generalization of Lemma 4.6, the laws of $\Pi^n\left(m_1^n\right), \Pi^n\left(m_2^n\right) - \Pi^n\left(m_1^n\right), \ldots, \Pi^n\left(m_k^n\right) - \Pi^n\left(m_{k-1}^n\right)$ converge weakly as $n \to \infty$ to the law of $\Pi(t_1), \Pi(t_2) - \Pi(t_1), \ldots, \Pi(t_k) - \Pi(t_{k-1})$ and the required independence follows. $\qquad\square$

Proof of (ii)   Note that the law of Π is uniquely determined by its finite-dimensional distributions. Thus it is the same as the law of the process given in the first part of the proposition by (4.3). Thus, in particular, (4.4) holds true and the random variables Z_k can be defined as follows: $Z_1 = \inf\{t : \Pi(t) = 1\}$, $Z_1 + \cdots + Z_n = \inf\{t : \Pi(t) = n\}$. $\qquad\square$

Proof of (iii)   This follows from (ii). Indeed,

$$\mathbb{E}\, e^{z\Pi(t)} = \sum_{k=0}^{\infty} e^{zk}\, \mathbb{P}\left(\Pi(t) = k\right) = \sum_{k=0}^{\infty} e^{-at} e^{zk} \frac{(at)^k}{k!}$$

$$= e^{-at} \sum_{k=0}^{\infty} \frac{\left(ate^z\right)^k}{k!} = e^{-at} \exp\{ate^z\}.$$

$\qquad\square$

Proof of (iv)   We have already shown that every Poisson process satisfies (4.4). Let Π be a $\mathbb{Z}_+$-valued Lévy process satisfying (4.4). Note that Π has a finite number of jumps on any finite interval, that its jumps are equal to 1 and that Π is constant between jumps. We have to show that there exists an $\alpha \ge 0$ such that, for any $t \ge s \ge 0$ and $k = 0, 1, \ldots$,

$$\mathbb{P}\left(\Pi(t) - \Pi(s) = k\right) = e^{-\alpha(t-s)} \frac{(\alpha(t-s))^k}{k!} .$$

Without any loss of generality we can assume that $s = 0$. Consider the increasing sequence $\left((t_0^n, \ldots, t_{2^n}^n)\right), t_k^n = (k/2^n)\, t, k = 0, 1, \ldots, 2^n$ of dyadic partitions of the interval $[0, t]$ and define $\zeta_k^n := \Pi\left(t_{k+1}^n\right) - \Pi\left(t_k^n\right), \widetilde{\zeta}_k^n := \zeta_k^n \wedge 1$, for $k = 0, 1, \ldots$, $2^n - 1$. If $A_n = \left\{\exists k \le 2^n - 1 : \zeta_k^n > 1\right\}$ then $A_n \supset A_{n+1}$. Since Π is a càdlàg process,

$$\lim_{n \to \infty} \mathbb{P}(A_n) = \mathbb{P}\left(\bigcap_{n=1}^{\infty} A_n\right) = 0.$$

Consequently, for all $k = 0, 1, \ldots,$

$$\mathbb{P}(\Pi(t) = k) = \lim_{n \to \infty} \mathbb{P}\big(\widetilde{\zeta}_0^n + \cdots + \widetilde{\zeta}_{2^n-1}^n = k\big)$$

$$= \lim_{n \to \infty} \binom{2^n}{k} \frac{p_n^k}{(1 - p_n)^k} (1 - p_n)^{2^n},$$

where $p_n = \mathbb{P}\big(\widetilde{\zeta}_1^n = 1\big) = \mathbb{P}(\zeta_1^n \geq 1)$. But $1 - p_n = \mathbb{P}\big(\Pi(t2^{-n}) = 0\big)$. Excluding the trivial case $\mathbb{P}(\Pi(t) = 0) = 1$, which corresponds to $\alpha = 0$, we can assume that $0 < \mathbb{P}(\Pi(t) = 0) = \gamma < 1$. But $\gamma = (1 - p_n)^{2^n}$ and $(2^n - j)p_n = (2^n - j)$ $\big(1 - \gamma^{1/2^n}\big) \to -\log \gamma$ for $j = 0, \ldots, k - 1$. Therefore

$$\mathbb{P}(\Pi(t) = k) = \frac{\gamma}{k!} (-\log \gamma)^k,$$

and the proof is completed by setting $e^{-\alpha t} = \gamma$. $\qquad \square$

The random variables Z_n appearing in (4.3) can be interpreted as waiting times for the consecutive occurrences of events such as the arrival of the next customer in a queue, the next car accident or the next call to a telephone exchange. The property articulated in the proposition attributes a lack of memory to the waiting-time mechanism. The value $\Pi(t)$ is the number of events that have occurred before or at time t.

4.2.2 Wiener processes

Definition 4.10 Let $q > 0$. A real-valued mean-zero Gaussian process $W = (W(t), t \geq 0)$ with continuous trajectories and covariance function

$$\mathbb{E}\, W(t)W(s) = (t \wedge s)\, q, \qquad t, s \geq 0,$$

is called a *Wiener process with diffusion q*. If the diffusion is equal to 1 then W is called *standard*.

Definition 4.11 Assume that $(\Omega, \mathcal{F}, (\mathcal{F}_t), \mathbb{P})$ is a filtered probability space and that W is a Wiener process in $\mathbb{R}^d$ adapted to $(\mathcal{F}_t)$. Then W is a *Wiener process with respect to* $(\mathcal{F}_t)$ or an $(\mathcal{F}_t)$-*Wiener process* if, for all $t, h \geq 0$, $W(t + h) - W(t)$ is independent of $\mathcal{F}_t$.

We have the following classical Lévy characterization of a Wiener process; see e.g. Kallenberg (2002).

Theorem 4.12 (Lévy) *A real-valued continuous process W is a standard Wiener process if and only if the process $\big(W^2(t) - t,\ t \geq 0\big)$ is a martingale with respect to its own filtration.*

It is well known (see Kallenberg (2002) or Remark 4.21 below) that any Wiener process has locally Hölder continuous trajectories for any exponent $\alpha < 1/2$. One can show, however, that it does not admit a Hölder continuous modification for an exponent $\alpha \geq 1/2$. In particular, it is not differentiable at any point.

We now pass to the construction of a Wiener process. The first rigorous proof of the existence of such a process was given by Wiener (1923). It was based on Daniell's method (see Daniell 1918) of constructing measures on infinite-dimensional spaces. In Paley and Wiener (1987) the Wiener process is constructed using Fourier series expansions and assuming only the existence of a sequence of independent, identically distributed, Gaussian random variables. Below we present a similar and elegant construction due to P. Lévy and Z. Ciesielski; see Lévy (1948) and Ciesielski (1961).

In the Lévy–Ciesielski construction, an essential role is played by the *Haar system* connected with a dyadic partition of the interval $[0, 1]$. Namely, set $h_0 \equiv 1$ and, for $2^n \leq k < 2^{n+1}$, set

$$h_k(t) = \begin{cases} 2^{n/2} & \text{if } \dfrac{k - 2^n}{2^n} \leq t < \dfrac{k - 2^n}{2^n} + \dfrac{1}{2^{n+1}}, \\[2mm] -2^{n/2} & \text{if } \dfrac{k - 2^n}{2^n} + \dfrac{1}{2^{n+1}} \leq t < \dfrac{k - 2^n}{2^n} + \dfrac{1}{2^n}, \end{cases}$$

$$h_k(1) = 0.$$

The system $(h_k, \, k = 0, 1, \ldots)$ forms an orthonormal basis of $L^2(0, 1)$.

Theorem 4.13 *Let $(X_k, \, k = 0, 1, \ldots)$ be a sequence of independent random variables with distribution $\mathcal{N}(0, 1)$ defined on a probability space $(\Omega, \mathcal{F}, \mathbb{P})$. Then, $\mathbb{P}$-a.s., the series*

$$\sum_{k=0}^{\infty} X_k(\omega) \int_0^t h_k(s) \, ds, \qquad t \in [0, 1],$$

converges uniformly on $[0, 1]$ and defines a Wiener process on $[0, 1]$.

4.3 Compound Poisson processes in a Hilbert space

Definition 4.14 Let ν be a finite measure on a Hilbert space U such that $\nu(\{0\}) = 0$. A *compound Poisson process* with the *Lévy measure* (also called the *jump intensity measure*) ν is a càdlàg Lévy process L satisfying

$$\mathbb{P}(L(t) \in \Gamma) = e^{-\nu(U)t} \sum_{k=0}^{\infty} \frac{t^k}{k!} \nu^{*k}(\Gamma), \qquad \forall t \geq 0, \ \Gamma \in \mathcal{B}(U). \tag{4.5}$$

In the formula above, we use the convention that v^0 is equal to the unit measure concentrated at $\{0\}$, that is, $v^0 = \delta_0$.

The theorem below provides the construction of a compound Poisson process with given v.

Theorem 4.15   *Let v be a finite measure supported on $U \setminus \{0\}$, and let $a = v(U)$.*

(i) *Let $Z_1, Z_2, \ldots$ be independent random variables with identical laws equal to $a^{-1}v$. In addition, let $(\Pi(t), \ t \geq 0)$ be a Poisson process with intensity a, independent of $Z_1, Z_2, \ldots$ Then*

$$L(t) = \sum_{j=1}^{\Pi(t)} Z_j \tag{4.6}$$

is a compound Poisson process with jump intensity measure v.

(ii) *Given a compound Poisson process L with jump intensity measure v, one can find a sequence of independent random variables $Z_1, Z_2, \ldots$ with identical laws equal to $a^{-1}v$ and a Poisson process $(\Pi(t), \ t \geq 0)$ with intensity a, independent of $Z_1, Z_2, \ldots$, such that (4.6) holds.*

(iii) *For $z \in \mathbb{C}$, $t \geq 0$ and $x \in U$,*

$$\mathbb{E}\, e^{z\langle x, L(t)\rangle_U} = \exp\left\{-t \int_U \left(1 - e^{z\langle x, y\rangle_U}\right) v(dy)\right\}.$$

Proof   Let L be given by (4.6). Then L has stationary independent increments, since Π does. To see that (4.5) holds, note that

$$\mathbb{P}(L(t) \in \Gamma) = \mathbb{P}(L(t) \in \Gamma \text{ and } \Pi(t) = 0) + \sum_{k=1}^{\infty} \mathbb{P}(L(t) \in \Gamma \text{ and } \Pi(t) = k).$$

Thus

$$\mathbb{P}(L(t) \in \Gamma) = e^{-at}\delta_{\{0\}}(\Gamma) + \sum_{k=1}^{\infty} \mathbb{P}(Z_1 + \cdots + Z_k \in \Gamma \text{ and } \Pi(t) = k)$$

$$= e^{-at}\delta_{\{0\}}(\Gamma) + \sum_{k=1}^{\infty} \mathbb{P}(Z_1 + \cdots + Z_k \in \Gamma)\mathbb{P}(\Pi(t) = k)$$

$$= e^{-at}\delta_{\{0\}}(\Gamma) + \sum_{k=1}^{\infty} a^{-k}v^{*k}(\Gamma)e^{-at}\frac{(at)^k}{k!}$$

$$= e^{-at}\sum_{k=0}^{\infty} \frac{t^k}{k!}v^{*k}(\Gamma).$$

Since the law of $L(0)$ is equal to $v^0 = \delta_0$, it follows that $L(0) = 0$. The process L is càdlàg by (4.6).

The proof of the second part of the theorem uses arguments from the proof of Proposition 4.9. Since L is right-continuous, its law is determined by finite-dimensional distributions. Thus we define Π by

$$\Pi(t) := \#\{s \leq t \colon \Delta L(s) = L(s) - L(s-) \neq 0\}.$$

By Proposition 4.9, Π is given by (4.3) with a properly chosen sequence (X_n) of independent exponentially distributed random variables. Let $\tau_k = X_1 + \cdots + X_k$. Then (4.3) can be written equivalently in the form $\Pi(t) = \sum_k \delta_{\tau_k}([0, t])$ for $t \geq 0$. Note that the sequence (Z_j) of random variables given by $Z_1 := L(\tau_1), \ldots, Z_k := L(\tau_k) - L(\tau_{k-1})$ has the desired properties.

The last part of the theorem can be shown using arguments from the proof of Proposition 4.9(iii). The details are left to the reader. $\qquad\square$

Let $\Delta L(t) := L(t) - L(t-)$. We define the *Poisson random measure corresponding to L* by the formula

$$\pi([0, t], \Gamma) := \#\{s \leq t \colon \Delta L(s) \in \Gamma\}, \qquad \Gamma \in \mathcal{B}(U \setminus \{0\}).$$

Note that

$$\pi([0, t], \Gamma) = \sum_{n=1}^{\Pi(t)} \delta_{Z_n}(\Gamma) = \sum_{n \colon Z_1 + \cdots + Z_n \leq t} \delta_{Z_n}(\Gamma).$$

Later (see Definition 6.1) we will introduce the concept of a Poisson random measure on an arbitrary measurable space $(E, \mathcal{E})$ with intensity measure λ. According to this general definition, the identity above means that π is a Poisson random measure on $[0, \infty) \times U$ with intensity measure $dt\nu(dx)$. The process

$$\widehat{\pi}([0, t], \Gamma) := \pi([0, t], \Gamma) - t\nu(\Gamma), \qquad t \geq 0, \ \Gamma \in \mathcal{B}(U \setminus \{0\}),$$

is called the *compensated Poisson random measure*. We have the following result.

Proposition 4.16

(i) *For each $\Gamma \in \mathcal{B}(U \setminus \{0\})$, $(\pi([0, t], \Gamma), t \geq 0)$ is a Poisson process with intensity $\nu(\Gamma)$.*

(ii) *If sets $\Gamma_1, \ldots, \Gamma_M$ are disjoint then the random variables $\pi([0, t], \Gamma_j)$, $j = 1, \ldots, M$, are independent.*

(iii) *For each $\Gamma \in \mathcal{B}(U \setminus \{0\})$, the process $(\widehat{\pi}([0, t], \Gamma), t \geq 0)$ is a martingale with respect to the filtration $\left(\overline{\mathcal{F}}_{t+}^{\pi}\right)$, where*

$$\mathcal{F}_t^{\pi} := \sigma\{\pi([0, s], \Gamma) \colon s \leq t, \ \Gamma \in \mathcal{B}(U)\}.$$

Proof By Proposition 4.9(iii), the proof of the first two parts is complete if we
can show that, for all $t \geq 0$, $z_1, \ldots, z_M \in \mathbb{C}$ and disjoint Borel sets $\Gamma_1, \ldots, \Gamma_M$,

$$\mathbb{E} \exp \left\{ \sum_{j=1}^{M} z_j \pi([0, t], \Gamma_j) \right\} = \exp \left\{ \sum_{j=1}^{M} v(\Gamma_j) t (e^{z_j} - 1) \right\}.$$

We can assume that $\{\Gamma_j\}$ is a partition of $U \setminus \{0\}$. Let $a = v(U)$. We have

$$\mathbb{E} \exp \left\{ \sum_{j=1}^{M} z_j \pi([0, t], \Gamma_j) \right\} = \mathbb{E} \exp \left\{ \sum_{n=1}^{\Pi(t)} \sum_{j=1}^{M} z_j \delta_{Z_n}(\Gamma_j) \right\}$$

$$= e^{-at} + \sum_{k=1}^{\infty} \mathbb{P}(\Pi(t) = k) \left(\mathbb{E} \exp \left\{ \sum_{j=1}^{M} z_j \delta_{Z_n}(\Gamma_j) \right\} \right)^k$$

$$= e^{-at} + \sum_{k=1}^{\infty} e^{-at} \frac{(at)^k}{k!} \left(\sum_{j=1}^{M} e^{z_j} \frac{v(\Gamma_j)}{a} \right)^k,$$

which gives the desired conclusion. Since $(\pi([0, t], \Gamma), t \geq 0)$ is a càdlàg process
with independent increments and $\mathbb{E} \pi([0, t], \Gamma) = t v(\Gamma)$, statement (iii) follows
from Proposition 3.25.   □

Remark 4.17 The proposition is true for the jump intensity measure of an arbitrary Lévy process and sets Γ that are separated from the origin, that is, satisfying
$\Gamma \cap \{y : |y|_U \leq r\} = \emptyset$ for r sufficiently small.

Proposition 4.18 *Let L be the compound Poisson process with jump intensity
measure v.*

(i) *The process L is integrable if and only if*

$$\int_U |y|_U v(dy) < \infty. \tag{4.7}$$

Moreover, if (4.7) holds then

$$\mathbb{E} L(t) = t \int_U y v(dy) \tag{4.8}$$

*and the compensated compound process $\widehat{L}(t) = L(t) - \mathbb{E} L(t), t \geq 0$, is a
martingale with respect to $\left(\overline{\mathcal{F}}_{t+}^L \right)$.*

(ii) *For all $z \in \mathbb{C}$, $t \geq 0$ and $x \in U$,*

$$\mathbb{E} e^{z \langle x, \widehat{L}(t) \rangle_U} = \exp \left\{ -t \int_U \left(1 - e^{z \langle x, y \rangle_U} + z \langle x, y \rangle_U \right) v(dy) \right\}.$$

(iii) *The process L, and hence $\widehat{L}$, is square integrable if and only if*

$$\int_U |y|_U^2 \nu(dy) < \infty. \tag{4.9}$$

Moreover $\mathbb{E} \left|\widehat{L}(t)\right|_U^2 = t \int_U |y|_U^2 \nu(dy)$ *and, for all* $x, \tilde{x} \in U$ *and* $t \geq 0$,

$$\mathbb{E} \left\langle \widehat{L}(t), x \right\rangle_U \left\langle \widehat{L}(t), \tilde{x} \right\rangle_U = t \int_U \langle x, y \rangle_U \langle \tilde{x}, y \rangle_U \nu(dy).$$

Proof Let $a = \nu(U)$. Then

$$\mathbb{E} |L(t)|_U = \sum_{k=1}^\infty \mathbb{E} \left| \sum_{j=1}^k Z_j \right|_U e^{-at} \frac{(at)^k}{k!}$$

$$= \sum_{k=1}^\infty \int_U \cdots \int_U \left| \sum_{j=1}^k y_j \right|_U \nu(dy_1) \cdots \nu(dy_k) e^{-at} \frac{t^k}{k!}$$

$$\leq \sum_{k=1}^\infty k \int_U |y|_U \nu(dy) a^{k-1} e^{-at} \frac{t^k}{k!} \leq t \int_U |y|_U \nu(dy).$$

Since

$$\mathbb{E} |L(t)|_U \geq \mathbb{E} \left(|Z_1|_U \, \chi_{\{\pi(t)=1\}} \right) = \mathbb{E} |Z_1|_U \, e^{-at} \, at = t \, e^{-at} \int_U |y|_U \nu(dy),$$

we have the desired equivalence. In the same way we obtain

$$\mathbb{E} L(t) = \sum_{k=1}^\infty \int_U \cdots \int_U \left(\sum_{j=1}^k y_j \right) \nu(dy_1) \cdots \nu(dy_k) e^{-at} \frac{t^k}{k!}$$

$$= \sum_{k=1}^\infty k \int_U y \nu(dy) (\nu(U))^{k-1} e^{-at} \frac{t^k}{k!}$$

$$= \sum_{k=1}^\infty \int_U y \nu(dy) a^{k-1} e^{-at} \frac{t^k}{(k-1)!} = \int_U y \nu(dy).$$

The martingale property of $\widehat{L}$ and (iii) follows from Proposition 3.25 and Theorem 4.15. In order to compute the second moment note that

$$\mathbb{E} |L(t)|_U^2 = \sum_{k=1}^\infty \mathbb{E} \left| \sum_{j=1}^k Z_j \right|_U^2 e^{-at} \frac{(at)^k}{k!}$$

$$= \sum_{k=1}^\infty \sum_{j,l=1}^k \mathbb{E} \langle Z_j, Z_l \rangle_U e^{-at} \frac{(at)^k}{k!}.$$

Now

$$\mathbb{E} |Z_j|_U^2 = \frac{1}{a} \int_U |y|_U^2 \nu(dy),$$

and if $j \neq l$ then

$$\mathbb{E} \langle Z_j, Z_l \rangle_U = \langle \mathbb{E} Z_j, \mathbb{E} Z_l \rangle_U = \frac{1}{a^2} \left| \int_U y\nu(dy) \right|_U^2 .$$

Hence

$$\mathbb{E} |L(t)|_U^2 = t \int_U |z|_U^2 \nu(dy) + t^2 \left| \int_U y\nu(dy) \right|_U^2 .$$

Let $x, \tilde{x} \in U$. Then

$$\mathbb{E} \langle L(t), x \rangle_U \langle L(t), \tilde{x} \rangle_U = \sum_{k=1}^{\infty} \sum_{j,l=1}^{k} \mathbb{E} \langle Z_l, x \rangle_U \langle Z_j, \tilde{x} \rangle_U \, e^{-at} \frac{(at)^k}{k!}$$

$$= t \int_U \langle y, x \rangle_U \langle y, \tilde{x} \rangle_U \nu(dy)$$

$$+ \sum_{k=2}^{\infty} k(k-1) \int_U \int_U \langle y, x \rangle_U \langle \tilde{y}, \tilde{x} \rangle_U \nu(dy)\nu(d\tilde{y}) \frac{1}{a^2} e^{-at} \frac{(at)^k}{k!}$$

$$= t \int_U \langle y, x \rangle_U \langle y, \tilde{x} \rangle_U \nu(dy) + t^2 \int_U \int_U \langle y, x \rangle_U \langle \tilde{y}, \tilde{x} \rangle_U \nu(dy)\nu(d\tilde{y}).$$

Hence, from (4.8),

$$\mathbb{E} \big(\widehat{L}(t), x \big)_U \big(\widehat{L}(t), \tilde{x} \big)_U = \mathbb{E} \langle L(t), x \rangle_U \langle L(t), \tilde{x} \rangle_U - \mathbb{E} \langle L(t), x \rangle_U \, \mathbb{E} \langle L(t), \tilde{x} z \rangle_U$$

$$= t \int_U \langle y, x \rangle_U \langle y, \tilde{x} \rangle_U \nu(dy).$$

$\square$

4.4 Wiener processes in a Hilbert space

It is convenient to start with the following general definition.

Definition 4.19 A mean-zero Lévy process W with continuous trajectories in U is called a *Wiener process*.

The following theorem gathers basic properties of Wiener processes taking values in a Hilbert space.

Theorem 4.20 *Let W be a Wiener process in U. Then W is Gaussian and square integrable. Moreover, for all $t_1, \ldots, t_n \geq 0$ and $x_1, \ldots, x_n \in U$, the random vector*

$(\langle W(t_1), x_1 \rangle_U, \ldots, \langle W(t_n), x_n \rangle_U)$ *has a normal distribution* $\mathcal{N}(0, [q_{i,j}])$, *where*

$$q_{i,j} = (t_i \wedge t_j)\langle Qx_i, x_j \rangle_U, \qquad i, j = 1, \ldots, n, \tag{4.10}$$

and Q is the covariance operator of W. Moreover, let $\{e_n\}$ be the orthonormal basis of U consisting of eigenvectors of the covariance operator Q of W, and let $\{\gamma_n\}$ be the corresponding eigenvalues. Then

$$W(t) = \sum_n W_n(t)e_n, \qquad t \geq 0, \tag{4.11}$$

where the real-valued Wiener processes

$$W_n(t) = \langle W(t), e_n \rangle_U, \qquad n \in \mathbb{N}, \tag{4.12}$$

are independent and have covariances

$$\mathbb{E}\, W_n(t)W_n(s) = (t \wedge s)\gamma_n, \tag{4.13}$$

and the series (4.11) converges $\mathbb{P}$-a.s. and in $L^2(\Omega, \mathcal{F}, \mathbb{P}; U)$.

Proof By Theorem 4.12 and Definition 3.30, W is Gaussian. Hence, by Theorem 3.31 it is square integrable, and its covariance operator Q is non-negative and of trace class;[1] thus (4.11) follows. Let W_n be given by (4.12). Then

$$\mathbb{E}\, W_n(t)W_m(s) = (t \wedge s)\langle Qe_n, e_m \rangle_U = \gamma_n \delta_{n,m}(t \wedge s).$$

Convergence $\mathbb{P}$-a.s. follows from the identity $x = \sum_n \langle x, e_n \rangle_U e_n$. To show convergence in $L^2(\Omega, \mathcal{F}, \mathbb{P}; U)$, note that

$$\mathbb{E}\left| \sum_{n=k}^N W_k(t)e_n \right|_U^2 = \sum_{n=k}^N \gamma_k t \to 0 \qquad \text{as } k, N \to \infty.$$

$\square$

Remark 4.21 Assume that Q belongs to the class $L_1^+(U)$ of non-negative trace-class operators on U. Then, using (4.11)–(4.13), we can construct a Gaussian process with covariance Q. In order to show that any such process has a continuous modification we will prove that for every m there is a constant C_m such that $\mathbb{E}\,|W(h)|_U^{2m} \leq C_m h^m$ for $h > 0$. By the Kolmogorov criterion this guarantees that W has Hölder-continuous trajectories with an arbitrary exponent $\alpha < 1/2$. To this end we fix m. We have

$$\mathbb{E}\,|W(t) - W(s)|_U^{2m} = \mathbb{E}\,|W(t-s)|_U^{2m} = \mathbb{E}\left(\sum_k W_k^2(t-s) \right)^m,$$

---

[1] Briefly, $Q \in L_1^+(U)$; see Appendix A.

where $W_k(t) = \langle W(t), e_k \rangle_U$ and $\{e_k\}$ is an orthonormal basis consisting of eigenvectors of Q. Let $\{\gamma_k\}$ be the corresponding sequence of eigenvalues. Since $W_k = 0$ for $\gamma_k = 0$, we can assume that $\gamma_k \neq 0$ for all k. Then

$$
\begin{aligned}
\mathbb{E} \, |W(h)|_U^{2m} &= \lim_{n \to \infty} \mathbb{E} \left(\sum_{k=1}^{n} W_k^2(h) \right)^m \\
&= \lim_{n \to \infty} \left(\prod_{k=1}^{n} \frac{1}{\sqrt{2\pi \gamma_j h}} \right) \int_{\mathbb{R}^n} \left(\sum_{k=1}^{n} \xi_k^2 \right)^m \exp \left\{ -\sum_{k=1}^{n} \frac{\xi_k^2}{2\gamma_k h} \right\} \mathrm{d}\xi_1 \cdots \mathrm{d}\xi_n \\
&= \lim_{n \to \infty} h^m \left(\prod_{k=1}^{n} \frac{1}{\sqrt{2\pi \gamma_j}} \right) \int_{\mathbb{R}^n} \left(\sum_{k=1}^{n} \eta_k^2 \right)^m \exp \left\{ -\sum_{k=1}^{n} \frac{\eta_k^2}{2\gamma_k} \right\} \mathrm{d}\eta_1 \cdots \mathrm{d}\eta_n \\
&= h^m \, \mathbb{E} \, |W(1)|_U^{2m}.
\end{aligned}
$$

Since Theorem 3.31 gives $\mathbb{E} \, |W(1)|_U^{2m} < \infty$ for $m \in \mathbb{N}$, we arrive at the desired estimate.

Remark 4.22   The reasoning above leads to the scaling property of a Wiener process; namely, for all $t, h \geq 0$, the laws of $W(th)$ and $t^{1/2} W(h)$ are identical.

Since $\langle W(t), x \rangle_U$ has the distribution $\mathcal{N}(0, t \langle Qx, x \rangle_U)$, we have

$$
\mathbb{E} \, \mathrm{e}^{\mathrm{i}\langle x, W(t) \rangle_U} = \exp \left\{ -\frac{t}{2} \langle Qx, x \rangle_U \right\}, \qquad t \geq 0, \ x \in U. \tag{4.14}
$$

4.5 Lévy–Khinchin decomposition

Assume that L is a càdlàg Lévy process on a Hilbert space U. Given a Borel set A separated from 0 (see Remark 4.17), write

$$
\pi_A(t) := \sum_{s \leq t} \chi_A(\Delta L(s)), \qquad t \geq 0.
$$

Note that the càdlàg property of L implies that π_A is $\mathbb{Z}_+$-valued. Clearly it is a Lévy process with jumps of size 1. Thus, by Proposition 4.9(iv), π_A is a Poisson process. Note that $\mathbb{E} \, \pi_A(t) = t \, \mathbb{E} \, \pi_A(1) = t\nu(A)$, where ν is a measure that is finite on sets separated from 0. Write

$$
L_A(t) := \sum_{s \leq t} \chi_A(\Delta L(s))\Delta L(s).
$$

Then L_A is a well-defined Lévy process. Our aim is to prove the following Lévy–Khinchin decomposition.

Theorem 4.23 (Lévy–Khinchin decomposition)

(i) *If ν is a jump intensity measure corresponding to a Lévy process then*

$$\int_U \left(|y|_U^2 \wedge 1\right) \nu(dy) < \infty. \tag{4.15}$$

(ii) *Every Lévy process has the following representation:*

$$L(t) = at + W(t) + \sum_{k=1}^{\infty} \left(L_{A_k}(t) - t \int_{A_k} y\nu(dy)\right) + L_{A_0}(t),$$

where $A_0 := \{x : |x|_U \geq r_0\}$, $A_k := \{x : r_k \leq |x|_U < r_{k-1}\}$, (r_k) is an arbitrary sequence decreasing to 0, W is a Wiener process, all members of the representation are independent processes and the series converges $\mathbb{P}$-a.s. uniformly on each bounded subinterval of $[0, \infty)$.

It follows from the proof, given below, that the processes

$$L_n(t) := L_{A_n}(t) - t \int_{A_n} y\nu(dy), \qquad t \geq 0, \tag{4.16}$$

are independent compensated compound Poisson processes. Hence we have the decomposition

$$L(t) = at + W(t) + \sum_{n=1}^{\infty} L_n(t) + L_0(t), \qquad t \geq 0, \tag{4.17}$$

of the Lévy process L, where the processes $W, L_n, n \geq 0$, and L_0 are independent, W is a Wiener process, L_0 is a compound Poisson process with jump intensity measure $\chi_{\{|y|_U \geq r_0\}}(y)\nu(dy)$ and each L_n is a compensated compound Poisson process with jump intensity measure

$$\chi_{\{r_{n+1} \leq |y|_U < r_n\}}(y)\nu(dy).$$

A similar representation theorem holds not only for Hilbert spaces but also for Banach spaces (see Kruglov 1984, Linde 1986, Tortrat 1967, 1968 and Section 4.10) and for some topological linear spaces such as the space of tempered distributions on Euclidean spaces; see Ustunel (1984). In Section 14.2 we will investigate basic properties of Lévy processes on the space of tempered distributions.

The proof of Theorem 4.23 follows Gikhman and Skorokhod (1974), Vol. II, and requires several lemmas. The proof of the first is found in Appendix F.

Lemma 4.24 *For any disjoint Borel sets $A_1, \ldots, A_m$ separated from zero, the processes $L_{A_1}, \ldots, L_{A_m}, L - L_{A_1}, \ldots, L - L_{A_m}$ are independent Lévy.*

Lemma 4.25 *For every Borel set A separated from 0 and for all $u \in U$,*

$$\mathbb{E} \exp\{i \langle u, L_A(t)\rangle_U\} = \exp\left\{-t \int_A \left(1 - e^{i\langle u,x\rangle_U}\right) \nu(dx)\right\}.$$

Proof By an easy limiting argument we can see that A is bounded. Given $\delta > 0$ let $A_1, \ldots, A_m$ be disjoint sets of diameters less than δ and such that $A = \bigcup_{k=1}^m A_k$. In addition let $x_k \in A_k$, $k = 1, \ldots, m$. Then

$$\left| L_A(t) - \sum_{k=1}^m x_k \pi_{A_k}(t) \right|_U \leq \sum_{k=1}^m \left| L_{A_k}(t) - x_k \pi_{A_k}(t) \right|_U$$

$$\leq \delta \sum_{k=1}^m \pi_{A_k}(t) = \delta \pi_A(t),$$

and therefore

$$\sum_{k=1}^m x_k \pi_{A_k}(t) \to L_A(t),$$

$\mathbb{P}$-a.s., as $\delta \to 0$. Consequently,

$$\mathbb{E} \exp \left\{ i \langle y, L_A(t) \rangle_U \right\} = \lim_{\delta \to 0} \mathbb{E} \exp \left\{ i \left\langle y, \sum_{k=1}^m x_k \pi_{A_k}(t) \right\rangle_U \right\}$$

$$= \lim_{\delta \to 0} \prod_{k=1}^m \mathbb{E} \exp \left\{ i \langle y, x_k \rangle_U \pi_{A_k}(t) \right\}$$

$$= \lim_{\delta \to 0} \prod_{k=1}^m \exp \left\{ -t \nu(A_k) \left(1 - e^{i \langle y, x_k \rangle_U} \right) \right\}$$

$$= \lim_{\delta \to 0} \exp \left\{ -t \sum_{k=1}^m \left(1 - e^{i \langle y, x_k \rangle_U} \right) \nu(A_k) \right\}$$

$$= \exp \left\{ -t \int_A \left(1 - e^{i \langle y, x \rangle_U} \right) \nu(dx) \right\}.$$

$\square$

Assume now that L_n, $n \in \mathbb{N}$, is given by (4.16).

Lemma 4.26 *If assumption (4.15) is satisfied then the series in (4.17) converges $\mathbb{P}$-a.s. uniformly on each bounded interval $[0, T]$.*

Proof By Proposition 4.18(iii) and assumption (4.15),

$$\mathbb{E} \left| \sum_{k=n_0}^n L_k(T) \right|_U^2 = T \int_{\{r_{n+1} \leq |y|_U < r_{n_0}\}} |y|_U^2 \nu(dy)$$

$$\leq T \int_{\{|y|_U < r_{n_0}\}} |y|_U^2 \nu(dy) < \infty.$$

Since L_k, $k \in \mathbb{N}$, is a martingale,

$$Z_{n_0,n}(t) := \left| \sum_{k=n_0}^{n} L_k(t) \right|_U^2, \qquad t \geq 0,$$

is a submartingale. Note that $Z_{n_0,n}$ is càdlàg. Thus, by Doob's inequality,

$$\mathbb{P}\left(\sup_{t \in [0,T]} Z_{n_0,m}(t) \geq \varepsilon \right) \leq \frac{\mathbb{E}\, Z_{n,m}(T)}{\varepsilon} = \frac{T}{\varepsilon} \int_{\{r_{n+1} \leq |y|_U < r_{n_0}\}} |y|_U^2 \, \nu(dy).$$

Therefore, the series converges in probability uniformly on $[0, T]$. Since the L_k, $k = 1, 2, \ldots$, are independent random elements in the space of all càdlàg functions with the supremum norm, the desired $\mathbb{P}$-a.s. convergence follows from Corollary 3.12. □

Proof of Theorem 4.23 We show first that condition (4.15) is satisfied. The Lévy process $\tilde{L} = L - L_0$ has jumps bounded by 1 and therefore, by Theorem 4.4, the process $\tilde{L}$ has finite second moments. Let A_k, $k = 1, \ldots, n$, be the sets appearing in the formulation of the theorem. Then for each n the processes

$$\tilde{L}(t) - \sum_{k=1}^{n} L_{A_k}(t), \qquad \sum_{k=1}^{n} L_{A_k}(t), \qquad t \geq 0,$$

are independent (Lévy) processes. Consequently, setting

$$\tilde{L}_n(t) := \left(\tilde{L}(t) - \sum_{k=1}^{n} L_{A_k}(t) \right) - \mathbb{E}\left(\tilde{L}(t) - \sum_{k=1}^{n} L_{A_k}(t) \right),$$

we obtain

$$\mathbb{E}\left| \tilde{L}(t) - \mathbb{E}\, \tilde{L}(t) \right|_U^2 = \mathbb{E}\left| \tilde{L}_n(t) \right|_U^2 + \mathbb{E}\left| \sum_{k=1}^{n} L_{A_k}(t) - \mathbb{E} \sum_{k=1}^{n} L_{A_k}(t) \right|_U^2.$$

Therefore, for every $n \in \mathbb{N}$,

$$\int_{\{r_{n+1} \leq |y|_U < r_0\}} |y|_U^2 \, \nu(dy) = \mathbb{E}\left| \sum_{k=1}^{n} \left(L_{A_k}(t) - \mathbb{E}L_{A_k}(t) \right) \right|_U^2$$

$$\leq \mathbb{E}\left| \tilde{L}(t) - \mathbb{E}\, \tilde{L}(t) \right|_U^2 < \infty,$$

and the first part of the theorem follows.

We now proceed to the proof of part (ii) of the theorem. The convergence of the series is a consequence of Lemma 4.26. Define

$$\tilde{W}(t) = L(t) - \sum_{k=1}^{\infty} \left(L_{A_k}(t) - t \int_{A_k} y\nu(dy) \right) - L_{A_0}(t), \qquad t \geq 0.$$

Then $\tilde{W}$ is a Lévy process with continuous trajectories. Thus $a := \mathbb{E}\tilde{W}(1)$ is finite and $W(t) := \tilde{W}(t) - at, t \geq 0$, is a mean-zero Lévy process with continuous trajectories. In particular, for each vector $u \in U$ the process $(\langle u, W(t) \rangle_U, t \geq 0)$ has continuous trajectories and therefore, by the Lévy characterization of real-valued Wiener processes, it is a Wiener process. Consequently, by the definition of the infinite-dimensional Gaussian distribution, W is a Gaussian Lévy process. $\square$

4.6 Lévy–Khinchin formula

The following result is a direct consequence of the Lévy–Khinchin decomposition. For a different proof see e.g. Parthasarathy (1967) or Linde (1986).

Theorem 4.27 (Lévy–Khinchin formula)

(i) *Given $a \in U$, $Q \in L_1^+(U)$ and a non-negative measure ν concentrated on $U \setminus \{0\}$ satisfying (4.15), there is a convolution semigroup (μ_t) of measures such that*

$$\int_U e^{i\langle x, y \rangle_U} \mu_t(\mathrm{d}y) = e^{-t\psi(x)}, \tag{4.18}$$

where

$$\psi(x) = -i\langle a, x \rangle_U + \tfrac{1}{2}\langle Qx, x \rangle_U$$
$$+ \int_U \left(1 - e^{i\langle x, y \rangle_U} + \chi_{\{|y|_U < 1\}}(y)i\langle x, y \rangle_U\right)\nu(\mathrm{d}y). \tag{4.19}$$

(ii) *Conversely, for each convolution semigroup (μ_t) of measures, there exist $a \in U$, $Q \in L_1^+(U)$ and a non-negative measure ν concentrated on $U \setminus \{0\}$ satisfying (4.15) in such a way that (4.18) holds with ψ defined by (4.19).*

Definition 4.28   Let L be a Lévy process and let (μ_t) be the family of its distributions. We call the measure ν appearing in (4.19) the *Lévy measure* or the *jump intensity measure* of L or (μ_t). We call the triple (a, Q, ν) the *characteristics* of L.

The Lévy–Khinchin formula gives the characteristic function of a Lévy process. It turns out that it is also useful for computing characteristic functionals of stochastic integrals. As an example of a simple application, we present the following result.

Corollary 4.29   *Let L be a real-valued Lévy process with exponent ψ, and let $f \colon \mathbb{R} \mapsto \mathbb{R}$. Assume that the Riemann–Stieltjes integrals $\int_0^t f(s)\,\mathrm{d}L(s)$ and*

$\int_0^t \psi\, (xf(s))\, ds$ *exist. Then*

$$\mathbb{E} \exp\left\{ ix \int_0^t f(s)\, dL(s) \right\} = \exp\left\{ -\int_0^t \psi(xf(s))\, ds \right\}, \qquad x \in \mathbb{R}.$$

Proof Let (s_j^n), $j = 0, \ldots k_n$, $n \in \mathbb{N}$, be a sequence of partitions on $[0, t]$ satisfying standard assumptions. Then

$$\mathbb{E} \exp\left\{ ix \int_0^t f(s)\, dL(s) \right\}$$

$$= \lim_{n \to \infty} \mathbb{E} \exp\left\{ ix \sum_{j=0}^{k_n - 1} f(s_j^n)(L(s_{j+1}^n) - L(s_j^n)) \right\}$$

$$= \lim_{n \to \infty} \prod_{j=0}^{k_n - 1} \mathbb{E} \exp\left\{ ixf(s_j^n)(L(s_{j+1}^n) - L(s_j^n)) \right\}$$

$$= \lim_{n \to \infty} \prod_{j=0}^{k_n - 1} \exp\left\{ \psi(xf(s_j^n))(s_{j+1}^n - s_j^n) \right\}$$

$$= \exp\left\{ -\int_0^t \psi(xf(s))\, ds \right\}.$$

$\square$

The assumptions on Riemann–Stieltjes integrability can be relaxed. This, however, requires some knowledge of stochastic integration; see Theorem 6.6.

4.7 Laplace transforms of convolution semigroups

In some situations it is more convenient to determine convolution semigroups of measures in terms of Laplace rather than Fourier transforms.

Theorem 4.30 *Let (μ_t) be a convolution semigroup of measures on a Hilbert space U, with exponent given by (4.18) and (4.19).*

(i) *Let $x \in U$. Then the Laplace transform $\int_U e^{-\langle x, y \rangle_U} \mu_t(dy)$ is finite for some $t > 0$ (equivalently for all $t > 0$) if and only if*

$$\int_{\{|y|_U \geq 1\}} e^{-\langle x, y \rangle_U} \nu(dy) < \infty. \tag{4.20}$$

(ii) *If (4.20) holds then*

$$\int_U e^{-\langle x, y \rangle_U} \mu_t(dy) = e^{-t\tilde{\psi}(x)}, \qquad \forall t > 0,$$

where

$$\tilde{\psi}(x) = \langle a, x \rangle_U - \tfrac{1}{2} \langle Qx, x \rangle_U + \tilde{\psi}_0(x),$$

$$\tilde{\psi}_0(x) = \int_U \left(1 - e^{-\langle x, y \rangle_U} - \langle x, y \rangle_U \chi_{\{|y|_U \leq 1\}} \right) \nu(dy).$$

Proof Let L be a Lévy process with distributions $\mu_t, t \geq 0$. Note that if X and Y are independent random variables then $\mathbb{E} \, e^{-X-Y} = \mathbb{E} \, e^{-X} \mathbb{E} \, e^{-Y}$ and the left-hand side is finite if and only if both terms on the right-hand side are finite. Thus we can assume that, in the Lévy–Khinchin decomposition of L, $a = 0$ and $W \equiv 0$. Denote by ν_0 the restriction of ν to the set $A_0 := \{y \in U : |y|_U \geq r_0\}$. Let (r_k), (A_k) and (L_{A_k}) be as in Theorem 4.23. Then, by (4.5),

$$\mathbb{E} \, e^{-\langle x, L_{A_0}(t) \rangle_U} = e^{-t\nu_0(U)} \sum_{k=0}^{\infty} \frac{t^k}{k!} \left(\int_U e^{-\langle x, y \rangle_U} \nu_0(dy) \right)^k$$

$$= \exp \left\{ -t \int_U \left(1 - e^{-\langle x, y \rangle_U} \right) \nu_0(dy) \right\}$$

and, since $\nu_0(U) < \infty$, $\mathbb{E} \, e^{-\langle x, L_{A_0}(t) \rangle_U} < \infty$ if and only if (4.20) holds. The fact that

$$\mathbb{E} \exp \left\{ - \left\langle x, \sum_{k=1}^{\infty} \left(L_{A_k}(t) - t \int_{A_k} y\nu(dy) \right) \right\rangle_U \right\} < \infty$$

follows directly from Theorem 4.4. We prefer to give a self-contained proof for the reader's convenience. Namely, for every $\tilde{x} \in U$,

$$\mathbb{E} \exp \left\{ - \left\langle \tilde{x}, \sum_{k=1}^{N} \left(L_{A_k}(t) - t \int_{A_k} y\nu(dy) \right) \right\rangle_U \right\}$$

$$= \exp \left\{ -t \int_U \left(1 - e^{-\langle \tilde{x}, y \rangle_U} - \langle \tilde{x}, y \rangle_U \chi_{[r_N, r_0)}(|y|_U) \right) \nu(dy) \right\}. \quad (4.21)$$

Since $\int_{\{|y|_U < r_0\}} |y|_U^2 \nu(dy) < \infty$, the right-hand side of (4.21) tends to the (finite) quantity

$$\exp \left\{ -t \int_{\{|y|_U < r_0\}} \left(1 - e^{-\langle \tilde{x}, y \rangle_U} - \langle \tilde{x}, y \rangle_U \right) \nu(dy) \right\}.$$

Therefore, by Fatou's lemma,

$$\mathbb{E} \exp \left\{ - \left\langle \tilde{x}, \sum_{k=1}^{\infty} \left(L_{A_k}(t) - t \int_{A_k} y\nu(dy) \right) \right\rangle_U \right\}$$

$$\leq \exp \left\{ -t \int_{\{|y|_U < r_0\}} \left(1 - e^{-\langle \tilde{x}, y \rangle_U} - \langle \tilde{x}, y \rangle_U \right) \nu(dy) \right\} < \infty. \quad (4.22)$$

In particular, (4.22) holds for $\tilde{x} = \gamma x$, $\gamma \geq 1$, which implies that the random variables

$$\exp\left\{-\left\langle x, \sum_{k=1}^{N}\left(L_{A_k}(t) - t\int_{A_k} y\nu(\mathrm{d}y)\right)\right\rangle_U\right\}, \qquad N \in \mathbb{N},$$

are uniformly integrable. Therefore

$$\mathbb{E}\exp\left\{-\left\langle x, \sum_{k=1}^{\infty}\left(L_{A_k}(t) - t\int_{A_k} y\nu(\mathrm{d}y)\right)\right\rangle_U\right\}$$

$$= \lim_{N\to\infty}\mathbb{E}\exp\left\{-\left\langle x, \sum_{k=1}^{N}\left(L_{A_k}(t) - t\int_{A_k} y\nu(\mathrm{d}y)\right)\right\rangle_U\right\}$$

$$= \exp\left\{-t\int_U \left(1 - e^{-\langle x,y\rangle_U} - \langle x, y\rangle_U \chi_{\{|y|_U < r_0\}}\right)\nu(\mathrm{d}y)\right\} < \infty,$$

which completes the proof. $\qquad\qquad\qquad\qquad\qquad\qquad\qquad\qquad\qquad\square$

If λ is a measure on $\mathbb{R}$ then we denote its Laplace transform by

$$\tilde{\lambda}(r) = \int_{-\infty}^{+\infty} e^{-r\xi}\,\lambda(\mathrm{d}\xi)$$

for those values of r for which the integral is finite. In particular, if λ has support on $[0, +\infty)$ then $\tilde{\lambda}(r)$ is defined at least for $r \geq 0$.

The theorem below can be proved in a similar way to the Lévy–Khinchin formula. A different and much shorter proof, based on the theory of C_0-semigroups, is given in Appendix E. The semigroup-theoretic method can also be used to prove the Lévy–Khinchin formula on $\mathbb{R}^d$; see Zabczyk (2003).

Theorem 4.31 *A family (λ_t) of measures on $[0, +\infty)$ is a convolution semigroup of measures if and only if their Laplace transforms $\tilde{\lambda}_t$ are of the form*

$$\tilde{\lambda}_t(r) = e^{-t\tilde{\psi}(r)}, \qquad \tilde{\psi}(r) = \gamma r + \int_0^{+\infty}\left(1 - e^{-r\xi}\right)\nu(\mathrm{d}\xi), \qquad r > 0,$$

where γ is a positive constant and ν is a non-negative measure on $(0, +\infty)$ satisfying

$$\int_0^1 \xi\,\nu(\mathrm{d}\xi) < \infty, \qquad \int_1^{+\infty}\nu(\mathrm{d}\xi) < \infty.$$

The family (λ_t) described in this theorem is called a *subordinator*. The theorem is concerned with measures on the half line $[0, +\infty)$; it is, however, possible to find Laplace transforms of some important families on $\mathbb{R}$.

Theorem 4.32   *Assume that v is a measure on $(0, +\infty)$ satisfying the following conditions:*

$$\int_0^1 \xi^2 \, v(d\xi) < \infty \quad and \quad \int_1^{+\infty} \xi v(d\xi) < \infty.$$

Then there exists a convolution semigroup (μ_t) of measures on $\mathbb{R}$ such that

$$\int_{\mathbb{R}} e^{ir\xi} \mu_t(d\xi) = e^{-t\psi(r)}, \qquad r \in \mathbb{R},$$

where the Lévy exponent is given by

$$\psi(r) = \int_0^{+\infty} \left(1 - e^{ir\xi} + ir\xi\right) v(d\xi), \qquad r \in \mathbb{R}.$$

Moreover, for all $r > 0$ and $t > 0$, $\int_{\mathbb{R}} e^{-r\xi} \mu_t(d\xi) < \infty$ and

$$\int_{\mathbb{R}} e^{-r\xi} \mu_t(d\xi) = e^{-t\widetilde{\psi}(r)},$$

where

$$\widetilde{\psi}(r) = \int_0^{+\infty} \left(1 - e^{-r\xi} - r\xi\right) v(d\xi).$$

Proof   The first part follows directly from the general Lévy–Khinchin theorem. Since

$$\int_0^{+\infty} \left(1 - e^{-r\xi} - r\xi\right) v(d\xi)$$
$$= \int_0^{+\infty} \left(1 - e^{-r\xi} - r\xi \chi_{[-1,1]}(\xi)\right) v(d\xi) - r \int_1^{+\infty} \xi v(d\xi),$$

the second part follows from Theorem 4.30.   □

Remark 4.33   Clearly any Lévy process L corresponding to a convolution semigroup of measures (λ_t) with support in $[0, +\infty)$ has increasing trajectories. The process L corresponding to the semigroup constructed in Theorem 4.32 has only positive jumps but, owing to a drift, it takes strictly negative values with positive probability.

4.7.1 Examples

We present some specific examples of convolution semigroups of measures important in applications.

Example 4.34 (Stable families of order $\beta \in (0, 1)$)   We are in the framework of Theorem 4.31. Assume that $\gamma = 0$ and $v(d\xi) = \xi^{-1-\beta} \, d\xi$ for $\beta \in (0, 1)$.

Then

$$\tilde{\psi}(r) = \int_0^{+\infty} \left(1 - e^{-r\xi}\right) \frac{d\xi}{\xi^{1+\beta}}, \qquad r > 0.$$

For $\alpha > 0$, the change of variable $\eta = r\xi$ gives

$$\tilde{\psi}(r) = \int_0^{+\infty} \left(1 - e^{-r\xi}\right) \frac{d\xi}{\xi^{1+\beta}} = \int_0^{+\infty} \left(1 - e^{-\eta}\right) \frac{r^\beta \, d\eta}{\eta^{1+\beta}} = r^\beta \tilde{\psi}(1).$$

Integrating by parts we obtain $\tilde{\psi}(1) = (1/\beta)\Gamma(1-\beta)$.

If $\beta = 1/2$ then there exists an explicit formula for the density of λ_t (see Feller 1971):

$$\lambda_t(d\xi) = \frac{t}{\sqrt{2\pi}} \frac{1}{\sqrt{\xi^3}} e^{-t^2/(2\xi)} \, d\xi.$$

Example 4.35 (Stable families of order $\beta \in (1, 2)$) Assume that $\beta \in (1, 2)$ and

$$\tilde{\psi}(r) = \int_0^{+\infty} \left(1 - e^{-r\xi} - r\xi\right) \frac{d\xi}{\xi^{1+\beta}}, \qquad r > 0.$$

Then, in the same way as in Example 4.34, we obtain $\tilde{\psi}(r) = r^\beta \tilde{\psi}(1)$, $r > 0$, where $\tilde{\psi}(1) = -[1/\beta(\beta-1)]\Gamma(2-\beta)$.

Here are examples of convolution semigroups of measures on $\mathbb{R}^d$.

Example 4.36 (α-stable families) Assume that $a = 0$, $Q = 0$ and the Lévy measure ν on $\mathbb{R}^d$ is of the form $\nu(d\xi) = c\,d\xi/|\xi|^{d+\alpha}$, where $c > 0$, $\alpha \in (0, 2)$. Then (4.15) is satisfied. In this case the exponent ψ is given by $\psi(\xi) = c_1|\xi|^\alpha$, $\xi \in \mathbb{R}^d$, where c_1 is a positive constant. This follows by considering

$$\psi(\xi) = \int_{\mathbb{R}^d} (1 - \cos\langle\xi, \eta\rangle) \frac{c\,d\eta}{|\eta|^{d+\alpha}}.$$

It is clear that ψ is invariant under rotation around 0. Moreover, if $r > 0$ then

$$\psi(r\xi) = \int_{\mathbb{R}^d} (1 - \cos\langle\xi, r\eta\rangle) \frac{c\,d\eta}{|\eta|^{d+\alpha}} = r^\alpha \psi(\xi).$$

The semigroups (infinitely divisible families) described above are called *α-stable rotationally invariant families*.

Example 4.37 (Symmetric Cauchy family on $\mathbb{R}^d$) This is the α-stable family with parameter $\alpha = 1$. Then (see Feller 1971),

$$\lambda_t(d\xi) = \frac{\Gamma((d+1)/2)}{\pi^{(d+1)/2}} \frac{t\,d\xi}{(|\xi|^2 + t^2)^{(d+1)/2}}.$$

We finish this section by describing how one can produce new convolution semigroups of measures on Hilbert spaces using subordinators. Namely, assume that $(\zeta_t,\ t \geq 0)$ is a convolution semigroup of measures on a Hilbert space U with exponent φ, that is, $\int_U e^{i\langle x,y \rangle_U} \zeta_t(dy) = e^{-t\varphi(x)}$. Let (λ_t) be a convolution semigroup of measures on $[0, +\infty)$ such that $\int_0^{+\infty} e^{-r\xi} \lambda_t(d\xi) = e^{-t\widetilde{\psi}(r)}$. Then, by direct computation, the family

$$\zeta_t^{(\lambda_t)} := \int_0^{+\infty} \zeta_s \, \lambda_t(ds), \qquad t \geq 0,$$

is a convolution semigroup of measures with exponent $\widetilde{\varphi}(x) = \widetilde{\psi}(\varphi(x))$, $x \in U$.

Example 4.38 Let $\varphi(x) = \frac{1}{2}\langle Qx, x \rangle_U$, $x \in U$, where Q is a non-negative trace-class operator. The corresponding infinitely divisible family corresponds to a Wiener process on U. It follows from the preceding arguments that for arbitrary $\alpha \in (0, 1)$ the function $\psi(x) = \frac{1}{2}\langle Qx, x \rangle_U^{\alpha}$, $x \in U$, is the exponent of an infinitely divisible family. It could be regarded as an infinite-dimensional version of a stable family.

4.8 Expansion with respect to an orthonormal basis

Let L be a Lévy process in U. Assume that $\{e_n\}$ is an orthonormal basis of U. Then

$$L(t) = \sum_n \langle L(t), e_n \rangle_U e_n = \sum_n L_n(t) e_n, \qquad t \geq 0. \tag{4.23}$$

It is clear that the processes L_n are real-valued càdlàg Lévy processes.

Theorem 4.39 *The series in (4.23) converges, in probability, uniformly in t on any compact interval* $[0, T]$.

Proof Let us fix a finite time interval $[0, T]$. If L is a compound Poisson process then its trajectories are piecewise constant and right-continuous and each takes only a finite number of values. Thus the result is true in this case. By Theorem 4.23 we can therefore assume that L is a square integrable martingale in U. Define $M_k(t) = \sum_{n=1}^k L_n(t) e_n$. For any $k \geq l$, the process $(M_k(t) - M_l(t),\ t \geq 0)$ is a square integrable martingale and thus the process $\left(|M_k(t) - M_l(t)|_U^2,\ t \geq 0 \right)$ is a submartingale. By Doob's submartingale inequality,

$$\mathbb{P}\left(\sup_{0 \leq t \leq T} |M_k(t) - M_l(t)|_U \geq c \right) \leq \frac{1}{c^2} \, \mathbb{E} \, |M_k(T) - M_l(T)|_U^2.$$

$\square$

Let (L_n) be a sequence of real-valued Lévy processes. Then the series in (4.23) converges $\mathbb{P}$-a.s. in U if and only if

$$\sum_n |L_n(t)|^2 < \infty, \qquad \mathbb{P}\text{-a.s.} \tag{4.24}$$

If (4.24) holds for all $t \geq 0$ then $\sum_n L_n(t)e_n$, $t \geq 0$, defines a process with independent increments. If it is stochastically continuous at 0 then it has a càdlàg modification.

4.8.1 Expansion with independent terms

Assume now that the L_n are independent real-valued Lévy processes. Each has the Lévy–Khinchin representation

$$\mathbb{E}\, e^{i\xi L_n(t)} = e^{-t\psi_n(\xi)}, \qquad t \geq 0,\ \xi \in \mathbb{R}. \tag{4.25}$$

For simplicity we assume that the L_n are pure jump processes in the sense that

$$\psi_n(\xi) = \int_{\mathbb{R}} \left(1 - e^{i\xi z} + \chi_{\{|z|<1\}}(z)i\xi z\right) \nu_n(dz). \tag{4.26}$$

The following theorem gives if and only if conditions under which the process $L := \left(\sum_n L_n(t)e_n,\ t \geq 0\right)$ takes values in U. The theorem also states that if the process does take values in U then its Lévy measure is concentrated on the union of the axes.

Theorem 4.40

 (i) *Assume that a U-valued Lévy process L is given by (4.23) with independent Lévy processes L_n having the representation (4.25), (4.26). Then*

$$\sum_n \int_{\mathbb{R}} (|z|^2 \wedge 1)\, \nu_n(dz) < \infty. \tag{4.27}$$

Moreover, L is square integrable if and only if

$$\sum_n \int_{\mathbb{R}} |z|^2\, \nu_n(dz) < \infty.$$

 (ii) *Conversely, if (4.25)–(4.27) hold and the L_n are independent Lévy processes then (4.23) defines a Lévy process on U. Moreover, the series in (4.23) converges, $\mathbb{P}$-a.s., uniformly in t on any compact interval.*

Proof   We can assume that $U = l^2$ and that $e_1 = (1, 0, \ldots), e_2 = (0, 1, 0, \ldots), \ldots$ Let $M_n(t) = \left(L_1(t), \ldots, L_n(t)\right)$, $t \geq 0$. Then M_n is a Lévy process on $\mathbb{R}^n$.

Moreover, for $x = (x_1, x_2, \ldots, x_n)$,

$$\mathbb{E}\, e^{i\langle x, M_n(t)\rangle} = \prod_{k=1}^{n} e^{-t\psi_k(x_k)} = \exp\left\{ -t \sum_{k=1}^{n} \psi_k(x_k) \right\}.$$

Consequently, the Lévy exponent Ψ_n of M_n is given by

$$\Psi_n(x_1, \ldots, x_n) = \sum_{k=1}^{n} \int \left(1 - e^{ix_k z_k} + \chi_{\{|z_k|<1\}} ix_k z_k \right) \nu_k(\mathrm{d}z_k).$$

Therefore the Lévy measure of M_n is concentrated on the axes of $\mathbb{R}^n$, and on the k-axis it is exactly ν_k. The jump measure ν for the process L is concentrated on the axes as well. Note that

$$\int_{l^2} \left(|x|_{l^2}^2 \wedge 1 \right) \nu(\mathrm{d}x) = \sum_{n=1}^{\infty} \int_{\mathbb{R}} \left(|z|^2 \wedge 1 \right) \nu_n(\mathrm{d}z),$$

and the first part follows by the Lévy–Khinchin theorem. The square integrability condition can be shown in a similar way. Convergence $\mathbb{P}$-a.s. follows from the Lévy–Ottaviani inequality. $\qquad\square$

The following examples are of some importance in mathematical finance; see Cont and Tankov (2004).

Example 4.41 (One-sided exponentially tempered stable coordinates)   As the Lévy measure ν_n of L_n, we take

$$\nu_n(\mathrm{d}r) = c_n \frac{e^{-\beta_n r}}{r^{1+\alpha_n}} \chi_{[0,\infty)}(r)\, \mathrm{d}r.$$

Then $L = \sum_n L_n e_n$ is U-valued and square integrable if and only if

$$\infty > \sum_n c_n \int_0^\infty e^{-\beta_n r} r^{1-\alpha_n}\, \mathrm{d}r = \sum_n \frac{c_n}{\beta_n^{2-\alpha_n}} \int_0^\infty e^{-r} r^{(2-\alpha_n)-1}\, \mathrm{d}r$$

$$= \sum_n \frac{c_n}{\beta_n^{2-\alpha_n}} \Gamma(2-\alpha_n).$$

In particular, if $\alpha_n = \alpha \in (0, 2)$ then the condition for square integrability reads $\sum_n c_n / \beta_n^{2-\alpha} < \infty$.

Example 4.42 (Two-sided exponentially tempered stable coordinates)   In this case

$$\frac{\mathrm{d}\nu_n}{\mathrm{d}z}(z) = c_n^- \frac{1}{|z|^{1+\alpha_n^-}} e^{-\beta_n^-|z|} \chi_{(-\infty,0)}(z) + c_n^+ \frac{1}{z^{1+\alpha_n^+}} e^{-\beta_n^+|z|} \chi_{(0,+\infty)}(z).$$

Thus if $\alpha_n^- = \alpha_n^+ = \alpha \in (0, 2)$ then the process L is a square integrable Lévy process with values in U if and only if

$$\sum_n \left(\frac{c_n^-}{(\beta_n^-)^{2-\alpha}} + \frac{c_n^+}{(\beta_n^+)^{2-\alpha}} \right) < \infty.$$

4.8.2 Expansion with uncorrelated terms

Assume now that a certain Lévy process L is square integrable with mean zero and covariance operator Q; see Definition 4.45 below. If L has an expansion (4.23) with independent Lévy processes L_n then $Qe_n = \lambda_n e_n$, where $\lambda_n = \mathbb{E}\, L_n^2(1)$, $n = 1, 2, \ldots$ Thus the e_n are eigenvectors of the operator Q. Conversely, assume that Q is the covariance operator of L with eigenvectors and eigenvalues $\{e_n\}$ and $\{\lambda_n\}$. If L has an expansion (4.23) then the Lévy processes L_n are in general not independent but only uncorrelated, that is, $\mathbb{E}L_n(t)L_m(s) = 0$ if $n \neq m$ and $\mathbb{E}\, L_n(t)L_n(s) = \lambda_n |t - s|$. One can easily calculate the Lévy–Khinchin exponent ψ_n of L_n in terms of the Lévy exponent ψ of L. In particular, if

$$\psi(x) = \tfrac{1}{2}\langle \tilde{Q}x, x \rangle_U + \int_U \left(1 - e^{i\langle x, y \rangle_U} \right) \nu(dy)$$

then $\psi_n(z) = \psi(ze_n)$ and the measure ν_n corresponding to L_n is the image of ν under the projection $y \mapsto \langle e_n, y \rangle_U$.

4.9 Square integrable Lévy processes

Let L be a Lévy process in U defined on a filtered probability space satisfying the usual conditions. We assume that for $t > s$ the increment $L(t) - L(s)$ is independent of $\mathcal{F}_s$.

Remark 4.43 Clearly, for all $t > s$, $L(t) - L(s)$ is independent of $\overline{\mathcal{F}}_s^L$. If L is right-continuous then the increment $L(t) - L(s)$ is independent of $\overline{\mathcal{F}}_{s+}^L$.

If L is integrable and of mean zero then, by Proposition 3.25, L is a martingale with respect to $(\mathcal{F}_t)$.

Assume that L is square integrable. Our first result provides exact forms for the mean and covariance of L. We denote by $L_1^+(U)$ the class of all symmetric non-negative-definite nuclear operators on U; see Appendix A.

Theorem 4.44 *There exist an $m \in U$ and a linear operator $Q \in L_1^+(U)$ such that, for all $t, s \geq 0$ and $x, y \in U$,*

$$\mathbb{E}\,\langle L(t), x\rangle_U = \langle m, x\rangle_U\, t,$$

$$\mathbb{E}\,\langle L(t) - mt, x\rangle_U \langle L(s) - ms, y\rangle_U = t \wedge s\,\langle Qx, y\rangle_U, \qquad (4.28)$$

$$\mathbb{E}\,|L(t) - mt|_U^2 = t\,\mathrm{Tr}\,Q.$$

Proof   Since

$$\mathbb{E}\,\langle L(t+s), x\rangle_U = \mathbb{E}\,\langle L(t+s) - L(s), x\rangle_U + \mathbb{E}\,\langle L(s), x\rangle_U$$
$$= \mathbb{E}\,\langle L(t), x\rangle_U + \mathbb{E}\,\langle L(s), x\rangle_U$$

and $t \mapsto \mathbb{E}\,\langle L(t), x\rangle_U$ is measurable, there is a mapping $m\colon U \mapsto \mathbb{R}$ such that $\mathbb{E}\,\langle L(t), x\rangle_U = m(x)\,t$ for all $x \in U$ and $t \geq 0$. Since $x \mapsto m(x)$ is linear and continuous, it has the desired form $m(x) = \langle m, x\rangle_U, x \in U$.

We pass to the covariance of L. Replacing $L(t)$ by $L(t) - mt$, we may assume that L is of mean zero. Let $s \geq 0$. Note that

$$U \times U \ni (x, y) \mapsto [x, y]_L := \mathbb{E}\,\langle L(s), x\rangle_U \langle L(s), y\rangle_U$$

is a symmetric non-negative-definite continuous bilinear form on U. Thus there is a symmetric non-negative-definite continuous linear operator $Q(s)$ such that $\langle Q(s)x, y\rangle_U = [x, y]_L$ for $x, y \in U$. Since, for any orthonormal basis $\{e_n\}$ of U,

$$\sum_n \langle Q(s)e_n, e_n\rangle_U = \sum_n \mathbb{E}\,\langle L(s), e_n\rangle_U^2 = \mathbb{E}\,|L(s)|_U^2 < \infty,$$

$Q(s)$ is nuclear and $\mathrm{Tr}\,Q(s) = \mathbb{E}\,|L(s)|_U^2$. Let $0 \leq s < t$, and let $x, y \in U$. Then

$$\mathbb{E}\,\langle L(t) - L(s), x\rangle_U \langle L(s), y\rangle_U = \mathbb{E}\,\langle L(t) - L(s), x\rangle_U \langle L(s) - L(0), y\rangle_U$$
$$= \mathbb{E}\,\langle L(t) - L(s), x\rangle_U\,\mathbb{E}\,\langle L(s) - L(0), y\rangle_U$$
$$= 0.$$

Hence

$$\mathbb{E}\,\langle L(t), x\rangle_U \langle L(s), y\rangle_U = \mathbb{E}\,\langle L(s), x\rangle_U \langle L(s), y\rangle_U = \langle Q(s)x, y\rangle_U,$$

and the proof is complete if we can show that $Q(s) = s\,Q(1)$ for $s \in I$. Since

$$\langle Q(s+h)x, y\rangle_U = \mathbb{E}\big(\langle L(s+h) - L(s) + L(s) - L(0),\ x\rangle_U$$
$$\times \langle L(s+h) - L(s) + L(s) - L(0),\ y\rangle_U\big)$$
$$= \mathbb{E}\,\langle L(s+h) - L(s), x\rangle_U \langle L(s+h) - L(s), y\rangle_U$$
$$+ \mathbb{E}\,\langle L(s) - L(0), x\rangle_U \langle L(s) - L(0), y\rangle_U$$
$$= \langle Q(h)x, y\rangle_U + \langle Q(s)x, y\rangle_U,$$

we have $Q(s + h) = Q(s) + Q(h)$. Since the functions $s \mapsto \langle Q(s)x, x \rangle_U, x \in U$, are increasing, they are measurable. Then, for all $x, y \in U$, the function

$$s \mapsto \langle Q(s)x, y \rangle_U = \tfrac{1}{4}\big(\langle Q(s)(x + y), x + y \rangle_U - \langle Q(s)(x - y), x - y \rangle_U \big)$$

is measurable. Consequently $\langle Q(s)x, y \rangle_U = s \langle Q(1)x, y \rangle_U$ for $x, y \in U$, which is the desired conclusion. $\qquad\square$

Definition 4.45 The vector m and the operator Q appearing in the theorem above are called the *mean* and the *covariance operator* of the process L, respectively.

Remark 4.46 Note that the covariance operator of the process L is the same as the covariance operator of $L(1)$.

Theorem 4.47

(i) *A Lévy process L on a Hilbert space U is square integrable if and only if its Lévy measure satisfies*

$$\int_U |y|_U^2 \nu(dy) < \infty. \tag{4.29}$$

(ii) *Assume (4.29). Let L have the representation (4.17), let Q_0 be the covariance operator of the Wiener part of L and let Q_1 be the covariance operator of the jump part. Then*

$$\langle Q_1 x, z \rangle_U = \int_U \langle x, y \rangle_U \langle z, y \rangle_U \nu(dy), \qquad x, z \in U,$$

$$\mathbb{E}\, L(t) = \left(a + \int_{\{|y|_U \geq r_0\}} y\nu(dy) \right) t,$$

and the covariance Q of L is equal to $Q_0 + Q_1$.

Proof Since the elements appearing in the decomposition (4.17) are independent, W is square integrable (see Theorem 4.20) and, since

$$\mathbb{E}\left| \sum_{n=1}^{\infty} L_n \right|_U^2 = \int_{\{|y|_U < r_0\}} |y|_U^2 \nu(dy) < \infty,$$

the process L is square integrable if and only if the compound Poisson process L_0 with jump intensity measure $\chi_{\{|y|_U \geq r_0\}} \nu(dy)$ is square integrable. Hence, by Proposition 4.18(iii), L is square integrable if and only if $\int_{\{|y|_U \geq r_0\}} |y|_U^2 \nu(dy) < \infty$. Since $\int_{\{|y|_U < r_0\}} |y|_U^2 \nu(dy) < \infty$, we obtain the desired conclusion. In order to compute the mean and the covariance operator of L we use (4.17) again. Clearly the Wiener part has mean zero. The part $\sum_{n=1}^{\infty} L_n(t), t \geq 0$, has mean zero as it is the sum of compensated, and hence mean-zero, processes. Thus by

Proposition 4.18(i)

$$\mathbb{E}\,L(t) = at + \mathbb{E}\,L_0(t) = at + t \int_{\{|y|_U \geq r_0\}} y\nu(\mathrm{d}y).$$

Let $\widehat{L}_0(t) = L_0(t) - \mathbb{E}\,L_0(t)$. Since $\widehat{L}_0$ and the L_n, $n = 0, 1, \ldots$, are mutually independent compensated compound Poisson processes, we have

$$\mathbb{E}\left\langle \widehat{L}_0(t) + \sum_{n=1}^{\infty} L_n(t), \ x \right\rangle_U \left\langle \widehat{L}_0(t) + \sum_{n=1}^{\infty} L_n(t), \ y \right\rangle_U$$

$$= \mathbb{E}\,\langle \widehat{L}_0(t), x\rangle_U\, \langle \widehat{L}_0(t), y\rangle_U + \sum_{n=1}^{\infty} \mathbb{E}\,\langle \widehat{L}_n(t), x\rangle_U\, \langle \widehat{L}_n(t), y\rangle_U,$$

and the desired identity follows from Proposition 4.18(iii). □

Remark 4.48 It follows from Theorem 4.47 that if a square integrable Lévy process does not have a Wiener part then its covariance operator coincides with the covariance of the jump intensity measure,

$$\langle Qx, y\rangle_U = \int_U \langle x, z\rangle_U \langle y, z\rangle_U \nu(\mathrm{d}z), \qquad x, y \in U.$$

4.9.1 Martingale property

Let us recall that the Lévy process L is adapted to $(\mathcal{F}_t)$ and, for $t > s$, the increment $L(t) - L(s)$ is independent of $\mathcal{F}_s$. Below, $x \otimes y(z) := \langle x, z\rangle_U y$ for $x, y, z \in U$; $L_1(U)$ denotes the space of all nuclear operators on U.

Theorem 4.49

(i) *If L is an integrable Lévy process with mean zero then L is a martingale with respect to $(\mathcal{F}_t)$.*

(ii) *If L is a mean-zero and square integrable process with covariance operator Q then the processes $|L(t)|_U^2 - t \operatorname{Tr} Q$ and $L(t) \otimes L(t) - t Q$, $t \in [0, \infty)$, are martingales with values in U and $L_1(U)$, respectively. Moreover, for all $x, y \in U$, $\langle L, x\rangle_U$ and $\langle L, y\rangle_U$ are square integrable martingales and*

$$\langle L(t), x\rangle_U \langle L(t), y\rangle_U - t\,\langle Qx, y\rangle_U, \qquad t \geq 0,$$

is a martingale.

Proof The first part follows from Proposition 3.25. In order to show part (ii) note that, for all $u, v \in U$,

$$\langle L(t), x\rangle_U \langle L(t), y\rangle_U$$

$$= \langle L(t) - L(s), x\rangle_U \langle L(t) - L(s), y\rangle_U + \langle L(t) - L(s), x\rangle_U \langle L(s), y\rangle_U$$

$$+ \langle L(t) - L(s), y\rangle_U \langle L(s), x\rangle_U + \langle L(s), x\rangle_U \langle L(s), y\rangle_U.$$

Hence, as $L(t) - L(s)$ is independent of $\mathcal{F}_s$ and has mean zero,

$$\mathbb{E}\big(\langle L(t), x\rangle_U \langle L(t), y\rangle_U - \langle L(s), x\rangle_U \langle L(s), y\rangle_U \big| \mathcal{F}_s\big)$$
$$= \mathbb{E}\,\langle L(t) - L(s), x\rangle_U \langle L(t) - L(s), y\rangle_U = (t-s)\,\langle Qx, y\rangle_U.$$

Let $\{e_n\}$ be an orthonormal basis of U. Then

$$|L(t)|_U^2 = \sum_n \langle L(t), e_n\rangle_U.$$

For every n, $\langle L(t), e_n\rangle_U^2 - t\,\langle Qe_n, e_n\rangle_U, t \geq 0$, is a martingale and therefore so is

$$|L(t)|_U^2 - t\sum_n \langle Qe_n, e_n\rangle_U = |L(t)|_U^2 - t\,\mathrm{Tr}\,Q.$$

$\square$

Remark 4.50   Theorem 4.49(ii) says that $\langle L, L\rangle_t = t\,\mathrm{Tr}\,Q$ and $\langle\langle L, L\rangle\rangle_t = tQ$, $t \geq 0$, where $\langle L, L\rangle$ and $\langle\langle L, L\rangle\rangle$ are the *angle bracket* and the *operator angle bracket* and will be defined for general square integrable martingales in Section 8.1.

4.9.2 Covariance operators in $L^2(\mathcal{O})$

In this subsection we give a condition under which an operator on $L^2(\mathcal{O})$ is nuclear, symmetric and non-negative-definite and therefore can be the covariance of a square integrable random element in $L^2(\mathcal{O})$.

Assume that Q is a symmetric non-negative-definite nuclear operator on the space $L^2(\mathcal{O}) = L^2(\mathcal{O}, \mathcal{B}(\mathcal{O}), \mathrm{d}\xi)$, where $\mathcal{O}$ is a subset of $\mathbb{R}^d$. Since Q is compact, then (see Appendix A) there is an orthonormal basis $\{e_n\}$ consisting of eigenvectors of Q. Let $\{\gamma_n\}$ be the corresponding set of eigenvalues. Then

$$Q^{1/2}\psi = \sum_n \sqrt{\gamma_n}\langle\psi, e_n\rangle_{L^2(\mathcal{O})}e_n, \qquad \psi \in L^2(\mathcal{O}),$$

is the square root of Q. Since $\mathrm{Tr}\,Q = \sum_n \gamma_n < \infty$, the operator $Q^{1/2}$ is Hilbert–Schmidt. Thus, by Proposition A.7, $Q^{1/2}$ is an integral operator. Let q_0 be the kernel of $Q^{1/2}$. Then

$$Q^{1/2}\psi(\xi) = \int_\mathcal{O} q_0(\eta, \xi)\psi(\eta)\,\mathrm{d}\eta, \qquad \xi \in \mathcal{O},\ \psi \in L^2(\mathcal{O}),$$

and

$$\int_\mathcal{O}\int_\mathcal{O} |q_0(\eta, \xi)|^2\,\mathrm{d}\xi\,\mathrm{d}\eta < \infty.$$

Since Q is non-negative-definite,

$$\int_\mathcal{O}\int_\mathcal{O} q_0(\eta, \xi))\psi(\eta)\psi(\xi)\,\mathrm{d}\xi\,\mathrm{d}\eta \geq 0, \qquad \psi \in L^2(\mathcal{O}).$$

Proposition 4.51 *Any symmetric non-negative-definite nuclear operator Q is an integral operator with kernel*

$$q(\eta, \xi) = \int_{\mathcal{O}} q_0(\eta, \zeta) q_0(\zeta, \xi) \, d\zeta, \qquad \xi, \eta \in \mathcal{O}.$$

Proof It is sufficient to show that, for each $\psi \in L^2(\mathcal{O})$, for almost all $\xi \in \mathcal{O}$,

$$Q\psi(\xi) = \int_{\mathcal{O}} q(\xi, \eta) \psi(\eta) \, d\eta.$$

Note that

$$Q\psi(\xi) = \int_{\mathcal{O}} q_0(\xi, \zeta) \left(\int_{\mathcal{O}} q_0(\zeta, \eta) \psi(\eta) \, d\eta \right) d\zeta.$$

Thus it is sufficient to justify the change in the order of integration. By the Schwarz inequality, for almost all $\xi \in \mathcal{O}$,

$$\int_{\mathcal{O}} \int_{\mathcal{O}} |q_0(\xi, \zeta) q_0(\zeta, \eta)| |\psi(\eta)| \, d\eta \, d\zeta$$

$$\leq \int_{\mathcal{O}} |\psi(\eta)| \left[\left(\int_{\mathcal{O}} |q_0(\xi, \zeta)|^2 \, d\zeta \right)^{1/2} \left(\int_{\mathcal{O}} |q_0(\zeta, \eta)|^2 \, d\zeta \right)^{1/2} \right] d\eta$$

$$\leq \left(\int_{\mathcal{O}} |\psi(\eta)|^2 \, d\eta \right)^{1/2} \left(\int_{\mathcal{O}} |q_0(\xi, \zeta)|^2 \, d\zeta \right)^{1/2} \left(\int_{\mathcal{O}} \int_{\mathcal{O}} |q_0(\zeta, \eta)|^2 \, d\zeta \, d\eta \right)^{1/2}$$

$$< \infty.$$

Thus the result follows. $\qquad\qquad\square$

The kernel q can be interpreted as the *spatial correlation* of the process L. In fact, for each $t > 0$, $L(t, \cdot)$ is a function-valued random variable. Thus $L(t, \xi)$ is a real-valued random variable for almost all $\xi \in \mathcal{O}$. Moreover,

$$\mathbb{E}\langle L(t), \phi \rangle_{L^2(\mathcal{O})} \langle L(t), \psi \rangle_{L^2(\mathcal{O})} = \mathbb{E} \left(\int_{\mathcal{O}} L(t, \xi) \phi(\xi) \, d\xi \right) \left(\int_{\mathcal{O}} L(t, \eta) \psi(\eta) \, d\eta \right).$$

Reasoning as in the proof of Proposition 4.51, one can change the order of integration and expectation to obtain

$$t \langle Q\phi, \psi \rangle_{L^2(\mathcal{O})} = \int_{\mathcal{O}} \int_{\mathcal{O}} \left(\mathbb{E} \, L(t, \xi) L(t, \eta) \right) \phi(\xi) \psi(\eta) \, d\xi \, d\eta.$$

Thus, for almost all $\xi, \eta \in \mathcal{O}$ and all $t \geq 0$, $q(\xi, \eta) = t^{-1} \mathbb{E} \, L(t, \xi) L(t, \eta)$.

The following theorem, whose proof can be found in Appendix A, gives a sufficient condition for a function q of two variables to be the kernel of a symmetric non-negative-definite nuclear operator on $L^2(\mathcal{O})$.

We recall that a continuous function $q \colon \mathcal{O} \times \mathcal{O} \mapsto \mathbb{R}$ is *non-negative-definite* if, for any $M \in \mathbb{N}$, $v_j \in \mathbb{R}$, $j = 1, \ldots, M$, and $\xi_j \in \mathcal{O}$, $j = 1, \ldots, M$,

$$\sum_{i,j=1}^{M} q(\xi_i, \xi_j) v_i v_j \geq 0.$$

We say that q is *symmetric* if $q(\eta, \xi) = q(\xi, \eta)$ for all $\xi, \eta \in \mathcal{O}$.

Theorem 4.52 *Let $\mathcal{O}$ be a bounded closed subset of $\mathbb{R}^d$, and let q be a symmetric non-negative-definite continuous function on $\mathcal{O} \times \mathcal{O}$. Then the operator $Q \colon L^2(\mathcal{O}) \mapsto L^2(\mathcal{O})$ given by*

$$Q\psi(\xi) = \int_{\mathcal{O}} q(\eta, \xi) \psi(\eta) \, d\eta, \qquad \psi \in L^2(\mathcal{O}), \ \xi \in \mathcal{O},$$

is nuclear, symmetric and non-negative-definite.

We now describe a general method of constructing kernels satisfying the assumptions of Theorem 4.52. Assume that $\big(Z(\xi), \xi \in \mathcal{O}\big)$ is a family of real random variables that have finite second moments and are mean-square continuous, that is, $\lim_{\eta \to \xi} \mathbb{E}\,|Z(\xi) - Z(\eta)|^2$ for $\xi \in \mathcal{O}$. Then $q(\xi, \eta) = \mathbb{E}\,Z(\xi)Z(\eta), \xi, \eta \in \mathcal{O}$, has the required properties. Indeed, its continuity follows from the identity

$$\mathbb{E}\,Z(\xi)Z(\eta) - \mathbb{E}\,Z(\xi_0)Z(\eta_0)$$
$$= \mathbb{E}\,(Z(\xi) - Z(\xi_0))(Z(\eta) - Z(\eta_0))$$
$$+ \mathbb{E}\,Z(\xi_0)(Z(\eta) - Z(\xi_0)) + \mathbb{E}\,Z(\eta_0)(Z(\xi) - Z(\eta_0)).$$

To prove that q is positive-definite, take $v_j \in \mathbb{R}$ and $\xi_j \in \mathcal{O}$, $j = 1, \ldots, M$. Then

$$\sum_{i,j=1}^{M} q(\xi_i, \xi_j) v_i v_j = \mathbb{E}\left(\sum_{i=1}^{M} Z(\xi_1) v_i\right)^2 \geq 0.$$

Example 4.53 Take $\mathcal{O} = [0, \pi]$ and $q(\xi, \eta) := \xi \wedge \eta$, $\xi, \eta \in [0, \pi]$. Note that q is the covariance of a standard Wiener process in $\mathbb{R}$. Write

$$Qx(\xi) := \int_0^{\pi} q(\xi, \eta) x(\eta) \, d\eta, \qquad x \in L^2(0, \pi).$$

Then Q has eigenvectors $e_n(\xi) = \sqrt{2/\pi} \sin\big((n + \tfrac{1}{2})\xi\big), \xi \in [0, \pi], n = 0, 1, \ldots,$ and the corresponding eigenvalues are $\lambda_n := \big(n + \tfrac{1}{2}\big)^{-2}$. Thus, by (4.23), any square integrable Lévy process L on $L^2(0, \pi)$ with covariance Q has an expansion

$$L(t, \xi) = \sum_{n=1}^{\infty} \frac{\tilde{L}_n(t)}{n + \tfrac{1}{2}} \sqrt{\frac{2}{\pi}} \sin\big((n + \tfrac{1}{2})\xi\big),$$

where the $\tilde{L}_n$ are uncorrelated normalized Lévy processes. Note that

$$\mathbb{E}\, L(t,\xi)L(s,\eta) = t \wedge s\, \xi \wedge \eta.$$

If, in particular, L is a Wiener process then it is often called a *Brownian sheet* on $[0, +\infty) \times [0, \pi]$.

Necessary and sufficient conditions for the shift-invariant kernel

$$q(\eta,\xi) = \phi(\xi - \eta), \qquad \xi,\eta \in \mathbb{R}^d, \tag{4.30}$$

to be continuous and non-negative-definite are given by the following theorem due to S. Bochner; see e.g. Feller (1971).

Theorem 4.54 (Bochner) *Assume that q is given by (4.30) for some $\phi \colon \mathbb{R}^d \mapsto \mathbb{R}$. Then q is continuous and non-negative-definite if and only if ϕ is the Fourier transform of a finite positive Borel measure on $\mathbb{R}^d$.*

As a consequence of Theorems 4.52 and 4.54 we have the following result. For more details see Feller (1971).

Example 4.55 The functions

$$q(\eta,\xi) = e^{-|\eta-\xi|^\alpha}, \qquad 0 < \alpha \le 2,$$

$$q(\eta,\xi) = \begin{cases} \dfrac{\sin\langle a, \eta - \xi\rangle}{\langle a, \eta - \xi\rangle}, & \eta \ne \xi, \\ 1, & \eta = \xi, \end{cases}$$

$$q(\eta,\xi) = \begin{cases} \dfrac{2}{|a|^2}\dfrac{1 - \cos\langle a, \eta - \xi\rangle}{|\eta - \xi|^2}, & \eta \ne \xi, \\ 1, & \eta = \xi, \end{cases}$$

where $a \in \mathbb{R}^d$, are kernels of non-negative-definite nuclear operators on $L^2(\mathcal{O})$.

4.10 Lévy processes on Banach spaces

We conclude this chapter with a few facts about Lévy processes about Banach spaces. Let us start from the observation that if (μ_t) is a convolution semigroup of measures on a Banach space B then $\mu = \mu_1$ is infinitely divisible.

Assume that ν is a finite measure on B. We denote by B^* the dual space. Define

$$\pi(\nu) := e^{-\nu(B)} \sum_{n=0}^\infty \frac{\nu^{*n}}{n!}.$$

Clearly, $\pi(\nu) = \mu_1$ for the convolution semigroup of measures of a compound Poisson process with intensity ν. We have the following simple result.

Proposition 4.56

(i) *The measure $\pi(\nu)$ is infinitely divisible. Moreover, for all $f \in B^*$,*

$$\int_B e^{if(x)} \pi(\nu)(dx) = e^{-\nu(B)} \sum_{n=0}^{\infty} \left(\int_B e^{if(x)} \nu(dx) \right)^n$$

$$= \exp \left\{ \int_B \left(e^{if(x)} - 1 \right) \right\} \nu(dx).$$

(ii) *Assume that $\int_B |x|_B \nu(dx) < \infty$. Let $m := \int_B x\nu(dx)$ and let $\widehat{\pi}(\nu) := \pi(\nu) * \delta_{-m}$. Then*

$$\int_B e^{if(x)} \widehat{\pi}(\nu)(dx) = \exp \left\{ \int_B \left(e^{if(x)} - 1 - if(x) \right) \nu(dx) \right\}.$$

Define

$$Tx := \begin{cases} x, & \text{if } |x|_B \leq 1, \\ x/|x|_B, & \text{if } |x|_B \geq 1. \end{cases}$$

Clearly, if ν has support in $\{x \in B : |x|_B > 1\}$ then $\nu \circ T^{-1}$ is concentrated on the sphere $\{x \in B : |x|_B = 1\}$. Let

$$K(f, x) := e^{if(x)} - 1 - if(Tx), \qquad x \in B.$$

Definition 4.57 A measure ν concentrated on $B \setminus \{0\}$ is called a *Lévy measure* or a *jump measure* if

(i) $\nu \{x : |x|_B > r\} < \infty$ for every $r > 0$,
(ii) $\int_{\{|x|_B \leq 1\}} f^2(x) \nu(dx) < \infty$ for every $f \in B^*$,
(iii) $\exp \left\{ \int_B K(f, x)\nu(dx) \right\}$, $f \in B^*$, is a characteristic functional of a probability measure (denoted by $\pi(\nu)$) on B.

Note that the integrals $\int_B K(f, x)\nu(dx)$, $f \in B^*$, are well defined, since

$$\int_{|x|_B \leq 1} K(f, x)\nu(dx) = \int_{\{|x|_B \leq 1\}} \left(e^{if(x)} - 1 - if(x) \right) \nu(dx).$$

Note that

$$\left| e^z - 1 - z \right| \leq \sum_{k=2}^{\infty} \frac{|z|^k}{k!} = |z|^2 \sum_{k=0}^{\infty} \frac{|z|^k}{(k+2)!} \leq \frac{|z|^2}{2} \sum_{k=0}^{\infty} \frac{|z|^k}{k!},$$

because $1/2k! \geq 1/(k+2)!$. Therefore $|e^z - z - 1| \leq \frac{1}{2}|z|^2 e^{|z|}$ for every $z \in \mathbb{C}$, and hence

$$\left| e^{if(x)} - 1 - if(x) \right| \leq \frac{1}{2} |f(x)|^2 e^{|f(x)|}.$$

One can formulate an equivalent definition by putting

$$K(f, x) = e^{if(x)} - 1 - if(x)\chi_{\{|x|_B \le 1\}}.$$

However, a normalization factor (such as $if(x)\chi_{\{|x|_B \le 1\}}$) is important. The following characterization of infinitely divisible measures is due to A. Tortrat; see Tortrat (1967, 1969), Linde (1986) or Kruglov (1984).

Theorem 4.58 (Tortrat) *A probability measure μ on a Banach space is infinitely divisible if and only if it is of the form $\mu = \delta_a * N * \pi(\nu)$, where $a \in B$, N is a symmetric Gaussian measure on B and $\pi(\nu)$ is a Lévy measure.*

With the help of the Itô–Nisio theorem one can derive the following result.

Corollary 4.59 *If (μ_t) is a convolution semigroup of measures on a Banach space B and $\mu_1 = \delta_a * N * \pi(\nu)$ then the corresponding Lévy process L is of the form*

$$L(t) = ta + W(t) + \sum_{n=1}^{\infty} L_n(t) + L_0(t), \tag{4.31}$$

where W is a Wiener process, the L_n are compound Poisson martingales and L_0 is a compound Poisson process with jump measure supported on $\{x : |x|_B > 1\}$. For each t the sum in (4.31) converges $\mathbb{P}$-a.s.

5

Lévy semigroups

This chapter is devoted to transition semigroups of Lévy processes and their generators. A family $S = (S(t), t \geq 0)$ of bounded linear operators on a Banach space $(B, |\cdot|_B)$ is called a C_0-*semigroup* if

  (i) $S(0)$ is the identity operator I,
  (ii) $S(t)S(s) = S(t+s)$ for all $t, s \geq 0$,
  (iii) $[0, \infty) \ni t \mapsto S(t)z \in B$ is continuous for each $z \in B$.

Assume that S is a C_0-semigroup on B. We say that an element $z \in B$ is in the *domain of the generator* of S if $\lim_{t \downarrow 0} t^{-1}(S(t)z - z) =: Az$ exists. The set of all such z is denoted by $D(A)$ and Az, $z \in D(A)$, is then a linear operator called the *generator* of S.

5.1 Basic properties

Let L be a Lévy process on a Hilbert space U, and let μ_t be the distribution of $L(t)$. Then (see Chapter 4) (μ_t) is a convolution semigroup of measures and L is a Markov process with transition function $P_t(x, \Gamma) = \mu_t(\Gamma - x)$. The corresponding semigroup is given by

$$P_t \varphi(x) = \int_U \varphi(x+y)\mu_t(\mathrm{d}y). \tag{5.1}$$

Denote by $C_b(U)$ and $UC_b(U)$ the spaces of all bounded continuous functions on U and all bounded and uniformly continuous functions on U, equipped with the supremum norm. The following result is due to Tessitore and Zabczyk (2001b).

Theorem 5.1 *Let (P_t) be defined on $C_b(U)$ by (5.1). Then (P_t) is a C_0-semigroup on $C_b(U)$ if and only if either (μ_t) is the convolution semigroup of measures of a compound Poisson process or $\mu_t = \delta_{\{0\}}$, $t \geq 0$.*

75

Proof   If (μ_t) corresponds to a compound Poisson process with intensity $\alpha > 0$ and jump measure ν then

$$P_t\varphi(x) = e^{-\alpha t} \sum_{n=0}^{\infty} \frac{(\alpha t)^n}{n!} \left(\int_U \varphi(x+y)\, \nu^{*n}(dy) \right), \qquad x \in U.$$

Hence

$$P_t\varphi(x) - \varphi(x) = \left(e^{-\alpha t} - 1\right)\varphi(x) + e^{-\alpha t} \sum_{n=1}^{\infty} \frac{(\alpha t)^n}{n!} \int_U \varphi(x+y)\, \nu^{*n}(dy)$$

and therefore

$$\sup_{x \in U} |P_t\varphi(x) - \varphi(x)| \le \left(1 - e^{-\alpha t}\right) \|\varphi\|_\infty + \left(e^{-\alpha t} \sum_{n=1}^{\infty} \frac{(\alpha t)^n}{n!} \right) \|\varphi\|_\infty.$$

Consequently, $\|P_t\varphi - \varphi\|_\infty \to 0$ as $t \downarrow 0$. If $P_t\varphi = \varphi$ then $\|P_t\varphi - \varphi\|_\infty = 0$.

Assume now that (μ_t) is not the convolution semigroup of any compound Poisson process and that $\mu_t \equiv \delta_0$ does not hold. Let (x_n) be a sequence of elements in U such that $|x_n - x_m|_U \ge 1, n \ne m$. Fix $\varepsilon \in (0, 1)$. By the assumptions on (μ_t), there is a sequence $t_n \downarrow 0$ such that $\mu_{t_n}\big(\{x : |x|_U \le \frac{1}{2}\}\big) \ge 1 - \varepsilon$. For each n, one can find $r_n < \frac{1}{2}$ such that $\mu_{t_n}\big(\{x : |x|_U \le r_n\}\big) \le \varepsilon$. Let φ_n be a continuous function on U taking values in $[0, 1]$ and such that $\varphi_n(x_n) = 1$ and $\varphi_n(x) = 0$ if $|x - x_n|_U \ge r_n$. Define

$$\varphi(x) = \sum_{m=1}^{\infty} \varphi_m(x), \qquad x \in U.$$

It is clear that $\varphi \in C_b(U)$ and $\varphi(x_m) = 1$ for $m = 1, 2, \dots$ Moreover,

$$P_{t_n}\varphi(x_n) = P_{t_n}\varphi_n(x_n) + \sum_{m \ne n} P_{t_n}\varphi_m(x_n)$$

$$\le \mu_{t_n}\left(\{x : |x|_U \le r_n\}\right) + \mu_{t_n}\left(\{x : |x|_U > \tfrac{1}{2}\}\right) \le 2\varepsilon.$$

Thus

$$\sup_x |P_{t_n}\varphi(x) - \varphi(x)| \ge |P_{t_n}\varphi(x_n) - \varphi(x_n)| = |1 - P_{t_n}\varphi(x_n)| \ge 1 - 2\varepsilon.$$

$\square$

For each $a \in U$ define the translation $\tau_a\varphi$ of a function φ by the formula $\tau_a\varphi(x) = \varphi(x+a), a \in U, x \in U$.

Definition 5.2   A semigroup P of continuous linear operators on $UC_b(U)$ is called *translation invariant* or *spatially homogeneous* if, for any $a \in U, t \ge 0$ and $\varphi \in UC_b(U)$, $P_t(\tau_a\varphi) = \tau_a(P_t\varphi)$.

The space $UC_b(U)$ is more convenient for treating the transition semigroups of Lévy processes, as the following theorem shows.

Theorem 5.3

(i) *If (P_t) is defined on $UC_b(U)$ by (5.1) then (P_t) is a C_0-semigroup on $UC_b(U)$.*

(ii) *A Markov semigroup[1] (P_t) on $UC_b(U)$ is translation invariant if and only if it is given by (5.1) for some convolution semigroup of measures.*

Proof Assume that $\varphi \in UC_b(U)$ and $t > 0$. Then

$$|P_t\varphi(x) - P_t\varphi(z)| = \left| \int_U (\varphi(x + y) - \varphi(z + y))\, \mu_t(\mathrm{d}y) \right|$$

$$\leq \int_U |\varphi(x + y) - \varphi(z + y)|\, \mu_t(\mathrm{d}y).$$

For every $\varepsilon > 0$ there exists a $\delta > 0$ such that if $|x - x'|_U < \delta$ then $|\varphi(x) - \varphi(x')| < \varepsilon$. Thus if $|x - z|_U < \delta$ then $|(x + y) - (z + y)|_U < \delta$ and $|\varphi(x + y) - \varphi(z + y)| < \varepsilon$. This gives $|P_t\varphi(x) - P_t\varphi(y)| < \varepsilon$ and consequently $P_t : UC_b(U) \mapsto UC_b(U)$. Let us observe now that for each $\varepsilon > 0$ there exists a $\delta > 0$ such that $|\varphi(x + y) - \varphi(x)| < \varepsilon$ for $|y|_U < \delta$ and all x. Therefore

$$|P_t\varphi(x) - \varphi(x)| \leq \varepsilon \int_{\{|y|_U \leq \delta\}} \mu_t(\mathrm{d}y) + 2\|\varphi\|_\infty \int_{\{|y|_U > \delta\}} \mu_t(\mathrm{d}y)$$

$$\leq \varepsilon + 2\|\varphi\|_\infty \int_{\{|y|_U > \delta\}} \mu_t(\mathrm{d}y).$$

Since (μ_t) converges weakly to δ_0 as $t \downarrow 0$, we have that $P_t\varphi \to \varphi$ uniformly.

To show the second part, note that if (5.1) is satisfied then

$$\tau_a(P_t\varphi)(x) = P_t(\tau_a\varphi)(x), \qquad \forall a, x \in U,\ t \geq 0,\ \varphi \in UC_b(U). \tag{5.2}$$

If (5.2) holds then, for all $a, x \in U, t \geq 0$, and $\varphi \in UC_b(U)$,

$$\int_U \varphi(y) P_t(x + a, \mathrm{d}y) = \int_U \varphi(y + a) P_t(x, \mathrm{d}y)$$

and consequently

$$\int_U \varphi(y) P_t(a, \mathrm{d}y) = \int_U \varphi(y + a) P_t(0, \mathrm{d}y).$$

Hence it is suffcient to show that $(P_t(0, \mathrm{d}y),\ t \geq 0)$ is a convolution semigroup of measures. Clearly $P_0(0, \cdot) = \delta_0$. The condition $P_t(0, \cdot) * P_s(0, \cdot) = P_{t+s}(0, \cdot)$ follows from the semigroup property of (P_t). Since (P_t) is C_0 on $UC_b(U)$, $P_t(x, \cdot)$ converges weakly to δ_x as $t \to 0$. Thus $P_t(0, \cdot)$ converges weakly to δ_0, which is

---

[1] That is, a semigroup defined by a transition probability; see Definition 1.7.

equivalent to

$$\lim_{t\downarrow 0} P_t(0, \{x : |x|_U > r\}) = 0, \qquad \forall\, r > 0.$$

$\square$

5.2 Generators

In this section we derive the form of the generator (on regular functions) for an arbitrary Lévy process on a Hilbert space U. However, we first consider some special processes.

5.2.1 Compound Poisson process

We will show that the generator of a compound Poisson process L is given by

$$A\varphi(x) = \int_U (\varphi(x+y) - \varphi(x))\nu(\mathrm{d}y), \qquad \varphi \in UC_b(U), \qquad (5.3)$$

where ν is the Lévy measure of L. Note that A is a bounded linear operator on $UC_b(U)$; see the proof of Theorem 5.3(i). Recall that the corresponding distributions μ_t are given by (4.5). To show that the operator A given by (5.3) is in fact the generator of the process L, we have to show that, for every $\varphi \in UC_b(U)$,

$$\left\| \frac{P_t\varphi - \varphi}{t} - A\varphi \right\|_\infty$$

$$= \sup_{x \in U} \left| \int_U (\varphi(x+y) - \varphi(x)) \left(\frac{1}{t}\mu_t(\mathrm{d}y) - \nu(\mathrm{d}y) \right) \right| =: J(t) \to 0$$

as $t \downarrow 0$. Since

$$\frac{1}{t}\mu_t - \nu = \frac{e^{-at}}{t}\delta_0 + (e^{-at} - 1)\nu + e^{-at} \sum_{k=2}^\infty \frac{t^{n-1}}{n!} \nu^{*n},$$

we have

$$J(t) \leq 2\|\varphi\|_\infty \left(\left| e^{-at} - 1 \right| \nu(U) + e^{-at} \sum_{n=2}^\infty \frac{t^{n-1}}{n!} \nu^{*n}(U) \right),$$

which gives the desired conclusion.

5.2.2 Uniform motion

Consider the deterministic process $L(t) = at$, where $a \in U$ is a fixed parameter. It is clearly a Lévy process and $\mu_t = \delta_{ta}, t \geq 0$. The generator A, acting on functions

$\varphi \in UC_b^1(U)$, satisfies

$$A\varphi(x) = \lim_{t \downarrow 0} \frac{1}{t}(\varphi(x + ta) - \varphi(x)) = \langle a, D\varphi(x)\rangle_U.$$

5.2.3 Wiener process

Consider a Wiener process W with covariance operator Q. Then $\mu_t = \mathcal{N}(0, tQ)$, $t > 0$. We will show that the domain of the generator contains $UC_b^2(U)$ and that

$$A\varphi(x) = \tfrac{1}{2}\mathrm{Tr}\, QD^2\varphi(x), \qquad \forall\, \varphi \in UC_b^2(U).$$

Indeed, let $\{e_n\}$ be an orthonormal basis of eigenvectors of Q and let $\{\gamma_n\}$ be the corresponding eigenvalues. Let $I = \{n: \gamma_n \neq 0\}$. Since W lives in the space

$$V := \left\{ \sum_{n \in I} x_n e_n : \sum_{n \in I} x_n^2 < \infty \right\},$$

we can assume that $U = V$, that is, $\gamma_n \neq 0$ for every n. On the product space $\mathbb{R}^{\mathbb{N}}$ consider the product measures

$$\mathcal{N}(0, tQ)(\mathrm{d}\mathbf{y}) = \bigotimes_n \frac{1}{\sqrt{2\pi\gamma_n t}} e^{-y_n^2/(2t\gamma_n)}\, \mathrm{d}\mathbf{y}, \qquad t \geq 0,$$

$$\mathcal{N}(0, I)(\mathrm{d}\mathbf{y}) = \bigotimes_n \frac{1}{\sqrt{2\pi}} e^{-\frac{y_n^2}{2}}\, \mathrm{d}\mathbf{y},$$

where $\mathrm{d}\mathbf{y} := \mathrm{d}y_1\, \mathrm{d}y_2 \cdots$. Then, for every $\varphi \in UC_b^2(U)$,

$$P_t\varphi(x) - \varphi(x) = \int_{\mathbb{R}^{\mathbb{N}}} \left(\varphi\left(x + \sum_n y_n e_n\right) - \varphi(x)\right) \mathcal{N}(0, tQ)(\mathrm{d}\mathbf{y})$$

$$= \int_{\mathbb{R}^{\mathbb{N}}} \left(\varphi\left(x + \sum_n \sqrt{\gamma_n t}\, y_n e_n\right) - \varphi(x)\right) \mathcal{N}(0, I)(\mathrm{d}\mathbf{y}).$$

Let $\partial_j \varphi := \langle D\varphi, e_j\rangle_U$ and $\partial_{j,k}^2 \varphi := \langle D^2\varphi[e_j], e_k\rangle_U$. Then

$$\varphi\left(x + \sum_n \sqrt{\gamma_n t}\, y_n e_n\right) - \varphi(x) = \int_0^1 \frac{\mathrm{d}}{\mathrm{d}s}\varphi\left(x + s\sum_n \sqrt{\lambda_n t}\, y_n e_n\right) \mathrm{d}s$$

$$= \int_0^1 \left\langle D\varphi\left(x + s\sum_n \sqrt{\gamma_n t}\, y_n e_n\right), \sum_n \sqrt{\gamma_n t}\, y_n e_n\right\rangle_U \mathrm{d}s$$

$$= \sum_j \int_0^1 \partial_j \varphi\left(x + s\sum_n \sqrt{\gamma_n t}\, y_n e_n\right)\sqrt{\gamma_j t}\, y_j\, \mathrm{d}s$$

$$= \sum_j \partial_j \varphi(x)\sqrt{\gamma_j t}\, y_j$$

$$+ \sum_{j,k} \int_0^1 \int_0^1 \partial_{j,k}^2 \varphi\left(x + s\rho \sum_n \sqrt{\gamma_n t}\, y_n e_n\right)\sqrt{\gamma_j t}\, y_j \sqrt{\gamma_k t}\, y_k s\, \mathrm{d}s\, \mathrm{d}\rho.$$

Since

$$\int_{\mathbb{R}^N} \sum_j \partial_j \varphi(x) \sqrt{\gamma_j t}\, y_j \mathcal{N}(0, I)(\mathrm{d}\mathbf{y}) = 0,$$

it follows that

$$\frac{P_t\varphi(x) - \varphi(x)}{t} = \sum_{j,k} \int_{\mathbb{R}^N} \int_0^1 \int_0^1 \partial^2_{j,k}\varphi\Big(x + s\rho \sum_n \sqrt{\gamma_n t}\, y_n e_i\Big)$$
$$\times \sqrt{\gamma_j}\, y_j \sqrt{\gamma_k}\, y_k \, \mathcal{N}(0, I)(\mathrm{d}\mathbf{y})s \, \mathrm{d}s \, \mathrm{d}\rho,$$

which converges uniformly in x as $t \to 0$ to

$$\frac{1}{2} \sum_{j,k} \int_{\mathbb{R}^N} \partial^2_{j,k}\varphi(x) \sqrt{\gamma_j}\, y_j \sqrt{\gamma_k}\, y_k \, \mathcal{N}(0, I)(\mathrm{d}\mathbf{y}) = \frac{1}{2} \sum_j \partial^2_{j,j}\varphi(x)\gamma_j$$
$$= \frac{1}{2} \operatorname{Tr} Q D^2 \varphi(x).$$

5.2.4 Arbitrary Lévy semigroup

Denote by $UC_b^1(U)$, $UC_b^2(U)$ the spaces of uniformly continuous bounded functions together with all derivatives up to order 1 and up to order 2, respectively. We have the following result.

Theorem 5.4 *Assume that (P_t) is the transition semigroup of a Lévy process L on a Hilbert space U with Lévy exponent (4.19). If $\varphi \in UC_b^2(U)$ then, for each, $x \in U$,*

$$\lim_{t \downarrow 0} \frac{1}{t}\big(P_t\varphi(x) - \varphi(x)\big) = A\varphi(x)$$
$$= \langle a, D\varphi(x) \rangle_U + \frac{1}{2} \operatorname{Tr} Q D^2 \varphi(x)$$
$$+ \int_U \big(\varphi(x+y) - \varphi(x) - \chi_{\{|y|_U < 1\}}(y)\langle D\varphi(x), y \rangle_U\big)\nu(\mathrm{d}y),$$

$$(5.4)$$

where the convergence is uniform in x.

Proof   Let (P_t^n) be the semigroup corresponding to the process

$$X_n(t) := at + W(t) + Z_n(t),$$

where W is a Wiener process with covariance Q and Z_n is a compound Poisson process independent of W and with Lévy measure

$$\nu_n(\mathrm{d}y) = \chi_{\{1/n \le |y|_U\}}(y)\nu(\mathrm{d}y).$$

The generator A_n of (P_t^n) is given by

$$A_n\varphi(x) = \langle a, D\varphi(x)\rangle_U + \tfrac{1}{2}\mathrm{Tr}\, QD^2\varphi(x)$$
$$+ \int_U \big(\varphi(x+y) - \varphi(x) - \langle D\varphi(x), y\rangle\big)\nu_n(dy), \qquad \varphi \in UC_b^1(U).$$
$$(5.5)$$

It is easily checked that, for $\varphi \in UC_b^2(U)$, $P_t\varphi$ and $P_t^n\varphi$ belong to $UC_b^2(U)$ and, for each n,

$$P_t^n\varphi(x) = \varphi(x) + \int_0^t A_n(P_s^n\varphi)(x)\,ds = \varphi(x) + \int_0^t P_s^n(A_n\varphi)(x)\,ds. \qquad (5.6)$$

It is sufficent to show that (5.6) also holds for (P_t) and A, so that

$$P_t\varphi(x) = \varphi(x) + \int_0^t P_s(A\varphi(x))\,ds, \qquad t \geq 0, \ x \in U. \qquad (5.7)$$

Note that, for $\varphi \in UC_b^2(U)$,

$$\varphi(x+y) = \varphi(x) + \langle D\varphi(x), y\rangle_U + \int_0^1\int_0^1 \sigma\langle D^2\varphi(x+\sigma sy)y, y\rangle_U\,d\sigma\,ds.$$

Therefore

$$A\varphi(x) = \langle a, D\varphi(x)\rangle_U + \tfrac{1}{2}\mathrm{Tr}\, QD^2\varphi(x)$$
$$+ \int_U\int_0^1\int_0^1 \sigma\langle D^2\varphi(x+\sigma sy)y, y\rangle_U\,d\sigma\,ds\,\nu(dy). \qquad (5.8)$$

The integral in (5.8) is well defined and $A\varphi \in UC_b(U)$.

Note that (5.8) also holds when A is replaced by the operator A_n. It follows that

$$|A_n\varphi(x) - A\varphi(x)| \leq \tfrac{1}{2}\|D^2\varphi\|_\infty \int_{\{|u|_U < 1/n\}} |u|_U^2\,\nu(dy)$$

and $A_n\varphi \to A\varphi$ uniformly. It is therefore enough to show that for any sequence $(\varphi_n) \subset UC_b(U)$, converging uniformly to $\varphi \in UC_b(U)$,

$$\sup_{0\leq s\leq 1}\sup_{x\in U} |P_s^n\varphi_n(x) - P_s\varphi(x)| \to 0.$$

But

$$\big|P_s^n\varphi_n(x) - P_s\varphi(x)\big| \leq \big|\mathbb{E}\big(\varphi_n(x+X_n(s)) - \varphi(x+X_n(s))\big)\big|$$
$$+ \big|\mathbb{E}\big(\varphi(x+X_n(s)) - \varphi(x+X(s))\big)\big|$$
$$\leq I_1^n(x, s) + I_2^n(x, s).$$

The sequence I_1^n converges uniformly to 0, because the sequence (φ_n) converges uniformly to φ. By the Lévy–Khinchin decomposition,

$$\sup_{s \le t} |X_n(s) - X(s)|_U \to 0, \qquad \mathbb{P}\text{-a.s.}$$

Since φ is uniformly continuous the result follows.   □

The idea of determining the generator of a Markov process by finite-dimensional approximations, as described above, can be applied to solutions of equations with Lévy noise. Note, however, that owing to the result of Nemirovskii and Semenov (1973) we know that the space $UC_b^2(U)$ is not dense in $UC_b(U)$ unless U is finite-dimensional.

6

Poisson random measures

In this chapter, the basic properties of Poisson random measures are established. They are used for the construction and moment estimates of Lévy processes.

6.1 Introduction

Let $\overline{\mathbb{Z}}_+ = \{0, 1, 2, \ldots, +\infty\}$ and let $(E, \mathcal{E})$ be a measurable space. We denote by $\mathcal{P}_{\overline{\mathbb{Z}}_+}(E)$ the space of all $\overline{\mathbb{Z}}_+$-valued measures on $(E, \mathcal{E})$ and consider the measurable space $(\mathcal{P}_{\overline{\mathbb{Z}}_+}(E), \mathcal{A})$, where $\mathcal{A}$ is the σ-field generated by the mappings $\mathcal{P}_{\overline{\mathbb{Z}}_+} \ni \rho \mapsto \rho(\Gamma) \in \overline{\mathbb{Z}}_+$, $\Gamma \in \mathcal{E}$.

Definition 6.1 Let λ be a σ-finite measure on $(E, \mathcal{E})$. A *Poisson random measure on E with intensity measure (or mean measure)* λ is a random element π in $\mathcal{P}_{\overline{\mathbb{Z}}_+}(E)$ such that for every $\Gamma \in \mathcal{E}$ the random variable $\pi(\Gamma)$ has the Poisson distribution $\mathcal{P}(\lambda(\Gamma))$ and for any disjoint $\Gamma_1, \ldots, \Gamma_M$ the random variables $\pi(\Gamma_1), \ldots, \pi(\Gamma_M)$ are independent.

Definition 6.2 Let π be a Poisson random measure on $(E, \mathcal{E})$ with intensity measure λ. Then we call $\widehat{\pi}(d\xi) := \pi(d\xi) - \lambda(d\xi)$ the *compensated Poisson measure*. Note that $\widehat{\pi}(A)$ is well defined if $\lambda(A) < \infty$.

Clearly the intensity measure uniquely determines the law of a Poisson random measure.

Remark 6.3 Since λ is σ-finite there is a disjoint partition $E_n \in \mathcal{E}$ such that $\lambda(E_n) < \infty$, $n \in \mathbb{N}$. Let $\lambda_n(\Gamma) = \lambda(\Gamma \cap E_n)$, $n \in \mathbb{N}$, $\Gamma \in \mathcal{E}$. Assume that π_n, $n \in \mathbb{N}$, are independent Poisson random measures and that, for every n, π_n corresponds to λ_n. Since the sum of a sequence of independent random variables with Poisson distributions also has a Poisson distribution, $\pi = \sum_n \pi_n$ is a Poisson random measure corresponding to λ. Conversely, given π we can define $\pi_n(\Gamma) = \pi(\Gamma \cap$

E_n), $\Gamma \in \mathcal{E}$. Then the π_n are independent Poisson random measures, π_n corresponds to λ_n and $\pi = \sum \pi_n$.

The following theorem ensures the existence of a Poisson random measure with a given intensity.

Theorem 6.4 *Let λ be a σ-finite measure on $(E, \mathcal{E})$. Let $E_n \in \mathcal{E}$ be a partition of E, indexed by a finite or countable set $\mathcal{I}$, satisfying $E = \bigcup_n E_n$, $\lambda(E_n) < \infty$ and $E_n \cap E_m = \emptyset$ for $n \neq m$. Let q_n and Λ_m^n, $n \in \mathcal{I}$, $m \in \mathbb{N}$, be sequences of independent random variables such that $q_n \in \mathcal{P}(\lambda(E_n))$ and $\mathbb{P}\left(\Lambda_m^n \in \Gamma\right) = \lambda(\Gamma \cap E_n)/\lambda(E_n)$ for $n \in \mathcal{I}$ and $m \in \mathbb{N}$. Then $\pi = \sum_{n \in \mathcal{I}} \sum_{m=1}^{q_n} \delta_{\Lambda_m^n}$ is a Poisson random measure on E with intensity measure λ.*

Proof Let (E_n), (q_n), (Λ_m^n) and π be given as above. Following Remark 6.3, we may assume that λ is finite and $\#\mathcal{I} = 1$. Thus $(E_n) = (E)$, $(q_n) = (q)$, and $(\Lambda_m^n) = (\Lambda_m)$. It is sufficient to show that, for any disjoint $\Gamma_1, \ldots, \Gamma_M \in \mathcal{E}$ and $y_1, \ldots, y_M > 0$,

$$\mathbb{E} \exp \left\{ -\sum_{l=1}^{M} y_l \pi(\Gamma_l) \right\} = \exp \left\{ \sum_{l=1}^{M} \lambda(\Gamma_l)(e^{-y_l} - 1) \right\}.$$

Indeed, if this holds then the multivariate Laplace transform of the vector $(\pi(\Gamma_1), \ldots, \pi(\Gamma_M))$ is equal to the product of the Laplace transforms of $\pi(\Gamma_1), \ldots, \pi(\Gamma_M)$. Let $\Gamma = E \setminus \bigcup_{l=1}^{M} \Gamma_l$. Since q and $\Lambda_1, \Lambda_2, \ldots$ are independent,

$$\mathbb{E} \exp \left\{ -\sum_{l=1}^{M} y_l \pi(\Gamma_l) \right\} = \mathbb{E} \exp \left\{ -\sum_{l=1}^{M} y_l \sum_{i=1}^{q} \delta_{\Lambda_i}(\Gamma_l) \right\}$$

$$= e^{-\lambda(E)} \sum_{k=0}^{\infty} \frac{\lambda(E)^k}{k!} \mathbb{E} \exp \left\{ -\sum_{l=1}^{M} \sum_{i \leq k} y_l \delta_{\Lambda_i}(\Gamma_l) \right\}$$

$$= e^{-\lambda(E)} \sum_{k=0}^{\infty} \frac{\lambda(E)^k}{k!} \left(\mathbb{E} \exp \left\{ -\sum_{l=1}^{M} y_l \delta_{\Lambda_1}(\Gamma_l) \right\} \right)^k$$

$$= e^{-\lambda(E)} \sum_{k=0}^{\infty} \frac{\lambda(E)^k}{k!} \left(\sum_{l=1}^{M} e^{-y_l} \frac{\lambda(\Gamma_l)}{\lambda(E)} + \frac{\lambda(\Gamma)}{\lambda(E)} \right)^k$$

$$= \exp \left\{ \sum_{l=1}^{M} (e^{-y_l} - 1)\lambda(\Gamma_l) \right\}.$$

$\square$

Let us consider a Poisson random measure π on $E = [0, \infty) \times S$, where $(S, \mathcal{S})$ is a measurable space. Assume that π is *stationary* in t, that is, for all $t \geq 0$ and $\Gamma \in \mathcal{B}([0, \infty)) \times S$ the random variables $\pi(\Gamma)$ and $\pi(\tau_t(\Gamma))$ have the same law. Here $\tau_t(s, \sigma) = (s + t, \sigma)$, $t, s \in [0, \infty)$ and $\sigma \in S$ are *translation operators*. Clearly,

π is stationary in t if and only if its intensity measure λ is of the form $\lambda(dt, d\sigma) = dt\,\nu(d\sigma)$, where ν is a non-negative σ-additive measure on S. Assume for now that ν is finite. Then $\Pi(t) = \pi([0, t] \times S)$, $t \geq 0$, is a Poisson process with intensity $a := \nu(S)$. Let (τ_j) be the sequence of jumps of Π. Let $r_1 = \tau_1, r_{j+1} = \tau_{j+1} - \tau_j$. Then the r_j are independent random variables with exponential distribution $\mathbb{P}(r_j > t) = e^{-at}$, $j = 1, \ldots, t \geq 0$. This leads us to the following result.

Theorem 6.5   *Let ν be a σ-finite measure on $(S, \mathcal{S})$, let $(S_n) \subset \mathcal{S}$ be a disjoint partition of S such that $\nu(S_n) < \infty$ for all n and let (ξ_j^n) and (r_j^n) be sequences of mutually independent random variables with distributions*

$$\mathbb{P}(\xi_j^n \in A) = \frac{\nu(A \cap S_n)}{\nu(S_n)}, \qquad A \in \mathcal{S}, \qquad \mathbb{P}(r_j^n > t) = e^{-\mu(S_n)t}, \qquad t \geq 0.$$

Let $\tau_k^n = r_1^n + \cdots + r_k^n$. Then $\pi = \sum_{n,j} \delta_{(\tau_j^n, \xi_j^n)}$ is a Poisson random measure with intensity measure $dt\,\nu(d\sigma)$.

6.2 Stochastic integral of deterministic fields

Let π be a Poisson random measure on $(E, \mathcal{E})$ with intensity measure λ. Theorems 6.4 and 6.5 suggest that π has the following structure:

$$\pi(A)(\omega) = \sum_{j=1}^{\infty} \delta_{\xi_k(\omega)}(A), \qquad \omega \in \Omega, \ A \in \mathcal{E},$$

for a properly chosen sequence (ξ_j) of random elements in E. Thus π may be identified with a random distribution of a countable number of points ξ_j and $\pi(A)$ is equal to the number of points in the set A. We can also integrate with respect to π. Assume that f is a real- or vector-valued function defined on E. We can write

$$\int_E f(\xi)\pi(d\xi) := \sum_{j=1}^{\infty} f(\xi_k),$$

provided that the series is convergent $\mathbb{P}$-a.s. We can also define the integral using the Bochner–Lebesgue scheme. Namely, let U be a Hilbert space. We say that $f \colon E \mapsto U$ is a simple function if it has the form

$$f = \sum_{j=1}^{n} u_j \, \chi_{A_j},$$

where $n \in \mathbb{N}$, $(u_j) \subset U$ and $(A_j) \subset \mathcal{E}$ are such that $A_j \cap A_k = \emptyset$, $j \neq k$, and $\pi(A_j) < \infty$, $\mathbb{P}$-a.s., or equivalently $\lambda(A_j) < \infty$. Define

$$\int_E f(\xi)\pi(d\xi) := \sum_{j=1}^{n} u_j \pi(A_j).$$

Note that

$$\left| \int_E f(\xi) \pi(\mathrm{d}\xi) \right|_U \leq \int_E |f(\xi)|_U \pi(\mathrm{d}\xi) \qquad (6.1)$$

and

$$\mathbb{E} \int_E |f(\xi)|_U \pi(\mathrm{d}\xi) = \int_E |f(\xi)|_U \lambda(\mathrm{d}\xi) < \infty.$$

By taking monotone limits, we define the integral first for all measurable real-valued non-negative f and then for arbitrary real-valued f, by splitting it into the difference of positive and negative parts. Having defined the integral of a real-valued function and established estimate (6.1), we can extend the integral, by a standard limiting argument, to vector-valued functions satisfying $\int_E |f(\xi)|_U \lambda(\mathrm{d}\xi) < \infty$. We will gather together basic properties of the integral in the following theorem. Here $f \colon E \mapsto U$ is a measurable mapping satisfying the above estimate.

Theorem 6.6

(i) *If $U = \mathbb{R}$ and f is non-negative then, for $t \geq 0$,*

$$\mathbb{E} \exp \left\{ -t \int_E f(\xi) \pi(\mathrm{d}\xi) \right\} = \exp \left\{ - \int_E (1 - e^{-tf(\xi)}) \lambda(\mathrm{d}\xi) \right\}.$$

(ii) *For $z \in U$,*

$$\mathbb{E} \exp \left\{ i \left\langle z, \int_E f(\xi) \pi(\mathrm{d}\xi) \right\rangle_U \right\} = \exp \left\{ - \int_E (1 - e^{i\langle z, f(\xi) \rangle_U}) \lambda(\mathrm{d}\xi) \right\}.$$

(iii) $\mathbb{E} \left| \int_E f(\xi) \pi(\mathrm{d}\xi) \right|_U < \infty$ *and* $\mathbb{E} \int_E f(\xi) \widehat{\pi}(\mathrm{d}\xi) = 0.$
(iv) *If* $\int_E |f(\xi)|_U^2 \lambda(\mathrm{d}\xi) < \infty$ *then*

$$\mathbb{E} \left| \int_E f(\xi) \widehat{\pi}(\mathrm{d}\xi) \right|_U^2 = \int_E |f(\xi)|_U^2 \lambda(\mathrm{d}\xi).$$

(v) *If functions $f_1, f_2, \ldots, f_M$ have disjoint supports then the random variables $\int_E f_1(\xi) \pi(\mathrm{d}\xi), \ldots, \int_E f_M(\xi) \pi(\mathrm{d}\xi)$ are independent.*

Proof We prove only (i), since the proofs of the other parts of the theorem use similar arguments. If $f = \sum_{j=1}^M \alpha_j \chi_{A_j}$, $A_j \cap A_k = \emptyset$, $j \neq k$, then

$$\mathbb{E} \exp \left\{ -t \int_E f(x) \pi(\mathrm{d}\xi) \right\} = \prod_{j=1}^M \mathbb{E} \exp \left\{ -t\alpha_j \pi(A_j) \right\}$$

$$= \prod_{j=1}^M \exp \left\{ -\lambda(A_j)(1 - e^{-t\alpha_j}) \right\},$$

and the formula holds. By monotone passage to the limit, the result follows for all non-negative measurable functions. $\qquad\square$

The theorem above gives a necessary and sufficient condition under which the integral $\int_E f(\xi)\widehat{\pi}(\mathrm{d}\xi)$ is square integrable and also an evaluation of its second moment. Below, we formulate a sufficient condition for the existence of its moments of order $p > 2$.

Proposition 6.7 *For each $p \in \mathbb{N}$ there exists a polynomial R_p of p variables such that, if $f : E \mapsto [0, +\infty)$ is measurable and $m_k := \int_E f^k(\xi)\lambda(\mathrm{d}\xi) < \infty$, $k = 1, \ldots, p$, then*

$$\mathbb{E}\left(\int_E f(\xi)\pi(\mathrm{d}\xi)\right)^p = R_p(m_1, \ldots, m_p).$$

Proof Let $I := \int_E f(\xi)\pi(\mathrm{d}\xi)$. Then

$$\phi(u) := \mathbb{E}\,\mathrm{e}^{-uI} = \exp\left\{-\int_E \left(1 - \mathrm{e}^{uf(\xi)}\right)\lambda(\mathrm{d}\xi)\right\} = \mathrm{e}^{-\varphi(u)}, \qquad u \ge 0.$$

By induction, for $u > 0$,

$$\frac{\mathrm{d}^k\phi}{\mathrm{d}u^k}(u) = \phi(u)\tilde{R}_k\left(\frac{\mathrm{d}\varphi}{\mathrm{d}u}(u), \quad \ldots, \quad \frac{\mathrm{d}^k\varphi}{\mathrm{d}u^k}(u)\right),$$

where $\tilde{R}_k$ is a polynomial of k variables. Since

$$\frac{\mathrm{d}^k\phi}{\mathrm{d}u^k}(0) = (-1)^k\,\mathbb{E}\,I^k = \tilde{R}_k\left(\int_E f(\xi)\lambda(\mathrm{d}\xi), \quad \ldots, \quad (-1)^{k+1}\int_E f^k(\xi)\lambda(\mathrm{d}\xi)\right),$$

the desired formula follows. $\qquad\square$

6.3 Application to construction of Lévy processes

As an application of the concept of the random measure we present a simple way of constructing Lévy processes, as promised in Chapter 4. This will give an additional interpretation of the Lévy–Khinchin formula. Let ψ be the Lévy exponent of a process. Recall (see (4.19) in Theorem 4.27) that

$$\psi(x) = -\mathrm{i}\langle a, x\rangle_U + \tfrac{1}{2}\langle Qx, x\rangle_U + \psi_0(x),$$

where

$$\psi_0(x) = \int_{\{|y|_U \ge 1\}} \left(1 - \mathrm{e}^{\mathrm{i}\langle x, y\rangle_U}\right)\nu(\mathrm{d}y)$$

$$+ \int_{\{|y|_U < 1\}} \left(1 - \mathrm{e}^{\mathrm{i}\langle x, y\rangle_U} + \mathrm{i}\langle x, y\rangle_U\right)\nu(\mathrm{d}y) \qquad (6.2)$$

and the Lévy measure ν is concentrated on $U \setminus \{0\}$ and satisfies

$$\int_{\{|y|_U < 1\}} |y|_U^2 \, \nu(\mathrm{d}y) + \nu\{y : |y|_U \geq 1\} < \infty.$$

Theorem 6.8 *Let π be a Poisson random measure on $[0, \infty) \times U$ with intensity measure $\mathrm{d}t \, \nu(\mathrm{d}x)$.*

(i) *The formula*

$$L(t) = \int_0^t \int_{\{|y|_U \geq 1\}} y\pi(\mathrm{d}s, \mathrm{d}y)$$

$$+ \lim_{\varepsilon \downarrow 0} \int_0^t \int_{\{\varepsilon \leq |y|_U < 1\}} y\widehat{\pi}(\mathrm{d}s, \mathrm{d}y), \qquad (6.3)$$

where the limit is in $L^2(\Omega, \mathcal{F}, \mathbb{P}; U)$, defines a process with independent increments and exponent ψ_0.

(ii) *The limit in (6.3) exists $\mathbb{P}$-a.s. uniformly on any time interval $[0, T]$ if ε tends to 0 sufficiently fast.*

(iii) *The process L has trajectories with bounded variation if and only if*

$$\int_{\{|y|_U < 1\}} |y|_U \nu(\mathrm{d}y) < \infty.$$

Proof of (i) In the proof, basically we follow Bertoin (1996). Consider the process

$$L_1(t) = \int_0^t \int_{\{|y|_U \geq 1\}} y\pi(\mathrm{d}s, \mathrm{d}y), \qquad t \geq 0.$$

It is well defined, provided that

$$\int_0^t \int_{\{|y|_U \geq 1\}} |y|_U \pi(\mathrm{d}s, \mathrm{d}y) < \infty, \qquad \mathbb{P}\text{-a.s.}$$

This condition is justified as follows. Let

$$\Omega_n := \{\omega \in \Omega : \pi([0, t] \times \{y : |y|_U \geq n\})(\omega) = 0\}.$$

Since $\pi([0, t] \times \{y : |y|_U \geq 1\}) < \infty$, $\Omega = \bigcup \Omega_n$. Thus the desired conclusion follows from the fact that, for $\omega \in \Omega_n$,

$$\int_0^t \int_{\{|y|_U \geq 1\}} |y|_U \pi(\mathrm{d}s, \mathrm{d}y)(\omega) \leq n\pi([0, t] \times \{y : |y|_U \geq 1\})(\omega) < \infty.$$

Let $0 = t_0 < t_1 < \cdots < t_M$. Then

$$L_1(t_j) - L_1(t_{j-1}) = \int_{(t_{j-1}, t_j] \times \{y : |y|_U \geq 1\}} y\pi(\mathrm{d}s, \mathrm{d}y),$$

and since the sets $(t_{j-1}, t_j] \times \{y \colon |y|_U \geq 1\}$, $j = 1, \ldots, M$, are disjoint the increments $L_1(t_j) - L_1(t_{j-1})$, $j = 1, \ldots, M$, are independent. By Theorem 6.6(ii), for $z \in U$,

$$
\begin{aligned}
\mathbb{E} \exp \left\{ i \langle z, L_1(t) \rangle_U \right\} &= \mathbb{E} \exp \left\{ i \left\langle z, \int_0^t \int_{\{|y|_U \geq 1\}} y \pi(ds, dy) \right\rangle_U \right\} \\
&= \exp \left\{ - \int_0^t \int_{\{|y|_U \geq 1\}} \left(1 - e^{i \langle z, y \rangle_U} \right) \nu(dy) \, ds \right\} \\
&= \exp \left\{ -t \int_{\{|y|_U \geq 1\}} \left(1 - e^{i \langle z, y \rangle_U} \right) \nu(dy) \right\}.
\end{aligned}
$$

Write, for $\varepsilon \in (0, 1)$,

$$
L_{2,\varepsilon}(t) := \int_0^t \int_{\{\varepsilon \leq |y|_U < 1\}} y \widehat{\pi}(ds, dy), \qquad t \geq 0.
$$

One can show easily that the process $L_{2,\varepsilon}$ is well defined, has independent increments and is independent of L_1. Finally,

$$
\mathbb{E} \exp \left\{ i \langle z, L_{2,\varepsilon}(t) \rangle_U \right\} = \exp \left\{ -t \int_{\{\varepsilon \leq |y|_U < 1\}} \left(1 - e^{i \langle z, y \rangle_U} + i \langle z, y \rangle_U \right) \nu(dy) \right\}.
$$

Since, for $0 < \eta < \varepsilon < 1$,

$$
\mathbb{E} |L_{2,\varepsilon}(t) - L_{2,\eta}(t)|_U^2 = t \int_{\{\eta \leq |y|_U < \varepsilon\}} |y|_U^2 \nu(dy),
$$

convergence in $L^2(\Omega, \mathcal{F}, \mathbb{P}; U)$ follows.

Proof of (ii) It follows from Theorem 6.6(iii), (iv) that $L_{2,\varepsilon}$ is a square integrable martingale. Thus by Doob's inequality, for all $T > 0$ and $c > 0$,

$$
\begin{aligned}
\mathbb{P} \left(\sup_{0 \leq t \leq T} |L_{2,\varepsilon}(t) - L_{2,\eta}(t)|_U \geq c \right) &\leq \frac{1}{c} \mathbb{E} |L_{2,\varepsilon}(T) - L_{2,\eta}(T)|_U \\
&= \frac{1}{c} \left(\mathbb{E} |L_{2,\varepsilon}(T) - L_{2,\eta}(T)|_U^2 \right)^{1/2} \\
&= \frac{\sqrt{T}}{c} \left(\int_{\{\eta \leq |y|_U < \varepsilon\}} |y|_U^2 \nu(dy) \right)^{1/2}.
\end{aligned}
$$

Consequently, there exist a process L_2 and a sequence $\varepsilon_n \downarrow 0$ such that

$$
\mathbb{P} \left(\lim_{n \to \infty} \sup_{0 \leq t \leq T} |L_{2,\varepsilon_n}(t) - L_2(t)|_U = 0 \right) = 1.
$$

$\square$

Proof of (iii) Let $\Delta L_t = L(t) - L(t-)$, $t \in (0, T]$. Then

$$\mathbb{E} \exp \left\{ - \sum_{0 < t \leq T} |\Delta L_t|_U \right\} = \mathbb{E} \exp \left\{ - \int_0^\infty \int_U f(t, x) \pi(\mathrm{d}t, \mathrm{d}x) \right\},$$

where $f(t, x) = \chi_{[0,T]}(t) |x|_U$. Consequently,

$$\mathbb{E} \exp \left\{ - \sum_{0 < t \leq T} |\Delta L_t|_U \right\} = \exp \left\{ -T \int_U \left(1 - e^{-|x|_U} \right) \nu(\mathrm{d}x) \right\}$$

Thus if $\int_U (1 - e^{-|x|_U}) \nu(\mathrm{d}x) = \infty$, or equivalently if $\int_{\{|x|_U < 1\}} |x|_U \nu(\mathrm{d}x) = \infty$, then $\sum_{0 < t \leq T} |\Delta L_t|_U = \infty$, $\mathbb{P}$-a.s. If $\int_U (1 - e^{-|x|_U}) \nu(\mathrm{d}x) < \infty$, or equivalently if $\int_{\{|x|_U < 1\}} |x|_U \nu(\mathrm{d}x) < \infty$, then

$$\mathbb{E} \sum_{0 < t \leq T} |\Delta L_t|_U \chi_{\{|L(t)|_U \leq M\}} = T \int_{\{|x|_U \leq M\}} |x|_U \nu(\mathrm{d}x) < \infty,$$

so L is of bounded variation. $\qquad\square$

6.4 Moment estimates in Banach spaces

In this section we prove a simple sufficient condition under which a Lévy process with representation (6.3) in a Hilbert space U takes values in a smaller Banach space B.

Proposition 6.9 *Let L be a Lévy process on U with representation (6.3), and let B be a Banach space continuously embedded into U. Assume that*

$$\int_U |x|_B^k \nu(\mathrm{d}x) < \infty \qquad for\ k = 1, \ldots, n.$$

Then L is a Lévy process in B and $\mathbb{E} |L(t)|_B^n < \infty$ for every $t \geq 0$.

Proof Let π be the Poisson measure corresponding to L. As $\int_U |x|_U \nu(\mathrm{d}x) < \infty$, we have

$$L(t) = \int_0^t \int_U x \pi(\mathrm{d}s, \mathrm{d}x) + t \int_{\{|x|_U \leq 1\}} x \nu(\mathrm{d}x).$$

Thus, as π is a non-negative measure,

$$|L(t)|_B \leq \int_0^t \int_U |x|_B \pi(\mathrm{d}s, \mathrm{d}x) + t \int_{\{|x|_U \leq\}} |x|_B \nu(\mathrm{d}x),$$

and consequently $\mathbb{E} |L(t)|_B^n < \infty$ by Proposition 6.7. What is left is the continuity in probability of L in B. But, in fact, the arguments above show the continuity in $L^n(\Omega, \mathcal{F}, \mathbb{P}; B)$. $\qquad\square$

7

Cylindrical processes and reproducing kernels

In this chapter, the concept of the reproducing kernel space is introduced. Then cylindrical Lévy processes are discussed and their reproducing kernel spaces are calculated.

7.1 Reproducing kernel Hilbert space

In this book we are mainly developing the L^2-theory of stochastic equations driven by Lévy processes, and therefore solutions of the equations will be square integrable processes in properly chosen state spaces. One can expect, therefore, that the covariance operator of the noise should play a fundamental role. Instead of the covariance, however, it is more convenient to study the reproducing kernel Hilbert space (RKHS) of the noise; see Definition 7.2 below. The RKHS is a certain Hilbert subspace of the noise space. Unlike the covariance operator, the RKHS is independent of the space on which the noise is considered; see Theorem 7.4. It contains all the information on the noise necessary for developing the L^2-theory.

We will define the RKHS of an arbitrary square integrable mean-zero random variable, and the RKHS of a square integrable mean-zero Lévy process L is then by definition the RKHS of $L(1)$.

Let Z be a square integrable random variable with mean zero in a Hilbert space U, and let $\langle \cdot, \cdot \rangle \colon U^* \times U \mapsto \mathbb{R}$ be the duality form. Let

$$R(x, y) := \mathbb{E} \langle x, Z \rangle_U \langle y, Z \rangle_U, \qquad x, y \in U,$$
$$K(x, y) := \mathbb{E} \langle x, Z \rangle \langle y, Z \rangle, \qquad x, y \in U^*.$$

Finally, let Q be the covariance operator of Z. Since Q is a nuclear self-adjoint operator on U, there is an orthonormal basis $\{e_n\}$ of U consisting of eigenvectors of Q; see Theorem 4.44 and Appendix A. Then $Q e_n = \gamma_n e_n$, $n \in \mathbb{N}$, where γ_n is

91

the eigenvalue corresponding to e_n. Since Q is non-negative-definite, $\gamma_n \geq 0$ for every n. The square root of Q is given by

$$Q^{1/2}x = \sum_n \langle x, e_n \rangle_U \gamma_n^{1/2} e_n, \qquad x \in U.$$

Generally, Q and $Q^{1/2}$ are not injective. However, we will denote by $Q^{-1/2}$ the *operator pseudo-inverse* to $Q^{1/2}$, that is, for $h \in Q^{1/2}(U)$,

$$Q^{-1/2}y = x \quad \text{if} \quad Q^{1/2}x = y \qquad \text{and} \qquad |x|_U = \inf\left\{|y|_U \colon Q^{1/2}y = x\right\}.$$
$$(7.1)$$

Proposition 7.1 *Let $(\mathcal{H}, \langle \cdot, \cdot \rangle_{\mathcal{H}})$ be a Hilbert space continuously embedded into U. Then the following conditions are equivalent.*

(i) *$\mathcal{H} = Q^{1/2}(U)$ and $\langle x, y \rangle_{\mathcal{H}} = \langle Q^{-1/2}x, Q^{-1/2}y \rangle_U$, $x, y \in \mathcal{H}$.*
(ii) *For any orthonormal basis $\{h_k\}$ of $\mathcal{H}$,*

$$R(x, y) = \sum_k \langle x, h_k \rangle_U \langle y, h_k \rangle_U, \qquad \forall x, y \in U.$$

(iii) *For any orthonormal basis $\{h_k\}$ of $\mathcal{H}$,*

$$K(x, y) = \sum_k \langle x, h_k \rangle \langle y, h_k \rangle, \qquad \forall x, y \in U^*.$$

Proof First we will prove the equivalence (ii) $\Leftrightarrow$ (iii). To this end let j denote a linear isomorphism between U and U^*. Thus $j \colon U \mapsto U^*$ and $\langle jx, y \rangle = \langle x, y \rangle_U$, $x, y \in U$. Let $\{h_k\}$ be an orthonormal basis of $\mathcal{H}$. Then, for $x, y \in U$,

$$R(x, y) = \mathbb{E}\langle x, Z \rangle_U \langle y, Z \rangle_U = \mathbb{E}\langle jx, Z \rangle \langle jy, Z \rangle = K(jx, jy)$$

and

$$\sum_k \langle x, h_k \rangle_U \langle y, h_k \rangle_U = \sum_k \langle jx, h_k \rangle \langle jy, h_k \rangle,$$

and hence

$$R(x, y) = \sum_k \langle x, h_k \rangle_U \langle y, h_k \rangle_U$$

if and only if

$$K(jx, jy) = \sum_k \langle jx, h_k \rangle \langle jy, h_k \rangle.$$

In order to show that (i) $\Rightarrow$ (ii), take an arbitrary orthonormal basis $\{h_k\}$ of $\mathcal{H}$. If (i) holds then there is an orthonormal basis $\{f_k\}$ of U such that $\{Q^{1/2}f_k\} \cup \{0\} =$

$\{h_k\} \cup \{0\}$. Consequently,

$$R(x, y) = \langle Qx, y \rangle_U = \langle Q^{1/2}x, Q^{1/2}y \rangle_U = \sum_k \langle Q^{1/2}x, f_k \rangle_U \langle Q^{1/2}y, f_k \rangle_U$$

$$= \sum_k \langle x, Q^{1/2}f_k \rangle_U \langle y, Q^{1/2}f_k \rangle_U = \sum_k \langle x, h_k \rangle_U \langle y, h_k \rangle_U.$$

To show that (ii) $\Rightarrow$ (i), assume that $\{h_k\}$ is an orthonormal basis of $\mathcal{H}$ such that

$$R(x, y) = \sum_k \langle x, h_k \rangle_U \langle y, h_k \rangle_U, \qquad \forall x, y \in U.$$

Let $\{f_k\}$ be an orthonormal basis of U, and let $B : U \mapsto \mathcal{H}$ be a bounded linear operator, uniquely determined by $Bf_k = h_k$ if $\dim \mathcal{H} = \infty$ and by

$$Bf_k = \begin{cases} h_k, & k \leq \dim \mathcal{H}, \\ 0, & k > \dim \mathcal{H}, \end{cases}$$

if $\dim \mathcal{H} < \infty$. Since $\mathcal{H}$ is continuously embedded into U one has $B \in L(U, U)$. For, denote the embedding by i, then

$$\left| B \sum_k x_k f_k \right|_U^2 \leq \|i\|_{L(\mathcal{H},U)}^2 \left| \sum_k x_k Bf_k \right|_{\mathcal{H}}^2 = \|i\|_{L(\mathcal{H},U)}^2 \sum_k x_k^2$$

$$\leq \|i\|_{L(\mathcal{H},U)}^2 \left| \sum_k x_k f_k \right|_U^2.$$

Clearly $\mathcal{H} = B(U)$ and $\langle x, y \rangle_{\mathcal{H}} = \langle B^{-1}x, B^{-1}y \rangle_U$, $x, y \in \mathcal{H}$, where B^{-1} stands for the pseudo-inverse operator. For all $x, y \in U$,

$$\langle Qx, y \rangle_U = R(x, y) = \sum_k \langle x, h_k \rangle_U \langle y, h_k \rangle_U = \sum_k \langle x, Bf_k \rangle_U \langle y, Bf_k \rangle_U$$

$$= \langle B^*x, B^*y \rangle_U = \langle BB^*x, y \rangle_U.$$

Thus $BB^* = Q$ and the desired conclusion follows from Douglas's theorem (see appendix section A.4). □

Definition 7.2 Let Z be a square integrable mean-zero random variable taking values in a Hilbert space U, and let Q be the covariance of Z. Then $\mathcal{H} := Q^{1/2}(U)$ is called the *reproducing kernel Hilbert space* (RKHS) of Z. Let L be a square integrable mean-zero Lévy process in U. Then its RKHS is the RKHS of $L(1)$ and therefore it is equal to $Q^{1/2}(U)$, where Q is the covariance operator of L.

Remark 7.3 Let Z be a square integrable mean-zero random variable in U, and let $\mathcal{H}$ be its RKHS. Then, since Q is nuclear, $Q^{1/2}$ is Hilbert–Schmidt.

Consequently the embedding $\mathcal{H} \hookrightarrow U$ is Hilbert–Schmidt, that is, for an arbitrary orthonormal basis $\{e_n\}$ of $\mathcal{H}$ one has $\sum_n |e_n|_U^2 < \infty$.

The next theorem shows that the RKHS of Z is independent of the space on which the random element Z is considered. More precisely, let $\mathcal{H}$ and $\tilde{\mathcal{H}}$ be the RKHSs of Z considered as a random variable on U and $\tilde{U}$, respectively; then $\mathcal{H} = \tilde{\mathcal{H}}$.

Theorem 7.4   *Assume that Z is a square integrable mean-zero random variable in a Hilbert space U, continuously embedded in a Hilbert space $\tilde{U}$. Let Q be the covariance operator of Z considered as a U-valued random variable, and let $\tilde{Q}$ be the covariance operator of Z, considered as a $\tilde{U}$-valued random variable. Then $Q^{1/2}(U) = \tilde{Q}^{1/2}(\tilde{U})$.*

Proof   It is sufficient to show that $\mathcal{H} = \tilde{\mathcal{H}}$. By the definition of the RKHS this immediately implies the identity of ranges of $Q^{1/2}$ and $\tilde{Q}^{1/2}$.

Let $\{e_n\}$ be an orthonormal basis of $\mathcal{H}$. By Proposition 7.1, to prove that $\mathcal{H} = \tilde{\mathcal{H}}$ it is sufficient to show that, for all $x, y \in \tilde{U}$,

$$\mathbb{E}\,\langle Z, x \rangle_{\tilde{U}} \langle Z, y \rangle_{\tilde{U}} = \sum_k \langle x, e_k \rangle_{\tilde{U}} \langle y, e_k \rangle_{\tilde{U}}. \tag{7.2}$$

We first show that (7.2) holds for $x, y \in U$. To do this note that $\langle \cdot, \cdot \rangle_{\tilde{U}} : U \times U \mapsto \mathbb{R}$ is a symmetric continuous bilinear form on U. Therefore there is a self-adjoint operator $B \in L(U, U)$ such that $\langle x, y \rangle_{\tilde{U}} = \langle x, By \rangle_U$ for all $x, y \in U$. Hence, for all $x, y \in U$,

$$\mathbb{E}\,\langle Z, x \rangle_{\tilde{U}} \langle Z, y \rangle_{\tilde{U}} = \mathbb{E}\,\langle Z, Bx \rangle_U \langle Z, By \rangle_U.$$

Thus by Proposition 7.1(ii),

$$\mathbb{E}\,\langle Z, x \rangle_{\tilde{U}} \langle Z, y \rangle_{\tilde{U}} = \sum_k \langle Bx, e_k \rangle_U \langle By, e_k \rangle_U = \sum_k \langle x, e_k \rangle_{\tilde{U}} \langle y, e_k \rangle_{\tilde{U}}.$$

By continuity (7.2) holds for all x, y in the closure $\overline{U}$ of U in $\tilde{U}$. We have the orthogonal decomposition $\tilde{U} = \overline{U} \oplus \overline{U}^\perp$. Since Z takes values in U and $\mathcal{H} \subset U$ we have $\langle Z, x \rangle_{\tilde{U}} = 0$ and $\langle x, e_k \rangle_{\tilde{U}} = 0$ for all $x \in \overline{U}^\perp$ and k. Hence (7.2) holds for all $x, y \in \tilde{U}$. $\square$

Example 7.5   Let $Z = (Z_n)$ be a sequence of independent square integrable mean-zero random variables. Let $\sigma_n^2 = \mathbb{E}\,Z_n^2$, $n = 1, 2, \ldots$, and let $U = l_\rho^2$ be a weighted l^2-space with positive weight $\rho = (\rho_n)$. Then $\mathbb{E}\,|Z|_U^2 = \sum_{n=1}^\infty \sigma_n^2 \rho_n$. Thus if $\mathbb{E}\,|Z|_U^2 < \infty$ then the covariance operator Q of Z is given by $Qx =$

$(\sigma_n^2 \rho_n x_n)$, $x = (x_n) \in l_\rho^2$. Moreover,

$$\mathcal{H} = Q^{1/2}(U) = \left\{ h = Q^{1/2}x \colon x \in l_\rho^2 \right\} = \left\{ (\sigma_n \gamma_n) \colon (\gamma_n) \in l^2 \right\}$$

and

$$\langle h, g \rangle_{\mathcal{H}} = \left\langle Q^{-(1/2)}h, Q^{-(1/2)}g \right\rangle_U = \sum_{n=1}^{\infty} \frac{h_n g_n}{\sigma_n^2},$$

and we see explicitly that in the example the reproducing kernel is independent of choice of weight.

The reproducing kernel has been defined thus far for square integrable random variables. As the following example indicates, one can extend the concept of the reproducing kernel to a non-square-integrable random variable Z taking values in a Hilbert space U, provided that there is a bigger $\tilde{U} \hookleftarrow U$ such that $\mathbb{E}|Z|_{\tilde{U}}^2 < \infty$.

Example 7.6 There exists a mean-zero random variable Z taking values in l_ρ^2, for arbitrary positive weight $\rho = (\rho_n)$, such that

$$\mathbb{E}|Z|_{l_\rho^2}^2 < \infty \text{ if and only if } \sum_{n=1}^{\infty} \rho_n < \infty, \qquad \mathcal{H} = l^2.$$

For, let $Z = (Z_n)$ be a sequence of independent real-valued random variables each taking values in $\{a_n, 0, -a_n\}$ and such that

$$\mathbb{P}(Z_n = a_n) = \mathbb{P}(Z_n = -a_n) = \tfrac{1}{2}p_n, \quad \mathbb{P}(Z_n = 0) = 1 - p_n, \quad n = 1, \ldots$$

Then $\mathbb{E}\, Z_n^2 = \sigma_n^2 := p_n a_n^2$. Assume that $\sigma_n^2 = 1$, $n = 1, \ldots$, and $\sum_{n=1}^{\infty} p_n < \infty$. By the Borel–Cantelli lemma, $Z_n = 0$, $\mathbb{P}$-a.s., for n sufficiently large. Thus $Z \in l_\rho^2$ for an arbitrary positive weight ρ. By Example 7.5, the reproducing kernel space $\mathcal{H} = l^2$. It is also clear that $\mathbb{E}|Z|_{l_\rho^2}^2 < \infty$ if and only if $\sum_{n=1}^{\infty} \rho_n < \infty$.

7.1.1 Determination of reproducing kernels

In the two propositions below, $\mathcal{H}$ and U are Hilbert spaces. We assume that $\mathcal{H}$ is densely embedded into U. Then, under identification of $\mathcal{H}$ with $\mathcal{H}^*$, we have $U^* \hookrightarrow \mathcal{H}^* \equiv \mathcal{H} \hookrightarrow U$ and we can treat U^* as a subspace of $\mathcal{H}$. We denote by $\langle \cdot, \cdot \rangle$ the duality bilinear form on $U^* \times U$. Clearly $\langle x, y \rangle = \langle x, y \rangle_{\mathcal{H}}$, $x \in U^*$, $y \in \mathcal{H}$.

Proposition 7.7 *Let Z be a square integrable mean-zero random variable in U. Assume that $\mathbb{E}\,\langle x, Z \rangle \langle y, Z \rangle = \langle x, y \rangle_{\mathcal{H}}$ for all $x, y \in U^*$. Then $\mathcal{H}$ is the RKHS of Z.*

Proof Let $\{e_k\}$ be an arbitrary orthonormal basis of $\mathcal{H}$. Then, by the Parseval identity,

$$\langle x, y \rangle_{\mathcal{H}} = \sum_k \langle x, e_k \rangle_{\mathcal{H}} \langle y, e_k \rangle_{\mathcal{H}}.$$

Since, for $x, y \in U^*$, $\langle x, e_k \rangle_{\mathcal{H}} = \langle x, e_k \rangle$ and $\langle y, e_k \rangle_{\mathcal{H}} = \langle y, e_k \rangle$ we have

$$\mathbb{E}\,\langle x, Z \rangle \langle y, Z \rangle = \langle x, y \rangle_{\mathcal{H}} = \sum_k \langle x, e_k \rangle \langle y, e_k \rangle,$$

which by Proposition 7.1 completes the proof. $\qquad\qquad\square$

Example 7.8 Let X be a mean-zero real-valued random variable that is independent of a random variable Y uniformly distributed in $[0, 1]$. Define $Z := X \delta_Y$. Let U be the Sobolev space $W_0^{-1,2}$, dual to the Hilbert space $W_0^{1,2}$ of all absolutely continuous functions x on $[0, 1]$, such that $x(0) = 0$ and

$$|x|^2_{W_0^{1,2}} := \int_0^1 \left| \frac{dx}{d\xi} \right|^2 d\xi < \infty.$$

The space U contains all Dirac measures δ_a, $a \in (0, 1)$. Note that if $x, y \in W_0^{1,2}$ then $(x, Z) = X x(Y)$ and

$$\mathbb{E}\,(x, Z)(y, Z) = \left(\mathbb{E}\,X^2\right) \int_0^1 x(\xi) y(\xi) \, d\xi.$$

Thus the space $\mathcal{H}$ is identical with the space of all square integrable functions on $[0, 1]$ with scalar product

$$\langle x, y \rangle_{\mathcal{H}} = \left(\mathbb{E}\,X^2\right) \int_0^1 x(\xi) y(\xi) \, d\xi.$$

Proposition 7.9 *Assume that the embedding $\mathcal{H} \hookrightarrow U$ is Hilbert–Schmidt. Let $Z \colon \mathcal{H} \mapsto L^2(\Omega, \mathcal{F}, \mathbb{P})$ be a linear operator such that $\mathbb{E}\,(Zx)^2 = c\,|x|^2_{\mathcal{H}}$ and $\mathbb{E}\,Zx = 0$ for $x \in \mathcal{H}$. Then there is a unique square integrable mean-zero random variable $\tilde{Z}$ in U such that*

$$Zx = \langle x, \tilde{Z} \rangle, \qquad \forall x \in U^*. \tag{7.3}$$

Moreover, $\mathcal{H}$ is the RKHS of $\tilde{Z}$.

Proof Let $\{e_n\}$ be an orthonormal basis of $\mathcal{H}$. We assume that $\{e_n\} \subset U^*$. Such a basis exists since U^* is dense in $\mathcal{H}$. Then

$$\mathbb{E}\left| \sum_{j=n}^{n+m} (Ze_j)e_j \right|^2_U = \sum_{j,k=n}^{n+m} \left(\mathbb{E}\,Ze_j\,Ze_k\right)\langle e_j, e_k \rangle_U = \sum_{j,k=n}^{n+m} \langle e_j, e_k \rangle_{\mathcal{H}} \langle e_j, e_k \rangle_U$$

$$= \sum_{j=n}^{n+m} |e_j|^2_U \to 0 \qquad \text{as } n, m \to \infty.$$

Thus the series $\sum_n (Ze_n)e_n$ converges in $L^2(\Omega, \mathcal{F}, \mathbb{P}; U)$. We will show that its limit $\tilde{Z}$ has the desired properties. To this end note that, for any $x \in U^*$,

$$\langle x, \tilde{Z} \rangle = \sum_n (Ze_n)(x, e_n) = \sum_n (Ze_n)\langle x, e_n \rangle_{\mathcal{H}}.$$

Since Z is a continuous linear operator on $\mathcal{H}$,

$$\sum_n (Ze_n)\langle x, e_n \rangle_{\mathcal{H}} = Z\left(\sum_n \langle e_n, x \rangle_{\mathcal{H}} e_n\right) = Zx$$

and hence (7.3) holds. The uniqueness of $\tilde{Z}$ is obvious. The second part follows directly from Proposition 7.7. □

Later we will identify $\tilde{Z}$ with Z and write Z instead of $\tilde{Z}$.

Example 7.10 Given a square integrable mean-zero real-valued Lévy process L, define $Zx = \int_0^1 x(\xi) \, dL(\xi)$, $x \in L^2(0, 1)$. Then

$$\mathbb{E} |Zx|^2 = \mathbb{E} L^2(1) \int_0^1 |x(\xi)|^2 \, d\xi.$$

Thus, if $U = W_0^{-1,2}$ (see Example 7.8) then Z can be identified with the distributional derivative $dL/d\xi$.

7.1.2 Cylindrical Wiener process

Let $(\Omega, \mathcal{F}, (\mathcal{F}_t), \mathbb{P})$ be a filtered probability space, and let $\mathcal{H}$ be a Hilbert space.

Definition 7.11 An $(\mathcal{F}_t)$-*adapted cylindrical Wiener process on* $\mathcal{H}$ is a linear (in the second variable) mapping $W: [0, \infty) \times \mathcal{H} \mapsto L^2(\Omega, \mathcal{F}, \mathbb{P})$ satisfying the following conditions.

(i) For all $t \geq 0$ and $x \in \mathcal{H}$, $\mathbb{E} |W(t, x)|^2 = t |x|_{\mathcal{H}}^2$.
(ii) For each $x \in \mathcal{H}$, $(W(t, x), t \geq 0)$ is a real-valued $(\mathcal{F}_t)$-adapted Wiener process.

Usually the filtration $(\mathcal{F}_t)$ is specified and we just say that W is a cylindrical Wiener process on $\mathcal{H}$.

Lemma 7.12 *If W is a cylindrical Wiener process then, for all $t \geq s \geq 0$ and $x, y \in \mathcal{H}$, $\mathbb{E} W(t, x)W(s, y) = (t \wedge s) \langle x, y \rangle_{\mathcal{H}}$.*

Proof   Assume that $t \geq s \geq 0$. Then

$$
\begin{aligned}
\mathbb{E}\, W(t, x)W(s, y) &= \mathbb{E}\,\mathbb{E}\big((W(t, x) - W(s, y))W(s, y)\big|\mathcal{F}_s\big) \\
&\quad + \mathbb{E}\, W(s, x)W(s, y) \\
&= \mathbb{E}\, W(s, x)W(s, y) \\
&= \tfrac{1}{4}\mathbb{E}\left((W(s, x) + W(s, y))^2 - (W(s, x) - W(s, y))^2\right) \\
&= \tfrac{1}{4}\mathbb{E}\left((W(s, x + y))^2 - (W(s, x - y))^2\right) \\
&= \tfrac{1}{4}s\left(|x + y|_{\mathcal{H}}^2 - |x - y|_{\mathcal{H}}^2\right) = s\langle x, y\rangle_{\mathcal{H}}.
\end{aligned}
$$

$\square$

As in the previous section, let U be a Hilbert space such that the embedding $\mathcal{H} \hookrightarrow U$ is dense and Hilbert–Schmidt. We identify U^* with a subspace of $\mathcal{H}$ and denote by $\langle \cdot, \cdot \rangle$ the bilinear form on $U^* \times U$. Recall that $\langle x, y \rangle = \langle x, y \rangle_{\mathcal{H}}$ for $x \in U^*$ and $y \in \mathcal{H}$. As a simple consequence of Propositions 7.7 and 7.9 we have the following result.

Theorem 7.13

(i) *If W is a cylindrical Wiener process on $\mathcal{H}$ then there is a U-valued Wiener process, which we will denote also by W, such that*

$$
\langle x, W(t) \rangle = W(t, x), \qquad t \geq 0,\ x \in U^*. \tag{7.4}
$$

Moreover, the RKHS of W is equal to $\mathcal{H}$.

(ii) *Conversely, if W is a Wiener process in U with RKHS equal to $\mathcal{H}$ then (7.4) defines a cylindrical Wiener process on $\mathcal{H}$.*

Remark 7.14   Assume that W is a cylindrical Wiener process in $\mathcal{H}$. Let $\{e_n\}$ be an orthonormal basis of $\mathcal{H}$. Let $W_n(t) := W(t, e_n)$. Then (W_n) is a sequence of independent standard real-valued Wiener processes. Let U be a Hilbert space such that the embedding $\mathcal{H} \hookrightarrow U$ is Hilbert–Schmidt. Then $W(t) = \sum_n W_n(t)e_n$, $t \geq 0$, where the series converges in $L^2(\Omega, \mathcal{F}, \mathbb{P}; U)$.

The concept of a cylindrical Wiener process is closely related to that of space–time white noise. Loosely speaking, the latter is the time derivative of a cylindrical Wiener process; see Theorem 7.16 below. Assume that $(E, \mathcal{E})$ is a measurable space and that λ is a σ-finite measure on E. Write $\mathcal{E}_{\mathrm{fin}} := \{A \in \mathcal{E} : \lambda(A) < \infty\}$.

Definition 7.15   The mapping $\mathfrak{X} \colon \mathcal{E}_{\mathrm{fin}} \mapsto L^2(\Omega, \mathcal{F}, \mathbb{P})$ is called *a (Gaussian) white noise on E* if for all $A_1, \ldots, A_n \in \mathcal{E}_{\mathrm{fin}}$ the random vector $(\mathfrak{X}(A_1), \ldots, \mathfrak{X}(A_n))$ is Gaussian with mean zero and covariance matrix $Q = [q_{i,j}]$, where $q_{i,j} = \lambda(A_i \cap A_j)$, $i, j = 1, \ldots, n$.

Given a domain $\mathcal{O}$ in $\mathbb{R}^d$, let $E = \mathcal{O} \times [0, \infty)$ and let $\lambda = d\xi \, dt$. Then the white noise on E is called *space–time white noise on $\mathcal{O}$*. The following theorem establishes a relation between the concepts of a cylindrical Wiener process on $L^2(\mathcal{O})$ and a space–time white noise on E. In this theorem, we assume a knowledge of stochastic integration with respect to a real-valued Wiener process.

Theorem 7.16 *Let W be a cylindrical Wiener process on $L^2(\mathcal{O})$. Let $\{e_n\}$ be an orthonormal basis of $L^2(\mathcal{O})$, and let $W_n(t) = W(t, e_n)$. Then*

$$\mathcal{X}(A) = \sum_n \int_0^\infty \int_{\mathcal{O}} \chi_A(\xi, t) e_n(\xi) \, d\xi \, dW_n(t), \qquad A \in \mathcal{E}_{\text{fin}},$$

defines a space–time white noise on $\mathcal{O}$.

Proof Since the W_n are independent and χ_A and e_n are non-random, the stochastic integrals

$$\mathcal{X}_n(A) := \int_0^\infty \left(\int_{\mathcal{O}} \chi_A(t, \xi) e_n(\xi) \, d\xi \right) dW_n(t), \qquad n \in \mathbb{N},$$

are independent Gaussian random variables in $\mathbb{R}$. Thus, for $A, B \in \mathcal{E}_{\text{fin}}$,

$$\mathbb{E}\, \mathcal{X}(A)\mathcal{X}(B) = \sum_n \int_0^\infty \int_{\mathcal{O}} \chi_A(t, \xi) e_n(\xi) \, d\xi \int_{\mathcal{O}} \chi_B(\xi, t) \, d\xi \, dt$$

$$= \int_0^\infty \int_{\mathcal{O}} \chi_{A \cap B}(\xi, t) \, d\xi \, dt,$$

and the desired conclusion follows. $\qquad \square$

Remark 7.17 By the theorem above, a space–time white noise on $\mathcal{O}$ can be treated as a time derivative of a cylindrical Wiener process on $L^2(\mathcal{O})$.

Remark 7.18 Note that a cylindrical Wiener process on U has covariance equal to the identity operator, whereas every Wiener process taking values in U has a nuclear covariance.

Remark 7.19 Assume that W is a cylindrical Wiener process on $L^2(\mathcal{O})$, where $\mathcal{O} := \{\xi = (\xi_1, \ldots, \xi_d) : \xi_j \in [0, a]\}$. For any $t \geq 0$ and $\xi \in \mathcal{O}$, define

$$\mathcal{W}(t, \xi_1, \ldots, \xi_d) := W\big(t, \chi_{\{[0, \xi_1] \times \cdots \times [0, \xi_d]\}}\big).$$

Then $\mathcal{W}$ is a Gaussian random field on $[0, \infty) \times [0, a] \times \cdots \times [0, a]$. Note that it is a real-valued Wiener process with respect to each parameter $t, \xi_1, \ldots, \xi_d$ when the others are fixed. The field $\mathcal{W}$ is usually called a *Brownian sheet*; see also Example 4.53. Later (see Remark 7.25) we will introduce its Poissonian analogue, the so-called Lévy sheet. Note that the process

$$\frac{\partial^d \mathcal{W}(t, \xi_1, \ldots, \xi_d)}{\partial \xi_1 \cdots \partial \xi_d}, \qquad t \geq 0,$$

where the derivatives are taken in a distributional sense (see the lemma below) is a cylindrical Wiener process.

Lemma 7.20 *Given a finite measure λ on $[0, a]^d$ write*

$$F(\xi_1, \ldots, \xi_d) := \lambda \left([0, \xi_1] \times \cdots \times [0, \xi_d]\right).$$

Then, for every $\psi \in C([0, a]^d) \cap C_c^\infty((0, a)^d)$,

$$(\psi, \lambda) := \int_{[0,a]^d} \psi(\xi_1, \ldots, \xi_d)\lambda(d\xi_1, \ldots, d\xi_d)$$

$$= (-1)^d \int_{[0,a]^d} \frac{\partial^d \psi}{\partial \xi_1 \cdots \partial \xi_d} F(\xi_1, \ldots, \xi_d)\, d\xi_1 \cdots d\xi_d.$$

Proof We have

$$(\psi, \mu) = (-1)^d \int_{[0,a]^d} \left(\int_{\xi_1}^a \cdots \int_{\xi_d}^a \frac{\partial^d \psi}{\partial \eta_1 \cdots \partial \eta_d}(\eta)\, d\eta \right) \lambda(d\xi)$$

$$= (-1)^d \int_{[0,a]^d} \int_{[0,a]^d} \chi_{[\xi_1,a]}(\eta_1) \cdots \chi_{[\xi_d,a]}(\eta_d) \frac{\partial^d \psi}{\partial \eta_1 \cdots \partial \eta_d}(\eta)\lambda(d\xi)\, d\eta$$

$$= (-1)^d \int_{[0,a]^d} \frac{\partial^d \psi}{\partial \eta_1 \cdots \partial \eta_d}(\eta)\lambda \left([0, \eta_1] \times \cdots \times [0, \eta_d]\right) d\eta.$$

$\square$

7.2 Cylindrical Poisson processes

For a possibly unbounded domain $\mathcal{O} \subset \mathbb{R}^d$, consider a Poisson random measure π on $[0, \infty) \times \mathcal{O} \times \mathbb{R}$ with intensity measure $dt\, \lambda(d\xi)\nu(d\sigma)$, where λ is a nonnegative Radon measure on $\mathcal{O}$. Usually λ is the Lebesgue measure ℓ_d. Let π be defined on a probability space $(\Omega, \mathcal{F}, \mathbb{P})$ with a filtration $(\mathcal{F}_t)$. We assume that π is adapted to $(\mathcal{F}_t)$, that is, $\pi([0, t] \times \Gamma)$ is $(\mathcal{F}_t)$-measurable for all $t \geq 0$ and $\Gamma \in \mathcal{B}(\mathcal{O} \times \mathbb{R})$. Finally, we assume that $\pi((s, t] \times \Gamma)$ is independent of $\mathcal{F}_s$ for all $0 \leq s \leq t$ and $\Gamma \in \mathcal{B}(\mathcal{O} \times \mathbb{R})$. We denote by $\mathcal{P}$ the σ-field of predictable sets in $[0, \infty) \times \Omega$. Generally ν may be an infinite measure, but we will assume that

$\nu(\{0\}) = 0$ and that

$$\int_{\mathbb{R}} \sigma^2 \wedge 1 \, \nu(d\sigma) < \infty. \tag{7.5}$$

The restriction of ν to each $\mathbb{R} \setminus [-a, a]$, $a > 0$, is a finite measure; the restriction to $[-a, a]$ may be infinite. Let $\widehat{\pi}$ be a compensated Poisson measure. In this book we consider stochastic PDEs driven by a measure-valued process Z defined, informally, by

$$Z(t, d\xi) := \int_0^t \int_{\{|\sigma| \leq 1\}} \sigma \widehat{\pi}(ds, d\xi, d\sigma) + \int_0^t \int_{\{|\sigma| \geq 1\}} \sigma \pi(ds, d\xi, d\sigma) \tag{7.6}$$

or, if

$$\int_{\mathbb{R}} \sigma^2 \nu(d\sigma) < \infty, \tag{7.7}$$

by

$$Z(t, d\xi) := \int_0^t \int_{\mathbb{R}} \sigma \widehat{\pi}(ds, d\xi, d\sigma), \qquad t \geq 0. \tag{7.8}$$

Given a compactly supported continuous function ϕ, written $\phi \in C_c(\mathcal{O})$, we define (see Proposition 7.21 below) the real-valued process

$$Z(t, \phi) = \int_{\mathcal{O}} \phi(\xi) Z(t, d\xi).$$

Then (see Proposition 7.21 and Theorem 7.22) $Z(\cdot, \phi)$ is a Lévy process such that

$$\mathbb{E} \, e^{iZ(t,\phi)} = \exp \left\{ -t \int_{\mathcal{O}} \left[\int_{\mathbb{R}} \left(1 - e^{i\sigma\phi(\xi)} + i\sigma \chi_{(0,1]}\phi(\xi) \right) \nu(d\sigma) \right] \lambda(d\xi) \right\}.$$

To simplify the exposition we will assume (7.7) and (7.8). The assumption of (7.7) does not exclude the possibility that

$$\int_0^t \int_{\mathcal{O}} \int_{\mathbb{R}} |\phi(x)| |\sigma| \, ds \, \lambda(d\xi) \nu(d\sigma) = \infty.$$

Thus $Z(t, \phi)$ may not be defined directly by Theorem 6.6(iii) as the stochastic integral with respect to π of the deterministic integrand $(t, \xi, \sigma) \mapsto \phi(\xi)\sigma$.

If π has representation $\pi(t) = \sum_{\tau_n \leq t} \delta_{\xi_n, \sigma_n}$ then, still informally,

$$Z(t, d\xi) = \sum_{\tau_n \leq t} \sigma_n \delta_{\xi_n}(d\xi) - t \left(\int_{\mathbb{R}} \sigma \nu(d\sigma) \right) \lambda(d\xi).$$

We interpret σ_n as the amount of energy introduced into the system at time τ_n. The random variables ξ_n represent the sources of energy. More formally, we define Z as a Lévy process in a Hilbert space U such that the embedding $L^2(d\lambda) :=$

$L^2(\mathcal{O}, \mathcal{B}(\mathcal{O}), \lambda(\mathrm{d}\xi)) \hookrightarrow U$ is dense and Hilbert–Schmidt. We show that Z can be regarded as a square integrable mean-zero Lévy process in U with RHKS $L^2(\mathrm{d}\lambda)$.

Note that, for all $t > 0$, $\varepsilon > 0$ and $\phi \in C_c(\mathcal{O})$,

$$\int_0^t \int_{\mathcal{O}} \int_{\{|\sigma| \geq \varepsilon\}} |\phi(\xi)||\sigma| \, \mathrm{d}s \, \lambda(\mathrm{d}\xi) \nu(\mathrm{d}\sigma) < \infty.$$

Thus by Theorem 6.6 the stochastic integral

$$Z_\varepsilon(t, \phi) := \int_0^t \int_{\mathcal{O}} \int_{\{|\sigma| \geq \varepsilon\}} \phi(\xi) \sigma \widehat{\pi}(\mathrm{d}s, \, \mathrm{d}\xi, \, \mathrm{d}\sigma)$$

is well defined. Note that, for all $0 < \eta < \varepsilon$,

$$\mathbb{E} \left| Z_\varepsilon(t, \phi) - Z_\eta(t, \phi) \right|^2 = t \int_{\{\eta \leq |\sigma| < \varepsilon\}} \sigma^2 \nu(\mathrm{d}\sigma) \int_{\mathcal{O}} |\phi(\xi)|^2 \lambda(\mathrm{d}\xi).$$

Thus $Z_\varepsilon(t, \phi)$ is a Cauchy sequence in $L^2(\Omega, \mathcal{F}, \mathbb{P})$ as $\varepsilon \downarrow 0$. We denote its limit by $Z(t, \phi)$. In the proposition below we gather together the main properties of the process Z. The proof follows directly from the definition of Z and Theorem 6.6.

Proposition 7.21

(i) *For each* $\phi \in C_c(\mathcal{O})$, $(Z(t, \phi), \, t \geq 0)$ *is a Lévy process with respect to* $(\mathcal{F}_t)$.

(ii) *For all* $t \geq 0$ *and* $\phi \in C_c(\mathcal{O})$, $\mathbb{E} \, Z(t, \phi) = 0$ *and*

$$\mathbb{E} |Z(t, \phi)|^2 = t \int_{\mathbb{R}} \sigma^2 \nu(\mathrm{d}\sigma) \int_{\mathcal{O}} |\phi(\xi)|^2 \lambda(\mathrm{d}\xi).$$

(iii) *For all* $t \geq 0$ *and* $\phi \in L^2(\mathrm{d}\lambda)$, $\mathbb{E} \, \exp\{\mathrm{i}Z(t, \phi)\} = \exp\{-t\Psi(\phi)\}$, *where*

$$\Psi(\phi) = \int_{\mathcal{O}} \int_{\mathbb{R}} \left(1 - \mathrm{e}^{\mathrm{i}\phi(\xi)\sigma} + \mathrm{i}\phi(\xi)\sigma\right) \lambda(\mathrm{d}\xi)\nu(\mathrm{d}\sigma). \qquad (7.9)$$

Proof   The first two parts follows directly from the definition of Z and Theorem 6.6. To see (iii) note that

$$\mathbb{E} \, \mathrm{e}^{\mathrm{i}Z(t,\phi)} = \lim_{\varepsilon \downarrow 0} \mathbb{E} \, \mathrm{e}^{\mathrm{i}Z_\varepsilon(t,\phi)}$$

$$= \lim_{\varepsilon \downarrow 0} \exp \left\{ -t \int_{\mathcal{O}} \int_{\{|\sigma| \geq \varepsilon\}} \left(1 - \mathrm{e}^{\mathrm{i}\phi(\xi)\sigma} + \mathrm{i}\phi(\xi)\sigma\right) \lambda(\mathrm{d}\xi)\nu(\mathrm{d}\sigma) \right\}.$$

Since

$$C := \sup_{y \in \mathbb{R}} \frac{\left|1 - \mathrm{e}^{\mathrm{i}y} + \mathrm{i}y\right|}{|y|} < \infty,$$

it follows that

$$\int_{\mathcal{O}}\int_0^\infty \left| e^{-\phi(\xi)\sigma} - 1 - i\phi(\xi)\sigma \right| \lambda(\mathrm{d}\xi)\nu(\mathrm{d}\sigma)$$

$$\leq C \int_{\mathbb{R}} \sigma^2 \nu(\mathrm{d}\sigma) \int_{\mathcal{O}} |\phi(\xi)|^2 \lambda(\mathrm{d}\xi) < \infty.$$

$\square$

Clearly $C_c(\mathcal{O}) \ni \phi \to Z(t,\phi) \in L^2(\Omega,\mathcal{F},\mathbb{P})$ is linear and isometric if $C_c(\mathcal{O})$ is equipped with the norm $|\phi|_t := a(t)|\phi|_{L^2(\mathrm{d}\lambda)}$, where $a(t) = \sqrt{t \int_{\mathbb{R}} \sigma^2 \nu(\mathrm{d}\sigma)}$. Thus, as $C_c(\mathcal{O})$ is dense in $L^2(\mathrm{d}\lambda)$, $Z(\cdot,t)$ can be uniquely extended to a linear operator acting from $L^2(\mathrm{d}\lambda)$ to $L^2(\Omega,\mathcal{F},\mathbb{P})$. By Proposition 7.9 there is a unique U-valued process, denoted also by Z, such that

$$Z(t,\phi) = (\phi, Z(t)), \qquad \forall \phi \in U^*. \tag{7.10}$$

Thus we have the following consequence of Propositions 7.7 and 7.9.

Theorem 7.22 *The process Z given by (7.10) is a square integrable Lévy martingale with RKHS equal to $L^2(\mathrm{d}\lambda)$.*

Definition 7.23 We call Z an *impulsive cylindrical process* on $L^2(\mathrm{d}\lambda)$ with *jump size intensity* ν.

Definition 7.24 The time derivative[1] $\partial Z/\partial t$ of the impulsive cylindrical process Z on $L^2(\mathrm{d}\lambda)$ is called an *impulsive white noise* with *jump position intensity* λ and *jump size intensity* ν.

Remark 7.25 Let $\lambda = \ell_d$ be Lebesgue measure on $\mathcal{O} := [0,a]^d$. For $t \geq 0$ and $\xi \in \mathcal{O}$, define $\mathcal{Z}(t,\xi_1,\ldots,\xi_d) := Z(t,\chi_{\{[0,\xi_1]\times\cdots\times[0,\xi_d]\}})$. Then

$$\mathbb{E}\, e^{iy\mathcal{Z}(t,\xi_1,\ldots,\xi_d)} = \exp\left\{ -t\xi_1\cdots\xi_d \int_{\mathbb{R}} (1 - e^{iy\sigma} + iy\sigma)\nu(\mathrm{d}\sigma) \right\}, \qquad y \in \mathbb{R}.$$

Therefore, with respect to each parameter $t, \xi_1, \ldots, \xi_d$, with the others fixed, the process $\mathcal{Z}$ is, up to a multiplicative constant, a Lévy process with the same jump measure ν. This is why $\mathcal{Z}$ is called a *Lévy sheet*. For the Gaussian analogue of a Lévy sheet see Remark 7.19. Note that

$$\frac{\partial^d \mathcal{Z}(t,\xi_1,\ldots,\xi_d)}{\partial\xi_1\cdots\partial\xi_d}, \qquad t \geq 0,$$

is an impulsive cylindrical process.

---

[1] In a distributional sense; see Lemma 7.20.

Example 7.26   The following stable white-noise process was introduced in Mytnik (2002). We define the process using the framework that has just been developed; see also Example 4.35. Given $\beta \in (1, 2)$ set

$$\nu_\beta(d\sigma) := \frac{1}{\sigma^{1+\beta}} \, \chi_{[0,+\infty)}(\sigma) \, d\sigma.$$

Let π_β be a Poisson random measure on $[0, +\infty) \times \mathcal{O} \times [0, +\infty)$, with intensity $dt \, d\xi \, \nu_\beta(d\sigma)$. Write

$$Z_\beta(t, \cdot) := \int_0^t \int_0^{+\infty} \sigma \widehat{\pi}(ds, d\xi, d\sigma).$$

Note that, for every Borel set $A \subset \mathcal{O}$ with $\ell_d(A) < \infty$,

$$\mathbb{E} \, e^{-\alpha Z_\beta(t, A)} = \lim_{\varepsilon \downarrow 0} \mathbb{E} \exp \left\{ -\alpha \int_0^t \int_\mathcal{O} \int_\varepsilon^{+\infty} \chi_A(\xi) \sigma \widehat{\pi}(ds, d\xi, d\sigma) \right\}$$

$$= \lim_{\varepsilon \downarrow 0} \mathbb{E} \exp \left\{ -\alpha \int_0^t \int_\mathcal{O} \int_\varepsilon^{+\infty} \chi_A(\xi) \sigma \pi(ds, d\xi, d\sigma) \right\}$$

$$\times \exp \left\{ \alpha t \ell_d(A) \int_\varepsilon^{+\infty} \sigma \nu_\beta(d\sigma) \right\}.$$

By Theorem 6.6(i),

$$\mathbb{E} \exp \left\{ -\alpha Z_\beta(t, A) \right\} = \lim_{\varepsilon \downarrow 0} \exp \left\{ -\int_0^t \int_\mathcal{O} \int_\varepsilon^{+\infty} (1 - e^{-\alpha \chi_A(\xi) \sigma}) \, ds \, d\xi \, \nu_\beta(d\sigma) \right\}$$

$$\times \exp \left\{ \alpha t \ell_d(A) \int_\varepsilon^{+\infty} \sigma \nu_\beta(d\sigma) \right\}$$

$$= \lim_{\varepsilon \downarrow 0} \exp \left\{ -t \ell_d(A) \int_\varepsilon^{+\infty} (1 - e^{-\alpha \sigma} - \alpha \sigma) \nu_\beta(d\sigma) \right\}$$

$$= \exp \left\{ -t \ell_d(A) \int_0^{+\infty} (1 - e^{-\alpha \sigma} - \alpha \sigma) \nu_\beta(d\sigma) \right\}.$$

By Example 4.35, $\mathbb{E} \, e^{-\alpha Z_\beta(t, A)} = \exp\{t \ell_d(A) c_\beta \alpha^\beta\}, \alpha > 0$. Thus $\left(Z_\beta(t, A), \, t \geq 0 \right)$ is a β-stable process with positive jumps.

In Mueller (1998) an uncompensated β-stable impulsive white noise was introduced for $\beta \in (0, 1)$.

Example 7.27   Let $\beta \in (0, 1)$. Let π be a Poisson random measure on $[0, +\infty) \times \mathcal{O} \times [0, +\infty)$ with intensity $dt \, d\xi \, \nu_\beta(d\sigma)$, where

$$\nu_\beta(d\sigma) = \sigma^{-1-\beta} \chi_{[0,+\infty)}(\sigma) \, d\sigma.$$

Define

$$Z_\beta(t, \cdot) = \int_0^t \int_0^{+\infty} \sigma \pi(\mathrm{d}s, \mathrm{d}\xi, \mathrm{d}\sigma), \qquad t \geq 0.$$

This quantity is again a measure-valued Lévy process. Using the results of Example 4.34 we obtain $\mathbb{E} \, \mathrm{e}^{-\alpha Z_\beta(t,A)} = \exp\{-t\ell_d(A)c_\beta \alpha^\beta\}, \alpha > 0$. Thus $\left(Z_\beta(t, A), \, t \geq 0\right)$ is a β-stable increasing process. In particular, if $\mathcal{O} = [0, 1]$ then $\mathcal{Z}_\beta(t, \xi) := Z_\beta(t, [0, \xi]))$ satisfies $\mathbb{E} \, \mathrm{e}^{-\alpha \mathcal{Z}_\beta(t,\xi)} = \exp\{-t\xi C_\beta \alpha^\beta\}$.

7.3 Compensated Poisson measure as a martingale

In this section we show that compensated Poisson measures can be treated as infinite-dimensional martingales. Namely, let $(E, \mathcal{E})$ be a measurable space, and let π be a Poisson random measure on $[0, \infty) \times E$ with intensity measure $\mathrm{d}t \, \mu(\mathrm{d}\xi)$. In applications (see Chapter 12 and the previous section), $E = \mathcal{O} \times S$, where $\mathcal{O}$ is a region in $\mathbb{R}^d$.

For $t > 0$ and $\psi \in L^1(E, \mathcal{E}, \mu) \cap L^2(E, \mathcal{E}, \mu)$, write

$$\widehat{\pi}(t, \psi) := \int_0^t \int_E \psi(\xi) \widehat{\pi}(\mathrm{d}s, \mathrm{d}\xi).$$

By Theorem 6.6(iii), (iv) we have that $\widehat{\pi}(t, \psi)$ is well defined and has mean zero and that

$$\mathbb{E} \left(\widehat{\pi}(t, \psi) \right)^2 = t \, |\psi|_{L^2(E,\mathcal{E},\mu)}^2. \tag{7.11}$$

By Theorem 6.6(v), it is a Lévy process and hence a Lévy square integrable real-valued martingale. Let $\Gamma_1, \Gamma_2 \in \mathcal{E}$ be of finite μ-measure. Then, again by Theorem 6.6(iii), (v),

$$\mathbb{E} \, \widehat{\pi}(t, \chi_{\Gamma_1}) \widehat{\pi}(t, \chi_{\Gamma_2}) = \mathbb{E} \left(\widehat{\pi}(t, \chi_{\Gamma_1 \cap \Gamma_2}) \right)^2 = t \, \mu(\Gamma_1 \cap \Gamma_2)$$
$$= t \langle \chi_{\Gamma_1}, \chi_{\Gamma_2} \rangle_{L^2(E,\mathcal{E},\mu)}. \tag{7.12}$$

See Section 8.7 for more details.

Using (7.12), we can extend the definition of $\widehat{\pi}(t, \psi)$ to all $\psi \in L^2(E, \mathcal{E}, \mu)$, to obtain a family $\widehat{\pi}(\cdot, \psi)$ of square integrable Lévy martingales indexed by the elements of $L^2(E, \mathcal{E}, \mu)$. Moreover, by (7.12),

$$\mathbb{E} \, \widehat{\pi}(t, \psi) \widehat{\pi}(t, \phi) = t \langle \psi, \phi \rangle_{L^2(E,\mathcal{E},\mu)}, \qquad \forall \, \psi, \phi \in L^2(E, \mathcal{E}, \mu).$$

The theorem below follows from this together with Propositions 7.7 and 7.9.

Theorem 7.28 *The compensated Poisson random measure $\widehat{\pi}$ on $[0, \infty) \times E$ can be identified with a square integrable Lévy martingale (denoted also by $\widehat{\pi}$) in any*

Hilbert space U containing $\mathcal{H} := L^2(E, \mathcal{E}, \mu)$ *with a Hilbert–Schmidt embedding. Moreover,* $\mathcal{H}$ *is the RKHS of* $\widehat{\pi}$.

One should be aware that $\widehat{\pi}$ is identified with a corresponding process by a duality procedure. Note, in particular, that increasing the intensity measure μ decreases the RKHS of $\widehat{\pi}$ and consequently increases the class of spaces on which $\widehat{\pi}$ lives. In the limit $\mu = \infty$, we obtain $\mathcal{H} = \{0\}$.

8

Stochastic integration

In this chapter the concept of the stochastic integral with respect to a Hilbert-space-valued martingale is introduced and a fundamental isometric formula is established. The space of integrands is determined in terms of the martingale covariance. Stochastic integrals of predictable fields with respect to Poisson measures are treated as well.

Throughout this chapter $(U, \langle \cdot, \cdot \rangle_U)$ is a Hilbert space and $(\Omega, \mathcal{F}, (\mathcal{F}_t), \mathbb{P})$ is a filtered probability space satisfying the usual conditions. The space of all càdlàg square integrable martingales in U with respect to $(\mathcal{F}_t)$ is denoted by $\mathcal{M}^2(U)$.

8.1 Operator-valued angle bracket process

Let $M, N \in \mathcal{M}^2(U)$. Denote by $\langle M, N \rangle$ the unique predictable process, with trajectories having a locally bounded variation, for which

$$\langle M(t), N(t) \rangle_U - \langle M, N \rangle_t, \qquad t \geq 0,$$

is a martingale. By the Doob–Meyer decomposition, the process $\langle M, N \rangle$ always exists (see Remark 3.46) and is called the *angle bracket*.

In this section we introduce the so-called *operator angle bracket* $\langle\langle M, N \rangle\rangle$, and in Theorem 8.2 we will show the absolute continuity of the operator angle bracket with respect to the angle bracket. The relevant density is called the *martingale covariance*. It plays an important role in the construction of the stochastic integral with respect to M. In particular it appears in the fundamental isometric formula.

Denote by $L_1(U)$ the space of all nuclear operators on U equipped with the nuclear norm; see Appendix A. Then $L_1(U)$ is a separable Banach space. Recall that, given $x, y, z \in U$, $x \otimes y (z) = \langle y, z \rangle_U x$. It is easy to show that $x \otimes y \in L^1(U)$ and $\|x \otimes y\|_{L_1(U)} = |x|_U |y|_U$. We denote by $L_1^+(U)$ the subspace of $L_1(U)$ consisting of all self-adjoint non-negative nuclear operators. If $M \in \mathcal{M}^2(U)$ then the process $(M(t) \otimes M(t), \ t \geq 0)$ is an $L_1(U)$-valued right-continuous process such that

$$\mathbb{E} \, \|M(t) \otimes M(t)\|_{L_1(U)} = \mathbb{E} \, |M(t)|_U^2 \leq \mathbb{E} \, |M(T)|_U^2 < \infty, \qquad t \geq 0.$$

We will need the following result.

Lemma 8.1 *Assume that $M, N \in \mathcal{M}^2(U)$. There exists a predictable process $(q(s), \ s \geq 0)$ such that*

$$\langle M, N \rangle_t = \int_0^t q(s) \, \mathrm{d}\big(\langle M, M \rangle_s + \langle N, N \rangle_s\big).$$

Proof Let $T < \infty$. It is enough to show that, almost surely, the measure on $[0, T]$ corresponding to the process $\langle M, N \rangle$ of bounded variation is absolutely continuous with respect to the sum of the measures induced by $\langle M, M \rangle$ and $\langle N, N \rangle$. For fixed $s \in [0, T]$ and an arbitrary real a, the process

$$\langle M + aN, M + aN \rangle_t - \langle M + aN, M + aN \rangle_s, \qquad t \in [s, T]$$

is the angle bracket corresponding to $(M(t) + aN(t), \ t \in [s, T])$. Consequently, for all $a \in \mathbb{R}$,

$$
\begin{aligned}
\langle M + aN, M &+ aN \rangle_t - \langle M + aN, M + aN \rangle_s \\
&= a^2 \big(\langle N, N \rangle_t - \langle N, N \rangle_s\big) + 2a\big(\langle M, N \rangle_t - \langle M, N \rangle_s\big) \\
&\quad + \big(\langle M, M \rangle_t - \langle M, M \rangle_s\big) \geq 0,
\end{aligned}
$$

and hence

$$\big(\langle M, N \rangle_t - \langle M, N \rangle_s\big)^2 \leq \big(\langle M, M \rangle_t - \langle M, M \rangle_s\big)\big(\langle N, N \rangle_t - \langle N, N \rangle_s\big)$$

or, equivalently,

$$\big|\langle M, N \rangle_t - \langle M, N \rangle_s\big| \leq \big(\langle M, M \rangle_t - \langle M, M \rangle_s\big)^{1/2} \big(\langle N, N \rangle_t - \langle N, N \rangle_s\big)^{1/2}.$$

Hence

$$\big|\langle M, N \rangle_t - \langle M, N \rangle_s\big| \leq \tfrac{1}{2}\big(\big(\langle M, M \rangle_t + \langle N, N \rangle_t\big) - \big(\langle M, M \rangle_s + \langle N, N \rangle_s\big)\big).$$

Thus we have shown that on each subinterval of $[0, T]$ the total variation of the measure corresponding to $\langle M, N \rangle$ is smaller than the total variation corresponding to $\langle M, M \rangle + \langle N, N \rangle$. In particular, if a Borel set $\Gamma \subset [0, T]$ is of measure zero with respect to $\mathrm{d}(\langle M, M \rangle + \langle N, N \rangle)$, it is of measure zero with respect to $\mathrm{d}\langle M, N \rangle$, and this implies the required absolute continuity.

To prove the predictability of q we use a real-analysis result. If μ and ν are two finite non-negative measures on $[0, \infty)$ and μ is absolutely continuous with respect to ν then, for ν-almost all $t > 0$,

$$\frac{\mathrm{d}\mu}{\mathrm{d}\nu}(t) = \liminf_{r \uparrow t} \frac{\mu((r, t])}{\nu((r, t])},$$

where the limit is taken with respect to $r < t$, r rational. Since the limit inferior of predictable processes is predictable the result follows. $\qquad\square$

We have the following basic result, mentioned in the introduction to this chapter, from which the isometric formula follows.

Theorem 8.2 *Let $M \in \mathcal{M}^2(U)$. Then there is a unique right-continuous $L_1^+(U)$-valued increasing predictable process $(\langle\!\langle M, M \rangle\!\rangle_t, \ t \geq 0)$ such that $\langle\!\langle M, M \rangle\!\rangle_0 = 0$ and the process $\big(M(t) \otimes M(t) - \langle\!\langle M, M \rangle\!\rangle_t, \ t \geq 0\big)$ is an $L_1(U)$-valued martingale. Moreover, there exists a predictable $L_1^+(U)$-valued process $(Q_t, \ t \geq 0)$ such that*

$$\langle\!\langle M, M \rangle\!\rangle_t = \int_0^t Q_s \, \mathrm{d}\langle M, M \rangle_s, \qquad \forall t \geq 0. \tag{8.1}$$

Proof Let $\{e_k\}$ be an orthonormal basis of U. Then

$$M(t) = \sum_k \langle M(t), e_k \rangle_U e_k, \qquad t \geq 0.$$

Write $M^k(t) := \langle M(t), e_k \rangle_U$. Since

$$\mathbb{E} \, |M(t)|_U^2 = \mathbb{E} \sum_k \big(M^k(t)\big)^2 = \sum_k \mathbb{E} \big(M^k(t)\big)^2 < \infty, \qquad t \geq 0,$$

it follows that $M^k \in \mathcal{M}^2$ for every k. Let $\langle M^k, M^j \rangle$ be the corresponding angle bracket processes. As a candidate for an operator-valued angle bracket, we take

$$\langle\!\langle M, M \rangle\!\rangle_t := \sum_{k,j} e_k \otimes e_j \, \langle M^k, M^j \rangle_t, \qquad t \geq 0. \tag{8.2}$$

Thus the infinite matrix $(\langle M^k, M^l \rangle_t)$ is a representation of the operator $\langle\!\langle M, M \rangle\!\rangle_t$ in the basis $\{e_k\}$. The problem lies in the convergence of the series.

To see that (8.2) defines an $L_1^+(U)$-valued process, we first prove that for all t, k, j,

$$|\langle M^k, M^l \rangle_t|^2 \leq \langle M^k, M^k \rangle_t \langle M^j, M^j \rangle_t$$

To see this note that, for every $a \in \mathbb{R}$,

$$0 \leq \langle M^k + aM^j, M^k + aM^j \rangle_t = a^2 \langle M^j, M^j \rangle_t + 2a \langle M^j, M^k \rangle_t + \langle M^k, M^k \rangle_t,$$

which gives the desired conclusion. We now show that the series on the right-hand side of (8.2) converges in $L^1(\Omega, \mathcal{F}, \mathbb{P}; L_{(HS)}(U, U))$. The Hilbert–Schmidt norm of the right-hand side is equal to

$$\left(\sum_k \left| \sum_j \langle M^k, M^j \rangle_t e_j \right|_U^2 \right)^{1/2} = \left(\sum_{j,k} \langle M^j, M^k \rangle_t^2 \right)^{1/2}$$

$$\leq \left(\sum_{j,k} \langle M^j, M^k \rangle_t \langle M^k, M^k \rangle_t \right)^{1/2} \leq \sum_j \langle M^j, M^j \rangle_t.$$

Since

$$\mathbb{E} \sum_j \langle M^j, M^j \rangle_t = \mathbb{E} |M(t)|_U^2 < \infty,$$

the $L^1(\Omega, \mathcal{F}, \mathbb{P}; L_{(HS)}(U, U))$-convergence of the series follows. Thus

$$\langle\langle M, M \rangle\rangle = \left(\langle\langle M, M \rangle\rangle_t, \ t \geq 0 \right)$$

is a well-defined process taking values in the space of Hilbert–Schmidt operators on U. It is symmetric and non-negative. Note that for $0 \leq s \leq t < \infty$ the operator $\langle\langle M, M \rangle\rangle_t - \langle\langle M, M \rangle\rangle_s$ is also non-negative. Consequently,

$$\left\| \langle\langle M, M \rangle\rangle_t - \langle\langle M, M \rangle\rangle_s \right\|_{L_1(U)} = \mathrm{Tr} \left\{ \langle\langle M, M \rangle\rangle_t - \langle\langle M, M \rangle\rangle_s \right\}$$

$$= \sum_j \left\{ \langle M^j, M^j \rangle_t - \langle M^j, M^j \rangle_s \right\}$$

and

$$\mathbb{E} \left\| \langle\langle M, M \rangle\rangle_t - \langle\langle M, M \rangle\rangle_s \right\|_{L_1(U)} = \mathbb{E} \left(|M(t)|_U^2 - |M(s)|_U^2 \right) < \infty.$$

This shows that $\langle\langle M, M \rangle\rangle$ is an $L_1^+(U)$-valued predictable increasing process. To show that it is also right-continuous and can be represented in the form (8.2) we will apply Lemma 8.1. Namely, it follows from the lemma that for any pair k, j there exists a predictable process $(q^{k,j}(t), \ t \geq 0)$ such that

$$\langle M^k, M^j \rangle_t = \int_0^t q^{k,j}(s) \, \mathrm{d}\langle M, M \rangle_s = \int_0^t q^{k,j}(s) \, \mathrm{d} \sum_l \langle M^l, M^l \rangle_s.$$

Thus

$$\langle\langle M, M \rangle\rangle_t = \int_0^t Q_s \, \mathrm{d}\langle M, M \rangle_s,$$

where

$$Q_s = \sum_{k,j} e_k \otimes e_j \, q^{k,l}(s), \qquad s \geq 0,$$

is a predictable process with values in $L_1^+(U)$. $\qquad\qquad \square$

Definition 8.3   We call the $L_1^+(U)$-valued process Q satisfying (8.1) the *martingale covariance* of M.

The main ingredient in the construction of the stochastic integral with respect to a square integrable U-valued martingale is the following result.

Proposition 8.4   *Let* $M \in \mathcal{M}^2(U)$. *Then, for any vectors* $x, y \in U$ *and any* $0 \le s \le t \le u \le v < \infty$,

$$\mathbb{E}\big(\langle M(t) - M(s), x \rangle_U \langle M(t) - M(s), y \rangle_U \,\big|\, \mathcal{F}_s\big)$$
$$= \mathbb{E}\left(\int_s^t \langle Q_r x, y \rangle_U \, \mathrm{d}\langle M, M \rangle_r \,\Big|\, \mathcal{F}_s\right)$$

and

$$\mathbb{E}\big(\langle M(t) - M(s), x \rangle_U \langle M(u) - M(v), y \rangle_U \,\big|\, \mathcal{F}_u\big) = 0.$$

Proof   The first formula is obtained as follows:

$$\mathbb{E}\big(\langle M(t) - M(s), \, x \rangle_U \langle M(t) - M(s), \, y \rangle_U \,\big|\, \mathcal{F}_s\big)$$
$$= \mathbb{E}\big(\langle M(t), x \rangle_U \langle M(t), y \rangle_U \,\big|\, \mathcal{F}_s\big) - \langle M(s), x \rangle_U \langle M(s), y \rangle_U$$
$$= \mathbb{E}\big(\langle M(t) \otimes M(t)x, \, y \rangle_U \,\big|\, \mathcal{F}_s\big) - \langle M(s) \otimes M(s)x, \, y \rangle_U$$
$$= \big\langle \mathbb{E}\big(M(t) \otimes M(t) - M(s) \otimes M(s) \,\big|\, \mathcal{F}_s\big)x, \, y \big\rangle_U$$
$$= \big\langle \mathbb{E}\big(\langle\langle M, M \rangle\rangle_t - \langle\langle M, M \rangle\rangle_s \,\big|\, \mathcal{F}_s\big)x, \, y \big\rangle_U$$
$$= \mathbb{E}\left(\int_s^t \langle Q_r x, y \rangle_U \, \mathrm{d}\langle M, M \rangle_r \,\Big|\, \mathcal{F}_s\right).$$

The second formula follows from the martingale property. □

8.2 Construction of the stochastic integral

To deal with stochastic equations one needs the concept of the stochastic integral, $I_t^M(\Psi) := \int_0^t \Psi(s) \, \mathrm{d}M(s)$, where $M \in \mathcal{M}^2(U)$ and $\Psi(s, \omega)$ are operators from U to another Hilbert space H. As for real-valued martingales, first we define the stochastic integral for simple processes Ψ. Then, in the next section, we extend the class of integrands using the isometric formula (8.3) below. The isometric formula in the general case appeared for the first time in Métivier and Pistone (1975). We will denote by Q the martingale covariance of M introduced in Definition 8.3.

Definition 8.5   Let $L(U, H)$ be the Banach space of continuous linear operators from U into H. An $L(U, H)$-valued stochastic process Ψ is said to be *simple* if there exist a sequence of non-negative numbers $t_0 = 0 < t_1 < \cdots < t_m$, a sequence of operators $\Psi_j \in L(U, H)$, $j = 1, \ldots, m$, and a sequence of events $A_j \in \mathcal{F}_{t_j}$, $j = 0, \ldots, m - 1$, such that

$$\Psi(s) = \sum_{j=0}^{m-1} \chi_{A_j} \chi_{(t_j, t_{j+1}]}(s) \Psi_j, \qquad s \geq 0.$$

We shall denote by $\mathcal{S} := \mathcal{S}(U, H)$ the class of all simple processes with values in $L(U, H)$. For a simple process Ψ, we set

$$I_t^M(\Psi) := \sum_{j=0}^{m-1} \chi_{A_j} \Psi_j \big(M(t_{j+1} \wedge t) - M(t_j \wedge t) \big), \qquad t \geq 0.$$

Let $L_{(HS)}(U, H)$ be the space of all Hilbert–Schmidt operators from U into H equipped with the Hilbert–Schmidt norm $\| \cdot \|_{L_{(HS)}(U,H)}$. We prove the isometric formula first for simple processes.

Proposition 8.6   *For any simple process Ψ,*

$$\mathbb{E} \left| I_t^M(\Psi) \right|_H^2 = \mathbb{E} \int_0^t \left\| \Psi(s) Q_s^{1/2} \right\|_{L_{(HS)}(U,H)}^2 \mathrm{d}\langle M, M \rangle_s, \qquad t \geq 0. \tag{8.3}$$

Proof   By Definition 8.5,

$$\mathbb{E} \left| I_t^M(\Psi) \right|_H^2 = \mathbb{E} \left| \sum_{j=0}^{m-1} \chi_{A_j} \Psi_j \big(M(t_{j+1} \wedge t) - M(t_j \wedge t) \big) \right|_H^2$$

$$= \mathbb{E} \sum_{k,j=0}^{m-1} \chi_{A_k} \chi_{A_j} K_{k,j},$$

where

$$K_{k,j} := \Big\langle \Psi_k \big(M(t_{k+1} \wedge t) - M(t_k \wedge t) \big), \ \Psi_j \big(M(t_{j+1} \wedge t) - M(t_j \wedge t) \big) \Big\rangle_H.$$

By Proposition 8.4, if $k > j$ then

$$\mathbb{E} \big(\langle M(t_{k+1} \wedge t) - M(t_k \wedge t), x \rangle_U \langle M(t_{j+1} \wedge t) - M(t_j \wedge t), y \rangle_U | \mathcal{F}_{t_k} \big)$$
$$= 0, \tag{8.4}$$

and if $k = j$ we have

$$\mathbb{E} \big(\langle M(t_{k+1} \wedge t) - M(t_k \wedge t), x \rangle_U \langle M(t_{j+1} \wedge t) - M(t_j \wedge t), y \rangle_U | \mathcal{F}_{t_k} \big)$$
$$= \mathbb{E} \left(\int_{t_j}^{t_{j+1}} \langle Q_s x, y \rangle_U \mathrm{d}\langle M, M \rangle_s \Big| \mathcal{F}_{t_j} \right). \tag{8.5}$$

Let $\{e_l\}$ be an orthonormal basis of H and $k \leq j$. We have

$$
\begin{aligned}
I_{kjl} &:= \mathbb{E}\big(\langle \Psi_k(M(t_{k+1} \wedge t) - M(t_k \wedge t)), e_l \rangle_H \\
&\quad \times \langle \Psi_j(M(t_{j+1} \wedge t) - M(t_j \wedge t)), e_l \rangle_H \big| \mathcal{F}_{t_k}\big) \\
&= \mathbb{E}\big(\langle (M(t_{k+1} \wedge t) - M(t_k \wedge t)), \Psi_k^* e_l \rangle_U \\
&\quad \times \langle (M(t_{l+1} \wedge t) - M(t_j \wedge t)), \Psi_j^* e_l \rangle_U \big| \mathcal{F}_{t_k}\big).
\end{aligned}
$$

By (8.4) and (8.5), $\mathbb{E} I_{kjl} \chi_{A_k} \chi_{A_j} = 0$ if $k > j$ and

$$
\mathbb{E} I_{jjl} \chi_{A_j} = \mathbb{E} \chi_{A_j} \int_{t_j}^{t_{j+1}} \langle Q_s \Psi_j^* e_l, \Psi_j^* e_l \rangle_U \, d\langle M, M \rangle_s.
$$

But (see Appendix A),

$$
\sum_l \langle Q_s \Psi_j^* e_l, \Psi_j^* e_l \rangle_U = \sum_l \big| Q_s^{1/2} \Psi_j^* e_l \big|_U^2 = \big\| Q_s^{1/2} \Psi_j^* \big\|_{L_{(HS)}(H,U)}^2.
$$

Since $\big\| Q_s^{1/2} \Psi_j^* \big\|_{L_{(HS)}(H,U)}^2 = \big\| \Psi_j Q_s^{1/2} \big\|_{L_{(HS)}(U,H)}^2$, the required identity follows by elementary reasoning. $\qquad\square$

Let $T < \infty$. Equip the class of all simple processes $\mathcal{S} = \mathcal{S}(U, H)$ with the seminorm

$$
\|\Psi\|_{M,T}^2 := \mathbb{E} \int_0^T \big\| \Psi(s) Q_s^{1/2} \big\|_{L_{(HS)}(U,H)}^2 \, d\langle M, M \rangle_s. \tag{8.6}
$$

We may identify Ψ with Φ if $\|\Psi - \Phi\|_{M,T} = 0$. Let $\mathcal{L}_{M,T}^2(H)$ be the completion of $(\mathcal{S}, \| \cdot \|_{M,T})$. In the next section we give an explicit construction of $\mathcal{L}_{M,T}^2(H)$. The norm on $\mathcal{L}_{M,T}^2(H)$ will be denoted by $\| \cdot \|_{M,T}$. Let $\mathcal{L}_{M,T,U}^2(H)$ be the class of all $L(U, H)$-valued processes belonging to $\mathcal{L}_{M,T}^2(H)$. Note that for $\Psi \in \mathcal{L}_{M,T,U}^2(H)$ the $\mathcal{L}_{M,T}^2(H)$-norm is given by (8.6).

Theorem 8.7

(i) *For any $t \in [0, T]$, there is a unique extension of I_t^M to a continuous linear operator, denoted also by I_t^M, from $(\mathcal{L}_{M,T}^2(H), \| \cdot \|_{M,T})$ into $L^2(\Omega, \mathcal{F}, \mathbb{P}; H)$. Moreover, for any $\Psi \in \mathcal{L}_{M,T}^2(H)$, $\mathbb{E} \big| I_T^M(\Psi) \big|_H^2 = \|\Psi\|_{M,T}^2$.*

(ii) *For all $\Psi \in \mathcal{L}_{M,T}^2(H)$ and $0 \leq s \leq t \leq T$, we have $\chi_{(s,t]}\Psi \in \mathcal{L}_{M,T}^2(H)$ and*

$$
\mathbb{E} \big| I_t^M(\Psi) - I_s^M(\Psi) \big|_H^2 = \|\chi_{(s,t]}\Psi\|_{M,T}^2 \leq \|\Psi\|_{M,T}^2.
$$

(iii) *For any $\Phi \in \mathcal{L}_{M,T}^2(H)$, $\big(I_t^M(\Phi), \, t \in [0, T]\big)$ is an H-valued martingale. It is square integrable and mean-square continuous, and $I_0^M(\Psi) = 0$.*

(iv) *For any* $\Phi, \Psi \in \mathcal{L}^2_{M,T,U}(H)$ *and any* $t \in [0, T]$,

$$\langle I^M(\Psi), I^M(\Phi) \rangle_t = \int_0^t \left\langle \Psi(s)Q_s^{1/2}, \Phi(s)Q_s^{1/2} \right\rangle_{L_{(HS)}(U,H)} \mathrm{d}\langle M, M \rangle_s.$$

and

$$\langle\langle I^M(\Psi), I^M(\Psi) \rangle\rangle_t = \int_0^t \Psi(s)Q_s\Psi(s)^* \, \mathrm{d}\langle M, M \rangle_s.$$

(v) *Let A be a bounded linear operator from H into a Hilbert space V. Then, for every $\Phi \in \mathcal{L}^2_{M,T}(H)$, $A\Phi \in \mathcal{L}^2_{M,T}(V)$ and $AI^M(\Phi) = I^M(A\Phi)$.*

Proof   The first two assertions follow from the linearity of I^M_t on $\mathcal{S}$ and from (8.3) in Proposition 8.6 for $\Psi \in \mathcal{S}$. Clearly, $I^M(\Psi)$ is square integrable and $I^M_0(\Psi) = 0$. In order to prove mean-square continuity, we need to show that

$$\lim_{s \to t} \| \chi_{(s,t]}\Psi \|_{M,T} = 0. \tag{8.7}$$

To do this, consider the family of linear operators $U(s) \colon \Psi \mapsto \chi_{(s,t]}\Psi$ from $\mathcal{L}^2_{M,T}(H)$ into $\mathcal{L}^2_{M,T}(H)$. We have

$$\sup_s \| U(s) \|_{L(\mathcal{L}^2_{M,T}(H), \mathcal{L}^2_{M,T}(H))} \le 1.$$

Thus, since (8.7) holds on a dense subspace, say $\mathcal{S}$, it holds on the whole space by a Banach–Steinhaus argument.

It is enough to check the martingale property and the identities in (iv) for simple Ψ, Φ. By additivity it is sufficent to note, and we leave this to the reader, that

$$\mathbb{E}\left(\chi_A\Psi(M(t \wedge t_2) - M(t \wedge t_1)) \big| \mathcal{F}_s\right) = \chi_A\Psi\left((M(s \wedge t_2) - M(s \wedge t_1))\right)$$

for all $0 \le s \le t \le T, 0 \le t_1 \le t_2 \le T, A \in \mathcal{F}_{t_1}$ and $\Psi \in L(U, H)$. The last assertion of the theorem clearly holds for simple Φ and therefore for all Φ by standard limiting arguments. $\square$

8.3 Space of integrands

The aim of this section is to characterize the space $\mathcal{L}^2_{M,T}(H)$ of admissible integrands. We assume that M is a U-valued right-continuous square integrable martingale with martingale covariance $(Q_s, s \ge 0)$. We denote by $\mathcal{P}_{[0,T]}$ the σ-field of predictable sets in $\Omega \times [0, T]$. In the proof of the lemma below, we will use the following classical selection theorem from Kuratowski and Ryll-Nardzewski (1965); see also Dynkin and Yushkevich (1978) and Wagner (1977).

Theorem 8.8 (Kuratowski–Ryll-Nardzewski) *Assume that K is a compact metric space. Let $(\tilde{\Omega}, \tilde{\mathcal{F}}, \tilde{\mathbb{P}})$ be a probability space, and let $\psi \colon K \times \tilde{\Omega} \mapsto \mathbb{R}$ be a function such that, for every $x \in K$, $\psi(x, \cdot)$ is measurable and, for every $\tilde{\omega} \in \tilde{\Omega}$, $\psi(\cdot, \tilde{\omega})$ is continuous. Then there exists a K-valued random variable $\xi \colon \tilde{\Omega} \mapsto K$ such that*

$$\psi(\xi(\tilde{\omega}), \tilde{\omega}) = \sup_{x \in K} \psi(x, \tilde{\omega}), \qquad \forall \tilde{\omega} \in \tilde{\Omega}.$$

We need the following lemma.

Lemma 8.9 *There are predictable real-valued processes $\gamma_n = \gamma_n(t, \omega)$ and predictable U-valued processes $g_n = g_n(t, \omega)$, $n \in \mathbb{N}$, such that*

$$Q_t(\omega) = \sum_n \gamma_n(\omega, t) g_n(\omega, t) \otimes g_n(\omega, t), \qquad t \geq 0, \ \omega \in \Omega,$$

and $\langle g_n(\omega, t), g_m(\omega, t) \rangle_U = \delta_{n,m}$ for $t \geq 0$, $\omega \in \Omega$ and $n, m \in \mathbb{N}$.

Proof Given $T < \infty$, we apply the Kuratowski–Ryll-Nardzewski theorem to $\tilde{\Omega} = \Omega \times [0, T], \tilde{\mathcal{F}} = \mathcal{P}_{[0,T]}$ and $\tilde{\mathbb{P}} = T^{-1} \mathbb{P} \otimes dt$, to the set $K := \{x \in U \colon |x|_U \leq 1\}$ endowed with the weak topology and to the function

$$\psi(x, \omega, t) = \psi(x, \tilde{\omega}) = \langle Q_t(\omega)x, x \rangle_U.$$

Then there is a U-valued process $g_1(\omega, t), t \geq 0, \omega \in \Omega$, such that

$$\langle Q_t(\omega)g_1(\omega, t), \ g_1(\omega, t) \rangle_U = \sup_{|y|_U \leq 1} \langle Q_t(\omega)y, y \rangle_U = \gamma_1(\omega, t).$$

Using the random variables $\gamma_1, \ldots, \gamma_n$ and $g_1, \ldots, g_n$ we repeat this procedure for the operator-valued process

$$Q_t(\omega) - \sum_{j=1}^n \gamma_j(\omega, t) g_j(\omega, t) \otimes g_j(\omega, t).$$

$\square$

Let $\mathcal{H}$ be a Hilbert space with an orthonormal basis $\{e_n\}$, and let $\tilde{T}(\omega, t)$ be the unique continuous linear operator from $Q_t^{1/2}(\omega)U$ into $\mathcal{H}$ satisfying

$$\tilde{T}(\omega, t)\sqrt{\gamma_n(\omega, t)} \, g_n(\omega, t) = e_n, \qquad n \in \mathbb{N}.$$

On each space $([0, T] \times \Omega, \mathcal{P}_{[0,T]})$, $T \geq 0$, we introduce a σ-finite measure[1]

$$\mu_M(d\omega, dt) = d\langle M, M \rangle_t(\omega) \mathbb{P}(d\omega).$$

---

[1] Called the *Doléans measure* of $|M|_U^2$; see Métivier (1982).

In this book we are concerned mostly with martingales that are Lévy processes. In this case $\langle M, M \rangle_t = ct$, and hence $\mu_M(\mathrm{d}t, \mathrm{d}\omega) = c\,\mathrm{d}t\,\mathbb{P}(\mathrm{d}\omega)$.

Theorem 8.10 *For* $\Psi \in L^2(\Omega \times [0, T], \mathcal{P}_{[0,T]}, \mu_M; L_{(HS)}(\mathcal{H}, H))$, *it follows that*

$$\mathcal{L}_{M,T}^2(H) = \left\{ \Psi \circ \tilde{T} : \ \Psi \in L^2(\Omega \times [0, T], \mathcal{P}_{[0,T]}, \mu_M; L_{(HS)}(\mathcal{H}, H)) \right\} \quad (8.8)$$

and

$$\left\| \Psi \circ \tilde{T} \right\|_{M,T}^2 = \int_\Omega \int_0^T \|\Psi(\omega, t)\|_{L_{(HS)}(\mathcal{H}, H)}^2 \mu_M(\mathrm{d}\omega, \mathrm{d}t). \quad (8.9)$$

Proof Let us denote by $\mathcal{X}$ the space defined by the right-hand side of (8.8). To show that $\mathcal{S} \subset \mathcal{X}$, take $0 \le s < t \le T$, $\Phi \in L(U, H)$ and $A \in \mathcal{F}_s$. We have to find a $\Psi \in L^2(\Omega \times [0, T], \mathcal{P}_{[0,T]}, \mu_M; L_{(HS)}(\mathcal{H}, H))$ such that

$$\|\chi_A \chi_{(s,t]} \Phi - \Psi \circ \tilde{T}\|_{M,T} = 0.$$

By the definition of $\| \cdot \|_{M,T}$,

$$\|\chi_A \chi_{(s,t]} \Phi - \Psi \circ \tilde{T}\|_{M,T}^2$$
$$= \mathbb{E} \int_0^T \left\| \chi_A \chi_{(s,t]} \Phi \circ Q_r^{1/2} - \Psi(r) \circ \tilde{T}(r) \circ Q_r^{1/2} \right\|_{L_{(HS)}(U,H)}^2 \mathrm{d}\langle M, M \rangle_r.$$

Note that Ψ as defined by $\Psi(\omega, r) = 0$ for $(\omega, r) \notin A \times (s, t]$ and

$$\Psi(\omega, r)e_n = \sqrt{\gamma_n(\omega, r)} \Phi g_n(\omega, r) \qquad \text{for } (\omega, r) \in A \times (s, t]$$

has the desired properties. The fact that $\mathcal{S}$ is dense in $\mathcal{X}$ follows from Lemma 8.13 below. It will play an essential role in the proof of the stochastic Fubini theorem; see Section 8.5.

We now prove (8.9). It suffices to note that

$$\|\Psi\|_{L_{(HS)}(\mathcal{H}, H)} = \left\| \Psi \circ \tilde{T}(\omega, t) \circ Q_t^{1/2}(\omega) \right\|_{L_{(HS)}(U,H)}$$

for $\Psi \in L(\mathcal{H}, H)$ and $(\omega, t) \in \Omega \times [0, T]$. Since $\{g_n(\omega, t)\}$ is an orthonormal basis of U consisting of eigenvectors of $Q_t^{1/2}(\omega)$,

$$\left\| \Psi \circ T(\omega, t) \circ Q_t^{1/2}(\omega) \right\|_{L_{(HS)}(U,H)}^2 = \sum_n \left| \Psi \circ T(\omega, t) \circ Q_t^{1/2}(\omega) g_n(\omega, t) \right|_H^2$$

$$= \sum_n \left| \Psi \circ T(\omega, t) \sqrt{\gamma_n(\omega, t)} g_n(\omega, t) \right|_H^2$$

$$= \sum_n |\Psi e_n|_H^2 = \|\Psi\|_{L_{(HS)}(\mathcal{H}, H)}^2.$$

$\square$

8.4 Local properties of stochastic integrals

Recall (see subsection 3.7.3) that $\Sigma_{[0,T]}$ denotes the class of all Markov stopping times τ such that $\mathbb{P}(\tau \leq T) = 1$. The following result plays an important role in the investigation of local properties of stochastic integrals.

Proposition 8.11 *Assume that* $\mu_M(d\omega, dt) = d\langle M, M\rangle_t(\omega)\mathbb{P}(d\omega)$ *is absolutely continuous with respect to* $dt\,\mathbb{P}(d\omega)$. *Let* $X \in \mathcal{L}^2_{M,T}(H)$ *and* $A \in \mathcal{F}_T$ *be such that* $X = 0$ *on* $[0, T] \times A$. *Then*

$$\int_0^t X(s)\,dM(s)(\omega) = 0, \qquad \forall t \in [0, T], \ \mathbb{P}\text{-a.s. on } A.$$

Proof By Theorem 8.7(v), $\langle u, I^M_t(X)\rangle_H = I^M_t(\langle X, u\rangle_H)$ for every $u \in H$. We may therefore assume that $H = \mathbb{R}$. Since the process X can be approximated by a sequence of bounded processes $(X \wedge m) \vee (-m)$, $m = 0, 1, \ldots$, vanishing on $[0, T] \times A$ we may assume that X is bounded. Now, any bounded process can be approximated in $\mathcal{L}^2_{M,T}$ by processes with continuous trajectories,

$$X_n(t) := n \int_{(t-1/n)\vee 0}^t X(s)\,ds, \qquad t \in [0, T].$$

Here we make use of the absolute continuity of μ_M with respect to $dt\,\mathbb{P}(d\omega)$. Since X_n also vanishes on $[0, T] \times A$, we have reduced the proof to the case of bounded and continuous X. Note that

$$Y_n := X(0)\chi_0 + \sum_{k=0}^n X(Tk/n)\chi_{(T(k-1)/n, TK/n]}, \qquad n \in \mathbb{N},$$

is an approximation sequence of simple processes vanishing on $[0, T] \times A$. Since for each Y_n

$$\int_0^t Y_n(s)\,dM(s)(\omega) = 0, \qquad \forall (t, \omega) \in [0, T] \times A,$$

the desired conclusion follows. $\qquad\qquad\square$

Proposition 8.12 *Assume that* $M(t) = 0$ *on* $[0, T] \times A$ *for some* $A \in \mathcal{F}_T$. *Then, for every* $X \in \mathcal{L}^2_{M,T}(H)$,

$$\int_0^t X(s)\,dM(s)(\omega) = 0, \qquad \forall t \in [0, T], \ \mathbb{P}\text{-a.s. on } A.$$

Proof The result follows immediately for simple X. It therefore follows for all $X \in \mathcal{L}^2_{M,T}(H)$ by approximation. $\qquad\qquad\square$

8.5 Stochastic Fubini theorem

In this section, M is an U-valued square integrable martingale with respect to a filtration $(\mathcal{F}_t)$ and λ is a finite positive measure on a measurable space $(E, \mathcal{E})$. Recall that $\mu_M(d\omega, dt) = d\langle M, M\rangle_t(\omega)\mathbb{P}(d\omega)$ and (see Theorem 8.10) that

$$\left\|\Psi \circ \tilde{T}\right\|_{M,T}^2 = \int_\Omega \int_0^T \|\Psi(\omega, t)\|_{L_{(HS)}(\mathcal{H}, H)}^2 \mu_M(d\omega, dt)$$

for $\Psi \in L^2(\Omega \times [0, T], \mathcal{P}_{[0,T]}, \mu_M; L_{(HS)}(\mathcal{H}, H))$.

Lemma 8.13   *Assume that*

$$\Psi \in L^1(E, \mathcal{E}, \lambda; L^2(\Omega \times [0, T], \mathcal{P}_{[0,T]}, \mu_M; L_{(HS)}(\mathcal{H}, H))).$$

Then there is a sequence of mappings $\Psi_n\colon \Omega \times [0, T] \times E \mapsto L_{(HS)}(U, H)$ *of the form*

$$\Psi_n(\omega, t, x) = \sum_{j=1}^{J_n} \sum_{k=1}^{K_n} \sum_{l=1}^{L_n} a_{k,l}^n H_j \psi_k(x) \chi_{(s_{k,l}^n, t_{k,l}^n]}(t) \chi_{F_{k,j}^n}(\omega),$$

where the $a_{k,l}^n$ *are constants, the* H_j *are Hilbert–Schmidt operators from* U *into* H, *the* ψ_k *are bounded real-valued* $\mathcal{E}$-*measurable functions and the* $(s_{k,l}^n, t_{k,l}^n]$ *are subintervals of* $[0, T]$ *and* $F_{k,l}^n \in \mathcal{F}_{s_{k,l}^n}$, *such that*

$$\lim_{n \to \infty} \int_E \left\|\Psi(\cdot, \cdot, x) \circ \tilde{T}(\cdot, \cdot) - \Psi_n(\cdot, \cdot, x)\right\|_{M,T} \lambda(dx) = 0.$$

Proof   Observe that

$$\|\Psi(\cdot, \cdot, x)\|_{M,T}^2 = \sum_m \int_\Omega \int_0^T \gamma_m(\omega, t)|\Psi(\omega, t, x)e_m|_H^2 \mu_M(d\omega, dt).$$

Therefore, since

$$\sum_m \gamma_m(\omega, t) = \|Q_t(\omega)\|_{L_1(U)} < \infty,$$

we may assume that there exist $m_0 \in \mathbb{N}$ and $M < \infty$ such that $\Psi(\omega, t, x)e_m = 0$ for all $m \geq m_0$ and ω, t, x and

$$\sum_{j=1}^{m_0-1} |\Psi(\omega, t, x)e_j|_H^2 \leq M, \qquad \forall\, \omega, t, x.$$

Then $\Psi(\omega, t, x) \circ \tilde{T}(\omega, t) \in L_{(HS)}(U, H)$ and

$$\left\|\Psi(\omega, t, x) \circ \tilde{T}(\omega, t)\right\|_{L_{(HS)}(U,H)}^2 \leq M.$$

In particular,

$$\Psi \circ \tilde{T} \in \mathcal{V} := L^2\big(\Omega \times [0, T] \times E, \; \mathcal{P}_{[0,T]} \times \mathcal{E}, \; \mu_M \otimes \lambda; \; L_{(HS)}(U, H)\big).$$

Denote the norm on $\mathcal{V}$ by $||| \cdot |||$. Let $\{H_j\}$ be an orthonormal basis of $L_{(HS)}(U, H)$, let $\{f_k\}$ be an orthonormal basis of $L^2(\Omega \times [0, T], \mathcal{P}_{[0,T]}, \mu_M)$ and let $\{\psi_l\}$ be an orthonormal basis of $L^2(E, \mathcal{E}, \lambda)$. Clearly $\{H_j f_k \otimes \psi_l\}$ is an orthonormal basis of $\mathcal{V}$. Then

$$\lim_{n \to \infty} |||\Psi_n^1 - \Psi \circ \tilde{T}||| = 0,$$

where

$$\Psi_n^1(\omega, t, x) = \sum_{j,l=1}^n b_{j,k,l} H_j f_k(\omega, t) \psi_l(x)$$

and the $b_{j,k,l}$ are the coefficients of $\Psi \circ \tilde{T}$ in the basis $\{H_j f_k \otimes \psi_l\}$. Therefore the proof of the lemma is complete as soon as we can show that each f_k can be approximated by a sequence of simple real-valued processes. To do this, denote by $\mathfrak{G}$ the class of all sets of the form

$$\Gamma = \bigcup_{m=1}^l (s_m, t_m] \times F_m, \qquad F_m \in \mathcal{F}_{s_m}.$$

Since the linear combinations of the characteristic functions of sets from $\mathcal{P}_{[0,T]}$ are dense in $L^2(\Omega \times [0, T], \mathcal{P}_{[0,T]}, \mu_M)$, we need to show only that for all $A \in \mathcal{P}_{[0,T]}$ and $\varepsilon > 0$ there is a $\Gamma \in \mathfrak{G}$ such that

$$\mu_M\big((A \setminus \Gamma) \cup (\Gamma \setminus A)\big) \le \varepsilon.$$

Let $\tilde{\mathcal{P}}_{[0,T]}$ be the class of all $A \in \mathcal{P}_{[0,T]}$ that can be approximated in the above sense by elements from $\mathfrak{G}$. Then $\mathfrak{G} \subset \tilde{\mathcal{P}}_{[0,T]} \subset \mathcal{P}_{[0,T]} = \sigma(\mathfrak{G})$. Hence it is sufficent to observe that $\tilde{\mathcal{P}}_{[0,T]}$ is a σ-field. This follows from the Dynkin π–λ theorem (see Theorem 3.2), because $\mathfrak{G}$ is a π-system and $\tilde{\mathcal{P}}_{[0,T]}$ is a λ-system. $\qquad\qquad\square$

We can now formulate the main result of the present section.

Theorem 8.14 (Stochastic Fubini) *Assume that*

$$\Psi \in L^1\big(E, \mathcal{E}, \lambda; L^2(\Omega \times [0, T], \mathcal{P}_{[0,T]}, \mu_M; L_{(HS)}(\mathcal{H}, H))\big).$$

Then, $\mathbb{P}$-a.s.,

$$\int_E \left(\int_0^T \Psi(t, x) \circ \tilde{T}(t) \, dM(t) \right) \lambda(dx) = \int_0^T \left(\int_E \Psi(t, x) \circ \tilde{T}(t) \lambda(dx) \right) dM(t).$$

Proof We recall arguments from Da Prato and Zabczyk (1992a). From Theorem 8.10 it follows that for λ-almost all $x \in E$, the predictable process $\Psi(\cdot, \cdot, x) \circ \tilde{T}(\cdot, \cdot)$ is integrable with respect to M. We have to show that there is an $\mathcal{F}_T \times \mathcal{E}$-measurable version of the integral

$$I(\omega, x) := \int_0^T \Psi(\omega, t, x) \circ \tilde{T}(t) \, dM(\omega, t),$$

that for $\mathbb{P}$-almost all $\omega \in \Omega$, the integral is Bochner integrable with respect to λ and that

$$\int_E I(\omega, x) \lambda(dx) = \int_0^T J(t, \omega) \, dM(\omega, t), \qquad \mathbb{P}\text{-a.s.,}$$

where J is a predictable version of the Bochner integral

$$J(\omega, t) := \int_E \Psi(\omega, t, x) \circ \tilde{T}(t) \lambda(dx).$$

Let (Ψ_n) be the approximating sequence constructed in Lemma 8.13. The theorem clearly holds for each Ψ_n. By standard use of Chebyshev's inequality and the Borel–Cantelli lemma, it follows from Lemma 8.13 that there is a set $E_0 \in \mathcal{E}$ such that $\lambda(E \setminus E_0) = 0$ and a subsequence (Ψ_{n_m}) such that, for all $x \in E_0$ and $m \geq m(x)$,

$$\left\| \Psi(\cdot, \cdot, x) \circ \tilde{T}(\cdot, \cdot) - \Psi_{n_m}(\cdot, \cdot, x) \right\|_{M,T} \leq 2^{-m}.$$

Consequently, if

$$I(\omega, x) := \lim_{m \to \infty} \int_0^T \Psi_{n_m}(\omega, t, x) \, dM(\omega, t)$$

on the set where the limit exists and $I(\omega, x) = 0$ otherwise then, for any $x \in E_0$, I is an $\mathcal{F}_T \times \mathcal{E}$-measurable modification of

$$\int_0^T \Psi(\omega, t, x) \circ \tilde{T}(\omega, t) \, dM(\omega, t).$$

Note that

$$\mathbb{E} \left| \int_E \int_0^T \Psi(t, x) \circ \tilde{T}(t) \, dM(t) \lambda(dx) - \int_E \int_0^T \Psi_{n_m}(t, x) \, dM(t) \lambda(dx) \right|_H$$

$$\leq \int_E \mathbb{E} \left| \int_0^T \left(\Psi(t, x) \circ \tilde{T}(t) - \Psi_{n_m}(t, x) \right) dM(t) \right|_H \lambda(dx)$$

$$\leq \int_E \left\| \Psi(\cdot, \cdot, x) \circ \tilde{T}(\cdot, \cdot) - \Psi_{n_m}(\cdot, \cdot, x) \right\|_{M,T} \lambda(dx) \to 0.$$

Moreover,

$$
\mathbb{E}\left|\int_0^T \int_E \Psi(t,x) \circ \tilde{T}(t)\lambda(\mathrm{d}x)\,\mathrm{d}M(t) - \int_0^T \int_E \Psi_{n_m}(t,x)\,\mathrm{d}M(t)\,\lambda(\mathrm{d}x)\right|_H
$$

$$
\leq \mathbb{E}\left|\int_0^T \int_E \left(\Psi(t,x)\circ\tilde{T}(t) - \Psi_{n_m}(t,x)\right)\lambda(\mathrm{d}x)\,\mathrm{d}M(t)\right|_H
$$

$$
\leq \left\|\int_E \left(\Psi(\cdot,\cdot,x)\circ\tilde{T}(\cdot,\cdot) - \Psi_{n_m}(\cdot,\cdot,x)\right)\lambda(\mathrm{d}x)\right\|_{M,T}
$$

$$
\leq \int_E \|\Psi(\cdot,\cdot,x)\circ\tilde{T}(\cdot,\cdot) - \Psi_{n_m}(\cdot,\cdot,x)\|_{M,T}\,\lambda(\mathrm{d}x),
$$

and the result follows. $\qquad\square$

8.6 Stochastic integral with respect to a Lévy process

This section is concerned with stochastic integration with respect to a Lévy process. First we consider the L^2-theory of stochastic integration with respect to square integrable Lévy martingales, and then stochastic integration in L^p-spaces. In the latter case we first consider the stochastic integral with respect to a Wiener process. The target space, where the stochastic integral takes values, will be a space of the type $L^p(E, \mathcal{E}, \lambda)$ with $p \geq 2$. For establishing the space–time regularity of a solution to an SPDE driven by a Wiener process, the case of large p will be of special interest. Finally, we consider the case of integration with respect to a Poisson random measure, which can be treated as a Lévy process taking values in a properly chosen Hilbert space. This is an impulsive analogue of space–time Gaussian white noise. The fact that its quadratic variation process is a purely atomic measure enables us to develop the theory of integration with a target space of the type $L^p(E, \mathcal{E}, \lambda)$, $p < 2$. The fact that p is small will be an obstacle to investigating the space regularity of a solution to the equation driven by a Poisson random measure, but it will be very useful for establishing the existence of solutions to SPDEs considered on a domain $\mathcal{O} \subseteq \mathbb{R}^d$ with $d > 1$.

8.6.1 Square integrable integrators

It is convenient to introduce a special class of martingales satisfying the following condition:

$$
\exists\, Q \in L_1^+(U)\colon \forall t \geq s \geq 0, \quad \langle\langle M, M\rangle\rangle_t - \langle\langle M, M\rangle\rangle_s \leq (t-s)Q. \tag{8.10}
$$

By Theorem 8.2, (8.10) may be stated equivalently as follows: $(\langle\langle M, M\rangle\rangle_s, \; s \geq 0)$ is absolutely continuous and

$$Q_s = \frac{\mathrm{d}}{\mathrm{d}s}\langle\langle M, M\rangle\rangle_s \leq Q, \qquad \forall\, s \geq 0, \; \mathbb{P}\text{-a.s.} \tag{8.11}$$

The following lemma gives a motivation for the assumption in (8.11).

Lemma 8.15 *Assume that Q and R are non-negative operators on a Hilbert space V and that $R \leq Q$. If Φ is a linear operator from V into a Hilbert space H then $\left\|\Phi R^{1/2}\right\|_{L_{(HS)}(V,H)} \leq \left\|\Phi Q^{1/2}\right\|_{L_{(HS)}(V,H)}.$*

Proof Let $\{e_k\}$ and $\{f_l\}$ be orthonormal bases of V and H, respectively. Then

$$\left\|\Phi R^{1/2}\right\|_{L_{(HS)}(V,H)}^2 \leq \sum_k \left|\Phi R^{1/2} e_k\right|_H^2 = \sum_l \left|R^{1/2}\Phi^* f_l\right|_V^2$$

$$= \sum_l \langle R^{1/2}\Phi^* f_l, R^{1/2}\Phi^* f_l\rangle_V \sum_l \langle R\Phi^* f_l, \Phi^* f_l\rangle_V$$

$$= \sum_l \langle Q\Phi^* f_l, \Phi^* f_l\rangle_V = \left\|\Phi Q^{1/2}\right\|_{L_{(HS)}(V,H)}^2.$$

$\square$

Let $Q^{-1/2}$ be the pseudo-inverse operator given by (7.1) and let us equip $\mathcal{H} := Q^{1/2}(U)$ with the scalar product $\langle\psi, \varphi\rangle_{\mathcal{H}} = \langle(Q^{-1/2})\psi, (Q^{-1/2})\varphi\rangle_U$. Let

$$\mathbf{L}_{\mathcal{H},T}^2 := L^2(\Omega \times [0,T], \mathcal{P}_{[0,T]}, \mathbb{P}\,\mathrm{d}t; L_{(HS)}(\mathcal{H}, H)).$$

A straightforward application of Theorems 8.7 and 8.10 yields:

Proposition 8.16 *Assume (8.10). Then $\mathcal{L}_{M,T}^2 \supseteq \mathbf{L}_{\mathcal{H},T}^2(H)$ and, for every $X \in \mathbf{L}_{\mathcal{H},T}^2(H)$,*

$$\mathbb{E}\left|\int_0^t X(s)\,\mathrm{d}M(s)\right|_H^2 \leq \mathbb{E}\int_0^t \|X(s)\|_{L_{(HS)}(\mathcal{H},H)}^2\,\mathrm{d}s.$$

In the most important case, where M is a Lévy process, $(Q_t, \; t \geq 0)$ does not depend on t and ω, and $\langle M, M\rangle_t = c\,t$. In fact, by Theorem 4.47 and Remark 7.3, $\langle M, M\rangle_t = t\,\mathrm{Tr}\,Q$ and $\langle\langle M, M\rangle\rangle_t = t\,Q$ for $t \geq 0$, yielding $Q_t = Q/\mathrm{Tr}\,Q$, where Q is the covariance operator of M. It follows that $\mathcal{H}$ as defined above is the reproducing kernel Hilbert space of M. Note that $\tilde{T}(\omega, t) = I$, with I the identity operator. Thus we have the following consequence of Theorems 8.7 and 8.10.

Corollary 8.17 *If Q is constant then*

$$\mathcal{L}_{M,T}^2(H) = \mathbf{L}_{\mathcal{H},T}^2(H) = L^2(\Omega \times [0,T], \mathcal{P}_{[0,T]}, \mathbb{P}\,\mathrm{d}t; L_{(HS)}^2(\mathcal{H}, H)).$$

Moreover, for $\Psi \in \mathcal{L}_{M,T}^2(H)$, *the stochastic integral is a square integrable martingale with*

$$\mathbb{E}\left|I_T^M(\Psi)\right|_H^2 = \mathbb{E}\int_0^T \|\Psi(s)\|_{L_{(HS)}(\mathcal{H},H)}^2 \, \mathrm{d}s,$$

$$\langle\!\langle I^M(\Psi), I^M(\Psi)\rangle\!\rangle_t = \int_0^t \Psi(s)\Psi(s)^* \, \mathrm{d}s, \qquad t \in [0, T],$$

$$\langle I^M(\Psi), I^M(\Psi)\rangle_t = \int_0^t \|\Psi(s)\|_{L_{(HS)}(\mathcal{H},H)}^2 \, \mathrm{d}s, \qquad t \in [0, T].$$

8.6.2 General integrators

Recall (see Theorem 4.23) that an arbitrary Lévy process L taking values in a Hilbert space U can be written in the form

$$L(t) = mt + M(t) + P(t), \qquad t \geq 0, \tag{8.12}$$

where $m \in U$, M is a square integrable martingale (and also a Lévy process) in U and P is a compound (not necessarily square integrable) Poisson process with Lévy measure μ_P.

We assume that M and P are defined on a filtered probability space $(\Omega, \mathcal{F}, (\mathcal{F}_t), \mathbb{P})$, M is a martingale with respect to the filtration $(\mathcal{F}_t)$, P is $(\mathcal{F}_t)$-adapted and for all $t, h \geq 0$ the increment $P(t + h) - P(t)$ is independent of $\mathcal{F}_t$. Finally, we assume that $\mu_M(\mathrm{d}\omega, \mathrm{d}t) = \mathrm{d}\langle M, M\rangle_t(\omega)\mathbb{P}(\mathrm{d}\omega)$ is absolutely continuous with respect to $\mathrm{d}t\,\mathbb{P}(\mathrm{d}\omega)$.

We consider a finite time interval $[0, T]$. The aim of this subsection is to define the stochastic integral

$$I(t) := \int_0^t \Phi(s)\,\mathrm{d}s + \int_0^t \Psi_1(s)\,\mathrm{d}M(s) + \int_0^t \Psi_2(s)\,\mathrm{d}P(s), \qquad t \in [0, T], \tag{8.13}$$

where Φ, Ψ_1, Ψ_2 are operator-valued processes. Clearly, this will cover, in particular, the definition of the stochastic integral $\int_0^t \Psi(s)\,\mathrm{d}L(s)$ with respect to the Lévy process given by (8.12).

Recall that $P(t) = \sum_{j=1}^{\Pi(t)} Z_j$, where the Z_j are independent random variables in V with distribution

$$\mathbb{P}\left(Z_j \in \Gamma\right) = \frac{\mu_P(\Gamma)}{\mu_P(V)}, \qquad j \in \mathbb{N},$$

and Π is a Poisson process with intensity $\mu_P(V)$.

Assume that an increasing sequence (V_m) of bounded measurable subsets of V is given. For any $m > 0$, we write

$$\tau_m := \inf\{t \geq 0 \colon P(t) - P(t-) \notin V_m\} = \inf\{j \in \mathbb{N} \colon Z_j \notin V_m\}. \qquad (8.14)$$

We assume that the sequence (τ_m) increases to ∞.

Lemma 8.18 *If $V = \bigcup V_m$ then the above hypothesis holds true.*

Proof   The sequence (τ_m) is clearly non-decreasing. It follows that

$$\mathbb{P}(\tau_m > n) \geq \mathbb{P}(Z_1 \in V_m, Z_2 \in V_m, \ldots, Z_n \in V_m)$$
$$\geq (\mathbb{P}(Z_1 \in V_m))^n \to 1 \qquad \text{as } m \uparrow \infty.$$

$\square$

Define Z_j^m by $Z_j^m := Z_j \chi_{V_m}(Z_j)$. Define $P_m(t) := \sum_{j=1}^{\Pi(t)} Z_j^m$, $t \geq 0$. Then

$$P(t) = P_m(t) \qquad \text{on the set } \{t \leq \tau_m\}. \qquad (8.15)$$

Each P_m is clearly a compound Poisson process in V with Lévy measure $\mu_m(\Gamma) = \mu_P(\Gamma \cap V_m)$. Moreover, each P_m is square integrable, since it takes values in a bounded set V_m. Let

$$u_m := \int_V z \mu_m(\mathrm{d}z) = \int_{V_m} z \mu_P(\mathrm{d}z). \qquad (8.16)$$

Then the processes $(M_m(t) := P_m(t) - t u_m, \ t \geq 0)$, $m \in \mathbb{N}$, are square integrable mean-zero Lévy processes with covariances

$$Q_m = \int_{U_m} z \otimes z \mu_P(\mathrm{d}z), \qquad m \in \mathbb{N}. \qquad (8.17)$$

That is,

$$\langle Q_m u, v \rangle_V = \int_{U_m} \langle z, u \rangle_V \langle z, u \rangle_V \mu_P(\mathrm{d}z), \qquad u, v \in V.$$

Let $\mathcal{H}_m = Q_m^{1/2}(V)$. Note that $(\mathcal{H}_m)$ is an increasing sequence. Recall that $\mathcal{L}_{M,T}^2(H)$ denotes the space of all processes Ψ such that the stochastic integral $\int_0^t \Psi(s)\,\mathrm{d}M(s)$, $t \in [0, T]$, is square integrable in H.

We will define the stochastic integral (8.13) for integrands Φ, Ψ_1 and Ψ_2 satisfying the following conditions.

(H1) $\Phi \colon \Omega \times [0, T] \mapsto H$ is measurable and

$$\mathbb{P}\left(\int_0^T \chi_{[0,\tau_m]}(t)|\Phi(t)|_H\,\mathrm{d}t < \infty\right) = 1, \qquad \forall\, m \in \mathbb{N}.$$

(H2) For each $m \in \mathbb{N}$, $\Psi_1 \chi_{[0,\tau_m]} \in \mathcal{L}_{M,T}^2(H)$.

(H3) For each $m \in \mathbb{N}$,

$$\chi_{[0, \tau_m]} \Psi_2 \in L^2 \left(\Omega \times [0, T], \mathcal{P}_T, \mathbb{P} \, dt; L_{(HS)}(\mathcal{H}_m, H) \right),$$
$$(\omega, t) \mapsto \chi_{[0, \tau_m(\omega)]}(t) \Psi_2(t) u_m \text{ is measurable}$$

and

$$\mathbb{P} \left(\int_0^T \chi_{[0, \tau_m]}(t) |\Psi_2(t) u_m|_H \, dt < \infty \right) = 1.$$

Note that (H1)–(H3) imply that, for each m,

$$I_m(t) := \int_0^t \chi_{[0, \tau_m]}(s) \left(\Phi(s) + \Psi_2(s) u_m \right) ds$$
$$+ \int_0^t \chi_{[0, \tau_m]}(s) \Psi_1(s) \, dM(s)$$
$$+ \int_0^t \chi_{[0, \tau_m]}(s) \Psi_2(s) \, dM_m(s), \qquad t \in [0, T],$$

is a well-defined process having a càdlàg modification. As usual, we take the càdlàg version of I_m. The following result is a simple consequence of Propositions 8.11 and 8.12.

Lemma 8.19 *Let I_m, $m \in \mathbb{N}$ be defined as above. Then, for every $t \in [0, T]$ and all $m, n \in \mathbb{N}$ with $n \geq m$, $(I_m(t) - I_n(t)) \chi_{[0, \tau_m]}(t) = 0$, $\mathbb{P}$-a.s.*

By definition, the stochastic integral (8.13) is a uniquely determined càdlàg process satisfying $I(\omega, t) = I_m(\omega, t)$ for $\omega \in \Omega$ and $t \in [0, \tau_m(\omega) \wedge T]$. By Lemma 8.19, the process I is well defined.

8.7 Integration with respect to a Poisson random measure

As in Section 7.3, consider a measurable space $(E, \mathcal{E})$ and a Poisson random measure on $[0, \infty) \times E$ with intensity measure $dt \, \mu(d\xi)$. By Theorem 7.28, the compensated measure $\widehat{\pi}$ (see Definition 6.2) can be treated as a square integrable Lévy martingale in any Hilbert space U containing $\mathcal{H} := L^2(E, \mathcal{E}, \mu)$ with a Hilbert–Schmidt embedding. Moreover, $\mathcal{H}$ is the RKHS of $\widehat{\pi}$.

In this section we are concerned with the so-called L^p, $p \in [1, 2]$, theory of integration with respect to $\widehat{\pi}$. Starting with simple fields, we define the stochastic integrals with respect to π and $\widehat{\pi}$ of random fields $X = X(t, \xi)$ on $[0, \infty) \times E$. At the end of the section, we will relate these concepts to the stochastic integral with respect to Hilbert-space martingales.

It is worth noting that in the application to SPDEs (see Chapter 12) we will be concerned with the convolution-type integral

$$\int_0^t \int_O \int_S G(t-s, x, y) g(Y(s, y), \sigma) \widehat{\pi}(ds, dy, d\sigma).$$

An extension to such types of integrand is provided in subsection 8.8.1.

First we define the integral of a simple field. To this end, let us denote by $\mathcal{E}_{\text{fin}}$ the set of all $A \in \mathcal{E}$ such that $\mu(A) < \infty$. A field X is said to be *simple* if

$$X = \sum_{j=0}^k X_j \chi_{(t_j, t_{j+1}]} \chi_{A_j}, \tag{8.18}$$

where $0 = t_0 < \cdots < t_k < t_{k+1} < \infty$, $A_1, \ldots, A_k \in \mathcal{E}_{\text{fin}}$ and the X_j are bounded and $\mathcal{F}_{t_j}$-measurable. We denote by $\mathcal{L}_0$ the set of all simple fields on $[0, \infty) \times E$.

For a simple field X, we set

$$\int_0^t \int_E X(s, \xi) \pi(ds, d\xi) := \sum_{j=0}^k X_j \pi\big((t_j \wedge t, t_{j+1} \wedge t] \times A_j\big).$$

We will often write $I_t^\pi(X)$ instead of $\int_0^t \int_E X(s, \xi) \pi(ds, d\xi)$. Note that

$$\mathbb{E} I_t^\pi(X) = \mathbb{E} \sum_{j=0}^k X_j \pi\big((t_j \wedge t, t_{j+1} \wedge t] \times A_j\big)$$

$$= \sum_{j=0}^k \mathbb{E} X_j \, \mathbb{E} \, \pi\big((t_j \wedge t, t_{j+1} \wedge t] \times A_j\big)$$

$$= \sum_{j=0}^k \mathbb{E} X_j (t_{j+1} \wedge t - t_j \wedge t) \mu(A_j)$$

$$= \mathbb{E} \int_0^t \int_E X(s, \xi) \, ds \, \mu(d\xi).$$

This yields the following result.

Lemma 8.20 *If X is a simple field then*

$$\mathbb{E} \big| I_t^\pi(X) \big| \leq \mathbb{E} I_t^\pi(|X|) = \int_0^t \int_E \mathbb{E} |X(s, \xi)| \, ds \, \mu(d\xi).$$

Let X be given by (8.18). Then, by definition,

$$I_t^{\widehat{\pi}}(X) := \sum_{j=0}^k X_j \left(\pi\big((t_j \wedge t, t_{j+1} \wedge t] \times A_j\big) - (t_{j+1} \wedge t - t_j \wedge t) \mu(A_j) \right).$$

We will often write $I_t^{\widehat{\pi}}(X) = \int_0^t \int_E X(s, \xi) \widehat{\pi}(ds, d\xi)$.

Lemma 8.21 *For a simple field X, the integral $I_t^{\widehat{\pi}}(X)$, $t \geq 0$, is a martingale having all moments finite and quadratic variation*

$$\left[I^{\widehat{\pi}}(X), I^{\widehat{\pi}}(X)\right]_t = I_t^{\pi}(X^2), \qquad t \geq 0.$$

Proof Let X be given by (8.18). We have to show that, for all $t > r$, $\mathbb{E}\left(I_t^{\widehat{\pi}}(X)|\mathcal{F}_r\right) = I_r^{\widehat{\pi}}(X)$. Clearly it is suffcent to show this for $r = t_i$, $i \leq k$, and $t = t_{k+1}$. We have

$$\mathbb{E}\left(I_{t_{k+1}}^{\widehat{\pi}}(X)|\mathcal{F}_{t_i}\right) - I_{t_i}^{\widehat{\pi}}(X)$$

$$= \sum_{j=i}^{k} \mathbb{E}\left(X_j\left(\pi\left((t_j, t_{j+1}] \times A_j\right) - (t_{j+1} - t_j)\mu(A_j)\right)|\mathcal{F}_{t_i}\right)$$

$$= \sum_{j=i}^{k} \mathbb{E}\left(R_j|\mathcal{F}_{t_i}\right),$$

where

$$R_j := X_j\left(\pi\left((t_j, t_{j+1}] \times A_j\right) - (t_{j+1} - t_j)\mu(A_j)\right).$$

Thus, it is suffcent to show that $\mathbb{E}\left(R_j|\mathcal{F}_{t_i}\right) = 0$ for $i \leq j$. We have

$$\mathbb{E}\left(R_j|\mathcal{F}_{t_i}\right) = \mathbb{E}\left(\mathbb{E}\left(R_j|\mathcal{F}_{t_j}\right)|\mathcal{F}_{t_i}\right)$$

$$= \mathbb{E}\left(X_j\,\mathbb{E}\left(\pi\left((t_j, t_{j+1}] \times A_j\right) - (t_{j+1} - t_j)\mu(A_j)|\mathcal{F}_{t_j}\right)|\mathcal{F}_{t_i}\right).$$

Since $\pi((t_j, t_{j+1}] \times A_j)$ is independent of $\mathcal{F}_{t_j}$, $i \leq j$,

$$\mathbb{E}\left(\pi\left((t_j, t_{j+1}] \times A_j\right) - (t_{j+1} - t_j)\mu(A_j)|\mathcal{F}_{t_j}\right)$$

$$= \mathbb{E}\left(\pi\left((t_j, t_{j+1}] \times A_j\right) - (t_{j+1} - t_j)\mu(A_j)\right) = 0,$$

which gives the desired conclusion. The integrability of the integral follows easily, as the X_j are bounded random variables and all moments of $\pi((t_j, t_{j+1}] \times A_j)$ are finite.

To prove the formula for the quadratic variation $[I^{\widehat{\pi}}(X), I^{\widehat{\pi}}(X)]$, we may assume that X is of the form $X = X_1 \chi_{(t_1, t_2]} \chi_A$, where X_1 is $\mathcal{F}_{t_1}$-measurable. We have

$$I_t^{\widehat{\pi}}(X) = X_1\left(\pi\left((t_1 \wedge t, t_2 \wedge t] \times A\right) - (t_2 \wedge t - t_1 \wedge t)\mu(A)\right).$$

Let us take a partition $\{t_j^n\}$ of $[0, \infty)$ satisfying $\sup_j \left(t_{j+1}^n - t_j^n\right) \to 0$ as $n \uparrow \infty$ and $t_j^n \uparrow \infty$. Let us fix t and write

$$\Delta_j^n := \pi\left((t_1 \wedge t \wedge t_{j+1}^n, t_2 \wedge t \wedge t_{j+1}^n] \times A\right) - \pi\left((t_1 \wedge t \wedge t_j^n, t_2 \wedge t \wedge t_j^n] \times A\right).$$

Note that $(\pi((0, t] \times A), t \geq 0)$ is a Poisson process. It is therefore piecewise constant with jumps of size 1. Thus, for every $\omega \in \Omega$ and n sufficiently large,

$\Delta_j^n \in \{0, 1\}$. Hence

$$\left[I^{\widehat{\pi}}(X), I^{\widehat{\pi}}(X)\right]_t = \lim_{n\to\infty} \sum_j X_1^2 \left(\Delta_j^n\right)^2 = X_1^2 \lim_{n\to\infty} \sum_j \Delta_j^n$$

$$= X_1^2\, \pi\left((t_1 \wedge t, t_2 \wedge t] \times A\right) = I_t^\pi(X^2).$$

$\square$

The following lemma is taken from Saint Loubert Bié (1998).

Lemma 8.22 *Let $p \in [1, 2]$. Then there exists a constant c_p such that, for an arbitrary simple field X and $T < \infty$,*

$$\mathbb{E} \sup_{t\in[0,T]} \left|I_t^{\widehat{\pi}}(X)\right|^p \leq c_p \, \mathbb{E} \int_0^T \int_E |X(t,\xi)|^p \, dt\, \mu(d\xi).$$

Proof By the Burkholder–Davies–Gundy inequality and Lemma 8.21,

$$\mathbb{E} \sup_{t\in[0,T]} \left|I_t^{\widehat{\pi}}(X)\right|^p \leq c_p\, \mathbb{E}\left[I^{\widehat{\pi}}(X), I^{\widehat{\pi}}(X)\right]_T^{p/2} = c_p\, \mathbb{E}\left(I_T^\pi\left(X^2\right)\right)^{p/2}.$$

Let X be given by (8.18). Then

$$I_T^\pi\left(X^2\right) = \sum_{j=1}^k X_j^2 \pi\left((t_j \wedge T, t_{j+1} \wedge T] \times A_j\right).$$

Note that, since $p/2 \leq 1$,

$$\left(\sum_j x_j m_j\right)^{p/2} \leq \sum_j (x_j m_j)^{p/2} \leq \sum_j (x_j)^{p/2} m_j$$

for all $(x_1, \ldots, x_k) \in [0, \infty)^k$ and $(m_1, \ldots, m_k) \in \mathbb{Z}_+^k$. Thus since

$$\pi\left((t_j \wedge T, t_{j+1} \wedge T] \times A_j\right) \in \mathbb{Z}_+, \qquad j = 1, \ldots, k,$$

we have $\left(I_T^\pi\left(X^2\right)\right)^{p/2} \leq I_T^\pi\left(|X|^p\right)$. By Lemma 8.20,

$$\mathbb{E}\, I_T^\pi\left(|X|^p\right) = \int_0^T \int_S \mathbb{E}\, |X(t,\xi)|^p \, dt\, \mu(d\xi),$$

and the required inequality follows. $\square$

We will generalize these results to the class of predictable fields. Given $T < \infty$, we denote by $\mathcal{P}_{[0,T]}$ the σ-field of predictable sets in $[0, T] \times \Omega$. Write

$$\mathcal{L}_{\mu,T}^p := L^p\left([0, T] \times \Omega \times E, \mathcal{P}_{[0,T]} \otimes \mathcal{E}, dt\, \mathbb{P}\mu\right). \tag{8.19}$$

We have proved that, for each $p \in [1, 2]$ and each $t > 0$, the stochastic integral $I_t^{\widehat{\pi}}: \mathcal{L}_0 \mapsto L^p(\Omega, \mathcal{F}_t, \mathbb{P})$ is a bounded linear mapping if the class $\mathcal{L}_0$ of simple fields is equipped with the seminorm

$$\|X\|_{\mathcal{L}^p_{\mu,t}} = \left(\int_0^t \int_E \mathbb{E}\, |X(s,\xi)|^p \, ds \, \mu(d\xi) \right)^{1/p}.$$

The simple fields are dense in $\mathcal{L}^p_{\mu,T}$, yielding the following consequence of Lemmas 8.20–8.22.

Theorem 8.23

(i) *For $p \in [1, 2]$ and $t \in [0, T]$ there is a unique extension of the stochastic integral $I_t^{\widehat{\pi}}$ to a bounded linear operator, denoted also by $I_t^{\widehat{\pi}}$, from $\mathcal{L}^p_{\mu,t}$ into $L^p(\Omega, \mathcal{F}_t, \mathbb{P})$.*

(ii) *There is a unique extension of the mapping $\mathcal{L}_0 \ni X \mapsto I_t^\pi(X) \in L^1(\Omega, \mathcal{F}_t, \mathbb{P})$ to a bounded linear operator from $\mathcal{L}^1_{\mu,t}$ into $L^1(\Omega, \mathcal{F}_t, \mathbb{P})$. The value of this operator at X is given by*

$$\int_0^t \int_E X(s,\xi) \pi(ds, d\xi),$$

denoted $I_t^\pi(X)$.

(iii) *For $X \in \mathcal{L}^1_{\mu,T}$ and $0 \le s \le t \le T$,*

$$\mathbb{E}\left| I_t^{\widehat{\pi}}(X) - I_s^{\widehat{\pi}}(X) \right| \le c_1 \int_s^t \int_E \mathbb{E}\,|X(r,\xi)| \, dr \, \mu(d\xi)$$

and

$$\mathbb{E}\left| I_t^\pi(X) - I_s^\pi(X) \right| \le \int_s^t \int_E \mathbb{E}\,|X(r,\xi)| \, dr \, \mu(d\xi).$$

Hence the processes $I^{\widehat{\pi}}(X)$ and $I^\pi(X)$ admit predictable modifications.

(iv) *If $X \in \mathcal{L}^2_{\mu,T}$ then $\left(I_t^{\widehat{\pi}}(X), \, t \in [0, T] \right)$ is a square integrable martingale. Moreover, for $X, Y \in \mathcal{L}^2_{\mu,T}$ and $t \in [0, T]$, $\left[I^{\widehat{\pi}}(X), I^{\widehat{\pi}}(Y) \right]_t = I_t^\pi(XY)$.*

As for the case of simple fields, we write $\int_0^t \int_E X(s,\xi)\widehat{\pi}(ds, d\xi)$ instead of $I_t^{\widehat{\pi}}(X)$. In the majority of cases,

$$\int_0^t \int_E X(s,\xi) \pi(ds, d\xi) = \int_0^t \int_E X(s,\xi)\widehat{\pi}(ds, d\xi) + \int_0^t \int_E X(s,\xi) \, ds \, \mu(d\xi).$$

8.7.1 Comparing two integrals

Assume that $p = 2$. We will show that the stochastic integral with respect to a compensated Poisson measure, introduced above, can be regarded as a stochastic integral with respect to a square integrable martingale, described in Section 8.2. In fact, we will relate the integrand $X \in \mathcal{L}^2_{\mu,T}$ to $\tilde{X} \in \mathcal{L}^2_{\tilde{\pi},T}$ in such a way that $I_t^{\widehat{\pi}}(X) = \int_0^t \tilde{X}(s) \, d\widehat{\pi}(s)$.

For this purpose we regard $\widehat{\pi}(s, \cdot)$ as a U-valued random variable for a properly chosen Hilbert space U. Namely, we assume that U is a Hilbert space such that the embedding of the RKHS space $\mathcal{H} = L^2(E, \mathcal{E}, \mu) \hookrightarrow U$ is Hilbert–Schmidt. Additionally we assume that $\mathcal{H}$ is dense in U. Then, under the identification of $\mathcal{H}$ with its dual space, $U^* \hookrightarrow \mathcal{H} = \mathcal{H}^* \hookrightarrow U$. By Proposition 7.9, we identify $\widehat{\pi}(t)$ with the family $(\langle \psi, \widehat{\pi}(t) \rangle, \psi \in U^*)$, where $\langle \cdot, \cdot \rangle$ is the duality on $U^* \times U$. In Section 7.3 we started the construction by defining $\langle \psi, \widehat{\pi}(t) \rangle$ as the stochastic integral of the deterministic mapping. Thus, with the notation of Section 7.3,

$$\langle \psi, \widehat{\pi}(t) \rangle = \widehat{\pi}(t, \psi) = \int_0^t \int_E \psi(\xi) \widehat{\pi}(\mathrm{d}s, \mathrm{d}\xi), \qquad \psi \in \mathcal{H}.$$

Since

$$I_t^{\widehat{\pi}}(\psi) = \int_0^t \int_E \psi(\xi) \widehat{\pi}(\mathrm{d}s, \mathrm{d}\xi)$$

and, under the identification of ψ with an $(\mathcal{H}^* = \mathcal{H})$-valued process, $\langle \psi, \widehat{\pi}(t) \rangle = \int_0^t \psi \, \mathrm{d}\widehat{\pi}(s)$, it follows that, for a deterministic time-independent field X, $I_t^{\widehat{\pi}}(\psi) = \int_0^t \tilde{\psi}(s) \, \mathrm{d}\widehat{\pi}(s)$, where $\tilde{\psi} \in \mathcal{H}^* = L_{(HS)}(\mathcal{H}, \mathbb{R})$ is given by

$$\tilde{\psi}[\varphi] = \langle \psi, \varphi \rangle_{\mathcal{H}} = \int_E \psi(\xi) \varphi(\xi) \mu(\mathrm{d}\xi), \qquad \varphi \in \mathcal{H}.$$

Thus, for a simple field X, $I_t^{\widehat{\pi}}(X) = \int_0^t \tilde{X}(s) \, \mathrm{d}\widehat{\pi}(s)$, where $\tilde{X}$ is a simple process in $L_{(HS)}(\mathcal{H}, \mathbb{R})$ given by

$$\tilde{X}(s)[\varphi] = \int_E X(s)(\xi) \varphi(\xi) \mu(\mathrm{d}\xi), \qquad \varphi \in \mathcal{H}, s \geq 0. \qquad (8.20)$$

By approximation arguments we obtain the following result.

Proposition 8.24 *Given* $X \in \mathcal{L}_{\mu, T}^2$ *we denote by* $\tilde{X}$ *the* $L_{(HS)}(\mathcal{H}, \mathbb{R})$-*valued process defined by (8.20). Then* $\tilde{X} \in \mathcal{L}_{\widehat{\pi}, T}^2(\mathbb{R})$ *and*

$$I_t^{\widehat{\pi}}(X) = \int_0^t \tilde{X}(s) \, \mathrm{d}\widehat{\pi}(s), \qquad t \in [0, T].$$

From now on, we will not distinguish X from $\tilde{X}$ notationally.

8.8 L^p-theory for vector-valued integrands

Assume that M is a square integrable Lévy martingale in a Hilbert space U with RKHS $\mathcal{H}$. So far, we have seen how to integrate processes with values in the space of linear, possibly unbounded, operators from U or $\mathcal{H}$ into another Hilbert space H. A special role is played by the space of Hilbert–Schmidt operators. One may ask whether it is possible to develop a similar theory of stochastic integration in Banach spaces. Thus, given a Banach space B, we are looking for a subspace $\mathcal{R}$ of the space of linear operators from U to B such that, for a simple $\mathcal{R}$-valued process

$$\Psi = \sum_n \alpha_i \Psi_i \chi_{(t_i, t_{i+1}]},$$

where $\Psi_i \in \mathcal{R}$ and α_i is an $\mathcal{F}_{t_i}$-measurable real-valued bounded random variable, we have

$$\mathbb{E} \left| \int_0^T \Psi(s)\, \mathrm{d}M(s) \right|_B^q \le C_{T,q}\, \mathbb{E} \int_0^T \|\Psi(s)\|_{\mathcal{R}}^q\, \mathrm{d}s, \qquad T \ge 0, \qquad (8.21)$$

for some positive q. This, however, requires some geometrical properties of B; see Brzeźniak (1997) and Neidhardt (1978).

Let $(E, \mathcal{E})$ be a measurable space and let λ be a measure on $(E, \mathcal{E})$. We will establish some basic properties of stochastic integrals in the target space $B = L^p(E, \mathcal{E}, \lambda)$. We consider two cases, firstly where M is a compensated Poisson measure and $p \in [1, 2]$ and secondly where M is a Wiener process. The latter will be considered for $p \ge 2$. Our analysis of the first case will be based on Theorem 8.23. In the second case we will use the Burkholder–Davies–Gundy (BDG) inequality formulated in Theorem 3.49.

The results obtained here for a Wiener martingale W will be sufficient to study the regularity properties of solutions to SPDEs driven by W.

For brevity, we set $L^p := L^p(E, \mathcal{E}, \lambda)$. Let $K : U^* \times U^* \mapsto \mathbb{R}$ be the covariance form of M introduced in Section 7.1.

Let us denote by $R_{U,0}(\mathcal{H}, L^p)$ the class of all operators $\Psi : \mathcal{H} \mapsto L^p$ given by $(\Psi\psi)(x) = \langle \mathcal{Q}(x), \psi \rangle$, $x \in E$, $\psi \in \mathcal{H}$, where the kernel $\mathcal{Q}$ is a measurable mapping from E to U^* satisfying

$$\Delta_p(\mathcal{Q}) := \left(\int_E K \big(\mathcal{Q}(x), \mathcal{Q}(x)\big)^{p/2} \lambda(\mathrm{d}x) \right)^{1/p} < \infty.$$

Note that $R_{U,0}(\mathcal{H}, L^p)$, equipped with $\|\Psi\|_{R_{U,0}(\mathcal{H}, L^p)} := \Delta_p(\mathcal{Q})$, is a normed space. Here $\mathcal{Q}$ is such that $\Psi\psi(x) = \langle \mathcal{Q}(x), \psi \rangle$, λ-almost surely, for every $\psi \in \mathcal{H}$. Clearly, $\| \cdot \|_{R_{U,0}(\mathcal{H}, L^p)}$ is a seminorm. It is also a norm as, by Proposition 7.1,

$$K \big(\mathcal{Q}(x), \mathcal{Q}(x)\big) = \sum_k \langle \mathcal{Q}(x), e_k \rangle^2$$

for any orthonormal basis $\{e_k\}$ of $\mathcal{H}$. Thus $K(\mathcal{Q}(x), \mathcal{Q}(x)) = 0$ implies that $\langle \mathcal{Q}(x), e_k \rangle = 0$ for every k, and consequently the operator (on $\mathcal{H}$) corresponding to $\mathcal{Q}$ is equal to 0.

As in the case of stochastic integration in Hilbert spaces, we first define the stochastic integral in L^p for simple processes $\mathcal{S}\big(R_{U,0}(\mathcal{H}, L^p)\big)$ and then extend the integral to a properly chosen space of processes.

Definition 8.25 An $R_{U,0}(\mathcal{H}, L^p)$-valued stochastic process Ψ is said to be *simple* if there exist a sequence of non-negative numbers $t_0 = 0 < t_1 < \cdots < t_m$, a sequence of operators $\Psi_j \in R_{U,0}(\mathcal{H}, L^p)$, $j = 1, \ldots, m$ and a sequence of events $A_j \in \mathcal{F}_{t_j}$, $j = 0, \ldots, m - 1$, such that

$$\Psi(s) = \sum_{j=0}^{m-1} \chi_{A_j} \chi_{(t_j, t_{j+1}]}(s) \Psi_j, \qquad s \geq 0.$$

We shall denote by $\mathcal{S}\left(R_{U,0}(\mathcal{H}, L^p)\right)$ the class of all simple processes with values in $R_{U,0}(\mathcal{H}, L^p)$. For simple processes Ψ we set

$$I_t^M(\Psi) := \sum_{j=0}^{m-1} \chi_{A_j} \Psi_j \left(M(t_{j+1} \wedge t) - M(t_j \wedge t) \right), \qquad t \geq 0. \tag{8.22}$$

Note that, since $R_{U,0}(\mathcal{H}, L^p) \subset L(U, L^p)$ and M takes values in U, each term in the definition above is a well-defined random element in L^p.

8.8.1 Poisson case

In this subsection, $M = \widehat{\pi}$ is a compensated Poisson random measure. We denote by $\mathcal{L}_{\widehat{\pi}, T}^p(L^p)$ the space of all predictable processes $\Psi \colon \Omega \times [0, T] \mapsto R_{U,0}(\mathcal{H}, L^p)$ such that

$$\|\Psi\|_{\mathcal{L}_{\widehat{\pi}, T}^p(L^p)} := \left(\mathbb{E} \int_0^T \|\Psi(t)\|_{R_{U,0}(\mathcal{H}, L^p)}^p \, dt \right)^{1/p}.$$

Theorem 8.26 *Let $p \in [1, 2]$. Then, for all $t \geq 0$, there is a unique continuous extension of the stochastic integral $I_t^{\widehat{\pi}}$ to a continuous linear mapping acting from $\mathcal{L}_{\widehat{\pi}, t}^p(L^p)$ into $L^p(\Omega, \mathcal{F}_t, \mathbb{P}; L^p)$. Moreover, for all $T \in (0, \infty)$ and $\Psi \in \mathcal{L}_{\widehat{\pi}, T}^p(L^p)$, the process $\left(I_t^{\widehat{\pi}}(\Psi), t \in [0, T] \right)$ is continuous in probability and adapted and therefore admits a predictable modification.[2]*

Proof It is enough to show that there is a constant c such that, for any simple process Ψ,

$$\mathbb{E} \left| I_t^{\widehat{\pi}}(\Psi) \right|_{L^p}^p \leq c \, \mathbb{E} \int_0^T \|\Psi(t)\|_{R_{U,0}(\mathcal{H}, L^p)}^p \, dt.$$

We will derive this estimate from Theorem 8.23. To this end we fix a simple function Ψ of the form given in Definition 8.25 and write

$$Q(s, x, \omega) = \sum_{j=0}^{m-1} \chi_{A_j}(\omega) \chi_{(t_j, t_{j+1}]}(s) Q_j(x),$$

---

[2] We will always work with predictable versions of stochastic integral processes.

where Q_j is the kernel of Ψ_j. Then, by Proposition 8.24,

$$I_t^{\widehat{\pi}}(\Psi)(x) = \int_0^t Q(s, x) \, d\widehat{\pi}(s).$$

It now follows from Theorem 8.23 that

$$\mathbb{E}\left|I_t^{\widehat{\pi}}(\Psi)(x)\right|^p \leq c \, \mathbb{E}\int_0^t K\big(Q(s, x), Q(s, x)\big)^{p/2} \, ds.$$

$\square$

8.8.2 Wiener case

In this subsection $M = W$ is a Wiener process. For the definitions of simple processes with values in $R_{U,0}(\mathcal{H}, L^p)$ and the stochastic integral $I_t^W(\Psi)$, Ψ simple, see Definition 8.25 and identity (8.22).

Lemma 8.27 *For any $\Psi \in \mathcal{S}(R_{U,0}(\mathcal{H}, L^p))$, the stochastic integral $I^W(\Psi)$ is an L^p-valued $(\mathcal{F}_t)$-adapted mean-zero process with continuous trajectories. Moreover, there is a constant C such that*

$$\mathbb{E}\left|I_t^W(\Psi)\right|_{L^p}^p \leq C \, \mathbb{E}\left(\int_0^t \|\Psi(s)\|_{R_{U,0}(\mathcal{H}, L^p)}^2 \, ds\right)^{p/2},$$

for all $t \in [0, \infty)$ and $\Psi \in \mathcal{S}(R_{U,0}(\mathcal{H}, L^p))$.

Proof Let us fix $x \in E$. Then

$$I_t^W(\Psi)(x) = \sum_{j=1}^{m-1} \chi_{A_j} \langle Q_j(x), \, W(t_{j+1} \wedge t) - W(t_j \wedge t)\rangle,$$

where Q_j stands for the kernel of Ψ_j. Therefore $\big(I_t^W(\Psi)(x), t \geq 0\big)$ is a square integrable martingale with continuous trajectories. Moreover,

$$\langle I^W(\Psi)(x), I^W(\Psi)(x)\rangle_t = \sum_{j=1}^{m-1} \chi_{A_j} K\big(Q_j(x), Q_j(x)\big) \big(t_{j+1} \wedge t - t_j \wedge t\big)$$

$$= \int_0^t \eta(s, x) \, ds,$$

where $\eta(s, x) := \chi_{A_j} K(Q_j(x), Q_j(x))$ for $t \in (t_j, t_{j+1}]$. Now using the BDG inequality (see Theorem 3.49), we obtain

$$\mathbb{E}\left|I_t^W(\Psi)(x)\right|^p \leq C_p \, \mathbb{E}\langle I^W(\Psi)(x), I^W(\Psi)(x)\rangle_t^{p/2}$$

$$\leq C_p \, \mathbb{E}\left(\int_0^t \eta(s, x) \, ds\right)^{p/2}.$$

Since

$$\int_E \left(\int_0^t \eta(s, x) \, ds \right)^{p/2} \lambda(dx) = \left| \int_0^t \eta(s, \cdot) \, ds \right|_{L^{p/2}}^{p/2}$$

$$\leq \left(\int_0^t |\eta(s, \cdot)|_{L^{p/2}} \, ds \right)^{p/2}$$

$$\leq \left(\int_0^t \|\Psi(s)\|_{R_{U,0}(\mathcal{H}, L^p)}^2 \, ds \right)^{p/2},$$

we have

$$\mathbb{E} \left| I_t^W(\Psi) \right|_{L^p}^p = \mathbb{E} \int_E \left| I_t^W(\Psi)(x) \right|^p \lambda(dx)$$

$$\leq C_p \, \mathbb{E} \int_E \left(\int_0^t \eta(s, x) \, ds \right)^{p/2} \lambda(dx)$$

$$\leq C_p \, \mathbb{E} \left(\int_0^t \|\Psi(s)\|_{R_{U,0}(\mathcal{H}, L^p)}^2 \, ds \right)^{p/2},$$

which is the desired estimate. $\qquad\square$

Remark 8.28 Let $u \in U^*$ and let $\{e_k\}$ be an orthonormal basis of $\mathcal{H}$. Then, for every $h \in \mathcal{H}$,

$$\langle u, h \rangle^2 = \left(\sum_k \langle h, e_k \rangle_{\mathcal{H}} \langle u, e_k \rangle \right)^2$$

$$\leq \sum_k \langle h, e_k \rangle_{\mathcal{H}}^2 \sum_k \langle u, e_k \rangle^2 = |h|_{\mathcal{H}}^2 K(u, u).$$

Therefore, given $\Psi \in R_{U,0}(\mathcal{H}, L^p)$ with kernel $\mathcal{Q}$,

$$\|\Psi\|_{L(\mathcal{H}, L^p)}^p = \sup_{h \in \mathcal{H}: |h|_{\mathcal{H}} \leq 1} |\Psi h|_{L^p}^p = \sup_{h \in \mathcal{H}: |h|_{\mathcal{H}} \leq 1} \int_E \langle \mathcal{Q}(x), h \rangle^p \lambda(dx)$$

$$\leq \sup_{h \in \mathcal{H}: |h|_{\mathcal{H}} \leq 1} \int_E K(\mathcal{Q}(x), \mathcal{Q}(x))^{p/2} |h|^p \lambda(dx)$$

$$\leq \|\Psi\|_{R_{U,0}(\mathcal{H}, L^p)}^p.$$

Let $(R(\mathcal{H}, L^p), \|\cdot\|_{R(\mathcal{H}, L^p)})$ be the completion of $(R_{U,0}(\mathcal{H}, L_\rho^p), \|\cdot\|_{R_{U,0}(\mathcal{H}, L^p)})$ to a Banach space. By the estimate above, the space $R(\mathcal{H}, L^p)$ is continuously embedded into $L(\mathcal{H}, L^p)$.

Let us denote by $\mathcal{L}_{W,T}^p(L^p)$ the space of all predictable processes $\Psi \colon \Omega \times [0, T] \mapsto R(\mathcal{H}, L^p)$ such that

$$\|\Psi\|_{\mathcal{L}_{W,T}^p(L^p)} := \left[\mathbb{E} \left(\int_0^T \|\Psi(t)\|_{R(\mathcal{H}, L^p)}^2 \, dt \right)^{p/2} \right]^{1/p}.$$

Note the difference from the space $\mathcal{L}^p_{\hat{\pi},T}(L^p)$ defined in subsection 8.8.1. Since simple processes are dense in $\mathcal{L}^p_{W,T}(L^p)$ we have the following consequence of Lemma 8.27.

Theorem 8.29 *Let $p \in [2, \infty)$. Then, for all $t \geq 0$, there is a unique continuous extension of the stochastic integral I^W_t to a continuous linear mapping acting from $\mathcal{L}^p_{W,t}(L^p)$ to $L^p(\Omega, \mathcal{F}_t, \mathbb{P}; L^p)$. Moreover, for all $T \in (0, \infty)$ and $\Psi \in \mathcal{L}^p_{W,T}(L^p)$, the process $\left(I^W_t(\Psi), t \in [0, T]\right)$ is continuous in probability and adapted and therefore admits a predictable modification.*

In the general case, when the target space B is not necessarily of L^p-type, it is useful to introduce the concept of so-called radonifying operators. To do this, let us fix an orthonormal basis $\{e_k\}$ of $\mathcal{H}$. Let (β_k) be a sequence of independent real-valued random variables with distribution $\mathcal{N}(0, 1)$ defined on a probability space $(\Omega, \mathcal{F}, \mathbb{P})$.

Definition 8.30 A bounded linear operator $\Psi \colon \mathcal{H} \mapsto B$ is called *radonifying* if the series $\sum_{k=1}^{\infty} \beta_k \Psi e_k$ converges in $L^2(\Omega, \mathcal{F}, \mathbb{P}; B)$.

One can show that this definition does not depend on the choice of orthonormal basis $\{e_k\}$ or sequence (β_k). We use $\mathfrak{R}(\mathcal{H}, B)$ to denote the class of all radonifying operators from $\mathcal{H}$ into B. Given a linear operator Ψ from $\mathcal{H}$ into L^p, write

$$\|\Psi\|^2_{\mathfrak{R}(\mathcal{H}, B)} = \limsup_n \mathbb{E} \left| \sum_{k=1}^{n} \beta_k \Psi e_k \right|^2_B .$$

Then (see e.g. Neidhardt 1978) Ψ is radonifying if and only if $\|\Psi\|_{\mathfrak{R}(\mathcal{H}, B)}$ is finite. Note that $\mathfrak{R}(\mathcal{H}, B)$ equipped with the norm $\| \cdot \|_{\mathfrak{R}(\mathcal{H}, B)}$ is a Banach space and that if B is a Hilbert space then the spaces $\mathfrak{R}(\mathcal{H}, B)$ and $L_{(HS)}(\mathcal{H}, B)$ of the radonifying and Hilbert–Schmidt operators are equal. In this setting, $\| \cdot \|_{\mathcal{R}(\mathcal{H}, L^p)}$, which appears in the definition of $\mathcal{L}^p_{W,T}(L^p)$, and $\| \cdot \|_{\mathfrak{R}(\mathcal{H}, L^p)}$ are different norms! Using Remark 8.32 below they may be shown to be equivalent.

The lemma below provides a useful estimate for the radonifying norm of an operator given by a kernel. It is a reformulation of Brzeźniak and Peszat (1999), Proposition 2.1, and Peszat and Tindel (2007), Lemma 4.3.

Lemma 8.31 *Let $p \in [2, \infty)$. Then $R_{U,0}(\mathcal{H}, L^p) \subseteq \mathfrak{R}(\mathcal{H}, L^p)$. Moreover, there is a constant $C < \infty$ such that, for an arbitrary $\Psi \in R_{U,0}(\mathcal{H}, L^p)$,*

$$\|\Psi\|_{\mathfrak{R}(\mathcal{H}, L^p)} \leq C \|\Psi\|_{R_{U,0}(\mathcal{H}, L^p)}.$$

Proof Assume that Ψ is given by a kernel Q. Since for each x the real-valued random variable $\sum_{k=1}^{n} \beta_k \langle Q(x), e_k \rangle$ is Gaussian, there exists a constant C_1 depending only on p such that

$$\left(\mathbb{E} \left| \sum_{k=1}^{n} \beta_k \Psi e_k \right|_{L^p}^2 \right)^{p/2} = \left[\mathbb{E} \left(\int_E \left| \sum_{k=1}^{n} \beta_k \langle \mathcal{Q}(x), e_k \rangle \right|^p \lambda(dx) \right)^{2/p} \right]^{p/2}$$

$$\leq \mathbb{E} \int_E \left| \sum_{k=1}^{n} \beta_k \langle \mathcal{Q}(x), e_k \rangle \right|^p \lambda(dx)$$

$$\leq C_1 \int_E \left(\mathbb{E} \left| \sum_{k=1}^{n} \beta_k \langle \mathcal{Q}(x), e_k \rangle \right|^2 \right)^{p/2} \lambda(dx)$$

$$\leq C_1 \int_E \left| \sum_{k=1}^{n} \langle \mathcal{Q}(x), e_k \rangle^2 \right|^{p/2} \lambda(dx).$$

Therefore, by Proposition 7.1,

$$\|\Psi\|_{\mathfrak{R}(\mathcal{H},L^p)}^2 = \limsup_n \mathbb{E} \left| \sum_{k=1}^{\infty} \beta_k \Psi e_k \right|_{L^p}^2 \leq C_1^{2/p} \Delta_p(\mathcal{Q}),$$

which is the desired conclusion. $\qquad\square$

Remark 8.32 The lemma above provides an upper estimate for the radonifying norm. In fact (see Brzeźniak 1997), one can show that on $R_{U,0}(\mathcal{H}, L^p)$ the norms $\mathfrak{R}(\mathcal{H}, L^p)$ and $R_{U,0}(\mathcal{H}, L^p)$ are equivalent.

Remark 8.33 In a general theory of stochastic integration in Banach spaces the most suitable operator norm is the radonifying norm. Under a certain geometrical assumption on B one obtains (8.21) for all $q \geq 2$, whereas using the BDG inequality we succeeded in obtaining (8.21) only for $q = p$.

Part II

Existence and Regularity

9

General existence and uniqueness results

This chapter is devoted to the existence and uniqueness of solutions to an abstract stochastic evolution equation driven by a locally square integrable martingale. The concepts of weak and mild solutions are introduced and their equivalence is shown. Sufficient conditions for the existence and uniqueness of solutions are given. The Markov property of solutions is proved. Linear equations are treated in some detail.

9.1 Deterministic linear equations

Let us start from an abstract situation. Let A_0 be a densely defined linear operator on a Banach space B. Denote by D_0 the domain of A_0. Assume that the differential equation

$$\frac{dy}{dt} = A_0 y, \qquad y(0) = y_0 \in D_0, \tag{9.1}$$

has a unique solution $y(t)$, $t \geq 0$. Since the equation is linear the solution depends linearly on y_0: $y(t) = S(t)y_0, t \geq 0$, where $S(t)$ is a linear transformation from D_0 into B. One says that the Cauchy problem (9.1) is *well posed* on B if the operators $S(t)$, $t \geq 0$, have continuous extensions to B and, for each $z \in B$, $t \mapsto S(t)z$ is a continuous mapping. The mapping $t \mapsto S(t)z$ defined now for all $z \in B$ is called the *generalized solution* to (9.1). Clearly the operators $S(t)$, $t \geq 0$, satisfy the following conditions:

(i) $S(0) = I$ and $S(t)S(s) = S(t + s)$ for all $t, s \geq 0$;
(ii) $[0, \infty) \ni t \mapsto S(t)z \in B$ is continuous for each $z \in B$ or, equivalently,
$|S(t)z - z|_B \to 0$ as $t \downarrow 0$ for every $z \in B$.

Thus well-posed Cauchy problems lead to the concept of a C_0-semigroup, introduced earlier in Chapter 4 on Lévy processes. For convenience we repeat the definition here.

Definition 9.1   A family $S = (S(t), t \geq 0)$ of bounded linear operators on a Banach space $(B, | \cdot |_B)$ is called a C_0-*semigroup* if the conditions formulated above are satisfied.

Assume that S is a C_0-semigroup on B. One says that an element $z \in B$ is in the *domain of the generator* of S if $\lim_{t \downarrow 0} t^{-1} (S(t)z - z)$ exists. The set of all such z is denoted by $D(A)$. The limit, denoted by A, is called the *generator* of S. Clearly, in the case considered at the beginning of the section, $D(A) \supseteq D_0$ and A is an extension of A_0. Basic properties of semigroups are formulated in the following theorem. For its proof we refer the reader to e.g. Davies (1980), Engel and Nagel (2000), Pazy (1983), or Yosida (1965).

Theorem 9.2

(i) *If S is a C_0-semigroup on B then, for some ω and $M > 0$,*

$$|S(t)z|_B \leq e^{\omega t} M |z|_B, \qquad \forall z \in B, \ \forall t \geq 0.$$

(ii) *If a densely defined operator A generates a C_0-semigroup S then A is closed and, for any $z \in D(A)$ and $t > 0$,*

$$S(t)z \in D(A) \quad and \quad \frac{\mathrm{d}}{\mathrm{d}t} S(t)z = AS(t)z = S(t)Az.$$

The final part of the theorem shows that for $y_0 \in D(A)$ the function $t \mapsto S(t)y_0$ is a solution to the equation

$$\frac{\mathrm{d}y}{\mathrm{d}t}(t) = Ay(t), \qquad y(0) = y_0.$$

Consider, finally, the following non-homogeneous evolution equation:

$$\frac{\mathrm{d}y}{\mathrm{d}t}(t) = Ay(t) + \psi(t), \qquad y(0) = y_0 \in H, \tag{9.2}$$

where ψ is an H-valued function. If y_0 is in the domain of A and ψ is continuously differentiable then the unique solution of (9.2) is given by the so-called *variation-of-constants formula*,

$$y(t) = S(t)y_0 + \int_0^t S(t-s)\psi(s)\,\mathrm{d}s, \qquad t \geq 0. \tag{9.3}$$

This expression has a meaning for less regular functions ψ, for example, for $\psi \in L^1_{\mathrm{loc}}([0, \infty), \mathcal{B}([0, \infty)), \ell_1; H)$. In this case y, given by (9.3), is called the *mild solution* or *generalized solution* to (9.2).

A continuous function y is a *weak solution* to (9.2) if

$$\langle y(t), h \rangle_H = \langle y_0, h \rangle_H + \int_0^t \langle y(s), A^*h \rangle_H \, \mathrm{d}s + \int_0^t \langle \psi(s), h \rangle_H \, \mathrm{d}s$$

for all $t \geq 0$ and h from the domain of the adjoint operator A^*. One can show that the mild solution y satisfies (9.2) in the weak sense. In Section 9.3 we will show the equivalence of mild and weak solutions to more general stochastic equations.

9.2 Mild solutions

We are concerned with the equation

$$\mathrm{d}X = (AX + F(X)) \, \mathrm{d}t + G(X) \, \mathrm{d}M, \qquad X(t_0) = X_0, \tag{9.4}$$

where A, with domain $D(A)$, is the generator of a C_0-semigroup S on a Hilbert space H and M is a square integrable martingale taking values in a Hilbert space U. We assume that M is defined on a filtered probability space $(\Omega, \mathcal{F}, (\mathcal{F}_t)_{t \geq 0}, \mathbb{P})$ and satisfies (8.10), that is,

$$\exists Q \in L_1^+(U) \colon \forall t \geq s \geq 0, \quad \langle\!\langle M, M \rangle\!\rangle_t - \langle\!\langle M, M \rangle\!\rangle_s \leq (t - s)Q. \tag{9.5}$$

We set $\mathcal{H} := Q^{1/2}(U)$. The space $\mathcal{H}$ is considered with an induced scalar product. By Proposition 8.16 the space $\mathcal{L}_{T,M}^2(H)$ of integrable processes contains

$$\mathbf{L}_{\mathcal{H},T}^2(H) := L^2 \left(\Omega \times [0, T], \mathcal{P}_{[0,T]}, \mathbb{P} \, \mathrm{d}t; L_{(HS)}(\mathcal{H}, H) \right).$$

Moreover, in the case where M is a Lévy process on H, one can take $\mathcal{H}$ as the RKHS of M. Then $\mathcal{L}_{T,M}^2(H) = \mathbf{L}_{\mathcal{H},T}^2(H)$. The assumption of the square integrability of M is rather restrictive and does not cover certain important cases. In particular, it does not cover stochastic equations in finite dimensions driven by local martingales. These equations have unique solutions under Lipschitz-type conditions. In the important case where M is a Lévy process the square integrability condition can be removed. For more details see Section 9.7.

We assume that X_0 is an $\mathcal{F}_{t_0}$-measurable random variable in H. The non-linear components of (9.4) are usually not continuous on H. Generally one should assume that F and G are defined on some subspace (or even a subset) of H and take values in $\tilde{H}$ and $L_0(\mathcal{H}, \tilde{H})$ respectively, where $\tilde{H}$ contains H and $L_0(\mathcal{H}, \tilde{H})$ is the space of all (not necessarily continuous) operators from $\mathcal{H}$ into $\tilde{H}$. In order to cover the most natural cases (see, however, Proposition 9.7), we assume that $F \colon D(F) \mapsto H$ and $G \colon D(G) \mapsto L(\mathcal{H}, H)$ satisfy Lipschitz-type conditions:

(F) $D(F)$ is dense in H and there is a function $a\colon (0, \infty) \mapsto (0, \infty)$ satisfying $\int_0^T a(t)\,dt < \infty$ for all $T < \infty$ such that, for all $t > 0$ and $x, y \in D(F)$,

$$|S(t)F(x)|_H \le a(t)(1 + |x|_H),$$
$$|S(t)(F(x) - F(y))|_H \le a(t)|x - y|_H.$$

(G) $D(G)$ is dense in H and there is a function $b\colon (0, \infty) \mapsto (0, \infty)$ satisfying $\int_0^T b^2(t)\,dt < \infty$ for all $T < \infty$ such that, for all $t > 0$ and $x, y \in D(G)$,

$$\|S(t)G(x)\|_{L_{(HS)}(\mathcal{H}, H)} \le b(t)(1 + |x|_H),$$
$$\|S(t)(G(x) - G(y))\|_{L_{(HS)}(\mathcal{H}, H)} \le b(t)|x - y|_H.$$

We will occasionally need the following strengthening of (G):

(GI) Condition (G) holds if $S(t) = I, t \ge 0$.

Remark 9.3   Since the domains $D(F)$ and $D(G)$ are dense in H, conditions (F) and (G) imply that, for each $t > 0$, $S(t)F$ and $S(t)G$ have unique extensions to continuous mappings from H to H and from H to $L(\mathcal{H}, H)$, respectively. We also denote these extensions by $S(t)F$ and $S(t)G$. Clearly, for all $t > 0$ and $x, y \in H$,

$$|S(t)F(x)|_H \le a(t)(1 + |x|_H),$$
$$|S(t)(F(x) - F(y))|_H \le a(t)|x - y|_H \tag{9.6}$$

and

$$\|S(t)G(x)\|_{L_{(HS)}(\mathcal{H}, H)} \le b(t)(1 + |x|_H),$$
$$\|S(t)(G(x) - G(y))\|_{L_{(HS)}(\mathcal{H}, H)} \le b(t)|x - y|_H. \tag{9.7}$$

Remark 9.4   The function $t \mapsto \|S(t)\|_{L(H, H)}$ is bounded on any finite interval $[0, T]$. Thus, if $F\colon H \mapsto H$ and $G\colon H \mapsto L_{(HS)}(\mathcal{H}, H)$ are Lipschitz continuous then (F) and (G) are satisfied.

We are now able to formulate the precise definition of a solution to (9.4).

Definition 9.5   Let X_0 be a square integrable $\mathcal{F}_{t_0}$-measurable random variable in H. A predictable process $X\colon [t_0, \infty) \times \Omega \mapsto H$ is called a *mild solution to (9.4) starting at time t_0 from X_0* if

$$\sup_{t \in [t_0, T]} \mathbb{E}\,|X(t)|_H^2 < \infty, \qquad \forall\, T \in (t_0, \infty) \tag{9.8}$$

and

$$X(t) = S(t - t_0)X_0 + \int_{t_0}^t S(t - s)F(X(s))\,ds$$

$$+ \int_{t_0}^t S(t - s)G(X(s))\,dM(s), \qquad \forall\, t \ge t_0. \tag{9.9}$$

Of course, it is to be understood that (9.9) holds $\mathbb{P}$-a.s., that is, for any fixed t there is an $\Omega_t \in \mathcal{F}_t$ of full $\mathbb{P}$-measure such that (9.9) holds for all $\omega \in \Omega_t$.

Remark 9.6 The integrals appearing in (9.9) are well defined. The first is the Bochner integral introduced in Section 3.3. Recall that it is defined if the integrand is measurable and

$$I_F(t) := \int_{t_0}^{t} |S(t-s)F(X(s))|_H \, ds < \infty, \qquad \forall\, t \in (t_0, \infty).$$

Measurability follows from the continuity of $S(t-s)F$, while the estimate can be shown in the following way:

$$\mathbb{E}\, I_F(t) \leq \mathbb{E} \int_{t_0}^{t} a(t-s)(1+|X(s)|_H) \, ds$$

$$\leq \left(1 + \sup_{s \in [t_0, t]} \mathbb{E}\,|X(s)|_H\right) \int_{t_0}^{t} a(t-s) \, ds < \infty.$$

By Corollary 8.17, the stochastic integral is well defined provided that the integrand is predictable and

$$I_G(t) := \mathbb{E} \int_{t_0}^{t} \|S(t-s)G(X(s))\|^2_{L_{(HS)}(\mathcal{H}, H)} \, ds < \infty. \qquad (9.10)$$

Predictability follows from the predictability of X and the continuity of

$$[t_0, t) \times H \ni (s, x) \mapsto S(t-s)G(x) \in L_{(HS)}(\mathcal{H}, H),$$

whereas (9.10) can be derived from (G) and (9.8) in the following way:

$$I_G(t) \leq \mathbb{E} \int_{t_0}^{t} b^2(t-s)(1+|X(s)|)^2 \, ds$$

$$\leq 2\left(1 + \sup_{s \in [t_0, t]} \mathbb{E}\,|X(s)|^2_H\right) \int_{0}^{t} b^2(s) \, ds < \infty.$$

As we shall see, assumptions (F) and (G) allow us to prove the existence and uniqueness of the solution to (9.4). They are satisfied in many interesting cases, but in special situations they may be relaxed. In particular, consider the equation

$$dX = AX\,dt + dM, \qquad X(0) = X_0. \qquad (9.11)$$

Then, for (9.11), the assumption in (G) implies in particular that the reproducing kernel of the noise is included in the state space. This assumption is not satisfied, however, for a large class of linear equations having well-defined solutions. For instance, assume (still for (9.11)) that A is a self-adjoint operator on a Hilbert space H. Assume that there exists an orthonormal and complete basis $\{e_n\}$ of H such that

$$Ae_n = -\gamma_n e_n, \qquad n = 1, 2, \ldots, \gamma_n > 0, \gamma_n \uparrow +\infty. \qquad (9.12)$$

Then we can identify H with l^2 and A with the diagonal operator $A(u_n) = (-\gamma_n u_n)$, $(u_n) \in l^2$. Let $M = (M_n)$, where M_n, $n \in \mathbb{N}$, are independent mean-zero Lévy processes with exponents $\mathbb{E}\,e^{ix M_n(t)} = e^{-t\psi_n(x)}$. Here

$$\psi_n(x) = \tfrac{1}{2} q_n x^2 + \int_{\mathbb{R}} \left(1 - e^{ixy}\right) v_n(dy).$$

Equation (9.11) can be solved for each coordinate of $X = (X_n)$ separately. Namely,

$$X_n(t) = e^{-\gamma_n t} X_n(0) + \int_0^t e^{-\gamma_n(t-s)}\, dM_n(s), \qquad t \geq 0.$$

To simplify the notation, we assume that $X_n(0) = 0$ for all n. By Corollary 4.29,

$$\mathbb{E}\,e^{ix X_n(t)} = \exp\left\{-\int_0^t \psi_n\!\left(xe^{-\gamma_n s}\right) ds\right\}.$$

Let

$$\psi_t(x) := \sum_{n=1}^{\infty} \int_0^t \psi_n\!\left(x_n e^{-\gamma_n s}\right) ds, \qquad x = (x_1, x_2, \ldots) \in l^2.$$

Then X takes values in $H = l^2$ if and only if, for each t, $e^{-\psi_t(x)}$ is the characteristic function of a probability measure on l^2. For $x = (x_1, x_2, \ldots) \in l^2$,

$$\psi_t(x) = \tfrac{1}{2} \sum_{n=1}^{\infty} \int_0^t q_n x_n^2 e^{-2\gamma_n s}\, ds + \sum_{n=1}^{\infty} \int_0^t \int_{\mathbb{R}} \left(1 - \exp\left\{ix_n e^{-\gamma_n s} y\right\}\right) v_n(dy)\, ds$$

$$= \tfrac{1}{4} \sum_{n=1}^{\infty} q_n \left(1 - e^{-2\gamma_n t}\right)\gamma_n^{-1} x_n^2 + \sum_{n=1}^{\infty} \int_{\mathbb{R}} \left(1 - e^{ix_n y}\right) v_n^t(dy),$$

where

$$v_n^t(\Gamma) := \int_0^t v_n\{y\colon e^{-\gamma_n s} y \in \Gamma\}\, ds, \qquad \Gamma \in \mathcal{B}(\mathbb{R}).$$

Consequently (see also the proof of Theorem 4.13), ψ_t is of the form

$$\psi_t(x) = \tfrac{1}{2}\langle Q(t)x, x\rangle_{l^2} + \int_{l^2} \left(1 - e^{i\langle x, z\rangle_{l^2}}\right) v^t(dz), \qquad x \in l^2,$$

where

$$(Q(t)x)_n = \tfrac{1}{2}\left(1 - e^{-2\gamma_n t}\right)\frac{q_n}{\gamma_n} x_n, \qquad x \in l^2,\ n \in \mathbb{N},$$

and ν_t is a measure concentrated on the axes; on the n-axis it is exactly ν_n^t. Hence, by the Lévy–Khinchin theorem, $X(t)$ takes values in l^2 if and only if $Q(t)$ is of trace class and ν^t is a measure on l^2 satisfying

$$\int_{l^2} \left(|z|^2 \wedge 1 \right) \nu^t(dz) < \infty.$$

Therefore, after standard manipulations we obtain the following result.

Proposition 9.7 *The process X takes values in l^2 if and only if*

$$\sum_{n=1}^{\infty} \left[\frac{q_n}{\gamma_n} + \int_0^t \left(e^{-2\gamma_n s} \int_{-e^{\gamma_n s}}^{e^{\gamma_n s}} |y|^2 \nu_n(dy) + \nu_n\{y: |y| > e^{\gamma_n s}\} \right) ds \right] < \infty.$$

Note that the RKHS of M is contained in l^2 if and only if the sequence $q_n + \int_{+\infty}^{-\infty} |y|^2 \nu_n(dy)$ is bounded; see Example 7.5. This is a much stronger condition than that required in Proposition 9.7.

9.2.1 Comments on the concept of solutions

Assume that $F: \mathbb{R}^d \mapsto \mathbb{R}^d, G: \mathbb{R}^d \mapsto M(d \times n)$ and that M is a square integrable $\mathbb{R}^n$-valued martingale. In the classical theory of stochastic equations (see e.g. Protter 2004), one looks for càdlàg solutions to equations of the form

$$dy(t) = F(y(t-)) dt + G(y(t-)) dM(t), \qquad y(0) = a, t \in [0, T], \qquad (9.13)$$

in which the left limit $y(t-)$ of the solution, rather than the solution $y(t)$ itself, enters into the coefficients. Note that if y is a càdlàg solution to (9.13) then the equation is well defined and the process $(y(t-), t \in [0, T])$ is predictable.

The left limits in (9.13) may be replaced by the requirement of predictability, provided that the angle bracket of the integrator is absolutely continuous with respect to Lebesgue measure or, equivalently, (9.5) is satisfied. An example constructed in the section on linear equations shows that mild solutions are not in general càdlàg and therefore, in infinite dimensions, instead of considering equations with left limits on the right-hand side, we require only that the solution should be predictable.

In the present section we discuss the relation between equations in $\mathbb{R}^d$ with and without left limits. Let ν be a finite measure on $[0, T]$ that may also be random. Set $d = 1$ and consider the equations

$$y(t) = a + \int_{(0,t]} G(y(s-))\nu(ds), \qquad t \in [0, T], \qquad (9.14)$$

$$y(t) = a + \int_{(0,t]} G(y(s))\nu(ds), \qquad t \in [0, T]. \qquad (9.15)$$

Proposition 9.8 *Assume that G is a Lipschitz function. Then we have the following.*

(i) *Bounded solutions to (9.14) and (9.15), if they exist, are càdlàg.*
(ii) *Equation (9.14) always has a unique solution.*
(iii) *Equation (9.15) may have zero, one or many solutions.*
(iv) *Equations (9.14) and (9.15) may have unique but different solutions.*

Proof We may assume that $d = n = 1$. To prove (i) assume that $G(y(s-))$ and $G(y(s))$ are bounded functions of $s \in [0, T]$. Given a bounded function $g : [0, T] \mapsto \mathbb{R}$ write

$$\psi(t) = \int_{(0,t]} g(s)\nu(ds), \qquad t \in [0, T].$$

Then, for $t \in [0, T)$ and $h > 0$ with $t + h \le T$,

$$|\psi(t + h) - \psi(t)| = \left| \int_{(t,t+h]} g(s)\nu(ds) \right| \le \nu((t, t + h]) \sup_{s \in [0,T]} |g(s)|.$$

Consequently, $\psi(t + h) \to \psi(t)$ as $h \downarrow 0$ and right-continuity follows. If $h > 0$ and $t - h \ge 0$ then

$$\psi(t) = \int_{(0,t-h]} g(s)\nu(ds) + \int_{(t-h,t]} g(s)\nu(ds)$$

$$= \psi(t - h) + \int_{(t-h,t)} g(s)\nu(ds) + g(t)\nu(\{t\}).$$

Consequently, $\lim_{h \downarrow 0} \psi(t - h) = \psi(t) - g(t)\nu(\{t\})$.

To show (ii), one considers G with Lipschitz constant K and an interval $[0, T]$ such that $K\nu((0, T]) < 1$. Then the transformation

$$J(y)(t) = a + \int_{(0,t]} G(y(s-))\nu(ds), \qquad t \in [0, T],$$

is a contraction on the space of càdlàg functions on $[0, T]$ equipped with the supremum norm. Indeed,

$$|J(y)(t) - J(\tilde{y})(t)| = \left| \int_{(0,t]} \big(G(y(s-)) - G(\tilde{y}(s-)) \big) \nu(ds) \right|$$

$$\le K \sup_{s \in [0,T]} |y(s) - \tilde{y}(s)| \, \nu((0, T]).$$

Thus if $K\nu((0, T]) < 1$ then the contraction property follows. We can use the classical argument; see Protter (2004). Consider times $T_0 = 0 < T_1 < \cdots < T_{k-1} < T_k = T$ such that $K\nu((T_j, T_{j+1})) < 1$. We first show that, on the interval $(0, T_1)$, (9.14) has a unique solution and the solution is bounded on $[0, T_1)$. Existence and

uniqueness follow from the contraction principle of J on $[0, S]$ for every $S < T_1$.
Note that $|G(x)| \le |G(0)| + K|x|$ for $x \in \mathbb{R}$ and, therefore, for $t < T_1$,

$$y(t) = a + \int_{(0,t]} G(y(s-))\nu(\mathrm{d}s);$$

consequently,

$$|y(t)| \le |a| + |G(0)|\nu((0, t]) + K\nu((0, t]) \sup_{s \le t} |y(s)|$$

$$\le |a| + |G(0)|\nu((0, T_1]) + K\nu((0, T_1]) \sup_{s \le t} |y(s)|.$$

Then, for $\delta \in (0, T_1)$,

$$\sup_{t \le T_1 - \delta} |y(t)| \le |a| + |G(0)|\nu((0, T_1)) + K\nu((0, T_1)) \sup_{t \le T_1 - \delta} |y(t)|.$$

Hence

$$\sup_{t \le T_1 - \delta} |y(t)| \le \frac{|a| + |G(0)|\nu((0, T_1))}{1 - K\nu((0, T_1))},$$

and so the required boundedness follows. Since y is uniformly bounded on $(0, T_1)$,
it follows that $y(t)$ has a finite limit as $t \uparrow T_1$. At $t = T_1$ the equation becomes

$$y(T_1) = a + \int_{(0,T_1]} G(y(s-))\nu(\mathrm{d}s).$$

This defines y at time T_1. One can repeat the reasoning for other intervals $[T_i, T_{i+1}]$
to obtain the unique solution on $[0, T]$.

To see (iii) take $\nu = \delta_{\hat{t}}$ and $a = 0$. Then any solution to (9.15) satisfies $y(t) = 0$
for $t < \hat{t}$ and $y(t) = y(\hat{t})$ for $t \ge \hat{t}$. Moreover, $y(\hat{t}) = G(y(\hat{t}))$. Now, if $G(x) = x$
then there are infinitely many solutions, whereas if $G(x) = x + 1$ there are no
solutions. Clearly there is a unique solution if the function G has a unique fixed
point.

To show (iv) we take $\nu = k\delta_{\hat{t}}$, where $\hat{t} \in (0, T)$ is such that $Kk < 1$, and $a = 0$.
Then the solution y_1 to (9.14) satisfies $y_1(t) = 0$ for $t < \hat{t}$,

$$y_1(\hat{t}) = \int_{(0,\hat{t}]} G(0)\nu(\mathrm{d}s) = kG(0)$$

and $y_1(t) = kG(0)$ for $t > \hat{t}$. The solution y_2 of (9.15) satisfies $y_2(t) = 0$ for $t < \hat{t}$
and

$$y_2(\hat{t}) = \int_{(0,\hat{t}]} G(y_2(s))\nu(\mathrm{d}s) = kG(y_2(\hat{t})).$$

Since $kK < 1$ the equation $x = kG(x)$ has a unique solution, which, in general,
is different from $kG(0)$. □

The following proposition shows that the absolute continuity assumption (9.5) allows us to introduce the concept of predictable solutions, in a unique way, up to modification.

Proposition 9.9   *Assume that M is a square integrable martingale satisfying (9.5). Let ψ and $\tilde{\psi}$ be two predictable stochastically equivalent processes such that*

$$\mathbb{E} \int_0^t \left(\|\psi(s)\|^2_{L_{(HS)}(\mathcal{H},H)} + \|\tilde{\psi}(s)\|^2_{L_{(HS)}(\mathcal{H},H)} \right) ds < \infty.$$

Then the stochastic integrals

$$\int_0^t \psi(s) dM(s), \qquad \int_0^t \tilde{\psi}(s) dM(s), \qquad t \geq 0,$$

are also stochastically equivalent.

Proof   Since

$$\mathbb{E} \left| \int_0^t \left(\psi(s) - \tilde{\psi}(s) \right) dM(s) \right|^2_H \leq \mathbb{E} \int_0^t \|\psi(s) - \tilde{\psi}(s)\|^2_{L_{(HS)}(\mathcal{H},H)} ds$$

and $\mathbb{E} \|\psi(s) - \tilde{\psi}(s)\|^2_{L_{(HS)}(\mathcal{H},H)} = 0$ for all $s \in [0, T]$, the result follows.   □

Proposition 9.10   *Assume that transformation $G \colon H \mapsto L_{(HS)}(\mathcal{H}, H)$ is Lipschitz and that (9.5) is satisfied. Let y be a càdlàg solution to the equation*

$$dy(t) = G(y(t-)) dM(t), \qquad y(0) = a.$$

Then $\tilde{y}(t) := y(t-)$, $t \geq 0$, is equivalent to y and is a predictable solution to

$$dy(t) = G(y(t)) dM(t), \qquad y(0) = a.$$

Proof   It is sufficient to show the equivalence of y and $\tilde{y}$. But this follows because

$$\lim_{s \uparrow t} \mathbb{E} \int_s^t \|G(y(\sigma-))\|^2_{L_{(HS)}(\mathcal{H},H)} d\sigma = 0.$$

□

9.3 Equivalence of weak and mild solutions

This section is devoted to the concept of weak (in the sense of partial differential equations) solutions to (9.4). Denote by $D(A^*)$ the domain of the adjoint operator A^*.

Definition 9.11 Assume that (F) and (G) hold. Let $t_0 \geq 0$, and let X_0 be a square integrable $\mathcal{F}_{t_0}$-measurable random variable in H. We say that a predictable H-valued process $(X(t), \, t \geq t_0)$ is a *weak solution* to (9.4) if it satisfies (9.8) and, for all $a \in D(A^*)$ and $t \geq t_0$,

$$\langle a, X(t) \rangle_H = \langle a, X_0 \rangle_H + \int_{t_0}^t \langle A^*a, X(s) \rangle_H \, \mathrm{d}s$$

$$+ \int_{t_0}^t \langle a, F(X(s)) \rangle_H \, \mathrm{d}s + \int_{t_0}^t \langle G^*(X(s))a, \mathrm{d}M(s) \rangle_{\mathcal{H}}. \quad (9.16)$$

Remark 9.12 In the majority of specific cases the domain $D(A^*)$ is not given explicitly; it is therefore important to realize that it is sufficient to verify (9.16) for elements a from the so-called core of A^*. A set $D \subset D(A^*)$ is a *core* of A^* if it is a linearly dense set in the space $D(A^*)$ equipped with the *graph norm*

$$\|a\|_{D(A^*)} := \left(|a|_H^2 + |A^*a|_H^2 \right)^{1/2}, \qquad a \in D(A^*).$$

Very useful sufficient criteria for a linear subset D of $D(A^*)$ to be a core are that D is dense in H and invariant for the adjoint semigroup $(S^*(t), \, t \geq 0)$; see Davies (1980).

To cope with unbounded operators A and, consequently, to treat the term $\int_{t_0}^t AX(s) \, \mathrm{d}s$, one takes scalar products of (9.4) with elements of the domain of A^*. One then treats

$$\left\langle a, \int_{t_0}^t AX(s) \, \mathrm{d}s \right\rangle_H \qquad \text{as} \qquad \int_{t_0}^t \langle A^*a, X(s) \rangle_H \, \mathrm{d}s.$$

Since non-linear mappings may not be defined on the whole H, we need the following lemma.

Lemma 9.13 *Assume that (F) and (G) hold. Then for every $a \in D(A^*)$ there is a constant $c(a) < \infty$ such that*

$$|\langle a, F(x) \rangle_H| \leq c(a)(1 + |x|_H), \qquad \forall \, x \in D(F),$$
$$|\langle a, F(x) - F(y) \rangle_H| \leq c(a)|x - y|_H, \qquad \forall \, x, y \in D(F), \quad (9.17)$$

$$|G^*(x)a|_{\mathcal{H}} \leq c(a)(1 + |x|_H), \qquad \forall \, x \in D(G),$$
$$|(G^*(x) - G^*(y))a|_{\mathcal{H}} \leq c(a)|x - y|_H, \qquad \forall \, x, y \in D(G). \quad (9.18)$$

Proof Since the semigroup S generated by A is C_0, there are real constants $\gamma > 0$ and $M \geq 1$ such that $\|S(t)\|_{L(H,H)} \leq Me^{\gamma t}$ for $t \geq 0$. Let $x \in D(F)$. Then the mapping $t \mapsto e^{-2\gamma t} S(t)F(x)$ is Bochner integrable over $(0, \infty)$ and

$$\int_0^\infty e^{-2\gamma t} S(t)F(x) \, \mathrm{d}s = (-A + 2\gamma I)^{-1} F(x).$$

By (F),

$$\int_0^\infty e^{-2\gamma t} |S(t)F(x)|_H \, dt$$

$$= \int_0^1 e^{-2\gamma t} |S(t)F(x)|_H \, dt + \int_1^\infty e^{-2\gamma t} |S(t)F(x)|_H \, dt$$

$$\le (1 + |x|_H) \left(\int_0^1 a(t) \, dt + \int_1^\infty e^{-2\gamma t} \|S(t-1)\|_{L(H,H)} \, dt \, a(1) \right).$$

It follows that

$$\left| (-A + 2\gamma I)^{-1} F(x) \right|_H \le c_1 (1 + |x|_H),$$

where c_1 is a finite constant. Now let

$$a \in D(A^*) = D(-A^* + 2\gamma I) = (-A^* + 2\gamma I)^{-1}(H).$$

Thus there is a $b \in H$ such that $a = (-A^* + \gamma I)^{-1} b$. Consequently, for all $x \in D(F)$,

$$|\langle a, F(x) \rangle_H| = |\langle b, (-A + \gamma I)^{-1} F(x) \rangle_H| \le c_1 |b|_H (1 + |x|_H).$$

In the same way one can show that

$$|\langle a, F(x) - F(y) \rangle_H| \le c_1 |b|_H |x - y|_H, \qquad \forall x, y \in D(F),$$

which proves (9.17). To show (9.18), note that by (G), for all $u \in U$ and $x \in D(H)$,

$$\left| (-A + 2\gamma I)^{-1} G(x)u \right|_H \le \int_0^\infty e^{-2\gamma t} \|S(t)G(x)\|_{L_{(HS)}(\mathcal{H}, H)} \, dt \, |u|_{\mathcal{H}}$$

$$\le c_2 (1 + |x|_H)|u|_{\mathcal{H}},$$

where c_2 is a finite constant. Let $a := (-A^* + 2\gamma I)^{-1} b \in D(A^*)$. Then

$$|G^*(x)a|_{\mathcal{H}} = \sup_{|u|_{\mathcal{H}} \le 1} |\langle u, G^*(x)a \rangle_{\mathcal{H}}|$$

$$= \sup_{|u|_{\mathcal{H}} \le 1} |\langle (-A + 2\gamma I)^{-1} G(x)u, b \rangle_H|$$

$$\le c_2 |b|_H (1 + |x|_H).$$

The proof of the second estimate in (9.18) follows in the same way. $\square$

As a direct consequence of the lemma, we have the following result.

Corollary 9.14 *Assume that (F) and (G) hold. Then for every $a \in D(A^*)$ the mappings*

$$D(F) \ni x \mapsto \langle a, F(x) \rangle_H \in \mathbb{R}, \qquad D(G) \ni x \mapsto G^*(x)a \in \mathcal{H}$$

have unique continuous extensions to the mappings, denoted by $\langle a, F(\cdot)\rangle_H$ and $G^*(\cdot)a$, satisfying (9.17) and (9.18).

We will consider an integral of the form

$$\left\langle a, \int_{t_0}^{t} G(X(s))\, dM(s) \right\rangle_H = \int_{t_0}^{t} G^*(X(s))a\, dM(s)$$

$$=: \int_{t_0}^{t} \langle G^*(X(s))a,\, dM(s)\rangle_{\mathcal{H}},$$

where $a \in D(A^*)$. Using the identification $L_{(HS)}(\mathcal{H}, \mathbb{R}) = \mathcal{H}$, we identify the integrand with the $L_{(HS)}(\mathcal{H}, \mathbb{R})$-valued process

$$\big(G^*(X(s))a\big)^{\sim}(z) := \langle G^*(X(s))a, z\rangle_{\mathcal{H}}, \qquad z \in \mathcal{H}.$$

Let $\{e_n\}$ be an orthonormal basis of $\mathcal{H}$. Then

$$\big\|\big(G^*(X(s))a\big)^{\sim}\big\|^2_{L_{(HS)}(\mathcal{H}, \mathbb{R})} = \sum_n \big|\big(G^*(X(s))a\big)^{\sim} e_n\big|^2$$

$$\sum_n \langle G^*(X(s))a, e_n\rangle^2_{\mathcal{H}} = |G^*(X(s))a|^2_{\mathcal{H}}.$$

Theorem 9.15 *Assume that (F) and (G) hold. Then X is a mild solution if and only if X is a weak solution.*

Proof We first show that the weak solution is mild. We will divide the proof of this into two steps. We assume that X is a weak solution to (9.4). Without any loss of generality, we may assume that $t_0 = 0$. Let $a \in D(A^*)$, $g \in C^1([0, \infty); \mathbb{R})$ and

$$\psi(s) := F(X(s)), \qquad \phi(s) := G(X(s)). \tag{9.19}$$

Step 1 Our goal is to show that, for any predictable process $(z(s), s \geq 0)$ with trajectories in $C^1([0, \infty); D(A^*))$,

$$\langle z(t), X(t)\rangle_H = \langle z(0), X_0\rangle_H + \int_0^t \langle A^*z(s) + z'(s), X(s)\rangle_H\, ds$$

$$+ \int_0^t \langle z(s), \psi(s)\rangle_H\, ds + \int_0^t \langle z(s), \phi(s)\, dM(s)\rangle_H. \tag{9.20}$$

We prove (9.20) first for processes of the form $z(t) = g(t)a$, where $a \in D(A^*)$ and $g \in C^1([0, \infty); \mathbb{R})$.

Using the integration-by-parts formula to be established in Proposition 9.16 below, we obtain

$$d\langle z(s), X(s)\rangle_H = d(g(s)\langle a, X(s)\rangle_H)$$
$$= g'(s)\langle a, X(s)\rangle_H \, ds + g(s)\big(\langle A^*a, X(s)\rangle_H \, ds$$
$$+ \langle a, \psi(s)\rangle_H \, ds + \langle a, \phi(s) \, dM(s)\rangle_H\big)$$
$$= \langle A^*z(s) + z'(s), X(s)\rangle_H \, ds + \langle z(s), \psi(s)\rangle_H \, ds$$
$$+ \langle z(s), \phi(s) \, dM(s)\rangle_H.$$

Now, using the fact that, for any $T > 0$, mappings of the form $g(s)a$, $s \in [0, T]$, where $a \in D(A^*)$ and $g \in C^1([0, T]; \mathbb{R})$, are linearly dense in $C^1([0, T]; D(A^*))$ and passing to the limit, we obtain (9.20) for all $z \in C^1([0, T]; D(A^*))$.

Step 2   Given $t > 0$ and $a \in D(A^*)$, define $z(s) := S^*(t-s)a$, $s \in [0, t]$. Then $z'(s) = -A^*z(s)$, $s \in [0, t]$. Hence

$$\langle a, X(t)\rangle_H = \langle S^*(t)a, X_0\rangle_H + \int_0^t \langle S^*(t-s)a, \psi(s)\rangle_H \, ds$$
$$+ \int_0^t \langle S^*(t-s)a, \phi(s) \, dM(s)\rangle_H.$$

Thus

$$\langle a, X(t)\rangle_H = \left\langle a, \ S(t)X_0 + \int_0^t S(t-s)\psi(s) \, ds\right\rangle_H$$
$$+ \left\langle a, \ \int_0^t S(t-s)\phi(s) \, dM(s)\right\rangle_H. \tag{9.21}$$

Since (9.21) holds for any a from the dense set $D(A^*)$, it holds for any $a \in H$. Consequently,

$$X(t) = S(t)X_0 + \int_0^t S(t-s)\psi(s) \, ds + \int_0^t S(t-s)\phi(s) \, dM(s).$$

We now prove that any mild solution is weak. We use the stochastic Fubini theorem, applied to $E = [0, T]$, Lebesgue measure and

$$\psi(r, s) := \chi_{[0,s]}(r)S(s-r)G(X(r)).$$

Assume for simplicity that

$$X(t) = \int_0^t S(t-s)G(X(s)) \, dM(s),$$

and write $\phi(s) := G(X(s))$. Let $a \in D(A^*)$. Then

$$\int_0^t \langle A^*a, X(s) \rangle_H \, ds = \int_0^t \left\langle A^*a, \int_0^t \chi_{[0,s]}(r)S(s-r)\phi(r)\,dM(r) \right\rangle_H \, ds$$

and, by the stochastic Fubini theorem,

$$\int_0^t \langle A^*a, X(s) \rangle_H \, ds$$

$$= \left\langle A^*a, \int_0^t \left(\int_0^t \chi_{[0,s]}(r)S(s-r)\phi(r)\,dM(r) \right) ds \right\rangle_H$$

$$= \left\langle A^*a, \int_0^t \left(\int_0^t \chi_{[0,s]}(r)S(s-r)\,ds \right) \phi(r)\,dM(r) \right\rangle_H.$$

Therefore

$$\int_0^t \langle A^*a, X(s) \rangle_H \, ds$$

$$= \left\langle A^*a, \int_0^t \left(\int_0^t \chi_{[0,s]}(r)S(s-r)\,ds \right) \phi(r)\,dM(r) \right\rangle_H$$

$$= \int_0^t \left\langle \int_0^t \chi_{[0,s]}(r)S^*(s-r)A^*a\,ds, \ \phi(r)\,dM(r) \right\rangle_H$$

$$= \int_0^t \left\langle \int_r^t S^*(s-r)A^*a\,ds, \ \phi(r)\,dM(r) \right\rangle_H$$

$$= \int_0^t \left\langle \int_r^t \left(\frac{d}{ds}S^*(s-r)a \right) ds, \ \phi(r)\,dM(r) \right\rangle_H$$

$$= \int_0^t \langle S^*(t-r)a - a, \ \phi(r)\,dM(r) \rangle_H$$

$$= \left\langle a, \int_0^t S(t-r)\phi(r)\,dM(r) \right\rangle_H - \left\langle a, \int_0^t \phi(r)\,dM(r) \right\rangle_H.$$

It follows that

$$\int_0^t \langle A^*a, X(s) \rangle_H \, ds = \langle a, X(t) \rangle_H - \left\langle a, \int_0^t \phi(r)\,dM(r) \right\rangle_H,$$

yielding

$$\langle a, X(t) \rangle_H = \int_0^t \langle A^*a, X(s) \rangle_H \, ds + \int_0^t \langle a, \phi(s)\,dM(s) \rangle_H.$$

$\square$

In the first step of the proof of the equivalence theorem we used a stochastic integration-by-parts formula. Its proof requires some facts about the Riemann–Stieltjes integral.

Proposition 9.16 *Assume that $g \in C^1([0, T], \mathbb{R})$ and l is a càdlàg function on* $[0, T]$. *Then the Riemann–Stieltjes integrals*

$$\int_0^T g(s)\,dl(s), \qquad \int_0^T g'(s)l(s)\,ds$$

exist and

$$\int_0^T g(s)\,dl(s) = g(T)l(T) - g(0)l(0) - \int_0^T g'(s)l(s)\,ds.$$

Before proving the proposition, we recall some properties of càdlàg functions.

Lemma 9.17 *Assume that l is a càdlàg function defined on $[0, T]$. Then*

(i) *the set $\{t \in [0, T]: l(t-) \neq l(t) = l(t+)\}$ is at most countable,*
(ii) *for each $\varepsilon > 0$, there exists a finite partition $0 = t_0 < t_1 < \cdots < t_n = T$ such that, for arbitrary $s, t \in [t_{k-1}, t_k)$, $|l(t) - l(s)| < \varepsilon$.*

Proof   In order to demonstrate (i), it is sufficient to show that for each n the set

$$S_n := \{t \in [0, T]: |l(t-) - l(t)| \geq 1/n\}$$

is finite. If this is not the case then there is a sequence (s_m) converging to s such that $|l(s_m-) - l(s_m)| \geq 1/n$. This leads to a contradiction, since l has left limits at each point.

In order to prove (ii), fix an ε and denote by τ the supremum of $t \in [0, T]$ for which there is a partition satisfying the desired property. Since $l(0) = l(0+)$, τ must be strictly positive. Since $l(\tau-)$ exists, the interval $[0, \tau]$ has the desired property and consequently $\tau = T$, as required. $\qquad \square$

Proof of Proposition 9.16   Let $0 = s_0 \leq t_0 < s_1 \leq t_1 \leq \cdots \leq s_{N-1} \leq t_{N-1} \leq s_N = T$, and let

$$I := \sum_{i=0}^{N-1} g(t_i)(l(s_{i+1}) - l(s_i)) = \sum_{i=1}^{N} l(s_i)g(t_{i-1}) - \sum_{i=0}^{N-1} l(s_i)g(t_i)$$

$$= -\sum_{i=1}^{N-1} l(s_i)(g(t_i) - g(t_{i-1})) + l(s_N)g(t_{N-1}) - l(s_0)g(t_0).$$

By the mean-value theorem, there exists a $\overline{t_i} \in [t_{i-1}, t_i]$ such that

$$
\begin{aligned}
I &= -\sum_{i=1}^{N-1} l(s_i) g'\left(\overline{t_i}\right) (t_i - t_{i-1}) + l(s_N) g(t_{N-1}) - l(s_0) g(t_0) \\
&= -\sum_{i=1}^{N-1} l\left(\overline{t_i}\right) g'\left(\overline{t_i}\right) (t_i - t_{i-1}) + l(T) g(t_{N-1}) - l(0) g(t_0) \\
&\quad + \sum_{i=1}^{N-1} \left(l\left(\overline{t_i}\right) - l(t_i) \right) g'\left(\overline{t_i}\right) (t_i - t_{i-1}) \\
&=: I_1 + I_2 + I_3 + I_4.
\end{aligned}
$$

It is clear that as the mesh of the partition converges to zero, $I_2 \to l(T) g(T)$ and $I_3 \to -l(0) g(0)$. Moreover, by the first part of the lemma and the Lebesgue dominated convergence theorem, $I_1 \to -\int_0^T l(s) g'(s) \, \mathrm{d}s$. Finally, by the second part of the lemma $I_4 \to 0$. $\qquad\square$

9.4 Linear equations

If (9.4) is linear, that is, if

$$
\mathrm{d}X = AX \, \mathrm{d}t + \mathrm{d}M, \qquad X(t_0) = X_0,
$$

then its mild solution is given by the formula

$$
X(t) = S(t - t_0) X_0 + Y(t),
$$

where

$$
Y(t) = \int_{t_0}^t S(t - s) \, \mathrm{d}M(s), \qquad t \geq t_0.
$$

In this section, we are interested in the regularity properties of the *stochastic convolution* Y. We will show that if M is a U-valued martingale and S is a semi-group of contractions on U then the process Y has a càdlàg version. We present two proofs, one based on the Kotelenez maximal inequalities (see Kotelenez 1987) and the other, due to Hausenblas and Seidler (2006), based on a dilation-type theorem. We also show that, in general, the stochastic convolution may not have a càdlàg version even if condition (G) is satisfied.

In the final subsection we show the existence of a càdlàg version of Y in the case of an analytic semigroup S and an M taking values in the domain $D\left((-A)^\alpha\right)$ of the generator.

The time continuity of solutions to equations driven by continuous martingales (in particular Wiener processes) is studied separately in Chapter 11.

9.4.1 The Kotelenez inequality

Let $M \in \mathcal{M}^2(U)$. Assume that $S(t, s), \tilde{S}(t, s), t \geq s$, are families of linear operators on U such that $S(s, s)$, $\tilde{S}(s, s)$ are each equal to the identity operator I and

$$\|S(t, s)\|_{L(U,U)} \leq e^{\beta(t-s)}, \quad \|\tilde{S}(t, s)\|_{L(U,U)} \leq e^{\beta(t-s)}, \quad t \geq s,$$

where $\beta \geq 0$ is a constant. In addition, let $S(\cdot, \cdot)$ be strongly continuous with respect to each variable separately; that is, for all t_0, s_0 and u,

$$\lim_{s \to s_0} S(s, t_0)u = S(s_0, t_0)u, \qquad \lim_{s \to s_0} \tilde{S}(s, t_0)u = \tilde{S}(s_0, t_0)u,$$

$$\lim_{t \to t_0} S(s_0, t)u = S(s_0, t_0)u, \qquad \lim_{t \to t_0} \tilde{S}(s_0, t)u = \tilde{S}(s_0, t_0)u,$$

where the limits are in the norm topology of U. Note that the process $\int_0^t S(t, s) \, dM(s), t \geq 0$, is well defined in U. We will require that, for arbitrary $t \geq s \geq 0$,

$$\int_0^s S(t, \sigma) \, dM(\sigma) = \tilde{S}(t, s) \int_0^s S(s, \sigma) \, dM(\sigma) = \tilde{S}(t, s)Y(s).$$

There are two important cases where the above identity holds. The first is where $\tilde{S}(t, s) = S(t, s)$ is an *evolution operator*, that is,

$$S(t, u) = S(t, s)S(s, u), \qquad t \geq s \geq u.$$

Note that if S is a semigroup then $S(t, s) := S(t - s)$ defines an evolution operator. The semigroup S satisfying $\|S(t)\|_{L(U,U)} \leq e^{\beta t}, t \geq 0$, is called a semigroup of *(generalized) contractions*.

The second case is where $\tilde{S}(t, s)$ is an evolution operator and $S(t, s) = \tilde{S}(t, t_k)$ for all $s \in (t_k, t_{k+1}]$ and for an increasing sequence t_k. For more details on the following result we refer the reader to Kotelenez (1982, 1984, 1987).

Proposition 9.18 (Kotelenez) *For any sequence* $t_0 = 0 < t_1 < \cdots < t_K = T < \infty$ *and for any constant* $c > 0$,

$$\mathbb{P}\left(\max_{1 \leq k \leq K} \left| \int_0^{t_k} S(t_k, \sigma) \, dM(\sigma) \right|_U > c \right) \leq \frac{1}{c^2} e^{2\beta T} \, \mathbb{E} \, |M(T)|_U^2.$$

Proof Write

$$Y(t) := \int_0^t S(t, s) \, dM(s), \qquad t \geq 0.$$

Then

$$\mathbb{P}\left(\max_{1 \le k \le K} |Y(t_k)|_U > c\right)$$

$$= \sum_{1 \le k \le K} \mathbb{P}\left(\bigcap_{j=1}^{k-1}\{|Y(t_j)|_U \le c\} \cap \{|Y(t_k)|_U > c\}\right)$$

$$\le \frac{1}{c^2} \sum_{1 \le k \le K} \mathbb{E}\left(|Y(t_k)|_U^2 \chi_{\{|Y(t_k)|_U > c\}} \prod_{j=1}^{k-1} \chi_{\{|Y(t_j)|_U \le c\}}\right).$$

Note that, for any k,

$$I_k := \mathbb{E}\left(|Y(t_k)|_U^2 \chi_{\{|Y(t_k)|_U > c\}} \prod_{j=1}^{k-1} \chi_{\{|Y(t_j)|_U \le c\}}\right)$$

$$\le \mathbb{E}\left(|Y(t_k)|_U^2 \prod_{j=1}^{k-1} \chi_{\{|Y(t_j)|_U \le c\}}\right) =: \tilde{I}_k.$$

Next,

$$Y(t_k) = \tilde{S}(t_k, t_{k-1}) \int_0^{t_{k-1}} S(t_{k-1}, s) \, dM(s) + \int_{t_{k-1}}^{t_k} S(t_k, s) \, dM(s)$$

$$= \tilde{S}(t_k, t_{k-1}) Y(t_{k-1}) + \int_{t_{k-1}}^{t_k} S(t_k, s) \, dM(s).$$

Let $\chi_k := \prod_{j=1}^{k-1} \chi_{\{|Y(t_j)|_U \le c\}}$. By the martingale property of stochastic integrals,

$$\tilde{I}_k = \mathbb{E}\left(|Y(t_k)|_U^2 \chi_k\right)$$

$$= \mathbb{E}\left(\left(\left|S(t_k, t_{k-1})Y(t_{k-1}) + \int_{t_{k-1}}^{t_k} S(t_k, s) \, dM(s)\right|_U^2\right) \chi_k\right)$$

$$= \mathbb{E}\left(\left(\left|S(t_k, t_{k-1})Y(t_{k-1})\right|_U^2 + \left|\int_{t_{k-1}}^{t_k} S(t_k, s) \, dM(s)\right|_U^2\right) \chi_k\right)$$

because the expectations of the cross terms disappear. Consequently, setting $J_{k-1} := \mathbb{E}\left(|Y(t_{k-1})|_U^2 \chi_k\right)$ we obtain

$$\tilde{I}_k \le e^{2\beta(t_k - t_{k-1})} J_{k-1} + e^{2\beta(t_k - t_{k-1})} \mathbb{E}\left|M(t_k) - M(t_{k-1})\right|_U^2.$$

Since $J_{k-1} + I_{k-1} = \tilde{I}_{k-1}$, we have

$$\tilde{I}_k \le \beta_k \left(\tilde{I}_{k-1} + \alpha_k\right),$$

where $\alpha_k := \mathbb{E} |M(t_k) - M(t_{k-1})|_U^2$ and $\beta_k := e^{2\beta(t_k - t_{k-1})}$. Thus, by iteration,

$$\sum_{k=1}^{K} I_k \leq \beta_K \alpha_K + \beta_K \beta_{K-1} \alpha_{K-1} + \cdots + \beta_K \beta_{K-1} \cdots \beta_1 \alpha_1.$$

Consequently,

$$\sum_{k=1}^{K} I_k \leq e^{2\beta T} (\alpha_1 + \cdots + \alpha_K) \leq e^{2\beta T} \mathbb{E} |M(T)|_U^2,$$

and the proof is complete. $\square$

Corollary 9.19 *For any $c > 0$ and a countable set $\mathcal{Q} \subset [0, T]$,*

$$\mathbb{P} \left(\sup_{t \in \mathcal{Q}} \left| \int_0^t S(t, \sigma) \, dM(\sigma) \right|_U > c \right) \leq \frac{1}{c^2} e^{2\beta T} \mathbb{E} |M(T)|_U^2.$$

9.4.2 Regular modifications

The following result of Kotelenez (1987) is a direct consequence of Proposition 9.18.

Theorem 9.20 *Assume that $M \in \mathcal{M}^2(U)$ and that $S(t, s), t \geq s \geq 0$, is an evolution operator on $[0, T]$ satisfying the generalized contraction principle $\|S(t, s)\|_{L(U,U)} \leq e^{\beta(t-s)}$ for all $t \geq s$. Then the process*

$$Y(t) := \int_0^t S(t, s) \, dM(s), \qquad t \in [0, T],$$

has a càdlàg version in U. If, in addition, the martingale M has continuous paths then Y has a continuous version.

Proof We restrict our considerations to the first assertion, as the other can be proved in a similar way. We may assume without any loss of generality that $T = 1$. For each $\sigma \in (i/2^k, (i+1)/2^k]$, $i = 0, 1, \ldots, 2^{k-1}$, set $\sigma_k = i/2^k$ and

$$Y^k(t) = \int_0^t S(t, \sigma_k) \, dM(\sigma) = \int_{(0,t]} S(t, \sigma_k) \, dM(\sigma).$$

For $t \in (i/2^k, (i+1)/2^k]$,

$$Y^k(t) = \int_{(i/2^k, t]} S\left(t, \frac{i}{2^k}\right) dM(\sigma) + \sum_{j=1}^{i} \int_{((j-1)/2^k, j/2^k]} S\left(t, \frac{j-1}{2^k}\right) dM(\sigma)$$

$$= S\left(t, \frac{i}{2^k}\right) \left[M(t) - M\left(\frac{i}{2^k}\right) \right] + \sum_{j=1}^{i} S\left(t, \frac{j-1}{2^k}\right) \left[M\left(\frac{j}{2^k}\right) - M\left(\frac{j-1}{2^k}\right) \right].$$

It is therefore clear that Y^k is a càdlàg process on the intervals $(i/2^k, (i+1)/2^k)$. Moreover, for $s = (i+1)/2^k$,

$$Y^k(s-)$$

$$= S\left(s, \frac{i}{2^k}\right)\left[M(s-) - M\left(\frac{i}{2^k}\right)\right] + \sum_{j=1}^{i} S\left(s, \frac{j-1}{2^k}\right)\left[M\left(\frac{j}{2^k}\right) - M\left(\frac{j-1}{2^k}\right)\right],$$

$$Y^k(s+)$$

$$= S\left(s, \frac{i+1}{2^k}\right)\left[M(s) - M\left(\frac{i+1}{2^k}\right)\right] + \sum_{j=1}^{i+1} S\left(s, \frac{j-1}{2^k}\right)\left[M\left(\frac{j}{2^k}\right) - M\left(\frac{j-1}{2^k}\right)\right]$$

$$= \sum_{j=1}^{i}\left(S\left(s, \frac{j-1}{2^k}\right)\left[M\left(\frac{j}{2^k}\right) - M\left(\frac{j-1}{2^k}\right)\right]\right) + S\left(s, \frac{i}{2^k}\right)\left[M\left(\frac{i+1}{2^k}\right) - M\left(\frac{i}{2^k}\right)\right],$$

$$Y^k(s)$$

$$= S\left(s, \frac{i}{2^k}\right)\left[M(s) - M\left(\frac{i}{2^k}\right)\right] + \sum_{j=1}^{i} S\left(s, \frac{j-1}{2^k}\right)\left[M\left(\frac{j}{2^k}\right) - M\left(\frac{j-1}{2^k}\right)\right].$$

Therefore $Y^k(s) = Y^k(s+)$ and the process Y^k is càdlàg. Assume that $m \geq k$. Then, for any $\sigma \geq 0$, we have $\sigma_m \geq \sigma_k$. It follows that $S(t, \sigma_k) = S(t, \sigma_m)S(\sigma_m, \sigma_k)$ and

$$Y^m(t) - Y^k(t) = \int_0^t (S(t, \sigma_m) - S(t, \sigma_k))\,\mathrm{d}M(\sigma)$$

$$= \int_0^t S(t, \sigma_m)(I - S(\sigma_m, \sigma_k))\,\mathrm{d}M(\sigma)$$

$$= \int_0^t S(t, \sigma_m)\,\mathrm{d}\widetilde{M}(\sigma),$$

where $\widetilde{M}(s) = \int_0^s [I - S(\sigma_m, \sigma_k)]\,\mathrm{d}M(\sigma)$. Without any loss of generality we may assume that the trajectories of all processes Y^k are càdlàg. Therefore, by the corollary to the Kotelenez inequality, for all $c > 0$,

$$\mathbb{P}\left(\sup_{0 \leq t \leq T} \left|Y^m(t) - Y^k(t)\right|_U > c\right) \leq \frac{1}{c^2}\,\mathrm{e}^{2\beta T}\,\mathbb{E}\left|\widetilde{M}(T)\right|_U^2$$

$$\leq \frac{1}{c^2}\,\mathrm{e}^{2\beta T}\,\mathbb{E}\int_0^T \left\|(I - S(\sigma_m, \sigma_k))Q_M^{1/2}\right\|_{L_{(HS)}(U,U)}^2\,\mathrm{d}\langle M, M\rangle_s.$$

Recall that Q is an operator appearing in (9.5). Since

$$\left\|(I - S(\sigma_m, \sigma_k))Q^{1/2}\right\|_{L_{(HS)}(U,U)}^2 \leq \left\|(I - S(\sigma_m, \sigma_k))\right\|_{L(U,U)}^2 \left\|Q^{1/2}\right\|_{L_{(HS)}(U,U)}^2$$

$$\leq \left(1 + \mathrm{e}^{\beta t}\right)^2 \left\|Q^{1/2}\right\|_{L_{(HS)}(U,U)}^2$$

and

$$\lim_{m,k\to\infty} \left\| (I - S(\sigma_m, \sigma_k))Q^{1/2} \right\|^2_{L_{(HS)}(U,U)} = 0,$$

we have

$$\lim_{m,k\to\infty} \mathbb{P}\left(\sup_{0\le t\le T} |Y^m(t) - Y^k(t)|_U > c \right) = 0.$$

By a standard application of the Borel–Cantelli lemma, one can find a subsequence (Y^{k_l}) that $\mathbb{P}$-a.s. converges uniformly to a càdlàg modification of Y. □

9.4.3 Regularization from the dilation theorem

In this section we sketch a method for establishing the regularity of stochastic convolutions based on the dilation theorem. We follow the ideas of Hausenblas and Seidler (2006). We start with the dilation theorem.

Definition 9.21 A mapping $S\colon \mathbb{R} \mapsto L(U, U)$ is said to be *positive-definite* if, for all $t_1, \ldots, t_N \in \mathbb{R}$ and $u_1, \ldots, u_N \in U$,

$$\sum_{j,k=1}^{N} \langle S(t_j - t_k)u_j, u_k \rangle_U \ge 0.$$

The following dilation theorem, due to S. Nagy, is of central importance in the theory of one-parameter semigroups on Hilbert spaces. Its proof can be found in Davies (1980) and Nagy and Foiaş (1970).

Theorem 9.22 (Nagy) *If S is a positive-definite $L(U, U)$-valued mapping with $S(0) = I$ then there exists a Hilbert space $\hat{U}$ containing U and a group $\hat{S}\colon \mathbb{R} \mapsto \hat{U}$ such that $P\hat{S}(t) = S(t)$, $t \in \mathbb{R}$, where P is the orthogonal projection of $\hat{U}$ onto U. Moreover, if S is strongly continuous then so is $\hat{S}$.*

We would like to apply the dilation theorem to the case where S is a semigroup. The following result holds; see e.g. Davies (1980).

Theorem 9.23 *If S is a C_0-semigroup of contractions on U and $S(t)$ is defined for $t < 0$ by $S(t) = S(-t)^*$ then S is a positive-definite $L(U, U)$-valued mapping.*

Assume that $M \in \mathcal{M}^2(U)$ and that S is a C_0-semigroup of contractions on U. Let $Q = (Q_s)$ be the martingale covariance of M and let Ψ be an $L(U, U)$-valued process such that $t \mapsto \Psi(t)u$ is predictable for every $u \in U$ and

$$\mathbb{E} \int_0^t \|\Psi(s)Q_s\|^2_{L_{(HS)}(U,U)}\, \mathrm{d}s < \infty, \qquad \forall\, t \ge 0.$$

Clearly, for each t, the process $s \mapsto S(t - s)\Psi(s)$ belongs to the space $\mathcal{L}^2_{M,t}$ of admissible integrands.

Theorem 9.24 *The process*

$$Y(t) := \int_0^t S(t - s)\Psi(s)\, \mathrm{d}M(s), \qquad t \geq 0,$$

has a càdlàg modification in U and there is a constant κ depending only on S such that, for all $T < \infty$ and $\alpha \in (0, 2)$,

$$\mathbb{P}\left(\sup_{t \in [0,T]} |Y(t)|_U \geq c\right) \leq \frac{\kappa}{c^2} \mathbb{E} \int_0^T \|\Psi(s)Q_s\|^2_{L_{(HS)}(U,U)}\, \mathrm{d}s \qquad (9.22)$$

and

$$\mathbb{E} \sup_{t \in [0,T]} |Y(t)|_U^\alpha \leq \frac{2\kappa}{2 - \alpha}\left(\mathbb{E}\int_0^T \|\Psi(s)Q_s\|^2_{L_{(HS)}(U,U)}\, \mathrm{d}s\right)^{2/\alpha}. \qquad (9.23)$$

Proof By Theorems 9.22 and 9.23, there exist a Hilbert space $\hat{U}$ and a group $\hat{S}$ of linear operators on $\hat{U}$ satisfying $P\hat{S} = S$, where P is an orthogonal projection. Hence $Y(t) = P\hat{S}(t)\hat{Y}(t)$, where

$$\hat{Y}(t) = \int_0^t \hat{S}(-s)\Psi(s)\, \mathrm{d}M(s).$$

Since $s \mapsto \hat{S}(-s)\Psi(s)$ belongs to $\mathcal{L}^2_{M,T}(\hat{U})$ for every $T < \infty$, we have $\hat{Y} \in \mathcal{M}^2(\hat{U})$ and hence, by the Doob theorem, $\hat{Y}$ has a càdlàg modification in $\hat{U}$. Since $\hat{S}$ is strongly continuous on $\hat{U}$, Y also has a càdlàg modification in U. Since $\hat{S}$ is unitary and P is a projection,

$$|Y(t)|_U = |P\hat{S}(t)\hat{Y}(t)|_U \leq |\hat{Y}(t)|_{\hat{U}}.$$

Hence, by Theorems 3.41 and 8.7(ii),

$$\mathbb{P}\left(\sup_{t \in [0,T]} |Y(t)|_U \geq c\right) \leq \mathbb{P}\left(\sup_{t \in [0,T]} |\hat{Y}(t)|_{\hat{U}} \geq c\right) \leq \frac{\mathbb{E}|\hat{Y}(T)|^2_{\hat{U}}}{c^2}$$

$$\leq \frac{1}{c^2} \mathbb{E} \int_0^T \|\hat{S}(-s)\Psi(s)Q_s\|^2_{L_{(HS)}(U,\hat{U})}\, \mathrm{d}s$$

$$\leq \frac{\kappa}{c^2} \mathbb{E} \int_0^T \|\Psi(s)Q_s\|^2_{L_{(HS)}(U,U)}\, \mathrm{d}s,$$

where $\sqrt{\kappa}$ is the norm of the inclusion $j : U \mapsto \hat{U}$. This proves (9.22). Using arguments from the proof of Theorem 3.41 one can derive (9.23) from (9.22). $\qquad\square$

9.4.4 A counterexample

We show here that conditions (F) and (G) do not imply in general the existence of a càdlàg solution, even if Z is a Lévy process.

Proposition 9.25 *There exist separable Hilbert spaces H and U, a generator A of a C_0-semigroup S on H and a U-valued square integrable mean-zero compound Poisson process Π such that the coefficients of the equation*

$$dX = AX\,dt + d\Pi, \qquad X(0) = 0, \tag{9.24}$$

satisfy assumptions (F) and (G) but the solution to (9.24) does not have a càdlàg modification in H.

Proof First of all note that $\Pi \in \mathcal{M}^2(U)$. Let H and U be infinite-dimensional Hilbert spaces such that H is densely embedded into U. Assume that $(S(t),\ t \geq 0)$ is a C_0-semigroup on H with the following properties:

(i) $\int_0^T \|S(s)\|^2_{L_{(HS)}(H,H)}\,ds < \infty$ for all $T > 0$;
(ii) for each $t > 0$, $S(t)$ has a continuous extension to U as an H-valued operator;
(iii) $\lim_{t\downarrow 0} |S(t)u|_H = \infty$ for all $u \in U \setminus H$.

Let Z be a U-valued square integrable mean-zero random variable with reproducing kernel H and distribution ν. Note that $\nu(U \setminus H) = 1$. If $(\Pi(t),\ t \geq 0)$ is a compound Poisson process with jump measure ν then the solution X of (9.24) is given by the formula

$$X(t) = \int_0^t S(t-\sigma)\,d\Pi(\sigma) = \sum_{\sigma_n < t} S(t-\sigma_n)Z_n,$$

where σ_n and Z_n are respectively the jump times and jump sizes of Π. It is clear that

$$\lim_{t\downarrow\sigma_1} |X(t)|_H = \lim_{t\downarrow\sigma_1} |S(t-\sigma_1)Z_1|_H = \infty,$$

and the result follows from Proposition 3.17. Taking into account Example 7.8, one can choose $H = L^2(0,1)$, $U = W_0^{-1,2}$, S equal to the heat semigroup on $L^2(0,1)$ and $Z = \eta\delta_\xi$. $\qquad\square$

Remark 9.26 Note that if W is a cylindrical Wiener process in $H = L^2(0,1)$, that is, $\mathcal{H}_W = L^2(0,1)$ and S is the heat semigroup on $L^2(0,1)$ then (see Da Prato and Zabczyk 1992a or Chapter 11) the equation

$$dX = AX\,dt + dW$$

has a solution with continuous trajectories in $L^2(0, 1)$. However, as was shown above, the equation driven by a Poisson process having the same RKHS does not have a càdlàg version.

Remark 9.27 It is shown in Fuhrman and Röckner (2000) that for any linear equation there exists a larger Hilbert space in which the solution has a càdlàg version.

9.4.5 The case of an analytic semigroup

Assume that the semigroup S generated by $(A, D(A))$ is analytic on a Banach space $(B, |\cdot|_B)$. For the convenience of the reader we present in appendix subsection B.1.2 the definition and main properties of analytic semigroups. In particular, we recall the concept of fractional powers of $(-A)$.

Proposition 9.28 *Assume that M is a square integrable martingale with càdlàg trajectories in $D(-A)^\alpha$ for some $\alpha \in [0, 1)$. Then*

$$Y(t) := \int_0^t S(t - s) \, \mathrm{d}M(s), \qquad t \geq 0,$$

has a càdlàg version in B.

Proof Using integration by parts we obtain

$$Y(t) = M(t) - \int_0^t (-A)S(t - s)M(s) \, \mathrm{d}s := M(t) - I(t),$$

provided that

$$a(t) := \int_0^t |(-A)S(t - s)M(s)|_B \, \mathrm{d}s < \infty.$$

Let $1 < p < 1/(1 - \alpha)$ and $q := (p + 1)/p$. Then

$$a(t) = \int_0^t \left| (-A)^{1-\alpha} S(t - s) (-A)^\alpha M(s) \right|_B \, \mathrm{d}s$$

$$\leq \left(\int_0^t \left\| (-A)^{1-\alpha} S(t - s) \right\|_{L(B,B)}^p \, \mathrm{d}s \right)^{1/p}$$

$$\times \left(\int_0^t |(-A)^\alpha M(s)|_B^q \, \mathrm{d}s \right)^{1/q}.$$

By subsection B.1.2 the function $t \mapsto \|(-A)^{1-\alpha} S(t)\|_{L(B,B)}^p$ is locally integrable. As $(-A)^\alpha M$ has càdlàg trajectories in B, a is locally bounded and, consequently, $\mathbb{P}$-a.s. the process Y has locally bounded trajectories in B and by an approximation procedure it has a càdlàg modification in B. $\qquad\square$

9.5 Existence of weak solutions

In this section we prove the following result. See Section 9.2 for conditions (F), (G) and (GI).

Theorem 9.29 (1) *Assume that conditions (F) and (G) are satisfied. Then the following hold.*

(i) *For all $t_0 \geq 0$ and an $\mathcal{F}_{t_0}$-measurable square integrable random variable X_0 in H there exists a unique (up to modification) solution $X(\cdot, t_0, X_0)$ of (9.4).*

(ii) $\forall \, 0 \leq t_0 \leq T < \infty \, \exists \, L < \infty$ *such that* $\forall \, x_1, x_2 \in H$,

$$\sup_{t \in [t_0, T]} \mathbb{E} \, |X(t, t_0, x_1) - X(t, t_0, x_2)|_H^2 \leq L \, |x_1 - x_2|_H^2.$$

(iii) $\forall \, 0 \leq t_0 \leq t, \, x \in H$, *the law of $X(t, t_0, x)$ does not depend on the choice of probability space or the process Z.*

(2) *If (F) and (GI) hold and S is a generalized contraction then the solution has a càdlàg version.*

Proof of (1) Given $0 \leq t_0 \leq T < \infty$ we denote by $\mathcal{X}_T$ the space of all predictable processes $Y \colon [t_0, T] \times \Omega \mapsto H$ such that

$$\|Y\|_T := \left(\sup_{t \in [t_0, T]} \mathbb{E} \, |Y(t)|_H^2 \right)^{1/2} < \infty.$$

Given $\beta \in \mathbb{R}$ and $Y \in \mathcal{X}_T$, write

$$\|Y\|_{T,\beta} := \left(\sup_{t \in [t_0, T]} e^{-\beta t} \, \mathbb{E} \, |Y(t)|_H^2 \right)^{1/2}.$$

Clearly $\mathcal{X}_T$ with the norm $\| \cdot \|_T = \| \cdot \|_{T,0}$ is a Banach space. Moreover, the norms $\| \cdot \|_{T,\beta}, \beta \in \mathbb{R}$ are equivalent.

Note that, from (F) and (G), for all $Y \in \mathcal{X}_T$ and $t \in [t_0, T]$ the integrals

$$I_F(Y)(t) := \int_{t_0}^t S(t - s) F(Y(s)) \, ds,$$

$$J_G(Y)(t) := \int_{t_0}^t S(t - s) G(Y(s)) \, dM(s) \tag{9.25}$$

are well defined. By Proposition 3.21 they have predictable versions since they are adapted and stochastically continuous. To prove their predictability one can also approximate the processes $I_F(Y)$ and $J_G(Y)$ by $S(1/n) I_F(Y)$ and $S(1/n) J_G(Y)$. From now on, we will always take predictable versions of $I_F(Y)$ and $J_G(Y)$. By the Banach fixed-point theorem it suffices to show that, for any $T < \infty$, there are $\beta \in \mathbb{R}$ and a constant $K_{\beta,T} < 1$ such that

$$\|I_F(Y) + J_G(Y) - I_F(V) - J_G(V)\|_{T,\beta} \le K_{\beta,T} \|Y - V\|_{T,\beta}, \qquad Y, Z \in \mathcal{X}_T. \tag{9.26}$$

To this end, we fix $Y, V \in \mathcal{X}_T$. Then

$$\|I_F(Y) + J_G(Y) - I_F(V) - J_G(V)\|_{T,\beta}^2 \le 2\,\|I_F(Y) - I_F(V)\|_{T,\beta}^2$$
$$+ 2\,\|J_G(Y) - J_G(V)\|_{T,\beta}^2.$$

Next, by (F),

$$\|I_F(Y) - I_F(V)\|_{T,\beta}^2$$
$$= \sup_{t \in [t_0,T]} e^{-\beta t}\, \mathbb{E} \left| \int_{t_0}^t S(t - s)\big(F(Y(s)) - F(V(s))\big)\, ds \right|_H^2$$
$$\le \sup_{t \in [t_0,T]} e^{-\beta t}\, \mathbb{E} \left(\int_{t_0}^t a(t - s)\,|Y(s) - V(s)|_H\, ds \right)^2.$$

By the Schwarz inequality,

$$\int_{t_0}^t a(t - s)|Y(s) - V(s)|_H\, ds$$
$$\le \left(\int_{t_0}^T a(s)\, ds \right)^{1/2} \left(\int_{t_0}^T |Y(s) - V(s)|_H^2\, a(t - s)\, ds \right)^{1/2}.$$

Hence, putting $c_1 := \int_{t_0}^T a(s)\, ds$, we obtain

$$\|I_F(Y) - I_F(V)\|_{T,\beta}^2 \le c_1 \sup_{t \in [t_0,T]} e^{-\beta t} \int_{t_0}^t a(t - s)\, \mathbb{E}\,|Y(s) - V(s)|_H^2\, ds$$
$$\le c_1 \sup_{t \in [t_0,T]} e^{-\beta t} \int_{t_0}^t a(t - s)\, e^{\beta s} e^{-\beta s}\, \mathbb{E}\,|Y(s) - V(s)|_H^2\, ds$$
$$\le c_1 \|Y - V\|_{T,\beta}^2 \sup_{t \in [t_0,T]} \int_{t_0}^t a(t - s) e^{-\beta(t-s)}\, ds.$$

Thus

$$\|I_F(Y) - I_F(V)\|_{T,\beta}^2 \le C_{1,\beta,T} \|Y - V\|_{T,\beta}^2$$

with

$$C_{1,\beta,T} = \int_{t_0}^T a(s)\, ds \int_{t_0}^T a(s) e^{-\beta s}\, ds.$$

Similarly, applying the isometric identity for stochastic integrals together with condition (G),

$$\|J_G(Y) - J_G(V)\|_{T,\beta}^2$$

$$= \sup_{t \in [t_0, T]} e^{-\beta t} \, \mathbb{E} \left| \int_{t_0}^t S(t-s)\big(G(Y(s)) - G(V(s))\big) dM(s) \right|_H^2$$

$$\leq \sup_{t \in [t_0, T]} e^{-\beta t} \, \mathbb{E} \int_{t_0}^t \big\| S(t-s)\big(G(Y(s)) - G(V(s))\big) \big\|_{L_{(HS)}(\mathcal{H}, H)}^2 ds$$

$$\leq \sup_{t \in [t_0, T]} e^{-\beta t} \int_{t_0}^t b^2(t-s) \, \mathbb{E} \, |Y(s) - V(s)|_H^2 \, ds$$

$$\leq \|Y - V\|_{T,\beta}^2 \int_{t_0}^T e^{-\beta s} \, b^2(s) \, ds.$$

Thus (9.26) follows with

$$K_{\beta,T} = \left(2\,C_{1,\beta,T} + 2 \int_{t_0}^T e^{-\beta s}\, b^2(s)\, ds \right)^{1/2}$$

and for the existence it suffices to observe that, for sufficiently large β, $K_{\beta,T} < 1$, which concludes the proof of (i).

In order to show (ii), note that, with the notation as above,

$$\|X(\cdot, t_0, x_1) - X(\cdot, t_0, x_2)\|_{\beta,T}^2$$

$$\leq 2 \sup_{t \in [t_0, T]} e^{-\beta t} |S(t)(x_1 - x_2)|_H^2$$

$$+ 2 \big\| I_F(X(\cdot, t_0, x_1)) + J_G(X(\cdot, t_0, x_1)) - I_F(X(\cdot, x_2)) - J_G(\cdot, x_2)) \big\|_{T,\beta}^2$$

$$\leq 2 \sup_{t \in [t_0, T]} e^{-\beta t} \|S(t)\|_{L(H,H)}^2 |x_1 - x_2|_H^2$$

$$+ 2K_{\beta,T}^2 \|X(\cdot, t_0, x_1) - X(\cdot, t_0, x_2)\|_{T,\beta}^2.$$

Choosing β large enough, we obtain $2K_{\beta,T}^2 < 1/2$. Then

$$\|X(\cdot, t_0, x_1) - X(\cdot, t_0, x_2)\|_{\beta,T}^2 \leq 4 \sup_{t \in [t_0, T]} e^{-\beta t} \|S(t)\|_{L(H,H)}^2 |x_1 - x_2|_H^2,$$

which gives the desired conclusion.

In order to show (iii) note, by the proof of (i), that the solution is a limit in $L^2(\Omega, \mathcal{F}, \mathbb{P}; H)$ of the sequences $(X_n(\cdot, t_0, x))$, where $X_0(t, t_0, x) = x$ and

$$X_{n+1}(t, t_0, x) = S(t - t_0)x + \int_{t_0}^t S(t-s)F(X_n(s, t_0, x))\, ds$$

$$+ \int_{t_0}^t S(t-s)G(X(s, t_0, x))\, dM(s)$$

for $n \geq 0$. Thus the claim follows from the easily verified fact that if the joint laws of (Ψ, M) and $(\tilde{\psi}, \tilde{M})$ are the same then the joint laws of

$$\left(\int_{t_0}^{t} \Psi(s) \, dM(s), \quad M(t), \, t \geq 0 \right), \quad \left(\int_{t_0}^{t} \tilde{\psi}(s) \, d\tilde{M}(s), \quad \tilde{M}(t), \, t \geq 0 \right)$$

are also the same. $\square$

Proof of (2)   Assume that X is a solution to (9.4) and define the process

$$N(t) := \int_0^t G(X(s)) \, dM(s), \qquad t \geq 0.$$

Then

$$\mathbb{E} \, |N(t)|_H^2 = \mathbb{E} \left| \int_0^t G(X(s)) \, dM(s) \right|_H^2$$

$$\leq \mathbb{E} \int_0^t \|G(X(s))\|_{L_{(HS)}(\mathcal{H}, H)}^2 \, ds \leq \mathbb{E} \int_0^t b^2(s) \big(1 + |X(s)|_H \big)^2 \, ds$$

$$\leq 2 \int_0^t b^2(s) \big(1 + \mathbb{E} \, |X(s)|_H^2 \big) \, ds < \infty,$$

by part (1) of the theorem. Therefore N is an H-valued square integrable martingale and the regularization theorem 9.24 implies that the process

$$Y(t) = \int_0^t S(t - s) \, dN(s) = \int_0^t S(t - s) G(X(s)) \, dM(s),$$

has a càdlàg version. $\square$

From now on we will write $X(t, x)$ instead of $X(t, 0, x)$.

9.6 Markov property

In this section we will be concerned with the *Markov property* of solutions to (9.4); see Section 3.4. Given $x \in H, 0 \leq s \leq t < \infty$, we denote by $X(t, s, x)$ the value at time t of the solution to (9.4) starting at time s from x. Define

$$P_{s,t} \varphi(x) := \mathbb{E} \, \varphi(X(t, s, x)), \qquad t \geq s \geq 0, \, \varphi \in B_b(H), \, x \in H, \qquad (9.27)$$

and, for all $t \geq s \geq 0, x \in H$ and $\Gamma \in \mathcal{B}(H)$,

$$P(s, x, t, \Gamma) := P_{s,t} \chi_{\Gamma}(x) = \mathbb{P} \, (X(t, s, x) \in \Gamma).$$

We call $P_{s,t}, 0 \leq s \leq t$ the *transition semigroup* and $P(s, x, t, \Gamma), 0 \leq s \leq t, x \in H, \Gamma \in \mathcal{B}(H)$, the *transition functions* corresponding to (9.4).

Theorem 9.30 *Assume that (F) and (G) hold and in addition that M is a Lévy process. Then, for all $0 \le s \le u \le t$, $\varphi \in B_b(H)$ and any square integrable $\mathcal{F}_s$-measurable random variable ζ,*

$$\mathbb{E}\big(\varphi(X(t, s, \zeta))\big|\mathcal{F}_u\big) = P_{u,t}\varphi(X(u, s, \zeta)), \qquad \mathbb{P}\text{-}a.s.$$

Proof By the uniqueness of solutions, $X(t, s, \zeta) = X(t, u, X(u, s, \zeta))$. It is therefore enough to show that, for any $\mathcal{F}_u$-measurable square integrable random variable η,

$$\mathbb{E}\big(\varphi(X(t, u, \eta))\big|\mathcal{F}_u\big) = P_{u,t}\varphi(\eta), \qquad \mathbb{P}\text{-}a.s. \tag{9.28}$$

Note that it is sufficient to show (9.28) for any $\varphi \in C_b(H)$. Indeed, if this is the case then it holds in particular for $\varphi = \chi_\Gamma$, where Γ is a closed subset of H. For any probability measure μ on a separable metric space E,

$$\mu(B) = \sup_{\Gamma = \bar{\Gamma} \subset B} \mu(\Gamma), \qquad B \in \mathcal{B}(E).$$

It follows that (9.28) holds for the characteristic function of any measurable set in H and, consequently, by a simple approximation argument, for any bounded measurable φ. Assume now that η is simple, that is, it takes only a finite number of values, so that we can write $\eta = \sum_{j=1}^N x_j \chi_{\Gamma_j}$ where

$$\{\Gamma_1, \ldots, \Gamma_N\} \subset \mathcal{F}_u, \qquad \bigcup_{j=1}^N \Gamma_j = \Omega, \qquad \Gamma_j \cap \Gamma_k = \emptyset \text{ for } j \ne k,$$

and $\{x_1, \ldots, x_N\} \subset H$. Then

$$X(t, u, \eta) = \sum_{j=1}^N X(t, u, x_j)\chi_{\Gamma_j}$$

and consequently, since $X(t, u, x_j)$ is independent of $\mathcal{F}_u$ and the functions χ_{Γ_j} are $\mathcal{F}_u$-measurable, we have

$$\mathbb{E}\big(\varphi(X(t, u, \eta))\big|\mathcal{F}_u\big) = \sum_{j=1}^N \mathbb{E}\left(\varphi(X(t, u, x_j))\chi_{\Gamma_j}\big|\mathcal{F}_u\right)$$

$$= \sum_{j=1}^N \left(\mathbb{E}\,\varphi(X(t, u, x_j))\right)\chi_{\Gamma_j}$$

$$= \sum_{j=1}^N P_{u,t}\varphi(x_j)\chi_{\Gamma_j} = P_{u,t}\varphi(\eta).$$

Now, if η is a square integrable $\mathcal{F}_u$-measurable random variable then there is a sequence (η_n) of simple random variables such that $\mathbb{E}\,|\eta - \eta_n|_H \to 0$ as $n \to \infty$.

Then

$$\mathbb{E}\big(\varphi(X(t, u, \eta_n)\big|\mathcal{F}_u\big) = P_{u,t}\varphi(\eta_n).$$

Since by Theorem 9.29 the solution depends continuously on the initial data and since φ is bounded and continuous, we may pass to the limit and obtain the desired conclusion. □

We have the following consequence of Theorem 9.30.

Proposition 9.31   *For arbitrary* $\varphi \in B_b(H)$, $\Gamma \in \mathcal{B}(H)$, $0 \leq s \leq u \leq t < \infty$ *and* $x \in H$,

$$P_{s,u}\big(P_{u,t}\varphi\big)(x) = P_{s,t}\varphi(x) \tag{9.29}$$

and

$$P(s, x, t, \Gamma) = \int_H P(s, x, u, \mathrm{d}y)P(u, y, t, \Gamma). \tag{9.30}$$

Proof   Clearly (9.30) follows from (9.29) by taking $\varphi = \chi_\Gamma$. To show (9.29) note that

$$P_{s,t}\varphi(x) = \mathbb{E}\,\varphi(X(t, s, x)) = \mathbb{E}\,\mathbb{E}\big(\varphi(X(t, s, , x))\big|\mathcal{F}_u\big)$$
$$= \mathbb{E}\,P_{u,t}\varphi(X(u, s, x)) = P_{s,u}\big(P_{u,t}\varphi\big)(x).$$

□

Since Z has stationary increments, we have the following result.

Proposition 9.32   *For all* $0 \leq s \leq t$, $P_{s,t} = P_{0,t-s}$. *In particular,*

$$P(s, x, t, \Gamma) = P(0, x, t - s, \Gamma), \qquad \forall x \in H, \ \forall \Gamma \in \mathcal{B}(H).$$

Proof   It is sufficient to show that for all $t \geq 0$ and $x \in H$ the processes $(X(h, 0, x), h \geq 0)$ and $(X(t + h, t, x), h \geq 0)$ have the same distribution. To see this note that

$$X(t + h, t, x) = S(h)x + \int_t^{t+h} S(t + h - s)F(X(s, t, x))\,\mathrm{d}s$$

$$+ \int_t^{t+h} S(t + h - s)G(X(s, t, x))\,\mathrm{d}M(s)$$

$$= S(h)x + \int_0^h S(h - s)F(X(t + s, t, x))\,\mathrm{d}s$$

$$+ \int_0^h S(h - s)G(X(t + s, t, x))\,\mathrm{d}M^t(s),$$

where $M^t(s) = M(t + s) - M(t)$. Clearly M^t and M have the same law. Thus, by Theorem 9.29(iii), $X(t + h, t, x)$ and $X(h, 0, x)$ have the same law. □

From now on, we write $X(t, x)$ instead of $X(t, 0, x)$, P_t instead of $P_{0,t}$ and $\mathbb{P}(t, x, \Gamma)$ instead of $P(t, x, 0, \Gamma)$.

Remark 9.33 By Theorem 9.30, $X(\cdot, x), x \in H$ is a family of Markov processes on the state space H with respect to the filtration $(\mathcal{F}_t)$. The corresponding transition semigroup is (P_t) and the transition probabilities are $P(t, x, \Gamma)$, $x \in H$, $t \geq 0$, $\Gamma \in \mathcal{B}(H)$. By Theorem 9.29(ii), the semigroup (P_t) is *Feller*, that is, for any $t \geq 0$, P_t transforms $C_b(H)$ into $C_b(H)$. In other words, *(9.4) defines a homogeneous Markov family satisfying the Feller property* (a *Feller family*).

9.7 Equations with general Lévy processes

In this section, we discuss the existence of a solution to the equation obtained from (9.4) by adding a Poissonian part that is not necessarily square integrable. Namely, we consider the problem

$$\mathrm{d}X = (AX + F(X))\,\mathrm{d}t + G(X)\,\mathrm{d}M + R(X)\,\mathrm{d}P, \qquad X(t_0) = X_0, \qquad (9.31)$$

where $(A, D(A))$, M, F and G are as in (9.4). We assume that P is a compound Poisson process on a Hilbert space V with (finite) Lévy measure μ and that the $R(x), x \in H$, are linear operators, usually unbounded, from V to H.

As in subsection 8.6.2, we introduce a sequence of stopping times

$$\tau_m := \inf\{t \geq 0 \colon P(t) - P(t-) \notin V_m\}, \qquad m \in \mathbb{N},$$

where (V_m) is an increasing sequence of bounded measurable subsets of V, such that $\bigcup V_n = V$. Note that $\tau_m \uparrow \infty$. Given m, define

$$P_m(t) := P(t \wedge \tau_m) = \sum_{s \leq t} \chi_{V_m}(P(s) - P(s-))(P(s) - P(s-)).$$

Then P_m is a compound Poisson process with Lévy measure μ_m, which is the restriction of μ to V_m. Let

$$u_m := \int_V z\mu_m(\mathrm{d}z) = \int_{V_m} z\mu(\mathrm{d}z).$$

Then $(M_m(t) := P_m(t) - tu_m, t \geq 0)$, $m \in \mathbb{N}$, are square integrable mean-zero Lévy processes. Moreover, each X_m has covariance

$$Q_m = \int_{V_m} z \otimes z\mu(\mathrm{d}z),$$

that is,

$$\langle Q_m u, v \rangle_V = \int_{V_m} \langle z, u \rangle_V \langle z, u \rangle_V \mu(dz), \qquad u, v \in V.$$

Let $\mathcal{H}_m = Q_m^{1/2}(V)$. Note that $(\mathcal{H}_m)$ is an increasing sequence.

We will assume that the nonlinear map R satisfies the following two conditions.

(R1) $D(R)$ is dense in H, and for each m there is a function
$b_m : (0, \infty) \mapsto (0, \infty)$ satisfying $\int_0^T b_m^2(t) \, dt < \infty$ for $T < \infty$ such that,
for all $t > 0$ and $x, y \in D(R)$,

$$\|S(t)R(x)\|_{L_{(HS)}(\mathcal{H}_m, H)} \le b_m(t)(1 + |x|_H),$$
$$\|S(t)(R(x) - R(y))\|_{L_{(HS)}(\mathcal{H}_m, H)} \le b_m(t) |x - y|_H.$$

(R2) For each m, there is a function $a_m : (0, \infty) \mapsto (0, \infty)$ satisfying

$$\int_0^T a_m(t) \, dt < \infty, \qquad \forall T < \infty,$$

such that, for all $t > 0$ and $x, y \in D(R)$,

$$|S(t)(R(x)u_m)|_H \le a_m(t)(1 + |x|_H),$$
$$\left|S(t)\big((R(x) - R(y))\, u_m\big)\right|_H \le a_m(t) |x - y|_H.$$

Given m, we consider the problem

$$\begin{aligned} dX_m &= (AX_m + F(X_m) + R(X_m)u_m) \, dt \\ &\quad + G(X_m) \, dM + R(X_m) \, dM_m \\ &= (AX_m + F(X_m)) \, dt + G(X_m) \, dM + R(X_m) \, dP_m, \qquad (9.32) \\ X_m(t_0) &= X_0. \end{aligned}$$

Note that the coefficients $F(\cdot) + R(\cdot)u_m$ and G, R satisfy assumptions (F) and (G) and hence that for each m there is a unique solution X_m to (9.32) satisfying $\sup_{t \in [0,T]} \mathbb{E} |X_m(t)|_H^2 < \infty$ for $T < \infty$.

Theorem 9.34 *For every $t \in [0, T]$ and all $m \le n$, $X_m(t) = X_n(t)$ $\mathbb{P}$-a.s. on $\{t \le \tau_m\}$. Moreover the process X given by $X(t) = X_m(t)$ for $t \le \tau_m$ is a weak solution to (9.31).*

Proof   We have

$$X_m(t) - X_n(t) = \int_0^t S(t-s)\big(F(X_m(s)) - F(X_n(s))\big)\,\mathrm{d}s$$

$$+ \int_0^t S(t-s)\big(G(X_m(s)) - G(X_n(s))\big)\,\mathrm{d}M(s)$$

$$+ \int_0^t S(t-s)\big(R(X_m(s)) - R(X_n(s))\big)\,\mathrm{d}M_m(s)$$

$$+ \int_0^t S(t-s)R(X_n(s))\,\mathrm{d}(M_m(s) - M_n(s))$$

$$+ \int_0^t S(t-s)\big(R(X_m(s))u_m - R(X_n(s))u_n\big)\,\mathrm{d}s.$$

Since on the set $\{t \le \tau_m\}$ we have $M_m(s) - M_n(s) = -(u_m - u_n)s$, $s \le t$, Proposition 8.11 yields

$$(X_m(t) - X_n(t))\chi_{\{t\le\tau_m\}}$$

$$= \int_0^t S(t-s)\big(F(X_m(s)) - F(X_n(s))\big)\,\mathrm{d}s\,\chi_{\{t\le\tau_m\}}$$

$$+ \int_0^t S(t-s)\big(G(X_m(s)) - G(X_n(s))\big)\,\mathrm{d}M(s)\,\chi_{\{t\le\tau_m\}}$$

$$+ \int_0^t S(t-s)\big(R(X_m(s)) - R(X_n(s))\big)\,\mathrm{d}M_m(s)\,\chi_{\{t\le\tau_m\}}$$

$$+ \int_0^t S(t-s)\big(R(X_m(s)) - R(X_n(s))\big)u_m\,\mathrm{d}s\,\chi_{\{t\le\tau_m\}}.$$

If $s \le t$ and $t \le \tau_m$ then $s \le \tau_m$. Consequently,

$$\left| \chi_{\{t\le\tau_m\}} \int_0^t S(t-s)\Psi(s)\,\mathrm{d}s \right|_H \le \left| \int_0^t S(t-s)\Psi(s)\chi_{\{s\le\tau_m\}}\,\mathrm{d}s \right|_H$$

and (see Proposition 8.11)

$$\left| \chi_{\{t\le\tau_m\}} \int_0^t S(t-s)\Psi(s)\,\mathrm{d}M(s) \right|_H \le \left| \int_0^t S(t-s)\Psi(s)\chi_{\{s\le\tau_m\}}\,\mathrm{d}M(s) \right|_H.$$

Thus

$$|X_m(t) - X_n(t)|_H \, \chi_{\{t \le \tau_m\}}$$

$$\le \int_0^t \left| S(t-s) \big(F(X_m(s)) - F(X_n(s)) \big) \right|_H \chi_{\{s \le \tau_m\}} \, \mathrm{d}s$$

$$+ \left| \int_0^t S(t-s) \big(G(X_m(s)) - G(X_n(s)) \big) \chi_{\{s \le \tau_m\}} \, \mathrm{d}M(s) \right|_H$$

$$+ \left| \int_0^t S(t-s) \big(R(X_m(s)) - R(X_n(s)) \big) \chi_{\{s \le \tau_m\}} \, \mathrm{d}M_m(s) \right|_H$$

$$+ \int_0^t \left| S(t-s) \big(R(X_m(s)) - R(X_n(s)) \big) u_m \right|_H \chi_{\{s \le \tau_m\}} \, \mathrm{d}s,$$

yielding

$$\mathbb{E} \, |X_m(t) - X_n(t)|_H^2 \, \chi_{\{t \le \tau_m\}}$$

$$\le 4 \, \mathbb{E} \left(\int_0^t \left| S(t-s) \big(F(X_m(s)) - F(X_n(s)) \big) \right|_H \chi_{\{s \le \tau_m\}} \, \mathrm{d}s \right)^2$$

$$+ 4 \, \mathbb{E} \left| \int_0^t S(t-s) \big(G(X_m(s)) - G(X_n(s)) \big) \chi_{\{s \le \tau_m\}} \, \mathrm{d}M(s) \right|_H^2$$

$$+ 4 \, \mathbb{E} \left| \int_0^t S(t-s) \big(R(X_m(s)) - R(X_n(s)) \big) \chi_{\{s \le \tau_m\}} \, \mathrm{d}M_m(s) \right|_H^2$$

$$+ 4 \, \mathbb{E} \left(\int_0^t \left| S(t-s) \big(R(X_m(s)) - R(X_n(s)) \big) u_m \right|_H \chi_{\{s \le \tau_m\}} \, \mathrm{d}s \right)^2.$$

Let $Y(s) := \mathbb{E} \, |X_m(s) - X_n(s)|_H^2 \, \chi_{\{s \le \tau_m\}}$. Using assumptions (F), (G) and (H1), (H2) we obtain, for $t \le T$,

$$Y(t) \le 4 \, \mathbb{E} \left(\int_0^t a(t-s) |X_m(s) - X_n(s)|_H \chi_{\{s \le \tau_m\}} \, \mathrm{d}s \right)^2$$

$$+ 4 \, \mathbb{E} \left(\int_0^t a_m(t-s) |X_m(s) - X_n(s)|_H \chi_{\{s \le \tau_m\}} \, \mathrm{d}s \right)^2$$

$$+ 4 \int_0^t \big(b^2(t-s) + b_m^2(t-s) \big) Y(s) \, \mathrm{d}s$$

$$\le C \int_0^t c(t-s) Y(s) \, \mathrm{d}s,$$

where

$$C = 4 \int_0^T \big(1 + a(t) + a_m(t) \big) \, \mathrm{d}t$$

and $c(t) = a(t) + a_m(t) + b^2(t) + b_m^2(t)$. Consequently, for any $\beta > 0$,

$$\sup_{t \leq T} e^{-\beta t} Y(t) \leq C_\beta \sup_{t \leq T} e^{-\beta t} Y(t),$$

where $C_\beta := C \int_0^T c(t) e^{-\beta t} \, dt$. Since $\int_0^T c(t) \, dt < \infty$, $C_\beta < 1$ for sufficiently large β and consequently $Y(t) = 0$, $t \in [0, T]$, as required. Since the processes X_m are weak solutions of (9.31) on the intervals $[0, T \wedge \tau_m]$, X is also a weak solution of (9.31). $\qquad\square$

The following theorem may now be proved in a similar way to Theorem 9.30.

Theorem 9.35 *Assume that M is a Lévy process and that P is independent of M. Then the solution X to (9.31) is a Markov process on H.*

9.8 Generators and a martingale problem

Let E be a Polish space, and let (P_t) be a transition semigroup defined on $B_b(E)$.

Definition 9.36 We say that $\phi \in C_b(E)$ belongs to the *domain $D(\mathcal{A})$ of the weak generator $\mathcal{A}$ of (P_t)* if for every $x \in E$ the limit

$$\lim_{t \downarrow 0} \frac{P_t \phi(x) - \phi(x)}{t} =: \mathcal{A}\phi(x) \tag{9.33}$$

exists, $\mathcal{A}\phi \in C_b(E)$ and

$$P_t \phi(x) = \phi(x) + \int_0^t P_s \mathcal{A}\phi(x) \, ds, \qquad x \in E. \tag{9.34}$$

Assume that (P_t) is the transition semigroup of the solution X of the equation

$$dX = (AX + F(X)) \, dt + G(X) \, dZ, \qquad X(0) = x \in H, \tag{9.35}$$

where A generates a C_0-semigroup on a Hilbert space H and Z is a square integrable Lévy martingale taking values in a Hilbert space U. Then the form of the generator $\mathcal{A}$ is well known in many specific cases. For instance, if Z is a Wiener process with covariance Q then, under appropriate conditions (see e.g. Da Prato and Zabczyk 1992a),

$$\mathcal{A}\phi(x) = \langle F(x), D\phi(x) \rangle_H + \langle x, A^* D\phi(x) \rangle_H$$
$$+ \tfrac{1}{2} \operatorname{Tr} G(x) Q^{1/2} D^2 \phi(x) \big(G(x) Q^{1/2} \big)^*.$$

Assume now that Z is a square integrable Lévy martingale in U of the form

$$Z(t) = \int_0^t \int_U y \widehat{\pi}(ds, dy), \tag{9.36}$$

where $\widehat{\pi}$ is a compensated Poisson random measure. One can conjecture that

$$\mathcal{A}\phi(x) = \langle F(x), D\phi(x)\rangle_H + \langle x, A^*D\phi(x)\rangle_H + \mathcal{A}_0\phi(x), \qquad (9.37)$$

where

$$\mathcal{A}_0\phi(x) = \int_U \big(\phi(x + G(x)y) - \phi(x) - \langle D\phi(x), G(x)y\rangle_H\big)\nu(dy),$$

and ν is the intensity of π. As an illustration of the applicability of Itô's lemma, we will show (9.37) for the so-called *cylinder functions*, that is, for functions ϕ of the form

$$\phi(x) = f(\langle h_1, x\rangle_H, \dots, \langle h_n, x\rangle_H), \qquad x \in H, \qquad (9.38)$$

where $f: \mathbb{R}^n \mapsto \mathbb{R}$ and $h_i \in D(A^*), i = 1, \dots, n$. The general case can be treated as in subsection 5.2.4. Note that if ϕ is given by (9.38) then

$$A^*D\phi(x) = \sum_{j=1}^n \frac{\partial f}{\partial \xi_j}(\langle h_1, x\rangle_H, \dots, \langle h_n, x\rangle_H)\, A^*h_j.$$

Proposition 9.37 *Assume that*

(i) $F: H \mapsto H$ *and* $G: H \mapsto L(U, H)$ *are Lipschitz and G is bounded,*
(ii) $f \in C_b^2(\mathbb{R}^n)$,
(iii) *Z is a square integrable U-valued martingale given by (9.36).*

Then ϕ given by (9.38) is in the domain $D(\mathcal{A})$ of the weak generator of the transition semigroup of the solution of (9.35) and $\mathcal{A}\phi$ is given by (9.37).

Proof To simplify the notation we assume that $n = 1$. We will show only that the limit in (9.33) exists and is equal to (9.37). If $h \in D(A^*)$ and X is the weak solution of (9.35) then

$$\langle h, X(t)\rangle_H = \langle h, x\rangle_H + \int_0^t \langle A^*h, X(s)\rangle_H\, ds$$

$$+ \int_0^t \langle h, F(X(s))\rangle_H\, ds + J_1(t), \qquad (9.39)$$

where

$$J_1(t) := \int_0^t \int_U \langle h, G(X(s-))y\rangle_H\, \widehat{\pi}(ds, dy).$$

Let $\xi(t) = \langle h, X(t)\rangle_H, \ t \geq 0$. By Itô's formula (see Appendix D),

$$\phi(X(t)) = f(\xi(t)) = f(\xi(0)) + \int_0^t f'(\xi(s-))\, d\xi(s) + J_2(t),$$

where

$$J_2(t) := \int_0^t \int_U \left(f(\xi(s-) + z) - f(\xi(s-)) - f'(\xi(s-))z \right) \pi_\xi(ds, dz)$$

and π_ξ is the Poisson jump measure corresponding to ξ. By (9.39), for non-negative predictable random fields ψ,

$$\int_0^t \int_R \psi(s, z)\pi_\xi(ds, dz) = \int_0^t \int_U \psi\left(s, \langle h, G(X(s-))y\rangle_H\right)\pi(ds, dy).$$

Thus

$$\phi(X(t)) - \phi(x) = \int_0^t f'(\xi(s-))\left(\langle A^*h, X(s)\rangle_H + \langle h, F(X(s))\rangle_H\right) ds$$
$$+ \int_0^t \int_U \Phi(s, y)\pi(ds, dy),$$

where

$$\Phi(s, y) := f\left(\xi(s-) + \langle h, G(X(s-))y\rangle_H\right) - f(\xi(s-))$$
$$- f'(\xi(s-))\langle h, G(X(s-))y\rangle_H.$$

We have

$$\int_0^t \int_U \Phi(s, y)\pi(ds, dy) = \int_0^t \int_U \Phi(s, y)\widehat{\pi}(ds, dy) + \int_0^t \int_U \Phi(s, y)ds\, \nu(dy).$$

Consequently,

$$\phi(X(t)) - \phi(x) = M(t) + \int_0^t \int_U \Phi(s, y) ds\, \nu(dy)$$
$$+ \int_0^t f'(\langle h, X(t)\rangle)\left(\langle A^*h, X(s)\rangle_H + \langle h, F(X(s))\rangle_H\right) ds$$

where M is a martingale, and hence

$$P_t\phi(x) - \phi(x) = \int_0^t \mathbb{E}\left(f'(\langle h, X(t)\rangle)\left(\langle A^*h, X(s)\rangle_H + \langle h, F(X(s))\rangle_H\right)\right) ds$$
$$+ \int_0^t \mathbb{E} \int_U \Phi(s, y)\nu(dy) ds,$$

which gives the desired conclusion. $\qquad\square$

9.8.1 Martingale problem

Let us start with the following general observation.

Proposition 9.38 *Assume that (P_t) is the transition semigroup of a Markov process X with respect to a filtration $(\mathcal{F}_t)$. If ϕ belongs to the domain of the weak generator $\mathcal{A}$ of (P_t) then the process*

$$M(t) := \phi(X(t)) - \int_0^t \mathcal{A}\phi(X(s))\,ds, \qquad t \ge 0, \qquad (9.40)$$

is a martingale.

Proof By Definition 3.22, for $t \ge s \ge 0$,

$$\mathbb{E}\,(M(t) - M(s)|\mathcal{F}_s) = \mathbb{E}\big(\phi(X(t)) - \phi(X(s))\big|\mathcal{F}_s\big) - \int_s^t \mathbb{E}\big(\mathcal{A}\phi(X(u))\big|\mathcal{F}_s\big)\,ds$$

$$= P_{t-s}\phi(X(s)) - \phi(X(s)) - \int_0^t P_{u-s}\mathcal{A}\phi(X(s))\,ds.$$

Consequently, the result follows from (9.34). □

Now let $(\mathcal{A}, D(\mathcal{A}))$ be a linear operator on $C_b(E)$, and let $\Lambda \subset D(\mathcal{A})$.

Definition 9.39 We say that a filtered probability space $(\Omega, \mathcal{F}, (\mathcal{F}_t), \mathbb{P})$ and an E-valued $(\mathcal{F}_t)$-adapted process X is a solution to the *martingale problem* $(\mathcal{A}, \Lambda)$ if X has measurable trajectories and, for every $\phi \in \Lambda$, the processes M defined by (9.38) are martingales with respect to $(\mathcal{F}_t)$.

Assume that $\mathcal{A}$ is given by (9.37). Then the solutions of the corresponding martingale problem are, in a sense, *generalized solutions* of (9.35) and exist under assumptions weaker than Lipschitz-type conditions. It is clear that the existence and uniqueness of the solution of any martingale problem depend on the family Λ. For small Λ the existence is easier to establish whereas for large Λ the uniqueness is easier to prove. An example of an interesting martingale problem is discussed in the following subsection.

9.8.2 Mytnik's equation

The following equation was studied in Mytnik (2002):

$$dX(t, \xi) = \Delta X(t, \xi)\,dt + (X(t-, \xi))^\delta\,dZ_\beta(t, \xi), \qquad \xi \in \mathbb{R}^d. \qquad (9.41)$$

Here Z_β, $\beta \in (1, 2)$, is the cylindrical process introduced in Example 7.26. Let us recall that

$$Z_\beta(t, d\xi) = \int_0^t \int_\mathbb{R} \sigma \widehat{\pi}(ds, d\xi, d\sigma),$$

where $\widehat{\pi}$ is the compensated Poisson measure on $[0, +\infty) \times \mathbb{R}^d \times [0, +\infty)$ with intensity $ds\, d\xi\, \sigma^{-(1+\beta)}\, d\sigma$. Assume that X is a weak solution of (9.41) and that $h \colon \mathbb{R}^d \mapsto \mathbb{R}$ and its first and second derivatives belong to any $L^p(\mathbb{R}^d)$, $p \geq 1$, spaces. Let $H = L^2(\mathbb{R}^d)$. Then

$$\langle h, X(t) \rangle_H - \langle h, x \rangle_H$$
$$= \int_0^t \langle \Delta h, X(s) \rangle_H \, ds + \int_0^t \int_{\mathbb{R}^d} \int_0^{+\infty} (X(s-, \xi))^\delta h(\xi) \sigma \widehat{\pi}(ds, d\xi, d\sigma).$$

If $f(z) = e^{-z}$, $z \geq 0$, then, applying Itô's formula to $f(\langle h, X(t) \rangle_H, t \geq 0$, in the same way as in the proof of Proposition 9.37, we find that the process

$$M(t) = e^{-\langle h, X(t) \rangle_H} - \int_0^t e^{-\langle h, X(s-) \rangle_H}$$
$$\times \int_{\mathbb{R}^d} \left[\Delta h(\xi) X(s, \xi) - c_\beta \big((X(s-, \xi))^\delta h(\xi) \big)^\beta \right] d\xi \, ds,$$
$$c_\beta = \frac{\Gamma(2-\beta)}{\beta(\beta-1)},$$

is a martingale. In Mytnik (2002) it is proved that under appropriate conditions on δ, β and d there exists a process X for which M is a martingale. One can arrive at the same definition of M deriving, formally, the formula for the generator of the solution of (9.41).

10

Equations with non-Lipschitz coefficients

In the present chapter, the Lipschitz-type conditions imposed on the coefficients F and G are relaxed. However, only problems with additive noise,

$$dX = (AX + F(X))\,dt + G\,dZ, \qquad X(0) = x \in H, \tag{10.1}$$

will be treated. In (10.1), Z is a square integrable mean-zero Lévy process with RKHS embedded into a Hilbert space U, G is a linear operator from U to H and F is an operator on H that may be non-linear and is not necessarily defined everywhere. The existence and uniqueness of solutions to (10.1) are proven under the condition that A and F are dissipative in type and that the solution to the linear problem

$$dY = AY\,dt + G\,dZ, \qquad Y(0) = 0,$$

is a càdlàg process in the domain of the map F. It is worth noting that the solution will be càdlàg as well.

10.1 Dissipative mappings

We start by recalling basic results on dissipative operators. For the proofs, we refer to subsection 5.5.1 in Da Prato and Zabczyk (1996) and Appendix D in Da Prato and Zabczyk (1992a). Let $(B, |\cdot|_B)$ be a separable Banach space. The *subdifferential* $\partial|x|_B$ of $|\cdot|_B$ at x is defined by the formula

$$\partial|x|_B := \left\{ x^* \in B^* : |x + y|_B - |x|_B \geq \langle x^*, y \rangle, \ \forall\, y \in B \right\}.$$

Proposition 10.1 *Let $u: [0, T] \mapsto B$ be a continuous function whose left derivative*

$$\frac{d^- u}{dt}(t_0) := \lim_{\substack{\varepsilon \to 0 \\ \varepsilon < 0}} \frac{u(t_0 + \varepsilon) - u(t_0)}{\varepsilon}$$

179

exists at $t_0 \in [0, T]$. Then the function $\gamma(t) = |u(t)|_B$, $t \in [0, T]$, is left differentiable at t_0 and

$$\frac{d^- \gamma}{dt}(t_0) = \min \left\{ \left\langle x^*, \frac{d^- u}{dt}(t_0) \right\rangle : x^* \in \partial |u(t_0)|_B \right\}.$$

Definition 10.2 A mapping $F \colon D(F) \subset B \mapsto B$ is said to be *dissipative* if for any $x, y \in D(F)$ there exists a $z^* \in \partial |x - y|_B$ such that $\langle z^*, F(x) - F(y) \rangle \leq 0$.

Example 10.3 If $B = H$ is a Hilbert space then $\partial |x|_H = \{x/|x|_H\}$ for $x \neq 0$, and F is dissipative if and only if $\langle x - y, F(x) - F(y) \rangle_H \leq 0$ for all $x, y \in D(F)$.

A dissipative mapping F is called *maximal dissipative* or *m-dissipative* if the image of $\lambda I - F$ is equal to the whole space B for some $\lambda > 0$ (and so for any $\lambda > 0$). Finally, a mapping F is called *almost m-dissipative* if $F + \eta I$ is m-dissipative for some $\eta \in \mathbb{R}$.

Assume that F is an m-dissipative mapping. Then its *Yosida approximations* F_α, $\alpha > 0$, are defined by

$$F_\alpha(x) = F(J_\alpha(x)) = \frac{1}{\alpha}(J_\alpha(x) - x), \qquad x \in B,$$

where $J_\alpha(x) = (I - \alpha F)^{-1}(x)$, $x \in B$. The basic properties of the operators F_α, J_α, $\alpha > 0$, are listed in the following two propositions; see Appendix D in Da Prato and Zabczyk (1992a).

Proposition 10.4 *Let $F \colon D(F) \mapsto B$ be an m-dissipative mapping on B. Then the following hold.*

(i) *For all $\alpha > 0$ and $x, y \in B$, $|J_\alpha(x) - J_\alpha(y)|_B \leq |x - y|_B$.*
(ii) *The mappings F_α, $\alpha > 0$, are dissipative and Lipschitz continuous:*

$$|F_\alpha(x) - F_\alpha(y)|_B \leq \frac{2}{\alpha}|x - y|_B, \qquad x, y \in B.$$

Moreover $|F_\alpha(x)|_B \leq |F(x)|_B$ for $x \in D(F)$.

(iii) $\lim_{\alpha \to 0} F_\alpha(x) = x$, $\forall x \in \overline{D(F)}$.

Proposition 10.5 *Let $B = H$ be a Hilbert space, let $F \colon D(F) \mapsto H$ be an m-dissipative mapping on H and let $\alpha, \beta > 0$. Then*

$$\langle x - y, F_\alpha(x) - F_\beta(y) \rangle_H \leq (\alpha + \beta)\big(|F_\alpha(x)|_H + |F_\beta(y)|_H\big)^2, \qquad x, y \in H.$$

10.1.1 Examples

Example 10.6 Let $(S, \mathcal{S}, \lambda)$ be a space with a measure. Assume that $B = L^p :=$ $L^p(S, \mathcal{S}, \lambda)$, where $p > 1$. If $x \neq 0$ then $\partial|x|_{L^p}$ consists of the unique element x^* of $L^q(S, \mathcal{S}, \mu)$, $q = p/(p-1)$, given by

$$x^*(\xi) = \left(|x|_{L^p}\right)^{1-p} |x(\xi)|^{p-2} x(\xi), \qquad \xi \in \mathcal{S}.$$

Example 10.7 Let $B = L^p$, $p > 1$, be as in Example 10.6. Let

$$F(x(\xi)) = f(x(\xi)), \qquad \xi \in \mathcal{S}, \tag{10.2}$$

where f is a decreasing function. Then F with domain

$$D(F) = \left\{ x \in L^p : \int_S |f(x(\xi))|^p \, \lambda(d\xi) < \infty \right\} \tag{10.3}$$

is m-dissipative.

Example 10.8 Let $B = L^p$, $p > 1$. Assume that, for some $\eta \in \mathbb{R}$, $f(x) + \eta x$, $x \in \mathbb{R}$, is decreasing. Then the mapping F given by (10.2) and (10.3) is almost m-dissipative. In particular, if

$$f(x) = -x^{2n+1} + a_{2n} x^{2n} + \cdots + a_0 \tag{10.4}$$

then the corresponding F is almost m-dissipative.

Example 10.9 If A is the infinitesimal generator of a C_0-semigroup $S(t)$, $t \geq 0$, on B then, by Proposition 10.11 below, A is almost m-dissipative if and only if, for some $\omega \in \mathbb{R}$, $\|S(t)\|_{L(B,B)} \leq e^{\omega t}$, $t \geq 0$, that is, if and only if S is a semigroup of generalized contractions; see subsection 9.4.1. For more details about the dissipativity of generators see the next subsection and Section 10.3.

Our final example deals with the Laplace operator. It is based on Proposition 9.4.5 from Da Prato and Zabczyk (1996); see also appendix section B.2.

Example 10.10 Let $d = 1, 2, \ldots$, and let

$$S(t)\psi(\xi) := \int_{\mathbb{R}^d} \mathfrak{g}(t)(\xi - \eta)\psi(\eta) \, d\eta = \mathfrak{g}(t) * \psi(\xi),$$

where $\mathfrak{g}(t)(\xi) = (4\pi t)^{d/2} e^{-|\xi|^2/(4t)}$ for $t > 0$ and $\xi \in \mathbb{R}^d$. Let $\kappa > 0$. Assume that λ is the measure on $(\mathbb{R}^d, \mathcal{B}(\mathbb{R}^d))$ given by $\lambda(d\xi) := e^{-\kappa|\xi|} \, d\xi$. Then $(S(t), \, t \geq 0)$ is a C_0-semigroup on $B = L^p(d\lambda) := L^p(\mathbb{R}^d, \mathcal{B}(\mathbb{R}^d), \lambda)$. Moreover, $\|S(t)\|_{L(B,B)} \leq \exp\{(\kappa^2/p)t\}$, $t \geq 0$. Similarly, if $\lambda(d\xi) := (1 + \kappa|\xi|^r)^{-1} \, d\xi$, where $\kappa > 0$

and $r > d$, then S is a C_0-semigroup on $B = L^p(d\lambda)$ and $\|S(t)\|_{L(B,B)} \leq \exp\{\kappa^{2/r}(\delta/p)t\}$, $t \geq 0$, where

$$\delta = \frac{r - d + 2}{r}(r - 1)^{2(1-1/r)}.$$

Thus, in both these cases, the Laplace operator (i.e. the generator of S) is almost m-dissipative.

10.1.2 Dissipativity of generators

We have the following result on m-dissipative generators.

Proposition 10.11 *Let $(A, D(A))$ be the generator of a C_0-semigroup S on a Banach space B. Then the following conditions are equivalent.*

(i) $\|S(t)\|_{L(B,B)} \leq 1$ *for all $t \geq 0$.*
(ii) *A is m-dissipative.*
(iii) $\langle x^*, Ax \rangle \leq 0$ *for all $x \in D(A)$ and some $x^* \in \partial|x|_B$.*
(iv) $\langle x^*, Ax \rangle \leq 0$ *for all $x \in D(A)$ and all $x^* \in \partial|x|_B$.*

Proof　We show first that (i) $\Rightarrow$ (iv). Note that, for all $x \in B$ and $x^* \in \partial|x|_B$,

$$0 \geq |S(t)x|_B - |x|_B = \left| x + (S(t)x - x) \right|_B - |x|_B \geq \langle x^*, S(t)x - x \rangle.$$

If also $x \in D(A)$ then $t \mapsto S(t)x$ is continuously differentiable and $S(t)x - x = \int_0^t S(s)Ax \, ds$. Consequently,

$$\left\langle x^*, \frac{1}{t} \int_0^t S(s)Ax \, ds \right\rangle \leq 0$$

and thus (iv) follows. To see that (iii) $\Rightarrow$ (i) we fix an $x \in D(A)$. Then, for all $t > 0$ and $x^* \in \partial|S(t)x|_B$,

$$\frac{d^-}{dt}|S(t)x|_B \leq \left\langle x^*, \frac{d^- S(t)x}{dt} \right\rangle$$

and hence, by (iii),

$$\frac{d^-}{dt}|S(t)x|_B \leq \langle x^*, AS(t)x \rangle \leq 0.$$

However, the function $t \mapsto |S(t)x|_B$ is absolutely continuous for all $x \in D(A)$ and therefore the function $t \mapsto |S(t)x|_B$ is non-increasing. As $|S(0)x|_B = |x|_B$, (i) follows. By the definition of maximal dissipativity, (ii) implies (iii). But if A is a generator then, by the Hille–Yosida theorem, for large enough $\lambda > 0$ the image of $\lambda I - A$ is equal to B, and thus (iii) implies (ii). □

10.2 Existence theorem

Let Z_A denote the *Lévy Ornstein–Uhlenbeck* process

$$Z_A(t) := \int_0^t S(t-s)G \, dZ(s), \qquad t \geq 0.$$

As in Da Prato and Zabczyk (1996), we make some assumptions on the operators A and F and on the process Z_A. In what follows, $(B, |\cdot|_B)$ is a reflexive Banach space continuously embedded into H as a dense Borel subset.

(H1) The operators A, F and their restrictions A_B, F_B to B are almost
  m-dissipative respectively in H and B. Moreover, $D(F) \supset B$ and F maps
  bounded sets in B into bounded sets of H.

(H2) The process Z_A is càdlàg in B and, for all $T > 0$, $\mathbb{P}$-a.s.,

$$\int_0^T |F(Z_A(t))|_B \, dt < \infty.$$

Definition 10.12 An adapted B-valued process X is said to be a *càdlàg mild solution to* (10.1) if it is càdlàg in B and satisfies, $\mathbb{P}$-a.s., the equation

$$X(t) = S(t)x + \int_0^t S(t-s)F(X(s)) \, ds + Z_A(t), \qquad t \geq 0.$$

In the definition it is required that $X(s) \in D(F)$ for $s \geq 0$, and consequently that $X(0) = x \in D(F)$.

Definition 10.13 If $x \in H$ and there exist a sequence $(x_n) \subset B$ and a sequence (X_n) of unique càdlàg mild solutions of (10.1) with $X_n(0) = x_n$, such that, $\mathbb{P}$-a.s., $|X_n(t) - X(t)|_H \to 0$ uniformly on each bounded interval then X is said to be a *generalized solution to* (10.1).

In the proof of our main existence result (see Theorem 10.14 below), we use the main ideas of the proof of Theorem 5.5.8 in Da Prato and Zabczyk (1996). But, in contrast with Da Prato and Zabczyk (1996), where Z is assumed to be Gaussian, here Z can be a general Lévy process. We also use a different approximating scheme.

Theorem 10.14 *Under assumptions (H1) and (H2), (10.1) has a unique càdlàg mild solution for any $x \in B$. For each $x \in H$, there exists a unique generalized solution to (10.1). Moreover, (10.1) defines Feller families on B and on H.*

Proof   Assume first that $x \in B$. Given a càdlàg $f: [0, \infty) \mapsto B$, consider the deterministic equation

$$y(t) = S(t)x + \int_0^t S(t-s)F(y(s) + f(s-))\,ds, \qquad t \geq 0,$$

which may be written formally as

$$\frac{dy}{dt}(t) = Ay(t) + F(y(t) + f(t-)), \qquad t \geq 0,$$
$$y(0) = x. \tag{10.5}$$

In the first equation above we take the left limit $f(t-)$ since, thanks to Proposition 10.1, we can estimate the left derivatives.

Note that if, for any càdlàg f, (10.5) has a unique solution then, denoting by y_A the solution to (10.5) with $f = Z_A$ and setting

$$X(t) = y_A(t) + Z_A(t), \qquad t \geq 0,$$

one arrives at the càdlàg mild solution to (10.1).

For $\alpha > 0$, $\beta > 0$ and sufficiently small η, denote by $(F + \eta)_\alpha$ and $(A + \eta)_\beta$ the Yosida approximations of the m-dissipative mappings $F + \eta$ and $A + \eta$:

$$(F + \eta)_\alpha := \frac{1}{\alpha}\big((I - \alpha(F + \eta))^{-1} - I\big),$$
$$(A + \eta)_\beta := \frac{1}{\beta}\big((I - \beta(A + \eta))^{-1} - I\big).$$

We will use the same notation for operators on H and for their restrictions to B. Denote by $y_{\alpha\beta}$ a solution to the approximating problem

$$\frac{d^-}{dt} y_{\alpha\beta}(t) = (A + \eta)_\beta y_{\alpha\beta}(t) + (F + \eta)_\alpha(y_{\alpha\beta}(t) + f(t-))$$
$$- 2\eta y_{\alpha\beta}(t) - \eta f(t-), \qquad t \geq 0,$$
$$y_{\alpha\beta}(0) = x \in B. \tag{10.6}$$

The left limit in $f(t-)$ is needed to prove that the solution $y_{\alpha\beta}$ satisfies the equation for all $t \geq 0$. The Yosida approximations are Lipschitz and therefore (10.6) has a unique continuous solution. We will show that

$$\lim_{\alpha \downarrow 0} \lim_{\beta \downarrow 0} y_{\alpha\beta}(t) = y(t), \qquad t \geq 0,$$

exists in H and defines the mild solution of the deterministic problem (10.5).

We assume that $\eta = 0$ to simplify the notation, but all the estimates can be performed when $\eta \neq 0$; compare also the proof of Theorem 16.6. We solve the equation

$$y_\alpha(t) = S(t)x + \int_0^t S(t-s)F_\alpha(y_\alpha(s) + f(s-))\,ds.$$

Since the transformation F_α is Lipschitz continuous both in H and in B and f is càdlàg in H and in B, there exists a solution, continuous in H and B. We show now that there exists a constant $c > 0$ such that, for all $\alpha > 0$ and $t \in [0, T]$, $|y_\alpha(t)|_B \le c$. Note that, for $\alpha > 0$ and $\beta > 0$,

$$y_\alpha(t) - y_{\alpha\beta} = S(t)x - S_\beta(t)x$$

$$+ \int_0^t \big(S(t-s) - S_\beta(t-s)\big) F_\alpha(y_\alpha(s) + f(s-))\,ds.$$

$$+ \int_0^t S_\beta(t-s)\big(F_\alpha(y_\alpha(s) + f(s-)) - F_\alpha(y_{\alpha\beta}(s) + f(s-))\big)\,ds.$$

Let M, ω and C_α be such that for all $t \ge 0$, $v, w \in B$ and $\beta > 0$,

$$\|S_\beta(t)\|_{L(B,B)} \le Me^{\omega t}, \qquad |F_\alpha(v) - F_\alpha(w)|_B \le C_\alpha|v - w|_B.$$

Then

$$|y_\alpha(t) - y_{\alpha\beta}(t)|_B$$

$$\le |S(t)x - S_\beta(t)x|_B + MC_\alpha \int_0^t e^{\omega(t-s)}|y_\alpha(s) - y_{\alpha\beta}(s)|_B\,ds$$

$$+ \int_0^t \big|\big(S_\beta(t-s) - S(t-s)\big)F_\alpha(y_\alpha(s) + f(s-))\big|_B\,ds. \qquad (10.7)$$

By the Hille–Yosida theorem (see Appendix B), it follows that $S_\beta(t)x \to S(t)x$ as $\beta \to 0$ uniformly in t on bounded intervals and in x on compact subsets of B. Consequently, the first and the third terms in (10.7) vanish uniformly on bounded intervals as $\beta \to 0$ and by Gronwall's lemma we obtain

$$\lim_{\beta \to 0} \sup_{t \le T} |y_\alpha(t) - y_{\alpha\beta}(t)|_B = 0, \qquad \forall\, T < \infty. \qquad (10.8)$$

From (10.6), by Proposition 10.1,

$$\frac{d^-}{dt}|y_{\alpha\beta}(t)|_B = \min\left\{\left\langle x^*, \frac{d^-}{dt} y_{\alpha\beta}(t)\right\rangle : x^* \in \partial|y_{\alpha\beta}(t)|_B\right\}$$

$$= \min\left\{\left\langle x^*, A_\beta y_{\alpha\beta}(t) + F_\alpha(y_{\alpha\beta}(t) + f(t-))\right\rangle : x^* \in \partial|y_{\alpha\beta}(t)|_B\right\}.$$

Both transformations A_β and F_α are m-dissipative. Therefore, by Definition 10.2, the linearity of A_β and Proposition 10.4(ii),

$$\frac{d^-}{dt}|y_{\alpha\beta}(t)|_B \le |F_\alpha(f(t-))|_B \le |F(f(t-))|_B$$

and consequently

$$|y_{\alpha\beta}(t)|_B \le |x|_B + \int_0^t |F(f(s-))|_B \, ds, \qquad t \ge 0. \tag{10.9}$$

By (10.8), for all $\alpha > 0$,

$$|y_\alpha(t)|_B \le |x|_B + \int_0^T |F(f(s-))|_B \, ds, \qquad t \in [0, T].$$

Similarly, for $t \in [0, T]$,

$$\frac{1}{2}\frac{d^-}{dt} |y_{\alpha\beta}(t) - y_{\gamma\beta}(t)|_H^2 = \left\langle \frac{d^-}{dt}(y_{\alpha\beta}(t) - y_{\gamma\beta}(t)), \; y_{\alpha\beta}(t) - y_{\gamma\beta}(t) \right\rangle_H$$

$$= \langle (A_\beta y_{\alpha\beta}(t) - A_\beta y_{\gamma\beta}(t))$$

$$+ \left(F_\alpha(y_{\alpha\beta}(t) + f(t-)) - F_\gamma(y_{\gamma\beta}(t) + f(t-)) \right), \; y_{\alpha\beta}(t) - y_{\gamma\beta}(t) \rangle_H$$

$$\le \langle F_\alpha(y_{\alpha\beta}(t) + f(t-)) - F_\gamma(y_{\gamma\beta}(t) + f(t-)), \; y_{\alpha\beta}(t) - y_{\gamma\beta}(t) \rangle_H$$

and so, using Proposition 10.5 and then Proposition 10.4(ii),

$$\frac{1}{2}\frac{d^-}{dt} |y_{\alpha\beta}(t) - y_{\gamma\beta}(t)|_H^2$$

$$\le (\alpha + \gamma) \left(\left| F_\alpha(y_{\alpha\beta}(t) + f(t-)) \right|_H + \left| F_\gamma(y_{\gamma\beta}(t) + f(t-)) \right|_H \right)^2$$

$$\le (\alpha + \gamma) \left(\left| F(y_{\alpha\beta}(t) + f(t-)) \right|_H + \left| F(y_{\gamma\beta}(t) + f(t-)) \right|_H \right)^2.$$

By (H1) and (10.9) there is a constant C such that

$$\frac{1}{2}\frac{d^-}{dt} |y_{\alpha\beta}(t) - y_{\gamma\beta}(t)|_H^2 \le C(\alpha + \gamma), \qquad t \in [0, T].$$

Consequently,

$$\tfrac{1}{2} |y_{\alpha\beta}(t) - y_{\gamma\beta}(t)|_H^2 \le C(\alpha + \gamma)T, \qquad t \in [0, T].$$

By (10.8),

$$\tfrac{1}{2} |y_\alpha(t) - y_\gamma(t)|_H^2 \le C(\alpha + \gamma)T, \qquad \alpha, \gamma > 0, \; t \in [0, T].$$

Thus $y_\alpha(t) \to y(t)$ in H, uniformly on $[0, T]$, as $\alpha \to 0$.

The proof that y is in fact a mild solution of (10.5) is given in Da Prato and Zabczyk (1996). By similar reasoning one can show that there exists a constant $c > 0$ such that, for all $x, \tilde{x} \in B$, the solutions $y(t, x), y(t, \tilde{x}), t \in [0, T]$, of (10.5) with initial conditions $x, \tilde{x}$ satisfy

$$\sup_{t \le T} |y(t, x) - y(t, \tilde{x})|_H \le c \, |x - \tilde{x}|_H.$$

This proves the existence of generalized solutions for any $x \in H$.

The fact that (10.1) defines Feller families on B and H follows from the continuous dependence of solutions on the initial data. □

10.3 Reaction–diffusion equation

In this section we present a typical example of (10.1), the so-called *stochastic reaction–diffusion equation*. Its long-time behavior is studied in Section 16.3.

Let $\mathcal{O}$ be an open bounded subset of $\mathbb{R}^d$. Given $p \in (1, \infty)$ write $L^p := L^p(\mathcal{O}, \mathcal{B}(\mathcal{O}), d\xi)$. Let $\mathcal{A} = \mathcal{A}(\xi, D)$ be a second-order elliptic differential operator,

$$\mathcal{A} = \mathcal{A}(\xi, D) = \sum_{i,j=1}^{d} a_{i,j}(\xi)\frac{\partial^2}{\partial\xi_i\partial x_j} + \sum_{i=1}^{d} b_i(\xi)\frac{\partial}{\partial\xi_i} + c(\xi)I, \qquad \xi \in \mathcal{O},$$

and let f be given by (10.4). Finally, let Z be a square integrable mean-zero Lévy process in a Hilbert space U with RKHS $\mathcal{H} \subset L^2$.

We are concerned with the existence of generalized solutions to the following problem:

$$dX(t, \xi) = \big(\mathcal{A}X(t, \xi) + f(X(t, \xi))\big)dt + dZ(t, \xi),$$
$$X(t, \xi) = 0 \qquad \text{for } \xi \in \partial\mathcal{O}, \tag{10.10}$$
$$X(0, \xi) = x(\xi) \qquad \text{for } \xi \in \mathcal{O}.$$

We assume that $\mathcal{O}$ and the coefficients $a_{i,j}$, b_j and c satisfy the assumptions formulated in Section 2.5. In particular we assume that the coefficients are of class $C_b^\infty(\mathcal{O})$ and that $\mathcal{A}$ is uniformly elliptic on $\overline{\mathcal{O}}$. Note that, in (10.10), $\mathcal{A}$ is considered in conjunction with the Dirichlet boundary operator $B\psi(\xi) = \psi(\xi)$, $\xi \in \partial\mathcal{O}$. Given $p \in (1, \infty)$ define

$$D(A_p) := \big\{\psi \in W^{2,p}(\mathcal{O})\colon \psi(\xi) = 0 \text{ for } \xi \in \partial\mathcal{O}\big\},$$
$$A_p\psi(\xi) = \mathcal{A}\psi(\xi), \qquad \psi \in D(A_p), \ \xi \in \mathcal{O}.$$

Then, see appendix section B.2, $(A_p, D(A_p))$ generates an analytic C_0-semigroup S_p on L^p. Since $S_q = S_p$ on $L^p \cap L^q$ we will often write A and S instead of A_p and S_p.

Theorem 10.15 *Let $p \geq 2n + 1$ and let $Z = W + L$, where W is a Wiener process and L is a Lévy process in L^2. Assume that*

$$W_A(t) = \int_0^t S(t - s)\,dW(s), \qquad t \geq 0,$$

has continuous trajectories in $L^{2p(2n+1)}$ and that there exist $\kappa > 0$ and $\alpha > 0$ such that L has càdlàg trajectories in $D\big((-A_{2p(2n+1)} - \kappa I)^\alpha\big)$. Then

(i) *for every $x \in L^{2p}$ there is a unique càdlàg solution to (10.10) in $B = L^{2p}$,*

(ii) *for every $x \in L^2$ there is a unique generalized solution to (10.10) in $H = L^2$.*

Moreover, (10.10) defines Feller families on L^{2p} and L^2.

Proof   We treat (10.10) as (10.1) with F the Nemytskii operator corresponding to f and $H = L^2$, $B = L^{2p}$. We have to verify the hypothesis of Theorem 10.14. By Example 10.8, F and F_B are almost m-dissipative. The fact that A is almost m-dissipative on H and B follows from Lemma 10.17 below. Therefore (H1) of Theorem 10.14 is satisfied. To see (H2) it is sufficient to show that L_A has càdlàg trajectories in $L^{2p(2n+1)}$. This follows from Proposition 9.28.   □

Remark 10.16   The assumption that W_A has continuous trajectories in the space $L^{2p(2n+1)}$ or even in L^∞ can be easily verified for many different cases; see subsections 12.2.2 and 12.4.3. For more details on stochastic reaction–diffusion equations with Wiener noise we refer the reader to e.g. Bally, Gyöngy and Pardoux (1994), Cerrai (2001), Da Prato and Zabczyk (1992b, 1996), Manthey and Zausinger (1999) and Peszat (1995).

Since the proof of the following result is rather long and technical it is given in appendix section B.6.

Lemma 10.17   *For all $p \geq 2$ the operator $(A_p, D(A_p))$ is almost m-dissipative on L^p.*

Example 10.18   Assume that $\mathcal{O} = (a, b) \subset \mathbb{R}$, that is, $\mathcal{O}$ is one-dimensional. Let $Z = W$ be a cylindrical Wiener process on L^2. Then W_A has continuous trajectories in $C\,(\overline{\mathcal{O}})$, and therefore (10.10) defines Feller families on L^2 and L^{2p}.

Example 10.19   Let $p \geq 2n + 1$ and let Z be a Lévy process in $L^2(\mathcal{O})$ with Lévy exponent (6.2). Assume that the Lévy measure ν satisfies

$$\int_{L^2(\mathcal{O})} |C_\alpha x|_{L^q}\, \nu(\mathrm{d}x) < \infty, \tag{10.11}$$

where $\alpha > 0$, $C_\alpha = \big(-A_{4p(n+1)} - \kappa I\big)^\alpha$ and $q = 2p(2n + 1)$. Then, by Proposition 6.9, Z satisfies the assumption of Theorem 10.15. For more details see the proof of Theorem 16.7. Since α is an arbitrary strictly positive number, the domain of C_α is known (see e.g. Lunardi 1995); instead of (10.11) it is sufficient to assume that there is a $\beta > 0$ such that

$$\int_{L^2(\mathcal{O})} |x|_{W^{\beta,q}(\mathcal{O})}\, \nu(\mathrm{d}x) < \infty,$$

where $W^{\beta,q}(\mathcal{O})$ is a fractional Sobolev space. For the definition of fractional Sobolev spaces see e.g. Lunardi (1995).

Example 10.20 Assume that Z is finite-dimensional and equal to $\sum_{k=1}^{N} f_k L_k$, where the L_k are real-valued Lévy square integrable martingales and the $f_k \in L^2(\mathcal{O})$ are sufficiently regular, for example $f_k \in C_c^1(\mathcal{O})$. Then the assumptions of the theorem are satisfied.

11

Factorization and regularity

The regularity of solutions to stochastic partial differential equations is of great importance. In general, when the integrator is a càdlàg process one cannot expect to establish regularity better than càdlàg. If the operator A in the drift term of an equation generates a generalized contraction then, under some additional assumptions, the càdlàg regularity of the solution is established by Theorem 9.20. In this chapter the so-called *factorization method*, introduced in Da Prato, Kwapień and Zabczyk (1987) (see also Zabczyk 1993) is introduced. This method is useful in many important cases for showing the time continuity of solutions to stochastic evolution equations driven by continuous processes. The method also leads to some maximal inequalities for stochastic convolutions with arbitrary semigroups; see Zabczyk (1993).

To motivate the method, the deterministic finite-dimensional case is considered first. This is then generalized to infinite dimensions. Finally, the regularity of the solutions is established.

11.1 Finite-dimensional case

A crucial aspect of the factorization method is the semigroup property of the *Liouville–Riemann operators* $I_{A,\alpha}$. Namely, given a matrix $A \in M(d \times d)$, a parameter $\alpha > 0$ and a function $\psi : (0, \infty) \mapsto \mathbb{R}^d$, we set

$$I_{A,\alpha}\psi(t) := \frac{1}{\Gamma(\alpha)} \int_0^t e^{A(t-s)}(t-s)^{\alpha-1}\psi(s)\,ds, \qquad t \geq 0,$$

where

$$\Gamma(\alpha) := \int_0^\infty t^{\alpha-1}e^{-t}\,dt, \qquad \alpha > 0,$$

is the *Euler Γ-function*.

Theorem 11.1 *For all α, $\beta > 0$ and $A \in M(d \times d)$, $I_{A,\alpha+\beta} = I_{A,\alpha} I_{A,\beta}$. In particular, for $\alpha \in (0, 1)$, we have $I_{A,1} = I_{A,\alpha} I_{A,1-\alpha}$.*

In the proof of the theorem we will use the following lemma.

Lemma 11.2 *For all $\beta > 0$, $\alpha > 0$,*

$$\int_0^1 (1 - z)^{\alpha-1} z^{\beta-1} \, dz = \frac{\Gamma(\alpha)\Gamma(\beta)}{\Gamma(\alpha + \beta)}. \tag{11.1}$$

Proof Consider the so-called γ-*distribution*

$$f_{a,\gamma}(t) := \frac{1}{\Gamma(\gamma)} a^\gamma t^{\gamma-1} e^{-at} \chi_{[0,\infty)}(t), \qquad t \in \mathbb{R}.$$

Denote by $*$ the *convolution operator*

$$f * g(t) = \int_{-\infty}^{+\infty} f(t - s) g(s) \, ds, \qquad t \in \mathbb{R}.$$

Then, for $t \geq 0$,

$$f_{a,\alpha} * f_{a,\beta}(t) = \frac{a^{\alpha+\beta}}{\Gamma(\alpha)\Gamma(\beta)} e^{-at} \int_0^t (t - s)^{\alpha-1} s^{\beta-1} \, ds$$

$$= f_{a,\alpha+\beta}(t) \frac{\Gamma(\alpha + \beta)}{\Gamma(\alpha)\Gamma(\beta)} t^{1-(\alpha+\beta)} \int_0^t (t - s)^{\alpha-1} s^{\beta-1} \, ds$$

$$= f_{a,\alpha+\beta}(t) \frac{\Gamma(\alpha + \beta)}{\Gamma(\alpha)\Gamma(\beta)} \int_0^1 (1 - z)^{\alpha-1} z^{\beta-1} \, dz.$$

Now, since $f_{a,\alpha}$ and $f_{a,\beta}$ are densities of probability measures, their convolution is also the density of a probability measure. Since $f_{a,\alpha+\beta}$ is also the density of a probability measure, the desired identity holds. □

Proof of Theorem 11.1 We have

$$I_{A,\alpha} I_{A,\beta} \psi(t)$$

$$= \frac{1}{\Gamma(\alpha)\Gamma(\beta)} \int_0^t e^{A(t-s)} (t - s)^{\alpha-1} \int_0^s e^{A(s-v)} (s - v)^{\beta-1} \psi(v) \, dv \, ds$$

$$= \frac{1}{\Gamma(\alpha)\Gamma(\beta)} \int_0^t e^{A(t-v)} \left(\int_v^t (t - s)^{\alpha-1} (s - v)^{\beta-1} \, ds \right) \psi(v) \, dv.$$

Thus we need to show that

$$\int_v^t (t - s)^{\alpha-1} (s - v)^{\beta-1} \, ds = \frac{\Gamma(\alpha)\Gamma(\beta)}{\Gamma(\alpha + \beta)} (t - v)^{(\alpha+\beta)-1}.$$

This is, however, a simple consequence of Lemma 11.2. □

Let $L^q(0, T; \mathbb{R}^d) := L^q \left((0, T), \mathcal{B}((0, T)), dt; \mathbb{R}^d \right).$

Theorem 11.3   *Assume that $A \in M(d \times d)$ and $1/q < \alpha < 1$. Then for every $T \in (0, \infty)$ there is a constant $c = c_{\alpha,q,A,T} < \infty$ such that, for every $\psi \in L^q(0, T; \mathbb{R}^d)$,*

$$|I_{A,\alpha}\psi(t) - I_{A,\alpha}\psi(s)| \leq c\,|t - s|^{\alpha - 1/q}\,|\psi|_{L^q(0,T;\mathbb{R}^d)}, \qquad t, s \in (0, T).$$

Proof   Let $0 \leq s \leq t \leq T$; then $I_{A,\alpha}\psi(t) - I_{A,\alpha}\psi(s) = J_1 + J_2 + J_3$, where

$$J_1 := \frac{1}{\Gamma(\alpha)} \int_s^t e^{A(t-v)}(t - v)^{\alpha - 1}\psi(v)\,\mathrm{d}v,$$

$$J_2 := \frac{1}{\Gamma(\alpha)} \int_0^s e^{A(t-v)}\left((t - v)^{\alpha - 1} - (s - v)^{\alpha - 1}\right)\psi(v)\,\mathrm{d}v,$$

$$J_3 := \frac{1}{\Gamma(\alpha)} \int_0^s \left(e^{A(t-v)} - e^{A(s-v)}\right)(s - v)^{\alpha - 1}\psi(v)\,\mathrm{d}v.$$

Let $p = q/(q - 1)$, so that $1/p + 1/q = 1$, and let

$$C_1 := \sup_{t \in [0,T]} \left\|e^{tA}\right\| (\Gamma(\alpha))^{-1},$$

where $\|\cdot\|$ stands for the operator norm on $L(\mathbb{R}^d, \mathbb{R}^d) \equiv M(d \times d)$. Note that $C_1 < \infty$. It follows that

$$
\begin{aligned}
|J_1| &\leq \frac{1}{\Gamma(\alpha)} \int_s^t \left\|e^{A(t-v)}\right\| |t - v|^{\alpha - 1}|\psi(v)|\,\mathrm{d}v \\
&\leq C_1 \left(\int_s^t (t - v)^{(\alpha - 1)p}\,\mathrm{d}v\right)^{1/p} \left(\int_s^t |\psi(v)|^q\,\mathrm{d}v\right)^{1/q} \\
&\leq C_1 \left(\int_0^{t-s} u^{(\alpha - 1)p}\,\mathrm{d}u\right)^{1/p} |\psi|_{L^q(0,T;\mathbb{R}^d)} \\
&\leq \frac{C_1(t - s)^{\alpha - 1/q}}{((\alpha - 1)p + 1)^{1/p}} |\psi|_{L^q(0,T;\mathbb{R}^d)}.
\end{aligned}
$$

To estimate J_2, note that

$$|J_2|^p \leq C_1^p \left(\int_0^s \left((s - v)^{\alpha - 1} - (t - v)^{\alpha - 1}\right)^p\,\mathrm{d}v\right) |\psi|_{L^q(0,T;\mathbb{R}^d)}^p.$$

For $a > b$ and $p > 1$, we have $(a - b)^p \leq a^p - b^p$. It follows that

$$
\begin{aligned}
|J_2|^p &\leq C_1^p \int_0^s \left((s - v)^{(\alpha - 1)p} - (t - v)^{(\alpha - 1)p}\right)\mathrm{d}v\,|\psi|_{L^q(0,T;\mathbb{R}^d)}^p \\
&\leq C_1^p \left(\frac{(t - s)^{(\alpha - 1)p+1}}{(\alpha - 1)p + 1} + \frac{s^{(\alpha - 1)p+1} - t^{(\alpha - 1)p+1}}{(\alpha - 1)p + 1}\right) |\psi|_{L^q(0,T;\mathbb{R}^d)}^p \\
&\leq C_1^p \frac{(t - s)^{(\alpha - 1)p+1}}{(\alpha - 1)p + 1} |\psi|_{L^q(0,T;\mathbb{R}^d)}^p
\end{aligned}
$$

and consequently

$$|J_2| \leq C_1 \frac{(t-s)^{\alpha-1/q}}{((\alpha-1)p+1)^{1/p}} |\psi|_{L^q(0,T;\mathbb{R}^d)}.$$

To estimate the last term, note that, for all $t, s \in (0, T]$,

$$\left\| e^{At} - e^{As} \right\| \leq |t-s| \sup_{r \in (0,T]} \left\| A e^{Ar} \right\|.$$

Let $C_2 := (\Gamma(\alpha))^{-1} \sup_{t \in (0,T]} \left\| A e^{At} \right\|$. Then

$$J_3 \leq C_2(t-s) \int_0^s (s-v)^{\alpha-1} |\psi(v)| \, dv$$

$$\leq C_2(t-s) \left(\int_0^T v^{(\alpha-1)p} \, dv \right)^{1/p} |\psi|_{L^q(0,T;\mathbb{R}^d)},$$

which completes the proof. $\qquad\square$

Let us denote by $C^\gamma([0, T]; \mathbb{R}^d)$ the space of all Hölder continuous functions $\psi : [0, T] \mapsto \mathbb{R}^d$ with exponent $\gamma \in (0, 1)$. On $C^\gamma([0, T]; \mathbb{R}^d)$ we consider the Hölder norm (see Section 2.2). The following result is a direct consequence of Theorem 11.3.

Corollary 11.4 *Assume that $A \in M(d \times d)$ and $1/q < \alpha < 1$. Then, for every $T \in (0, \infty)$, $I_{A,\alpha}$ is a bounded linear operator acting from $L^q(0, T; \mathbb{R}^d)$ into $C^{\alpha-1/q}([0, T]; \mathbb{R}^d)$.*

11.2 Infinite-dimensional case

11.2.1 General semigroups

Let A be the generator of a C_0-semigroup S on a separable Banach space B. Formally $S(t) = e^{At}$. Thus the infinite-dimensional analogues of the Liouville–Riemann operators introduced in the previous section have the form

$$I_{A,\alpha}\psi(t) := \frac{1}{\Gamma(\alpha)} \int_0^t (t-s)^{\alpha-1} S(t-s)\psi(s) \, ds. \tag{11.2}$$

Let $L^q(0, T; B) := L^q\big((0, T), \mathcal{B}((0, T)), dt; B\big)$, and let $C([0, T]; B)$ be the space of all continuous B-valued mappings equipped with the supremum norm.

Theorem 11.5

(i) *For all $\alpha > 0$ and $\beta > 0$, $I_{A,\alpha} I_{A,\beta} = I_{A,\alpha+\beta}$.*
(ii) *Assume that $1/q < \alpha < 1$. Then $I_{A,\alpha}$ is a bounded linear operator from $L^q(0, T; B)$ to $C([0, T]; B)$.*

Proof   The first assertion of the theorem can be proved in the same way as Theorem 11.1. To show the second, we fix a $T < \infty$. Let p be such that $1/p + 1/q = 1$. Since there are constants M and γ such that $\|S(t)\| \le Me^{\gamma t}$ for every $t \ge 0$,

$$
\left| \int_0^t (t - v)^{\alpha - 1} S(t - v) \psi(v) \, dv \right|_B
$$
$$
\le \left(\int_0^t (t - v)^{(\alpha - 1)p} \, dv \right)^{1/p} \left(\int_0^t |S(t - v) \psi(v)|_B^q \, dv \right)^{1/q}
$$
$$
\le Me^{\gamma t} \left(\frac{t^{(\alpha - 1)p + 1}}{(\alpha - 1)p + 1} \right)^{1/p} |\psi|_{L^q(0,t;B)}.
$$

Consequently, there is a constant $C < \infty$ such that

$$
\sup_{t \in [0,T]} |I_{A,\alpha} \psi(t)|_B \le C |\psi|_{L^q(0,T;B)}. \tag{11.3}
$$

Let $0 \le s \le t \le T$. Then

$$
\Gamma(\alpha) \left(I_{A,\alpha} \psi(t) - I_{A,\alpha} \psi(s) \right)
$$
$$
= \int_s^t v^{\alpha - 1} S(v) \psi(t - v) \, dv + \int_0^s v^{\alpha - 1} S(v)(\psi(t - v) - \psi(s - v)) dv.
$$

Hence, there is a constant C_1 such that

$$
|I_{A,\alpha} \psi(t) - I_{A,\alpha} \psi(s)|_B
$$
$$
\le C_1 \left(\int_s^t v^{(\alpha - 1)p} \, dv \right)^{1/p} |\psi|_{L^q(0,T;B)}
$$
$$
+ C_1 \left(\int_0^s v^{(\alpha - 1)p} \, dv \right)^{1/p} \left(\int_0^s |\psi(t - v) - \psi(s - v)|_B^q \, dv \right)^{1/q}.
$$

Therefore, $I_{A,\alpha} \psi \in C([0, T]; B)$ provided that $\psi \in C([0, T]; B)$. Since the space $C([0, T]; B)$ is dense in $L^q(0, T; B)$, the desired conclusion follows from (11.3). $\qquad \square$

11.2.2 Analytic semigroups

Denote by $C^\gamma([0, T]; B)$ the space of all Hölder continuous B-valued functions with exponent $\gamma \in (0, 1)$. As in the finite-dimensional case, the space $C^\gamma([0, T]; B)$ is equipped with the norm

$$
|\psi|_{C^\gamma([0,T];B)} := \sup_{t \in [0,T]} |\psi(t)|_B + \sup_{t,s \in [0,T], \, t \ne s} \frac{|\psi(t) - \psi(s)|_B}{|t - s|^\gamma}.
$$

In the finite-dimensional case we proved the Hölder continuity of $I_{A,\alpha}\psi$ for $\psi \in L^q(0, T; \mathbb{R}^d)$ with $1/q < \alpha < 1$. In the proof we used the fact that

$$\sup_{t\in[0,T]} \left\| Ae^{At} \right\| < \infty, \qquad \forall\, T < \infty. \tag{11.4}$$

In the infinite-dimensional case, the estimate (11.4) does not hold unless A is a bounded operator. However, Hölder continuity does hold if e^{tA} is an analytic semigroup. For these semigroups a condition slightly weaker than (11.4) is satisfied. For the definition of an analytic semigroup see appendix section B.1.2.

Theorem 11.6 *Assume that A generates an analytic C_0-semigroup. Let $q > 2$ and $\alpha \in (0, 1)$ be such that $1/q < \alpha$. Then, for any $\gamma \in (0, \alpha - 1/q)$, $I_{A,\alpha}$ is a bounded linear operator from $L^q(0, T; B)$ to $C^\gamma([0, T]; B)$.*

Proof Let $0 \leq s \leq t \leq T$. Let p be such that $1/p + 1/q = 1$; then

$$I_{A,\alpha}\psi(t) - I_{A,\alpha}\psi(s) = J_1 + J_2 + J_3,$$

where

$$J_1 := \frac{1}{\Gamma(\alpha)} \int_s^t S(t-v)(t-v)^{\alpha-1}\psi(v)\,dv,$$

$$J_2 := \frac{1}{\Gamma(\alpha)} \int_0^s S(t-v)\left((t-v)^{\alpha-1} - (s-v)^{\alpha-1}\right)\psi(v)\,dv,$$

$$J_3 := \frac{1}{\Gamma(\alpha)} \int_0^s \left(S(t-v) - S(s-v)\right)(s-v)^{\alpha-1}\psi(v)\,dv.$$

For the first two terms, we proceed as in the proof of Theorem 11.3; namely,

$$|J_1|_B \leq C_1 \frac{(t-s)^{\alpha-1/q}}{((\alpha-1)p+1)^{1/p}} \, |\psi|_{L^q(0,T;B)}$$

and

$$|J_2|_B \leq C_1 \frac{(t-s)^{\alpha-1/q}}{((\alpha-1)p+1)^{1/p}} \, |\psi|_{L^q(0,T;B)},$$

where $C_1 := \sup_{t\in[0,T]} \|S(t)\|_{L(B,B)} (\Gamma(\alpha))^{-1} < \infty$. Now the last term needs to be considered. We have

$$\left|(S(t-v) - S(s-v))\psi(v)\right|_B = \left| \int_{s-v}^{t-v} \frac{d}{du} S(u)\psi(v)\,du \right|_B$$

$$= \left| \int_{s-v}^{t-v} A S(u)\psi(v)\,du \right|_B.$$

By the analyticity of S it is clear that there is a constant C_2 such that $\|AS(u)\|_{L(B,B)}$ $\leq C_2/u$ for $u \in (0, T]$. Therefore, for all $\gamma \in (0, 1)$,

$$
\begin{aligned}
\big|(S(t-v) &- S(s-v))\psi(v)\big|_B \\
&\leq C_2\,|\psi(v)|_B \int_{s-v}^{t-v} \frac{1}{u}\,du = C_2\,|\psi(v)|_B \int_{s-v}^{t-v} \frac{1}{u^\gamma u^{1-\gamma}}\,du \\
&\leq C_2\,|\psi(v)|_B\,(s-v)^{-\gamma} \int_{s-v}^{t-v} \frac{1}{u^{1-\gamma}}\,du \\
&\leq C_2\,|\psi(v)|_B\,(s-v)^{-\gamma} \int_{s-v}^{t-v} \frac{1}{u^{1-\gamma}}\,du \\
&\leq \frac{C_2}{\gamma}|\psi(v)|_B\,(s-v)^{-\gamma}\,((t-v)^\gamma - (s-v)^\gamma) \\
&\leq \frac{C_2}{\gamma}|\psi(v)|_B\,(s-v)^{-\gamma}(t-s)^\gamma
\end{aligned}
$$

and, consequently,

$$
\begin{aligned}
J_3 &\leq \frac{C_2(t-s)^\gamma}{\gamma\,\Gamma(\alpha)} \int_0^s (s-v)^{\alpha-1-\gamma}|\psi(v)|_B\,dv \\
&\leq \frac{C_2(t-s)^\gamma}{\gamma\,\Gamma(\alpha)}|\psi|_{L^q(0,T;B)} \left(\int_0^s (s-v)^{(\alpha-1-\gamma)p}\,dv \right)^{1/p}.
\end{aligned}
$$

The proof now follows by noting that, for any $\gamma \in (0, \alpha - 1/q)$,

$$
\int_0^s (s-v)^{(\alpha-1-\gamma)p}\,dv < \infty.
$$

$\square$

In appendix subsection B.1.2 we recall the concept of the fractional power $(-A)^\gamma$ of an operator A. For the proof of the theorem below we can adapt the argument from the proof of Theorem 11.6. For a complete proof we refer the reader to Da Prato and Zabczyk (1996), Appendix A, but here we will assume that A generates an analytic semigroup on B satisfying (B.5). We equip the domain $D((-A)^\gamma)$ with the graph norm

$$
|x|_{D((-A)^\gamma)} := |x|_B + |(-A)^\gamma x|_B, \qquad x \in D((-A)^\gamma).
$$

Theorem 11.7   Let $\alpha \in (0, 1)$, $q \geq 1$ and $\gamma \in (0, 1)$ be such that $\alpha > \gamma + 1/q$. Then, for any $T < \infty$, $I_{A,\alpha}$ is a bounded linear operator from $L^q(0, T; B)$ to $C^{\alpha-\gamma-1/q}([0, T]; D((-A)^\gamma))$.

11.3 Applications to time continuity

In this section we will apply the factorization method to study the time continuity of solutions to the stochastic equation

$$dX = (AX + F(X)) \, dt + G(X) \, dW, \qquad X(0) = X_0; \qquad (11.5)$$

where A, with domain $D(A)$, is the generator of a C_0-semigroup S on a Hilbert space H, W is a Wiener process taking values in a Hilbert space U defined on a filtered probability space $(\Omega, \mathcal{F}, (\mathcal{F}_t), \mathbb{P})$ and X_0 is an $\mathcal{F}_0$-measurable random variable in H.

We denote by $\mathcal{H}$ the RKHS of W. We assume that the coefficients A, F and G satisfy assumptions (F) and (G) given in Section 9.2. This implies in particular that for any $t > 0$ the mappings $S(t)F$ and $S(t)G$ have unique extensions to continuous mappings from H to H and from H to $L_{(HS)}(\mathcal{H}, H)$, respectively.

Our goal is to show the existence of solutions that are continuous and, when the semigroup is analytic, Hölder continuous trajectories in H. Given $T < \infty, q \geq 1$ and $\gamma \in (0, 1)$ we denote by $\mathcal{C}_{T,q}$ and $\mathcal{C}_{T,q}^{\gamma}$ the classes of all processes $Y \colon \Omega \times [0, T] \mapsto H$ adapted and having respectively continuous and Hölder continuous trajectories in H such that

$$\|Y\|_{\mathcal{C}_{T,q}} := \left(\mathbb{E} \, |Y|_{C([0,T];H)}^q \right)^{1/q} < \infty,$$

$$\|Y\|_{\mathcal{C}_{T,q}^{\gamma}} := \left(\mathbb{E} \, |Y|_{C^{\gamma}([0,T];H)}^q \right)^{1/q} < \infty,$$

respectively.

Theorem 11.8 *Let q and $\alpha \in (0, 1/2)$ be such that $1/q < \alpha$. Assume that for each $T < \infty$ there are functions $a, b \colon (0, T] \mapsto (0, \infty)$ satisfying*

$$\int_0^T \left(a(t)t^{-\alpha} + b^2(t)t^{-2\alpha} \right) dt < \infty,$$

such that, for all $t \in (0, T]$ and $x, y \in H$,

$$|S(t)F(x)|_H \leq a(t)(1 + |x|_H),$$

$$\left| S(t)(F(x) - F(y)) \right|_H \leq a(t)|x - y|_H,$$

$$\|S(t)G(x)\|_{L_{(HS)}(\mathcal{H},H)} \leq b(t)(1 + |x|_H),$$

$$\|S(t)(G(x) - G(y))\|_{L_{(HS)}(\mathcal{H},H)} \leq b(t)|x - y|_H.$$

If $\mathbb{E} \, |X_0|_H^q < \infty$ then there is a unique solution X to (11.5) such that $X \in \mathcal{C}_{T,q}$ for $T < \infty$. If, moreover, S is analytic then $X \in \mathcal{C}_{T,q}^{\gamma}$ for every $T < \infty$ and for every exponent $\gamma \in (0, \alpha - 1/q)$.

Proof   Let us fix $T < \infty$. Denote by $\mathcal{Z}_{T,q}$ the class of all predictable processes $Y \colon \Omega \times [0,T] \mapsto H$ such that

$$\|Y\|_{\mathcal{Z}_{T,q}} := \left(\sup_{t \in [0,T]} \mathbb{E}\,|Y(t)|_H^q \right)^{1/q} < \infty.$$

Using the Banach fixed-point theorem as in the proof of Theorem 9.20, one can easily show the existence (and uniqueness) of a solution X to (11.5) in the class $\mathcal{Z}_{T,q}$. To control the qth moment of the stochastic term, the Burkholder–Davis–Gundy inequality may be used, as will be seen below.

Having established the existence of a solution X in the class $\mathcal{Z}_{T,q}$ we will show that in fact $X \in \mathcal{C}_{T,q}$ in the general case and $X \in \mathcal{C}_{T,q}^\gamma$, $\gamma \in (0, \alpha - 1/q)$, in the analytic case. To do this we will use factorization. We will concentrate on the stochastic term

$$I(t) := \int_0^t S(t-s)G(X(s))\,\mathrm{d}W(s), \qquad t \in [0,T];$$

the analysis of the term

$$J(t) := S(t)X_0 + \int_0^t S(t-s)F(X(s))\,\mathrm{d}s, \qquad t \in [0,T],$$

is left to the reader.

Write

$$Y(t) := \frac{1}{\Gamma(1-\alpha)} \int_0^t (t-s)^{-\alpha} S(t-s)G(X(s))\,\mathrm{d}W(s), \qquad t \in [0,T].$$

Note that

$$\int_0^t (t-s)^{-2\alpha} \mathbb{E}\,\|S(t-s)G(X(s))\|_{L_{(HS)}(\mathcal{H},H)}^2 \,\mathrm{d}s$$

$$\leq \int_0^t (t-s)^{-2\alpha} b^2(s)\,\mathrm{d}s \, \sup_{s \in [0,t]} \mathbb{E}\,(1+|X(s)|_H)^2 < \infty.$$

Thus Y is a well-defined process in H. In fact, $\int_0^T \mathbb{E}\,|Y(t)|_H^q\,\mathrm{d}t < \infty$ for $T > 0$. This follows by the Burkholder–Davis–Gundy inequality:

$$\int_0^T \mathbb{E}\,|Y(t)|_H^q\,\mathrm{d}t$$

$$\leq c_1 \int_0^T \left(\int_0^t (t-s)^{-2\alpha} \mathbb{E}\,\|S(t-s)G(X(s))\|_{L_{(HS)}(\mathcal{H},H)}^2\,\mathrm{d}s \right)^{q/2} \mathrm{d}t$$

$$\leq \int_0^T \left(\int_0^t (t-s)^{-2\alpha} b^2(s)\,\mathrm{d}s \right)^{q/2} \mathrm{d}t \, \sup_{s \in [0,T]} \mathbb{E}\,(1+|X(s)|_H)^q,$$

where c_1, depending only on q, appears in the BDG inequality. We show that

$$I = \Gamma(1)I_{A,\alpha}Y. \tag{11.6}$$

This relationship ensures that $I \in \mathcal{C}_{T,q}$ in the general case and $I \in \mathcal{C}_{T,q}^\gamma$ in the analytic case because, by Theorem 11.5, $I_{A,\alpha}$ is a bounded operator from $L^q(0, T; H)$ to $C([0, T]; H)$, $T \geq 0$. Thus, if (11.6) holds, I has continuous trajectories in H and

$$\sup_{t \in [0,T]} |I(t)|_H \leq C \left(\int_0^T |Y(t)|_H^q \, dt \right)^{1/q},$$

where the constant C is independent of X (see the proof of Theorem 11.5).

To show (11.6) we use the stochastic Fubini theorem in the same way as in the proof of Theorem 11.1. We have

$$\Gamma(\alpha)\Gamma(1 - \alpha)I_{A,\alpha}Y(t)$$
$$= \int_0^t S(t - s)(t - s)^{\alpha-1} \int_0^s S(s - v)(s - v)^{-\alpha} G(X(v)) \, dW(v) \, ds$$
$$= \int_0^t S(t - v) \left(\int_v^t (t - s)^{\alpha-1}(s - v)^{-\alpha} \, ds \right) G(X(v)) \, dW(v)$$
$$= \frac{\Gamma(\alpha)\Gamma(1 - \alpha)}{\Gamma(1)} \int_0^t S(t - v)G(X(v)) \, dW(v).$$

Assume now that S is analytic; then the desired Hölder continuity of I follows directly from Theorem 11.6 and (11.6). □

11.4 The case of an arbitrary martingale

Assume now that M is a square integrable càdlàg martingale taking values in a Hilbert space U. In this section, we discuss possible applications of the factorization method to the study of stochastic convolutions such as

$$Y(t) = \int_0^t S(t - s)\Psi(s) \, dM(s), \qquad t \geq 0,$$

where S is a C_0-semigroup on a Hilbert space H. We assume that, for $t \geq 0$, the process $(\omega, s) \mapsto S(t - s)\Psi(\omega, s) \in H$ belongs to the space of integrable processes $\mathcal{L}_{M,t}^2(H)$; see Section 8.3. Consequently, Y is a well-defined square integrable process in H. As we have already seen, it is not necessarily càdlàg; see Proposition 9.25. However, it might seem that the factorization method is based only on the stochastic Fubini theorem (valid in the general case) and on a certain (deterministic) integral representation of the Euler Γ-function. To see that this supposition fails, we assume that there is a constant $\alpha \in (0, 1/2)$ such that

$$Y_\alpha(t) := \frac{1}{\Gamma(1-\alpha)} \int_0^t (t-s)^{-\alpha} S(t-s)\Psi(s)\,dM(s)$$

is a well-defined square integrable predictable process in H; that is, for every t, $(\omega, s) \to S(t-s)\Psi(\omega, s) \in H$ belongs to $\mathcal{L}^2_{M,t}(H)$. Then with the generalized Liouville–Riemann operator $I_{A,\alpha}$ defined by (11.2), $Y(t) = \Gamma(1)I_{A,\alpha}(Y_\alpha)(t)$. We know that $I_{A,\alpha}$ is a bounded linear operator from $L^q(0,T;H)$ to $C([0,T];H)$ provided that $1/q < \alpha$, and hence for sufficiently large q. It remains to show that Y_α has trajectories in $L^q(0,T;H)$. To do this one can use the BDG inequality but, unlike the Wiener case, one obtains a condition expressed in terms of the quadratic variation $[M, M]$ of M. It turns out that in many cases one can show that $Y_\alpha \in L^q(0,T;H)$ only for $q < 2$. For more details we refer the reader to Section 8.7, which is on stochastic integration with respect to Poisson random measures.

12

Stochastic parabolic problems

The results of Chapter 9 will be applied here to the parabolic problem

$$
\begin{aligned}
\mathrm{d}X(t, \xi) = & \big(\mathcal{A}X(t, \xi) + f(\xi, X(t, \xi))\big)\mathrm{d}t \\
& + g(\xi, X(t, \xi))\,\mathrm{d}W(t, \xi) \\
& + \int_S g_1(\xi, \sigma, X(t, \xi))\widehat{\pi}_1(\mathrm{d}t, \mathrm{d}\xi, \mathrm{d}\sigma) \\
& + \int_S g_2(\xi, \sigma, X(t, \xi))\pi_2(\mathrm{d}t, \mathrm{d}\xi, \mathrm{d}\sigma), \qquad t > 0, \ \xi \in \mathcal{O}, \quad (12.1)
\end{aligned}
$$

with homogeneous boundary conditions

$$
B_j X(t, \xi) = 0, \ j = 1, \ldots, m \qquad \text{for } (t, \xi) \in (0, \infty) \times \partial\mathcal{O}, \qquad (12.2)
$$

and with the initial-value condition

$$
X(0, \xi) = x(\xi) \qquad \text{for } \xi \in \mathcal{O}. \qquad (12.3)
$$

In (12.1), $(S, \mathcal{S})$ is a measurable space, W is a cylindrical Wiener process on $L^2(\mathcal{O})$ and π_1, π_2 are Poisson random measures on $[0, \infty) \times \mathcal{O} \times S$ with Lévy measures $\mathrm{d}t \ \mathrm{d}\xi \ \nu_i(\mathrm{d}\sigma)$, $i = 1, 2$; see Section 7.2. Here ν_1 is an arbitrary non-negative measure on $(S, \mathcal{S})$. The assumptions on g_1, however, will guarantee the square integrability of the corresponding stochastic integral.

12.1 Introduction

As in Section 2.5, we assume that $\mathcal{A}$ is a uniformly elliptic operator of order $2m$ on a bounded region $\mathcal{O} \subset \mathbb{R}^d$ and that $\{B_j, \ j = 1, \ldots, m\}$ is a system of boundary operators. We assume that $\mathcal{O}$, $\mathcal{A}$ and $\{B_j\}$ satisfy the conditions formulated at the beginning of Section 2.5.

We assume further that ν_2 is a finite measure and that the following representation holds:

$$\pi_2 = \sum_j \delta_{t_j, \xi_j, \sigma_j}. \qquad (12.4)$$

Here ξ_k and σ_k are independent random variables; each ξ_j is uniformly distributed on $\mathcal{O}$ and each σ_j has distribution $\nu_2/\nu_2(S)$. Moreover, $t_j = r_1 + \cdots + r_j$, where the r_k are independent random variables, independent of ξ_j and σ_j, with exponential distribution[1]

$$\mathbb{P}\left(r_j > t\right) = e^{-t\ell_d(\mathcal{O})\nu_2(S)}. \qquad (12.5)$$

Note that π_2 has only isolated jumps and can be treated as a compound Poisson process in any Hilbert space V containing Dirac measures on $\mathcal{O} \times S$.

To simplify the exposition, we assume that W, π_1 and π_2 are independent. In this book we consider mostly mild solutions, but for better clarity we will take the weak form of (12.1)–(12.3) below; see (12.6).

Our first task is to rewrite the problem (12.1)–(12.3) in the abstract form (9.4) or (9.31) introduced in Chapter 9, where the general existence results were discussed. This will be done in the next section. Then, in subsections 12.1.2 and 12.1.3, we verify the existence of a solution by applying the general theorems 9.29 and 9.34. More precisely, we first develop the L^2-theory. We assume that $g_2 \equiv 0$ and treat (12.1) as a special case of the evolution equation (9.4). The existence of a mild solution X (in the sense of Definition 9.5) follows from Theorem 9.29. Next, we consider the equation with a non-vanishing term $g_2(\xi, X(t, \xi), \sigma)\pi_2(dt, d\xi, d\sigma)$. It turns out that in this case (12.1) is of the type (9.31), and the existence of a weak solution, in the sense of Definition 9.11, follows from Theorem 9.34. In Section 12.2 we consider the Gaussian case. Using the L^p-theory of integration with respect to Wiener processes, developed in subsection 8.8.2, and the factorization method we obtain the existence of a solution having continuous trajectories in the space $L^p(\mathcal{O})$, where $p \geq 2$. Finally, we discuss the space (and space–time) continuity of the solution. In Section 12.3 we set $g = 0$ and, consequently, obtain the pure jump case. We show the existence of a solution taking values in a certain $L^p(\mathcal{O})$-space for $p \in [1, 2)$. Special examples of (12.1), namely stochastic heat equations with Dirichlet or Neumann boundary conditions, are considered in the final section of the chapter.

We consider only equations on a bounded region with f, g, g_1 and g_2 satisfying certain Lipschitz conditions. Equations on $\mathcal{O} = \mathbb{R}^d$ and equations in which

---

[1] The quantity ℓ_d stands for Lebesgue measure on $\mathbb{R}^d$.

the noise enters through the boundary are considered in Chapters 14 and 15 respectively.

Some standard generalizations of the results presented here are possible. In particular, one can consider (12.1) driven by a Wiener process with nuclear covariance on $L^2(\mathcal{O})$ or, in particular, by a process taking values in a finite-dimensional space. Also, one can consider the case where the equation is driven by a cylindrical Wiener process on a space $L^2(\mathcal{O}, \mathcal{B}(\mathcal{O}), \lambda)$, with λ a given measure on $\mathcal{O}$. A much more ambitious task would be presented by equations with non-Lipschitz coefficients. In particular, equations where

$$f(\xi, r) = \sum_{j=0}^{2l+1} a_j(\xi) r^j$$

is a polynomial of odd order with a strictly negative leading coefficient a_{2l+1} are of great interest; see Example 10.8. They include the so-called reaction–diffusion equations. These equations with Gaussian noise were studied intensively in Da Prato and Zabczyk (1992b, 1996), Brzeźniak and Peszat (1999) and Cerrai (2001). In this book, we consider only stochastic reaction–diffusion equations with additive noise; see Section 10.3.

Stochastic parabolic equations with noise coefficients of the type $c|r|^\gamma$, $\gamma < 1$, appear naturally in some models of interacting-particle systems. So far, these equations have been treated as martingale problems; see Mytnik (2002), Mueller (1998) and Section 12.5.

12.1.1 Abstract form

Our goal is to write (12.1)–(12.3) in the standard form

$$dX = (AX + F(X)) \, dt + G(X) \, dM + R(X) \, dP, \qquad X(0) = x,$$

with properly chosen coefficients and state and noise spaces.

To this end, we take $H = L^2(\mathcal{O})$ as the state space. Next, for A we take the realization of $\mathcal{A}$ with the boundary conditions B_j; see appendix section B.2.

For the martingale part, we take $M := (W, \widehat{\pi}_1)$. Since W and $\widehat{\pi}_1$ are independent, the RKHS of M is the product of the RKHSs of W and $\hat{\pi}_1$. Thus, by Theorem 7.28, M is a square integrable Lévy martingale with RKHS equal to

$$\mathcal{H} = \mathcal{H}_W \times \mathcal{H}_{\widehat{\pi}_1} := L^2(\mathcal{O}) \times L^2(\mathcal{O} \times S, \mathcal{B}(\mathcal{O}) \times \mathcal{S}, d\xi \, \nu_1).$$

For the noise space U, we take any Hilbert space containing $\mathcal{H}$ with a Hilbert–Schmidt embedding. Clearly, the non-linear drift term F in the abstract formulation is the composition operator corresponding to f:

$$F(\psi)(\xi) := f(\xi, \psi(\xi)), \qquad \psi \in L^2(\mathcal{O}), \ \xi \in \mathcal{O}. \tag{12.6}$$

Before identifying the coefficients G and H it is useful to find the weak form of (12.1)–(12.3). Assume first that $g_2 \equiv 0$. A predictable $L^2(\mathcal{O})$-valued process X is a weak solution of (12.1)–(12.3) if

$$\sup_{t \le T} \mathbb{E} \, |X(t)|^2_{L^2(\mathcal{O})} < \infty, \qquad \forall \, T > 0, \tag{12.7}$$

and, for all $a \in D(A^*)$ and $t > 0$,

$$\begin{aligned}
\int_{\mathcal{O}} a(\xi) X(t, \xi) \, d\xi &= \int_{\mathcal{O}} a(\xi) x(\xi) \, d\xi \\
&\quad + \int_0^t \int_{\mathcal{O}} \left(A^* a(\xi) X(s, \xi) + a(\xi) f(s, X(s, \xi)) \right) d\xi \, ds \\
&\quad + \int_0^t a g(\cdot, X(s, \cdot)) \, dW(s) \\
&\quad + \int_0^t \int_{\mathcal{O}} \int_S a(\xi) g_1(\xi, \sigma, X(s, \xi)) \widehat{\pi}_1(ds, d\xi, d\sigma). \tag{12.8}
\end{aligned}$$

In the term involving W, $\Psi(s) := a g(\cdot, X(s, \cdot))$, $s \ge 0$ is an $L^2(\mathcal{O})$-valued or, under the proper identification, $\left(L^2(\mathcal{O}) \right)^* = L_{(HS)}(L^2(\mathcal{O}), \mathbb{R}) = L_{(HS)}(\mathcal{H}_W, \mathbb{R})$-valued process. Thus, assuming that $\Psi \in \mathcal{L}^2_{W,T}(\mathbb{R})$, $T > 0$, the integral is considered in the sense developed in Chapter 8. Similarly, in the Poisson term, $a(\xi) g_1(\xi, \sigma, X(s, \xi))$, $\xi \in \mathcal{O}$, $s \in [0, T]$, $\sigma \in S$, is understood as an element of $\mathcal{L}^2_{\widehat{\pi}, T}(\mathbb{R})$; see Proposition 8.24.

However (see (9.16)), the diffusion operator G should map $D(G) \subset L^2(\mathcal{O})$ into the space of linear operators from $\mathcal{H}$ to $L^2(\mathcal{O})$ in such a way that

$$\begin{aligned}
& \left\langle a, G(\psi)(\varphi, \varphi_1) \right\rangle_{L^2(\mathcal{O})} \\
&= \int_{\mathcal{O}} a(\xi) g(\xi, \psi(\xi)) \varphi(\xi) \, d\xi + \int_{\mathcal{O}} \int_S g_1(\xi, \sigma, \psi(\xi)) \varphi_1(\xi, \sigma) \, d\xi \, \nu_1(d\sigma).
\end{aligned}$$

Thus G is determined; namely, it is a generalized composition operator,

$$G(\psi)[(\varphi, \varphi_1)](\xi) := g(\xi, \psi(\xi)) \varphi(\xi) + \int_S g_1(\xi, \sigma, \psi(\xi)) \varphi_1(\xi, \sigma) \nu_1(d\sigma). \tag{12.9}$$

In order to identify the finite-variation term, let V be a Hilbert space containing the Dirac measures $\delta_{\xi, \sigma}$, $\xi \in \mathcal{O}$ and $\sigma \in S$. We assume that the mapping $j \colon \mathcal{O} \times S \ni (\xi, \sigma) \mapsto \delta_{\xi, \sigma} \in V$ is measurable. Then

$$P(t) := \sum_{t_j \le t} \delta_{\xi_j, \sigma_j},$$

where t_j, ξ_j, σ_j appear in (12.4), is a compound Poisson process in V with jump intensity measure $\nu_2 \circ j^{-1}$. The corresponding term in the definition of the weak solution is $\sum_{t_j \leq t} g_2(\xi_j, \sigma_j, X(t_j, \xi_j)) a(\xi_j)$. We would like to rewrite this in the form $\int_0^t R^*(X(s)) a \, dP(s)$. Thus

$$\langle R(\psi) \delta_{\xi, \sigma}, a \rangle_{L^2} = g_2(x, \sigma, \psi(\xi)) a(\xi). \tag{12.10}$$

If $g_2 \equiv 0$, and consequently the term in (9.31) containing the compound Poisson process vanishes, we treat (12.1) as (9.4) with the terms defined as above.

Remark 12.1 Let $\mathcal{W}$ be the Brownian sheet corresponding to W; see Remark 7.19. Using an approach based on the concept of multiparameter stochastic integration, as in Walsh (1986), we obtain

$$\int_0^t ag(\cdot, X(s, \cdot)) \, dW(s) = \int_0^t \int_{\mathcal{O}} a(\xi) g(\xi, X(s, \xi)) \mathcal{W}(ds, d\xi).$$

12.1.2 L^2-theory

In this section, we consider (12.1) with $g_2 \equiv 0$. We impose several conditions on the coefficients A, f and g_1. The first establishes a relation between the order of the operator A and the dimension of the region $\mathcal{O}$. Namely, we require

$$2m > d. \tag{12.11}$$

Remark 12.2 It is not difficult to prove that (12.11) is a necessary condition for the solution to the linear Gaussian problem

$$dX(t, \xi) = \mathcal{A}X(t, \xi) \, dt + dW(t, \xi), \qquad (t, \xi) \in (0, \infty) \times \mathcal{O},$$
$$B_j X(t, \xi) = 0, \qquad (t, \xi) \in (0, \infty) \times \partial\mathcal{O},$$

to be a function-valued process. In Section 12.3 we will see that in the pure jump case we can drop (12.11) and still obtain a function-valued solution, but this solution will take values in $L^p(\mathcal{O})$-spaces with $p < 2$.

Using the notation introduced in Section 2.4 we will formulate Lipschitz and linear-growth conditions on f, g and g_1. The main result of the present section is the following theorem.

Theorem 12.3 *Assume that (12.11) holds and that $g_2 \equiv 0$, $f \in \text{Lip}(2, \mathcal{O}, \ell_d)$ and $g \in \text{Lip}(\infty, \mathcal{O}, \ell_d)$. Assume that there are functions $l_1 \in L^2(S, \mathcal{S}, \nu_1)$ and $l_2 \in L^\infty(\mathcal{O})$ such that, for all $\xi \in \mathcal{O}$, $\sigma \in S$ and $r, u \in \mathbb{R}$,*

$$|g_1(\xi, \sigma, r)| \leq l_1(\sigma)(l_2(\xi) + |r|),$$
$$|g_1(\xi, \sigma, r) - g_1(\xi, \sigma, u)| \leq l_1(\sigma)|r - u|. \tag{12.12}$$

Then, for any $x \in L^2(\mathcal{O})$, there is a unique predictable $L^2(\mathcal{O})$-valued process X satisfying (12.1) such that, for all $T < \infty$, $\sup_{t \leq T} \mathbb{E} |X(t)|^2_{L^2(\mathcal{O})} < \infty$. Moreover, (12.1) defines a Feller family on $L^2(\mathcal{O})$.

Proof We will show that assumptions (F) and (G) of Section 9.2 are satisfied.

We first check (F). Note that in our case $D(F)$ is equal to the class of all $\psi \in L^2(\mathcal{O})$ such that $\xi \mapsto f(\xi, \psi(\xi))$ belongs to $L^2(\mathcal{O})$. Since $f \in \mathrm{Lip}\,(2, \mathcal{O}, \ell_d)$, we have $D(F) = L^2(\mathcal{O})$. Next, F is in fact a Lipschitz continuous mapping from $L^2(\mathcal{O})$ to $L^2(\mathcal{O})$, which clearly implies (F).

We now show (G). Note that $G \colon L^\infty(\mathcal{O}) \mapsto L(\mathcal{H}, L^2(\mathcal{O}))$. Thus we can take $L^\infty(\mathcal{O})$ for $D(G)$. Note that for $\psi \in L^2(\mathcal{O})$ and $t > 0$ the operator $S(t)G(\psi)$ is given by a kernel. In fact,

$$
\begin{aligned}
&S(t)G(\psi)[(\varphi, \varphi_1)](\xi) \\
&= \int_{\mathcal{O}} I(t, \psi, \xi, \eta)\varphi(\eta)\,\mathrm{d}\eta + \int_{\mathcal{O}} \int_S I_1(t, \psi, \xi, \eta, \sigma)\varphi_1(\eta, \sigma)\,\mathrm{d}\eta\,\nu_1(\mathrm{d}\sigma),
\end{aligned}
$$

where

$$
\begin{aligned}
I(t, \psi, \xi, \eta) &= \mathcal{G}(t, \xi, \eta)g(\eta, \psi(\eta)), \\
I_1(t, \psi, \xi, \eta, \sigma) &= \mathcal{G}(t, \xi, \eta)g_1(\eta, \sigma, \psi(\eta))
\end{aligned}
$$

and $\mathcal{G}$ is the Green function for $\mathcal{A}, \{B_j\}$. Thus, by Proposition A.7 from Appendix A,

$$
\begin{aligned}
&\|S(t)G(\psi)\|^2_{L_{(HS)}(\mathcal{H}, L^2(\mathcal{O}))} \\
&\leq 2 \int_{\mathcal{O}} \int_{\mathcal{O}} |I(t, \psi, \xi, \eta)|^2\,\mathrm{d}\xi\,\mathrm{d}\eta + 2 \int_{\mathcal{O}} \int_{\mathcal{O}} \int_S |I_1(t, \psi, \xi, \eta, \sigma)|^2\,\mathrm{d}\xi\,\mathrm{d}\eta\,\nu_1(\mathrm{d}\sigma).
\end{aligned}
$$

By the Aronson estimates (see Theorem 2.6) for an arbitrary T we can find a constant K such that, for all $t \in [0, T]$, $\eta \in \mathcal{O}$ and $\sigma \in S$,

$$
\int_{\mathcal{O}} |I(t, \psi, \xi, \eta)|^2\,\mathrm{d}\xi \leq Kt^{-d/2m} |g(\eta, \psi(\eta))|^2 \tag{12.13}
$$

and

$$
\int_{\mathcal{O}} |I_1(t, \psi, \xi, \eta, \sigma)|^2\,\mathrm{d}\xi \leq Kt^{-d/2m} |g_1(\eta, \sigma, \psi(\eta))|^2. \tag{12.14}
$$

By the assumptions on g and g_1, for an arbitrary T we can find a constant C such that, for all $t \in [0, T]$ and $\psi \in L^2(\mathcal{O})$,

$$
\|S(t)G(\psi)\|^2_{L_{(HS)}(\mathcal{H}, L^2(\mathcal{O}))} \leq Ct^{-d/2m} \left(1 + |\psi|^2_{L^2(\mathcal{O})}\right). \tag{12.15}
$$

The same arguments yield

$$\|S(t)(G(\psi_1) - G(\psi_2))\|^2_{L_{(HS)}(\mathcal{H}, L^2(\mathcal{O}))} \le C t^{-d/2m} |\psi_1 - \psi_2|^2_{L^2(\mathcal{O})}, \qquad (12.16)$$

for all $t \in [0, T]$ and $\psi_1, \psi_2 \in L^2(\mathcal{O})$, which, by (12.11), ensures (G). □

Remark 12.4 The assumptions that $f \in \text{Lip}(2, \mathcal{O}, \ell_d)$ and $g \in \text{Lip}(\infty, \mathcal{O}, \ell_d)$ and that g_1 satisfies (12.12) with $l_1 \in L^2(S, \mathcal{S}, \nu_1)$ and $l_2 \in L^\infty(\mathcal{O})$ can be weakened, but then the framework of Theorem 9.29 is exceeded. For the drift f it suffices to assume that $f \in \text{Lip}(q, \mathcal{O}, \ell_d)$, where q is such that

$$\int_0^T \|S(t)\|_{L(L^q(\mathcal{O}), L^2(\mathcal{O}))} \, \mathrm{d}t < \infty$$

for all, or equivalently for a certain, $T > 0$. By (B.15), it is sufficient to assume that

$$-\frac{d}{2m}\left(\frac{1}{q} - \frac{1}{2}\right) > -1.$$

This is necessary for the solution to the deterministic problem

$$\mathrm{d}X(t, \xi) = (AX(t, \xi) + f(\xi)) \, \mathrm{d}t, \qquad (t, \xi) \in (0, \infty) \times \mathcal{O},$$
$$B_j X(t, \xi) = 0, \qquad (t, \xi) \in (0, \infty) \times \partial\mathcal{O},$$

to be bounded in time in $L^2(\mathcal{O})$.

By (12.13), for g it is sufficient to assume that it belongs to the space Lip $(2, \mathcal{O}, \ell_d)$. Finally, for g_1 it suffices to assume that it satisfies (12.12) with $l_1 \in L^2(\mathcal{O})$; see (12.14).

12.1.3 Adding the finite-variation part

Let us add to the equation considered in the previous subsection (see Remark 12.2) a finite-variation term

$$\int_S g_2(\xi, \sigma, X(t, \xi)) \pi_2(\mathrm{d}t, \mathrm{d}\xi, \mathrm{d}\sigma);$$

equivalently, we could add to (9.4) the term $R(X(s)) \, \mathrm{d}P(s)$, where R is given by (12.10) and P is a compound Poisson process on the space V containing the Dirac measures on $\mathcal{O} \times S$.

Given a sequence $\{S_m\} \subset S$ we denote by $\mathcal{S}_m$ the restriction of $\mathcal{S}$ to S_m and by $\nu_{2,m}$ the restriction of ν_2 to S_m. Finally, we define

$$V_m := \{\delta_{\xi,\sigma} : \xi \in \mathcal{O}, \ \sigma \in S_m\}.$$

Theorem 12.5 *Let (12.11) hold, and let* f, g *and* g_1 *satisfy the assumptions of Theorem 12.3. Assume that* $S = \bigcup S_m$, *where* $\{S_m\} \subset S$ *is an increasing sequence such that the* V_m *are measurable and bounded in* V *and that for every* m *there are functions* $l_{1,m} \in L^2(S_m, S_m, \nu_{2,m})$ *and* $l_{2,m} \in L^\infty(\mathcal{O})$ *such that for all* $\xi \in \mathcal{O}$, $\sigma \in S_m$ *and* $r, u \in \mathbb{R}$,

$$|g_2(\xi, \sigma, r)| \le l_{1,m}(\sigma)(l_{2,m}(\xi) + |r|),$$
$$|g_2(\xi, \sigma, r) - g_2(\xi, \sigma, u)| \le l_{1,m}(\sigma)|r - u|.$$

Then for any $x \in L^2(\mathcal{O})$ *there is a unique weak solution to (12.1). Moreover, (12.1) defines a Feller family on* $L^2(\mathcal{O})$.

Proof One can show easily that, given m, the equation obtained from (12.1), by replacing π_2 with $\chi_{S_m}(\sigma)(\hat{\pi}_2(\mathrm{d}t, \mathrm{d}\xi, \mathrm{d}\sigma) - \mathrm{d}t\, \mathrm{d}\xi\, \nu_2(\mathrm{d}\sigma))$ and by adding to the drift f the term $f_m(\xi, r) = \int_{S_m} g_2(\xi, \sigma, r)\nu_2(\mathrm{d}\sigma)$, satisfies the assumptions of Theorem 12.3. □

12.2 Space–time continuity in the Wiener case

In this section we consider (12.1)–(12.3) in the case where the jump terms g_1 and g_2 vanish. We aim to prove the existence of solutions, regular in time and space, to the equation

$$\begin{aligned}
\mathrm{d}X(t, \xi) = \big(AX(t, \xi) + f(\xi, X(t, \xi))\big)\, \mathrm{d}t \\
+ g(\xi, X(t, \xi))\, \mathrm{d}W(t, \xi), \qquad t > 0, \quad \xi \in \mathcal{O}, \quad (12.17)
\end{aligned}$$

driven by a cylindrical Wiener process W on $L^2(\mathcal{O})$. We consider (12.17) with homogeneous boundary and initial-value conditions (12.2) and (12.3). We assume that f and g are Lipschitz continuous. To simplify the exposition, we assume further that f and g are measurable and satisfy the Lipschitz and linear-growth conditions

$$\begin{aligned}
\exists K < \infty: \forall \xi \in \mathcal{O}, \ \forall r, u \in \mathbb{R}: \\
|f(\xi, r)| + |g(\xi, r)| \le K(1 + |r|), \qquad\qquad (12.18) \\
|f(\xi, r) - f(\xi, u)| \\
+ |g(\xi, r) - g(\xi, u)| \le K|r - u|.
\end{aligned}$$

12.2.1 Time continuity in $L^p(\mathcal{O})$

Given $p, q \in [2, \infty)$ and $T < \infty$ we denote by $\mathfrak{Z}_{T,p,q}$ the class of all adapted processes $Y \colon \Omega \times [0, T] \mapsto L^p(\mathcal{O})$ with continuous trajectories in $L^p(\mathcal{O})$ and satisfying

$$\|Y\|_{3_{T,p,q}} := \left(\mathbb{E} \sup_{t\in[0,T]} |Y(t)|_{L^p(\mathcal{O})}^q \right)^{1/q} < \infty.$$

Our first result deals with the existence of a solution to (12.17), (12.2) and (12.3) with continuous trajectories in $L^p(\mathcal{O})$. We assume (12.11) and (12.18). Hence, by Theorem 12.3, for any $x \in L^2(\mathcal{O})$ there is a unique solution to (12.17), (12.2) and (12.3) satisfying (12.7).

Theorem 12.6 *Assume (12.11) and (12.18). Let X be a solution to (12.17), (12.2) and (12.3) starting from an $x \in L^2(\mathcal{O})$. Then the following hold.*

(i) *$X \in 3_{T,2,q}$ for all $T < \infty$ and $q \in [2, \infty)$.*
(ii) *Assume that $p \in (2, \infty)$ is such that there is an $\alpha \in (0, 1)$ satisfying*

$$\alpha > \frac{1}{p} \quad and \quad -\frac{d}{2m}\left(1 + \frac{2}{p} \wedge \left(1 - \frac{2}{p}\right)\right) - 2\alpha > -1. \quad (12.19)$$

If $x \in L^p(\mathcal{O})$ then $X \in 3_{T,p,p}$ for every $T < \infty$. Moreover (12.17), (12.2) and (12.3) define a Feller family on $L^p(\mathcal{O})$.

Proof Fix T. The first statement follows directly from Theorem 11.8, because our problem can be written in the form (11.5), with F and G as defined in subsection 12.1.1. Let $\alpha > 0$ be such that $-2\alpha - d/(2m) > -1$. Clearly such an α exists. It is enough to show that $X \in 3_{T,2,q}$ for q sufficiently large, say for $q: 1/q < \alpha$. To do this it suffices to check the hypothesis of Theorem 11.8. Since $F: L^2(\mathcal{O}) \to L^2(\mathcal{O})$ is a Lipschitz mapping, $S(t)F$ satisfies the assumptions if a is a constant function. Next, by (12.15) and (12.16), $S(t)G$ satisfies the assumptions with $b(t) = Ct^{-d/m}$.

We now prove the second assertion. Assume that α satisfies (12.19) and that $p: 1/p < \alpha$. We use the Banach contraction principle, the theory of stochastic integration in L^p-spaces developed in subsection 8.8.2 and the factorization established in Chapter 11.

First of all we need to estimate the second moment of the stochastic integral. The following lemma plays a crucial role. It provides analogues of (12.14) and (12.15) in terms of radonifying norms. Recall that $\mathcal{H} = L^2(\mathcal{O})$ is the RKHS of W. Assume that U is a Hilbert space such that $\mathcal{H} \hookrightarrow U$ is dense and Hilbert–Schmidt. Then the Wiener process W takes values in U. Recall that the spaces $R(\mathcal{H}, L^p(\mathcal{O}))$ and $R_{U,0}(\mathcal{H}, L^p(\mathcal{O}))$, the norms $\| \cdot \|_{R_{U,0}(\mathcal{H},L^p(\mathcal{O}))}$ and $\| \cdot \|_{R(\mathcal{H},L^p(\mathcal{O}))}$ and the space of integrable processes $\mathcal{L}_{W,T}^p(L^p(\mathcal{O}))$ were introduced in subsection 8.8.2.

Lemma 12.7 *For all $t > 0$ and $\psi \in L^p(\mathcal{O})$ the operator $\varphi \mapsto S(t)G(\psi)\varphi$ belongs to $R(\mathcal{H}, L^p(\mathcal{O}))$. Moreover, there is a constant C such that, for all $t \in [0, T]$*

and $\psi, \phi \in L^p(\mathcal{O})$,

$$\|S(t)G(\psi)\|^2_{R(\mathcal{H}, L^p(\mathcal{O}))} \leq Ct^\rho \left(1 + |\psi|^2_{L^p(\mathcal{O})}\right), \qquad (12.20)$$

$$\|S(t)(G(\psi) - G(\phi))\|^2_{R(\mathcal{H}, L^p(\mathcal{O}))} \leq Ct^\rho |\psi - \phi|^2_{L^p(\mathcal{O})}, \qquad (12.21)$$

where

$$\rho := -\frac{d}{2m}\left(1 + \frac{2}{p} \wedge \left(1 - \frac{2}{p}\right)\right).$$

Proof   Assume that the duality $\langle \cdot, \cdot \rangle$ on $U^* \times U$ is given by the scalar product on $\mathcal{H} = L^2(\mathcal{O})$. Therefore we identify the adjoint space U^* with a dense subspace of $\mathcal{H}$. Assume that $\psi \in C(\overline{\mathcal{O}})$. Let $\mathcal{G}$ be the Green function for $(\mathcal{A}, \{B_j\})$. Then

$$(S(t)G(\psi)\varphi)(\xi) = \int_{\mathcal{O}} \mathcal{Q}(t, \psi, \xi, \eta)\varphi(\eta)\, d\eta,$$

where the kernel is given by $\mathcal{Q}(t, \psi, \xi, \eta) := \mathcal{G}(t, \xi, \eta)g(\eta, \psi(\eta))$. By the Hölder inequality,

$$\langle \mathcal{Q}(t, \psi, \xi, \cdot), \mathcal{Q}(t, \psi, \xi, \cdot) \rangle = \int_{\mathcal{O}} |\mathcal{G}(t, \xi, \eta)|^2 |g(\eta, \psi(\eta))|^2\, d\eta$$

$$\leq \left(\int_{\mathcal{O}} |g(\eta, \psi(\eta))|^p\, d\eta\right)^{2/p} \left(\int_{\mathcal{O}} |\mathcal{G}(t, \xi, \eta)|^{2p/(p-2)}\, d\eta\right)^{(p-2)/p}.$$

By the Aronson estimates,

$$\int_{\mathcal{O}} |\mathcal{G}(t, \xi, \eta)|^{2p/(p-2)}\, d\eta$$

$$\leq C_1 \int_{\mathcal{O}} t^{-(d/2m)(2p/(p-2))} \exp\left\{-C_2\left(\frac{|\xi - \eta|^{2m}}{t}\right)^{1/(2m-1)}\right\} d\eta$$

$$\leq t^{-(d/2m)(p+2)/(p-2)} C_1 \int_{\mathbb{R}^d} t^{-d/2m} \exp\left\{-C_2\left(\frac{|\xi - \eta|^{2m}}{t}\right)^{1/(2m-1)}\right\} d\eta$$

$$\leq C_3 t^{-(d/2m)(p+2)/(p-2)}.$$

Taking into account the linear growth of g we obtain

$$\int_{\mathcal{O}} |g(\eta, \psi(\eta))|^p\, d\eta \leq C_4 (1 + |\psi|_{L^p})^p.$$

Summing up, we can find a constant C_5 such that, for all $t \in (0, T]$, $\xi \in \mathcal{O}$ and $\psi \in C(\overline{\mathcal{O}})$,

$$\langle \mathcal{Q}(t, \psi, \xi, \cdot), \mathcal{Q}(t, \psi, \xi, \cdot) \rangle \leq C_5 t^{-(d/2m)(p+2)/p} \left(1 + |\psi|^2_{L^p}\right).$$

Consequently, there is a constant C_6 such that, for all $t \in (0, t]$ and $\psi \in C(\overline{\mathcal{O}})$,

$$I := \left(\int_{\mathcal{O}} \langle \mathcal{Q}(t, \psi, \xi, \cdot), \mathcal{Q}(t, \psi, \xi, \cdot) \rangle^{p/2} \, d\xi \right)^{2/p} \le C_6 t^{-(d/2m)(p+2)/p} \left(1 + |\psi|^2_{L^p(\mathcal{O})} \right).$$

In order to estimate I we can also use the Jensen inequality. This will lead to a slightly different estimate. Namely, since $\mathcal{O}$ is of finite measure and $p \ge 2$,

$$I \le c_1 \left(\int_{\mathcal{O}} \int_{\mathcal{O}} |\mathcal{G}(t, \xi, \eta)|^p |g(\eta, \psi(\eta))|^p \, d\xi \, d\eta \right)^{2/p}$$
$$\le c_2 t^{-(d/2m)(2p-2)/p} \left(1 + |\psi|^2_{L^p(\mathcal{O})} \right).$$

Combining these two estimates on I we obtain

$$I \le C t^\rho \left(1 + |\psi|^2_{L^p(\mathcal{O})} \right),$$

where

$$\rho := -\frac{d}{2m} \left(1 + \frac{2}{p} \wedge \left(1 - \frac{2}{p} \right) \right).$$

Recall (see subsection 8.8.2) that for the case where the operator Ψ is given by a kernel $\mathcal{Q}(\xi) \in U^*, \xi \in \mathcal{O}$,

$$\|\Psi\|^2_{R_{U,0}(\mathcal{H}, L^p(\mathcal{O}))} = \left(\int_{\mathcal{O}} \langle \mathcal{Q}(\xi), \mathcal{Q}(\xi) \rangle^{p/2} \, d\xi \right)^{2/p}.$$

Since U^* is dense in $\mathcal{H}$ we can approximate $\mathcal{Q}(t, \psi, \xi, \cdot)$ by kernels taking values in U^*. Thus $S(t)G(\psi)$ belongs to the completion $R(\mathcal{H}, L^p(\mathcal{O}))$ of $R_{U,0}(\mathcal{H}, L^p(\mathcal{O}))$ and satisfies (12.20). Inequality (12.21) follows in the same way. Thus the proof of the lemma is complete. □

We now go back to the proof of Theorem 12.6(ii) and use the Banach contraction principle on the space $\mathfrak{Z} := \mathfrak{Z}_{T,p,p}$. On $\mathfrak{Z}$, we consider the family of equivalent norms

$$\|Y\|_\beta := \left(\mathbb{E} \sup_{t \in [0,T]} e^{-\beta t} |Y(t)|^p_{L^p(\mathcal{O})} \right)^{1/p}, \qquad \beta > 0.$$

Let

$$J_1(Y)(t) := S(t)x + \int_0^t S(t-s)F(Y(s)) \, ds,$$

$$J_2(Y)(t) := \int_0^t S(t-s)G(Y(s)) \, dW(s).$$

It is sufficient to show that $J_1, J_2 : \mathfrak{Z} \mapsto \mathfrak{Z}$ and that there is a β such that

$$\| J_i(Y) - J_i(Z) \|_\beta \le \tfrac{1}{3} \| Y - Z \|_\beta, \qquad i = 1, 2, \ \forall\, Y, Z \in \mathfrak{Z}.$$

We concentrate on the stochastic term, using the factorization method introduced in Chapter 11.

Recall that $I_{A,\alpha}$ is the infinite-dimensional Liouville–Riemann operator defined by (11.2). Since $1/p < \alpha < 1$ by Theorem 11.5, $I_{A,\alpha}$ maps the space $L^p(0, T; L^p(\mathcal{O}))$ into $C([0, T]; L^p(\mathcal{O}))$.

Let $Y \in \mathfrak{Z}$ and let $t \in [0, T]$. It follows from Lemma 12.7 that

$$\Psi : \Omega \times [0, t] \ni (\omega, s) \mapsto (t - s)^{-\alpha} S(t - s) G(Y(s, \omega)) \in R(\mathcal{H}, L^p(\mathcal{O}))$$

belongs to the space $\mathcal{L}^p_{W,t}(L^p(\mathcal{O}))$ of integrable processes and that

$$\mathbb{E} \left(\int_0^t (t - s)^{-2\alpha} \| S(t - s) G(Y(s)) \|^2_{R(\mathcal{H}, L^p(\mathcal{O}))} \, ds \right)^{p/2}$$

$$\le C_1 \left(1 + \| Y \|_0^p \right) \left(\int_0^t s^{-2\alpha + \rho} \, ds \right)^{p/2} \le C_2 (1 + \| Y \|_0)^p,$$

where ρ is given in Lemma 12.7, C_2 is a constant independent of Y and of $t \in [0, T]$ and $\| \cdot \|_0$ is the norm on $\mathfrak{Z}$ corresponding to $\beta = 0$. Let

$$I(Y)(t) = \frac{1}{\Gamma(1 - \alpha)} \int_0^t (t - s)^{-\alpha} S(t - s) G(Y(s)) \, dW(s), \qquad t \in [0, T].$$

By the estimate above and Theorem 8.23, $I(Y)$ is a well-defined process satisfying $\sup_{t \in [0,T]} \mathbb{E}\, |I(Y)(t)|^p_{L^p(\mathcal{O})} < \infty$. Thus, in particular, $I(Y)$ has trajectories in $L^p(0, T; L^p(\mathcal{O}))$. We have (see the proof of Theorem 11.8) $J_2(Y) = \Gamma(1) I_{A,\alpha}(I(Y))$ and consequently $J_2(Y) \in \mathfrak{Z}$. We now show that J_2 is contractive. In what follows, the C_j are constants independent of $t \in [0, T]$, $Y, Z \in \mathfrak{Z}$ and β. We have

$$\| J_2(Y) - J_2(Z) \|^p_\beta = \mathbb{E} \sup_{t \in [0,T]} e^{-\beta t} |J_2(Y)(t) - J_2(Z)(t)|^p_{L^p(\mathcal{O})}$$

$$\le C_1\, \mathbb{E} \sup_{t \in [0,T} e^{-\beta t} |I_{A,\alpha}(I(Y) - I(Z))(t)|^p_{L^p(\mathcal{O})}.$$

Let $\beta' = \beta/p$ and $q = p/(p - 1)$. Since

$$\left| \int_0^t (t - s)^{\alpha - 1} e^{-\beta'(t-s)} S(t - s) e^{-\beta' s} \big(I(Y)(s) - I(Z)(s) \big) ds \right|_{L^p(\mathcal{O})}$$

$$\le C_2 \left(\int_0^T t^{q(\alpha - 1)} e^{-q\beta' t} \, dt \right)^{1/q} \left(\int_0^T e^{-\beta s} |I(Y)(s) - I(Z)(s)|^p_{L^p(\mathcal{O})} \, ds \right)^{1/p}$$

we have

$$\|J_2(Y) - J_2(Z)\|_\beta \leq c(\beta) C_3 \left(\int_0^T e^{-\beta t}\, \mathbb{E}\, |I(Y)(t) - I(Z)(t)|^p_{L^p(\mathcal{O})}\, \mathrm{d}t \right)^{1/p}$$

where

$$c(\beta) := \left(\int_0^T t^{q(\alpha-1)} e^{-q\beta' t}\, \mathrm{d}t \right)^{1/q} \to 0 \qquad \text{as } \beta \uparrow \infty. \tag{12.22}$$

Now, by Lemma 12.7 and Theorem 8.23,

$$\mathbb{E}\, |I(Y)(t) - I(Z)(t)|^p_{L^p(\mathcal{O})} \leq C_4\, \mathbb{E} \sup_{s \in [0,t]} |Y(s) - Z(s)|^p_{L^p(\mathcal{O})}.$$

Thus

$$e^{-\beta t}\, \mathbb{E}\, |I(Y)(t) - I(Z)(t)|^p_{L^p(\mathcal{O})} \leq C_4\, \mathbb{E} \sup_{s \in [0,t]} e^{-\beta s} |Y(s) - Z(s)|^p_{L^p(\mathcal{O})}$$

$$\leq C_4\, \|Y - Z\|^p_\beta.$$

Summing up,

$$\|J_2(Y) - J_2(Z)\|_\beta \leq C_5 c(\beta) \|Y - Z\|_\beta,$$

and the desired estimate follows from (12.22). $\qquad\qquad\qquad\qquad\square$

12.2.2 Space continuity

Generally one cannot obtain solutions that are continuous in time in the space of all continuous functions $C(\overline{\mathcal{O}})$ on $\overline{\mathcal{O}}$. In fact $S(t)$ transforms $C(\overline{\mathcal{O}})$ into a certain smaller subspace depending on the boundary condition. Therefore, even the linear deterministic problem

$$\mathrm{d}X = \mathcal{A}X\, \mathrm{d}t, \qquad B_j X = 0, \qquad X(0) = x$$

does not have a space–time continuous solution for all $x \in C(\overline{\mathcal{O}})$.

We show below the existence of a solution to (12.17) that is bounded in time and continuous in space. For the space–time continuity of solutions, starting from $C_0(0, 1)$, to the stochastic heat equations on the interval $(0, 1)$ with Dirichlet boundary conditions we refer the reader to Remark 12.9 and subsection 12.4.3 below.

Let us equip $C(\overline{\mathcal{O}})$ with the supremum norm $\|\cdot\|_\infty$.

Theorem 12.8 *Assume (12.11) and (12.18). Let X be a solution to (12.17), (12.2) and (12.3). If $x \in C(\overline{\mathcal{O}})$ then*

$$\mathbb{E} \sup_{t \in [0,T]} \|X(t)\|^q_\infty < \infty, \qquad \forall\, T < \infty,\ \forall\, q > 1. \tag{12.23}$$

Proof   First note that for each sufficiently large p there is an $\alpha \in (0, 1)$ satisfying (12.19). By Theorem 12.6,

$$\mathbb{E} \sup_{t \in [0,T]} |X(t)|^p_{L^p(\mathcal{O})} < \infty, \qquad \forall\, T < \infty, \ \forall\, p > 1. \qquad (12.24)$$

Let us fix a $T < \infty$. We will show that the factorization and estimates (B.16) and (B.17) from Appendix B allow us to deduce (12.23) from (12.24). We have

$$X(t) = S(t)x + I_1(t) + I_2(t),$$

where

$$I_1(t) := \int_0^t S(t - s)F(X(s))\,\mathrm{d}s, \qquad I_2(t) := \int_0^t S(t - s)G(X(s))\,\mathrm{d}W(s).$$

By (B.16), $\sup_{t \in [0,T]} \|S(t)x\|_\infty < \infty$. Next, let $\alpha > 0$ and p be such that

$$\alpha - \frac{d}{2mp} > \frac{1}{p}, \qquad -2\alpha - \frac{d}{2m}\left(1 + \frac{2}{p} \wedge \left(1 - \frac{2}{p}\right)\right) > -1. \quad (12.25)$$

Let $I_{A,\alpha}$ be given by (11.2). We have $I_1(t) = \Gamma(1)I_{A,\alpha}(Y_1)(t)$ and $I_2(t) = \Gamma(1)I_{A,\alpha}(Y_2)(t)$, where

$$Y_1(t) := \frac{1}{\Gamma(1 - \alpha)} \int_0^t (t - s)^{-\alpha} S(t - s)F(X(s))\,\mathrm{d}s,$$

$$Y_2(t) := \frac{1}{\Gamma(1 - \alpha)} \int_0^t (t - s)^{-\alpha} S(t - s)G(X(s))\,\mathrm{d}W(s).$$

By (12.24), (12.25) and Lemma 12.7,

$$\mathbb{E} \int_0^T |Y_i(t)|^p_{L^p(\mathcal{O})}\,\mathrm{d}t < \infty, \qquad \forall\, p \geq p_0, \ i = 1, 2,$$

and the desired conclusion follows since, by (B.17), $I_{A,\alpha}$ maps $L^p(0, T; L^p(\mathcal{O}))$ to $L^\infty(0, T; C(\overline{\mathcal{O}}))$. $\square$

Remark 12.9   Given $t > 0$, define $\mathcal{C} := \overline{S(t)C(\overline{\mathcal{O}})}$. Then S is also a C_0-semigroup on $\mathcal{C}$. Hence, for every $x \in \mathcal{C}$, the solution is continuous in space and time. Moreover, for all $x \in C(\overline{\mathcal{O}})$ and $T_0 > 0$, $(X(t), t \geq T_0)$ is continuous in space and time.

12.3 The jump case

In this section we consider problem (12.1)–(12.3) in the case where the Wiener term vanishes. For simplicity we take the finite-variation term to be zero also.

Therefore we are concerned with the following SPDE:

$$dX(t, \xi) = \big(AX(t, \xi) + f(\xi, X(t, \xi))\big) \, dt$$
$$+ \int_S g_1(\xi, \sigma, X(t, \xi)) \widehat{\pi}_1(dt, d\xi, d\sigma), \qquad (12.26)$$

subject to the boundary and initial-value conditions (12.2) and (12.3). Recall that π_1 is a Poisson random measure on $[0, \infty) \times \mathcal{O} \times S$ with Lévy measure $dt \, d\xi \, \nu_1(d\sigma)$.

We will show that in this case we do not require (12.11). Thus, in particular, we can consider the second-order problem in all dimensions. However, we have to look for solutions taking values in $L^p(\mathcal{O})$-spaces with $p \in [1, 2)$.

Let $\mathcal{G}$ be the Green function for the system $\{\mathcal{A}, B_1, \dots, B_m\}$. Let us denote by $\mathcal{X}_{T,p}$ the space of all predictable fields $X \colon \Omega \times [0, T] \mapsto L^p(\mathcal{O})$ such that

$$\|X\|_{\mathcal{X}_{T,p}} := \sup_{t \in [0,T]} \left(\mathbb{E} \, |X(t)|^p_{L^p(\mathcal{O})} \right)^{1/p} < \infty.$$

Definition 12.10 We say that $X \in \mathcal{X}_{T,p}$ is an $\mathcal{X}_{T,p}$ *solution to* (12.26), (12.2) *and* (12.3) if it satisfies the following integral equation:

$$X(t, \xi) = \int_{\mathcal{O}} \mathcal{G}(t, \xi, \eta) x(\eta) \, d\eta + \int_0^t \int_{\mathcal{O}} \mathcal{G}(t - s, \xi, \eta) f(\eta, X(s, \eta)) \, d\eta \, ds$$
$$+ \int_0^t \int_{\mathcal{O}} \int_S \mathcal{G}(t - s, \xi, \eta) g_1(\eta, \sigma, X(s, \eta)) \widehat{\pi}_1(ds, d\eta, d\sigma).$$

In the definition above we have assumed that each integral is a well-defined process in $\mathcal{X}_{T,p}$, and the equation is understood as the equality of two elements of $\mathcal{X}_{T,p}$.

Theorem 12.11 *Let* $p \in [1, 2]$. *Assume that*

(i) $1 \leq p < 2m/d + 1$,
(ii) $f \colon \mathcal{O} \times \mathbb{R} \mapsto \mathbb{R}$ *and* $g_1 \colon \mathcal{O} \times S \times \mathbb{R} \mapsto \mathbb{R}$ *are measurable, and there are a constant* L *and a function* $l \in L^p(S, \mathcal{S}, \nu_1)$ *such that, for all* $u, v \in \mathbb{R}$, $\xi \in \mathcal{O}$ *and* $\sigma \in S$,

$$|f(\xi, u)| \leq L(1 + |u|),$$
$$|f(\xi, u) - f(\xi, v)| \leq L|u - v|,$$
$$|g_1(\xi, \sigma, u)| \leq l(\sigma)(1 + |u|),$$
$$|g_1(\xi, \sigma, u) - g_1(\xi, \sigma, v)| \leq l(\sigma)|u - v|.$$

Then for each $x \in L^p(\mathcal{O})$ *there exists a unique* $\mathcal{X}_{T,p}$ *solution to* (12.26), (12.2) *and* (12.3). *Moreover,* (12.26) *defines a Feller family on* $L^p(\mathcal{O})$.

Proof   Let us fix an $x \in L^p(\mathcal{O})$. We define a family $\| \cdot \|_\beta$, $\beta \in \mathbb{R}$, of norms on $\mathcal{X} := \mathcal{X}_{T,p}$ by

$$\|X\|_\beta := \sup_{t \in [0,T]} e^{-\beta t} \left(\mathbb{E} \,|X(t)|_{L^p(\mathcal{O})}^p \right)^{1/p}, \qquad X \in \mathcal{X}.$$

Note that each $\| \cdot \|_\beta$ is equivalent to the original norm $\| \cdot \|_{\mathcal{X}_{T,p}}$. Let

$$J(X)(t) := J_1(t) + J_2(X)(t) + J_3(X)(t),$$

where

$$J_1(t)(\xi) := \int_{\mathcal{O}} \mathcal{G}(t, \xi, \eta)x(\eta)\,d\eta,$$

$$J_2(X)(t)(\xi) := \int_0^t \int_{\mathcal{O}} \int_S \mathcal{G}(t - s, \xi, \eta)f(\eta, X(s, \eta))\,d\eta\,ds,$$

$$J_3(X)(t)(\xi) := \int_0^t \int_{\mathcal{O}} \int_S \mathcal{G}(t - s, \xi, \eta)g_1(\eta, \sigma, X(s, \eta))\widehat{\pi}_1(ds, d\eta, d\sigma).$$

We will show that $J \colon \mathcal{X} \mapsto \mathcal{X}$ and that J is a contraction for large enough β. This will complete the proof, on using the Banach fixed-point theorem.

By Theorem B.9, S is a C_0-semigroup on $L^p(\mathcal{O})$. Thus in particular there is a constant $C < \infty$ such that $\|S(t)\|_{L(L^p(\mathcal{O}), L^p(\mathcal{O}))} \le C$ for $t \in [0, T]$. Since $J_1(t) = S(t)x$ we have $J_1 \in \mathcal{X}$. Let $X \in \mathcal{X}$ and let F be the composition operator corresponding to f. Then

$$J_2(X)(t) = \int_0^t S(t - s)F(X(s))\,ds.$$

Note that $F \colon L^p(\mathcal{O}) \to L^p(\mathcal{O})$ is Lipschitz continuous. Then, with constants c_i independent of X,

$$\sup_{t \in [0,T]} \left(\mathbb{E}\,|J_2(X)(t)|_{L^p(\mathcal{O})}^p \right)^{1/p} \le c_1 \sup_{t \in [0,T]} \left(\mathbb{E}\,(1 + |X(t)|_{L^p(\mathcal{O})})^p \right)^{1/p}$$

$$\le c_2(1 + \|X\|_{\mathcal{X}}).$$

The process $J_2(X)$ is predictable since it is adapted and has continuous trajectories. As far as J_3 is concerned, by Theorems 8.23 and 2.6,[1]

$$\mathbb{E}\,|J_3(X)(t)|_{L^p(\mathcal{O})}^p$$

$$= \int_{\mathcal{O}} \mathbb{E} \left| \int_0^t \int_{\mathcal{O}} \int_S \mathcal{G}(t - s, \xi, \eta)g_1(\eta, \sigma, X(s, \eta))\widehat{\pi}_1(ds, d\eta, d\sigma) \right|^p d\xi$$

$$\le C_1 \int_{\mathcal{O}} \int_0^t \int_{\mathcal{O}} \int_S \mathbb{E}\,|\mathcal{G}(t - s, \xi, \eta)g_1(\eta, \sigma, X(s, \eta))|^p\,ds\,d\eta\,\nu_1(d\sigma)\,d\xi,$$

---

[1] Recall that g_m is given by (2.4).

which is less than or equal to

$$C_2 \int_{\mathcal{O}} \int_0^t \int_{\mathcal{O}} \int_S \mathbb{E} \, \mathfrak{g}_m^p(C_3(t-s), |\xi - \eta|) |g_1(\eta, \sigma, X(s, \eta))|^p \, ds \, d\eta \, \nu_1(d\sigma) \, d\xi.$$

Consequently,

$$\mathbb{E} \, |J_3(X)(t)|_{L^p(\mathcal{O})}^p$$

$$\leq C_4 \int_{\mathcal{O}} \int_0^t \int_{\mathcal{O}} \int_S (t-s)^{(1-p)d/2m} \mathfrak{g}_m(C_3(t-s), |\xi - \eta|) l^p(\sigma)$$

$$\times \left(1 + \mathbb{E} \, |X(s, \eta)|^p\right) \, ds \, d\eta \, \nu_1(d\sigma) \, d\xi$$

$$\leq C_5 \int_0^t \int_{\mathcal{O}} (t-s)^{(1-p)d/2m} \left(1 + \mathbb{E} \, |X(s, \eta)|^p\right) \, ds \, d\eta$$

$$\leq C_6 \int_0^t (t-s)^{(1-p)d/2m} \left(1 + \mathbb{E} \, |X(s)|_{L^p(\mathcal{O})}^p\right) \, ds$$

$$\leq C_7 \left(1 + \|X\|_{\mathcal{X}}^p\right) \int_0^t (t-s)^{(1-p)d/2m} \, ds.$$

Since $p < 1 + 2m/d$ implies that $(1-p)d/2m > -1$, it follows that

$$\mathbb{E} \, |J_3(X)(t)|_{L^p(\mathcal{O})}^p \leq C_8 \left(1 + \|X\|_{\mathcal{X}}^p\right)$$

and consequently that $\sup_{t \in [0,T]} \mathbb{E} \, |J_3(X)(t)|_{L^p(\mathcal{O})}^p < \infty$. Note that

$$\lim_{s \to t} \mathbb{E} \, |J_3(X)(s) - J_3(X)(t)|_{L^p(\mathcal{O})}^p = 0.$$

Hence, as the stochastic continuity of the adapted process implies the existence of a predictable version, we have $J_3 \colon \mathcal{X} \mapsto \mathcal{X}$.

Now let $X, Y \in \mathcal{X}$. Then

$$\|J(X) - J(Y)\|_\beta \leq \|J_2(X) - J_2(Y)\|_\beta + \|J_3(X) - J_3(Y)\|_\beta,$$

and, with $C := \sup_{t \in [0,T]} \|S(t)\|_{L(L^p(\mathcal{O}), L^p(\mathcal{O}))}^p$,

$$\|J_2(X) - J_2(Y)\|_\beta^p$$

$$= \sup_{t \in [0,T]} e^{-p\beta t} \, \mathbb{E} \left| \int_0^t S(t-s)\big(F(X(s)) - F(Y(s))\big) ds \right|_{L^p(\mathcal{O})}^p$$

$$\leq C \sup_{t \in [0,T]} \mathbb{E} \int_0^t e^{-p\beta(t-s)} e^{-p\beta s} \, |F(X(s)) - F(Y(s))|_{L^p(\mathcal{O})}^p \, ds$$

$$\leq C L^p \sup_{t \in [0,T]} \int_0^t e^{-p\beta(t-s)} e^{-p\beta s} \, \mathbb{E} \, |X(s) - Y(s)|_{L^p(\mathcal{O})}^p \, ds$$

$$\leq C L^p \|X - Y\|_\beta^p \int_0^T e^{-p\beta t} \, dt \leq \frac{C L^p}{p\beta} \|X - Y\|_\beta.$$

Finally, using the arguments from the estimate $\mathbb{E} \, |J_3(X)(t)|^p_{L^p(\mathcal{O})}$ we obtain

$$e^{-p\beta t} \, \mathbb{E} \, |J_3(X) - J_3(Y)|^p_{L^p(\mathcal{O})}$$

$$\leq Ce^{-p\beta t} \int_0^t (t-s)^{(1-p)d/2m} \, \mathbb{E} \, |X(t) - Y(t)|^p_{L^p(\mathcal{O})} \, ds$$

$$\leq C \int_0^t (t-s)^{(1-p)d/2m} e^{-p\beta(t-s)} e^{-p\beta s} \, \mathbb{E} \, |X(t) - Y(t)|^p_{L^p(\mathcal{O})} \, ds$$

$$\leq C \, \|X - Y\|^p_\beta \int_0^T e^{-p\beta t} t^{(1-p)d/2m} \, dt.$$

Consequently,

$$\|J_3(X) - J_3(Y)\|^p_\beta \leq C \, \|X - Y\|^p_\beta \int_0^T e^{-p\beta t} t^{(1-p)d/2m} \, dt.$$

Summing up, there is a constant C such that, for all $\beta \in \mathbb{R}$ and $X, Y \in \mathcal{X}$,

$$\|J(X) - J(Y)\|_\beta \leq CC_\beta \, \|X - Y\|_\beta,$$

where

$$C_\beta = \beta^{-1/p} + \left(\int_0^T e^{-p\beta t} t^{(1-p)d/2m} \, dt \right)^{1/p}.$$

Since $C_\beta \to 0$ as $\beta \to \infty$ we have

$$\|J(X) - J(Y)\|_\beta \leq \tfrac{1}{2} \|X - Y\|_\beta$$

for β large enough, which completes the proof. $\qquad\square$

12.3.1 Equations on the whole space

Although the present chapter is devoted mainly to the parabolic boundary problem on a bounded domain, we formulate here some results concerning the existence and uniqueness of solutions to the problem on the whole space,

$$dX(t, \xi) = \big(AX(t, \xi) + f(\xi, X(t, \xi))\big) \, dt$$

$$+ \int_S g_1(\xi, \sigma, X(t, \xi)) \widehat{\pi}_1(dt, d\xi, d\sigma), \qquad t > 0, \xi \in \mathbb{R}^d,$$

$$X(0, \xi) = x(\xi), \qquad \xi \in \mathbb{R}^d. \tag{12.27}$$

Analogously to (12.26), in the above problem π_1 is a Poisson random measure on $[0, \infty) \times \mathbb{R}^d \times S$ with Lévy measure $dt \, d\xi \, \nu_1(d\sigma)$.

One has to look for a solution in weighted L^p-spaces. We will formulate our result for the system L^p_ρ with exponential weights introduced in Section 2.3. A similar result holds for polynomial weights. As in the case of an equation on a bounded domain (see Theorem 12.11), we assume that $p \in [1, 2]$.

Let $\mathcal{G}$ be the Green function for $\mathcal{A}$. Let us denote by $\mathcal{X}_{T,p,\rho}$ the space of all predictable fields $X: \Omega \times [0, T] \mapsto L_\rho^p$ such that

$$\|X\|_{\mathcal{X}_{T,p,\rho}} := \sup_{t \in [0,T]} \left(\mathbb{E} \, |X(t)|_{L_\rho^p}^p \right)^{1/p} < \infty.$$

Definition 12.12 We say that $X \in \mathcal{X}_{T,p,\rho}$ is an $\mathcal{X}_{T,p,\rho}$ *solution to* (12.27) if it satisfies the integral equation

$$X(t, \xi) = \int_{\mathbb{R}^d} \mathcal{G}(t, \xi, \eta) x(\eta) \, d\eta + \int_0^t \int_{\mathbb{R}^d} \mathcal{G}(t - s, \xi, \eta) f(\eta, X(s, \eta)) \, d\eta \, ds$$

$$+ \int_0^t \int_{\mathbb{R}^d} \int_S \mathcal{G}(t - s, \xi, \eta) g(\eta, \sigma, X(s, \eta)) \widehat{\pi}_1(ds, d\eta, d\sigma).$$

Since one can prove the theorem below simply using arguments from the proof of Theorem 12.11, we will leave this to the reader. We note that the only differences are caused by the fact that constant functions on $\mathbb{R}^d$ are no longer integrable with respect to Lebesgue measure but belong to any L_ρ^p-space with strictly positive ρ.

Theorem 12.13 *Let $p \in [1, 2]$ and $\rho \in \mathbb{R}$. Assume that*

(i) $1 \leq p < 2m/d + 1$,
(ii) $f: \mathbb{R}^d \times \mathbb{R} \mapsto \mathbb{R}$ *and* $g: \mathbb{R}^d \times S \times \mathbb{R} \mapsto \mathbb{R}$ *are measurable and there are a constant L and functions $l \in L^p(S, \mathcal{S}, \nu_1)$ and $h \in L_\rho^p$ such that, for all $u, v \in \mathbb{R}$, $\xi \in \mathbb{R}^d$ and $\sigma \in S$,*

$$|f(\xi, u)| \leq L(h(\xi) + |u|),$$

$$|f(\xi, u) - f(\xi, v)| \leq L \, |u - v|,$$

$$|g(\xi, \sigma, u)| \leq l(\sigma)(h(\xi) + |u|),$$

$$|g(\xi, \sigma, u) - g(\xi, \sigma, v)| \leq l(\sigma)|u - v|.$$

Then for each $x \in L_\rho^p$ there exists a unique $\mathcal{X}_{T,p,\rho}$ solution to (12.27).

12.4 Stochastic heat equation

In this section we are concerned with the L^2-theory of the stochastic heat equation on a bounded interval $(b, c) \subset \mathbb{R}$. We will apply results from subsection 12.1.2. The existence of solutions in L^p-spaces for equations with purely jump noise can be derived easily from the results of Section 12.3. We will consider equations with Dirichlet and Neumann boundary conditions.

In this section W is a cylindrical Wiener process on $L^2(b, c)$ and Z is an impulsive cylindrical noise on $L^2(b, c)$ with jump size intensity ν; see Section 7.2. We assume that ν is supported on an interval $[-a, a]$ and that

$$\int_{-a}^{a} \sigma^2 \nu(d\sigma) < \infty. \tag{12.28}$$

Finally, to simplify the notation we assume that Z and W are independent.

12.4.1 Dirichlet boundary conditions

Consider the problem

$$
\begin{aligned}
\frac{\partial X}{\partial t}(t, \xi) &= \frac{\partial^2 X}{\partial \xi^2}(t, \xi) + f(X(t, \xi)) + g(X(t, \xi))\frac{\partial W}{\partial t}(t, \xi) \\
&\quad + h(X(t, \xi))\frac{\partial Z}{\partial t}(t, \xi), \qquad t > 0, \ \xi \in (b, c), \\
X(t, b) &= 0 = X(t, c), \qquad t > 0, \\
X(0, \xi) &= x(\xi), \qquad \xi \in (b, c),
\end{aligned}
\tag{12.29}
$$

where $f, g, h \colon \mathbb{R} \mapsto \mathbb{R}$. Then (12.29) is a special case of (12.1)–(12.3). Indeed, in this case

$$h(X(t, \xi))\frac{\partial Z}{\partial t}(t, d\xi) = \int_{-a}^{a} h(X(t, \xi))\sigma \widehat{\pi}(dt, d\xi, d\sigma).$$

As the operator A we take the *Laplace operator with Dirichlet boundary conditions*, that is, $A\psi(\xi) := d^2\psi(\xi)/d\xi^2$, $\psi \in D(A) := W^{2,2}(b, c) \cap W_0^{2,1}(b, c)$. Clearly, A is a differential operator of the type introduced in Section 2.5 with $(B\psi)(b) = \psi(b)$ and $(B\psi)(c) = \psi(c)$.

The following theorem follows directly from Theorem 12.3.

Theorem 12.14 *Assume that (12.28) holds and that f, g and h are Lipschitz continuous. Then for any $x \in L^2(b, c)$ there is a unique predictable $L^2(b, c)$-valued process X satisfying (12.29) such that, for all $T < \infty$,*

$$\sup_{t \in [0, T]} \mathbb{E}\,|X(t)|^2_{L^2(b, c)} < \infty.$$

Moreover, (12.29) defines a Feller family on $L^2(b, c)$.

Remark 12.15 Note that (12.29) is a special case of (9.4) with $H = L^2(c, b)$, $\mathcal{H} = L^2(c, b) \times L^2(c, b)$ and $G(\psi)(\varphi, \phi) = g(\psi)\varphi + h(\psi)\phi$. Note that $G(\psi) \in L(\mathcal{H}, L^2(b, c))$ if and only if $g(\psi)$ and $h(\psi)$ belong to $L^\infty(b, c)$. In fact one can show that, even if g and h are bounded, G maps $L^2(b, c)$ into $L_{(HS)}(\mathcal{H}, L^2(b, c))$

if and only if $g = 0 = h$. Of course, the existence follows from the much weaker facts that $S(t)G(\psi) \in L_{(HS)}\left(\mathcal{H}, L^2(b, c)\right)$ for all $t > 0$ and $\psi \in L^2(c, b)$ and that

$$\sup_{t \in (0,T]} t^{1/4} \|S(t)G(\psi)\|_{L_{(HS)}(\mathcal{H}, L^2(b,c))} \leq L\left(1 + |\psi|_{L^2(c,b)}\right),$$

$$\sup_{t \in (0,T]} t^{1/4} \|S(t)(G(\psi_1) - G(\psi_2))\|_{L_{(HS)}(\mathcal{H}, L^2(b,c))} \leq L |\psi_1 - \psi_2|_{L^2(c,b)}$$

where the constant L is independent of ψ, ψ_1 and ψ_2.

12.4.2 Neumann boundary conditions

Let us consider now the problem defined by the heat equation with Neumann boundary conditions:

$$\frac{\partial X}{\partial t}(t, \xi) = \frac{\partial^2 X}{\partial \xi^2}(t, \xi) + f(X(t, \xi)) + g(X(t, \xi)\frac{\partial W}{\partial t}(t, \xi)$$

$$+ h(X(t, \xi))\frac{\partial Z}{\partial t}(t, \xi), \qquad t > 0, \ \xi \in (b, c),$$

$$X(0, \xi) = x(\xi), \qquad \xi \in (b, c),$$

$$\frac{\partial X}{\partial \xi}(t, b) = 0 = \frac{\partial X}{\partial \xi}(t, c), \qquad t > 0. \tag{12.30}$$

The only difference from (12.29) is that now the operator A is defined as the *Laplace operator with Neumann boundary conditions*, that is,

$$A\psi(\xi) := \frac{d^2\psi}{d\xi^2}(\xi) \quad \text{and} \quad D(A) := \left\{\psi \in W^{2,2}(b, c) \colon \frac{d\psi}{d\xi}(b) = 0 = \frac{d\psi}{d\xi}(c)\right\}.$$

We have, therefore, the following result.

Theorem 12.16 *Assume that (12.28) holds and that f, g and h are Lipschitz continuous. Then for any $x \in L^2(b, c)$ there is a unique predictable $L^2(b, c)$-valued process X satisfying (12.30) such that, for each $T < \infty$,*

$$\sup_{t \in [0,T]} \mathbb{E} |X(t)|^2_{L^2(b,c)} < \infty.$$

Moreover, (12.30) defines a Feller family on $L^2(b, c)$.

12.4.3 Space–time continuity in the Wiener case

Let us consider the stochastic heat-equation problem

$$\frac{\partial X}{\partial t}(t,\xi) = \frac{\partial^2 X}{\partial \xi^2}(t,\xi) + f(X(t,\xi))$$

$$+ g(X(t,\xi))\frac{\partial W}{\partial t}(t,\xi), \qquad t > 0,\ \xi \in (b,c),$$

$$X(t,b) = 0 = X(t,c), \qquad t > 0, \tag{12.31}$$

$$X(0,x) = x(\xi), \qquad \xi \in (b,c),$$

driven by a cylindrical Wiener process W on $L^2(b,c)$. As in the case of (12.29) and (12.30) we assume that f and g are Lipschitz functions on $\mathbb{R}$.

For brevity we write $L^p = L^p(b,c)$. Recall that $C_0(b,c)$ is the space of all continuous functions on $[b,c]$ that vanish on the boundary. Then from Theorems 12.6 and 12.8 and Remark 12.9 we have the following regularity result.

Theorem 12.17 *Assume that f and g are Lipschitz continuous. Let X be the (unique) solution to (12.31) starting from $x \in L^2$. Then the following hold.*

(i) *X has continuous trajectories in L^2 and $\mathbb{E}\sup_{t\in[0,T]}|X(t)|^q_{L^2} < \infty$ for all $T < \infty$ and $q \geq 2$.*

(ii) *If $x \in L^p$ for some $p \in [2,\infty)$ then X has continuous trajectories in L^p and $\mathbb{E}\sup_{t\in[0,T]}|X(t)|^p_{L^p} < \infty$ for all $T < \infty$.*

(iii) *If $x \in C_0(b,c)$ then X has continuous trajectories in $C_0(b,c)$ and $\mathbb{E}\sup_{t\in[0,T]}|X(t)|^q_{C_0(b,c)} < \infty$ for all $T < \infty$ and $q \geq 2$.*

(iv) *If $x \in C([b,c])$ then X has bounded trajectories in $C([b,c])$ and $\mathbb{E}\sup_{t\in[0,T]}|X(t)|^q_{C([b,c])} < \infty$ for all $T < \infty$ and $q \geq 2$.*

Given $\gamma \in [0,1)$ we denote by $C_0^\gamma([0,T] \times [b,c])$ the space of all Hölder continuous functions $\psi \colon [0,T] \times [b,c] \mapsto \mathbb{R}$, with exponent γ, vanishing on $[0,T] \times \{b,c\}$. On $C_0^\gamma([0,T] \times [b,c])$ we consider the norm

$$\|\psi\|_{C_0^\gamma([0,T]\times[b,c])}$$

$$:= \sup_{(t,\xi)\in[0,T]\times[b,c]} |\psi(t,\xi)| + \sup_{t,s\in[0,T],\ t\neq s,\ \xi,\eta\in[b,c],\ \xi\neq\eta} \frac{|\psi(t,\xi) - \psi(s,\eta)|}{\left(|t-s|^2 + |\xi-\eta|^2\right)^{\gamma/2}}.$$

We denote by $C_0^\gamma(b,c)$ the space of all $\psi \in C^\gamma([b,c])$ vanishing on the boundary. Then, using the analyticity of the heat semigroup on L^p- and $C_0(b,c)$-spaces (see e.g. Lunardi 1995) and the factorization theorems 11.6 and 11.7, one can obtain the following result. For more details see e.g. Peszat (1995), Brzeźniak and Peszat (1999) or Peszat and Seidler (1998).

Theorem 12.18 *Assume that f and g are Lipschitz continuous, $\gamma < 1/4$ and $x \in C_0^{\gamma}(b, c)$. Then for every $T < \infty$ the solution X to (12.31) starting from x has a modification from $C_0^{\gamma}([0, T] \times [b, c])$. Moreover,*

$$\mathbb{E}\, \|X\|_{C_0^{\gamma}([0,T] \times [b,c])}^q < \infty, \qquad \forall\, T < \infty,\ \forall\, q \geq 2.$$

12.5 Equations with fractional Laplacian and stable noise

Let $Z_\beta(t, \cdot)$ be the stable white-noise process introduced in Examples 7.26 and 7.27. In Mueller (1998) and Mytnik (2002) the existence of a solution to the martingale problem (see Section 9.8) for the equation

$$dX(t, \xi) = -(-\Delta)^{\gamma} X(t, \xi)\, dt + (X(t, \xi))^{\delta}\, dZ_{\beta}(t, \xi) \tag{12.32}$$

was investigated. In (12.32), $\beta \in (0, 2) \setminus \{1\}$ $\gamma > 0$, $\delta > 0$ and $\xi \in \mathcal{O}$. Mytnik considered (12.32) on $\mathbb{R}^d$. This equation with $\delta = 1$ is a special case of the problem

$$dX(t, \xi) = -(-\Delta)^{\gamma} X(t, \xi)\, dt + g(X(t, \xi))\, dZ_{\beta}(t, \xi), \tag{12.33}$$

where $g \colon \mathbb{R} \mapsto \mathbb{R}$ is Lipschitz continuous. In this section we consider (12.33) on a bounded domain $\mathcal{O} \subset \mathbb{R}^d$, and we look for a solution in $L^p := L^p(\mathcal{O})$.

Theorem 12.19 *Let $\beta \in (1, 2)$ and $\gamma > 0$. Assume that g is Lipschitz continuous and $\beta < p < 2 \wedge (1 + 2\gamma/d)$. Then (12.33) defines a Feller family on L^p.*

Proof Recall that

$$Z_\beta(t, d\xi) = \int_0^t \int_0^{+\infty} \sigma \widehat{\pi}(ds, d\xi, d\sigma),$$

where π is a Poisson random measure with intensity $ds\, d\xi\, \sigma^{-1-\beta}\, d\sigma$. We decompose $Z_\beta(t, d\xi)$ into two parts, the martingale part

$$Z_{\beta,N}(t, d\xi) := \int_0^t \int_0^N \sigma \widehat{\pi}(ds, d\xi, d\sigma)$$

and the finite-variation part $\int_0^s \int_N^{+\infty} \sigma \widehat{\pi}(ds, d\xi, d\sigma)$. Thus we will treat (12.33) as an equation of the type (9.31). In particular, we will not be able to show that the solution has a moment of order p in L^p-space.

In order to solve (12.33) one should first fix N and solve separately

$$dX(t, \xi) = -(-\Delta)^{\gamma} X(t, \xi)\, dt + g(X(t, \xi))\, dZ_{\beta,N}(t, \xi). \tag{12.34}$$

In fact we will only sketch the proof of the existence of a solution to (12.34). Let $\mathcal{G}_\gamma$ be the Green function for $-(-\Delta)^{\gamma}$ (see appendix section B.5). Given $T < \infty$ we denote by $\mathcal{X}$ the class of all predictable fields $X \colon \Omega \times [0, T] \mapsto L^p$ such that

$\|X\|_{\mathcal{X}}^p := \sup_{t\in[0,T]} \mathbb{E}\,|X(t)|_{L^p}^p < \infty$. The proof can be based on the ideas used in the proof of Theorem 12.11. The most difficult part is to show that the stochastic convolution integral

$$J(X)(t)(\xi) := \int_0^t \mathcal{G}_\gamma(t-s,\xi,\eta)X(s,\eta)\,\mathrm{d}Z_{\beta,N}(s,\eta)$$

is a well-defined bounded linear operator from $\mathcal{X}$ to $\mathcal{X}$. Taking into account that $p - 1 - \beta > -1$, by Lemma B.23 we can find a constant C_1 independent of X such that

$$\int_{\mathcal{O}}\int_0^t\int_{\mathcal{O}}\int_0^N \mathbb{E}\,|\mathcal{G}_\gamma(t-s,\xi,\eta)X(s,\eta)|^p\,\mathrm{d}s\,\mathrm{d}\eta\,\mathrm{d}\xi\,\sigma^{p-1-\beta}\,\mathrm{d}\sigma$$

$$\leq C_1\|X\|_{\mathcal{X}}^p\int_0^t s^{(1-p)d/2\gamma}\,\mathrm{d}s.$$

Since $(1 - p)d/2\gamma > -1$ we conclude the proof by using Theorem 8.23(i). $\qquad\square$

Remark 12.20   A similar result can be stated for $\mathcal{O} = \mathbb{R}^d$ if L^2 is replaced by one of the weighted spaces L_ρ^2 or $\mathcal{L}_\rho^2$ introduced in Section 2.3. The required modification of the proof, which is made in a standard way, is left to the reader.

13

Wave and delay equations

In this chapter, stochastic wave equations on finite intervals and stochastic wave equations on $\mathbb{R}^d$, $d = 1, 2$, driven by an impulsive cylindrical process on $L^2(\mathbb{R}^d)$ are considered. It is shown that the general theory is applicable to stochastic delay equations.

13.1 Stochastic wave equation on $[0, 1]$

We start this section by recalling a generation result. Then we treat the wave equation driven by a process with a nuclear covariance. Finally, we consider equations driven by Gaussian and Poissonian space–time white noise.

13.1.1 Noise with nuclear covariance

Let $(A, D(A))$ be a negative-definite operator on a Hilbert space V. Then

$$H := \begin{pmatrix} D\left((-A)^{1/2}\right) \\ V \end{pmatrix}$$

with the scalar product

$$\left\langle \begin{pmatrix} x \\ y \end{pmatrix}, \begin{pmatrix} u \\ v \end{pmatrix} \right\rangle_H := \left\langle (-A)^{1/2}x, \ (-A)^{1/2}u \right\rangle_V + \langle y, v \rangle_V$$

is a real separable Hilbert space. By Lemma B.3, the operator

$$\mathcal{A} := \begin{pmatrix} 0 & I \\ A & 0 \end{pmatrix}, \qquad D(\mathcal{A}) := \begin{pmatrix} D(A) \\ D\left((-A)^{1/2}\right) \end{pmatrix} \tag{13.1}$$

generates a C_0-semigroup of contractions on H.

Consider the Cauchy problem for the wave equation on the interval [0, 1] with Dirichlet boundary conditions:

$$\frac{\partial^2 X}{\partial t^2}(t, \xi) = \frac{\partial^2 X}{\partial \xi^2}(t, \xi) + f(X(t, \xi)) + g_1(X(t, \xi))\frac{\partial Z_1}{\partial t}(t, \xi)$$

$$+ g_2(X(t, \xi))\frac{\partial Z_2}{\partial t}(t, \xi), \qquad t > 0, \ \xi \in (0, 1),$$

$$X(t, 0) = X(t, 1) = 0, \qquad t > 0,$$

$$X(0, \xi) = x(\xi), \qquad \xi \in (0, 1),$$

$$\frac{\partial X}{\partial t}(0, \xi) = y(\xi), \qquad \xi \in (0, 1).$$

(13.2)

In (13.2) we have assumed that Z_1 and Z_2 are independent Lévy processes taking values in $H = L^2 = L^2(0, 1)$ and that $f, g_1, g_2 \colon \mathbb{R} \mapsto \mathbb{R}$ are Lipschitz continuous. Thus the RKHS $\mathcal{H}$ of $M = (Z_1, Z_2)$ is a subspace of $L^2 \times L^2$ and the embedding $\mathcal{H} \hookrightarrow L^2 \times L^2$ is Hilbert–Schmidt. In the most interesting case Z_1 is a Wiener process and Z_2 is a Lévy jump process.

Our first goal will be to write (13.2) in the form (9.4), with properly chosen co-efficients. To this end we denote by A the Laplace operator with Dirichlet boundary conditions (see subsection 12.4.1). Set

$$\mathbf{X} := \begin{pmatrix} X \\ \dfrac{\partial X}{\partial t} \end{pmatrix} =: \begin{pmatrix} X \\ Y \end{pmatrix}.$$

Then

$$\frac{\partial \mathbf{X}}{\partial t}(t, \xi) = \begin{pmatrix} \dfrac{\partial X}{\partial t} \\ \dfrac{\partial Y}{\partial t} \end{pmatrix} = \begin{pmatrix} Y \\ \dfrac{\partial^2 X}{\partial t^2} \end{pmatrix}.$$

Write $V := D\left((-A)^{1/2}\right)$. We will write (13.2) as

$$d\mathbf{X} = (\mathcal{A}\mathbf{X} + \mathbb{F}(\mathbf{X}))\, dt + \mathbb{G}(\mathbf{X})(dZ_1, dZ_2), \qquad \mathbf{X}(0) = \begin{pmatrix} x \\ y \end{pmatrix}, \qquad (13.3)$$

on the Hilbert space

$$H := \begin{pmatrix} V \\ L^2 \end{pmatrix}$$

with $(\mathcal{A}, D(\mathcal{A}))$ given by (13.1) and with $\mathbb{F}$ and $\mathbb{G}$ defined by

$$\mathbb{F}\begin{pmatrix} u \\ v \end{pmatrix} = \begin{pmatrix} 0 \\ F(u) \end{pmatrix}, \qquad \mathbb{G}\begin{pmatrix} u \\ v \end{pmatrix}(\psi_1, \psi_2) = \begin{pmatrix} 0 \\ G(u)(\psi_1, \psi_2) \end{pmatrix},$$

$$F(u)(\xi) = f(u(\xi)),$$

$$G(u)[(\psi_1, \psi_2)](\xi) = g_1(u(\xi))\psi_1(\xi) + g_2(u(\xi))\psi_2(\xi).$$

Here (ψ_1, ψ_2) belongs to the RKHS of the driven process, that is, to the space $\mathcal{H} = Q_1^{1/2}(L^2) \times Q_2^{1/2}(L^2)$, where Q_1 and Q_2 are the covariance operators of Z_1 and Z_2, respectively. By Lemma B.3, $(\mathcal{A}, D(\mathcal{A}))$ generates a C_0-semigroup on H. We will show that $\mathbb{F} \colon H \mapsto H$ and $\mathbb{G} \colon H \mapsto L_{(HS)}(\mathcal{H}, H)$ are Lipschitz continuous. Taking into account that $\mathbb{F}$ and $\mathbb{G}$ do not depend on the second coordinate, it is sufficent to show that the mappings $F \colon V \mapsto L^2$,

$$G(\cdot)(Q_1^{1/2}, Q_2^{1/2}) \colon V \mapsto L_{(HS)}(L^2 \times L^2, L^2)$$

are Lipschitz continuous. Since F is a Lipschitz continuous mapping from L^2 to L^2, it is also Lipschitz continuous from V to L^2 because V is continuously embedded into L^2. To show the Lipschitz continuity of the diffusion coefficient it is sufficient to prove that, given a Hilbert–Schmidt operator Q and a Lipschitz continuous function g, the mapping

$$\tilde{G}(\cdot)Q^{1/2} \colon V \mapsto L_{(HS)}(L^2, L^2)$$

defined by $\tilde{G}(u)Q^{1/2}\psi = g(u)Q^{1/2}\psi$ is Lipschitz continuous. To do this, fix an orthonormal basis $\{e_n\}$ of L^2. Then, for $u \in L^\infty := L^\infty([0, 1], \mathcal{B}([0, 1]), \ell_1)$,

$$\sum_n \left| \tilde{G}(u)Q^{1/2}e_n \right|_{L^2}^2 = \sum_{n \in \mathbb{N}} \int_0^1 \left| g(u(\xi))(Q^{1/2}e_n)(\xi) \right|^2 \mathrm{d}\xi$$

$$\leq |g(u)|_{L^\infty}^2 \left\| Q^{1/2} \right\|_{L_{(HS)}(L^2, L^2)}^2.$$

Similarly, for $u, v \in L^\infty$,

$$\sum_n \left| (\tilde{G}(u) - \tilde{G}(v))Q^{1/2}e_n \right|_{L^2}^2 \leq |g(u) - g(v)|_{L^\infty}^2 \left\| Q^{1/2} \right\|_{L_{(HS)}(L^2, L^2)}^2.$$

Thus the proof of the Lipschitz property of $\tilde{G}(\cdot)Q^{1/2}$ is complete if we can show that the mapping $V \ni u \mapsto N_g(u) \in L^\infty$ is Lipschitz.[1] Note (see e.g. Theorem 15.6) that $V = W_0^{1,2}(0, 1)$. Next, $W_0^{1,2}(0, 1) \hookrightarrow C([0, 1])$. This follows from the fact that for every $\psi \in W_0^{1,2}(0, 1)$ we have $\psi(\xi) = \int_0^\xi \psi'(\eta) \, \mathrm{d}\eta$ and $\psi' \in L^2$. Now, since g

---

[1] N_g is the composition operator corresponding to g.

is Lipschitz, the mapping

$$C([0, 1]) \ni u \mapsto N_g(u) \in C([0, 1]) \hookrightarrow L^\infty$$

is also Lipschitz. We have shown that conditions (F) and (G) formulated in Section 9.2 are satisfied. Thus we have the following consequence of the abstract existence and uniqueness theorem.

Theorem 13.1 *Assume that* $f, g_1, g_2 \colon \mathbb{R} \to \mathbb{R}$ *are Lipschitz continuous and that* Z_1 *and* Z_2 *are independent Lévy processes on* L^2 *with a nuclear covariance operator. Let A be the Laplace operator on* L^2 *with Dirichlet boundary conditions. Then, for any initial value*

$$\mathbf{x} = \begin{pmatrix} x \\ y \end{pmatrix} \in H = \begin{pmatrix} D\left((-A)^{1/2}\right) \\ L^2 \end{pmatrix},$$

there is a unique solution to (13.2). Moreover, for any $T < \infty$ *there is a constant* K *such that for all* $\mathbf{x}_0, \mathbf{x}_1 \in H$ *the solutions* $\mathbf{X}(\cdot, \mathbf{x}_0)$ *and* $\mathbf{X}(\cdot, \mathbf{x}_1)$ *starting from* $\mathbf{x}_0$ *and* $\mathbf{x}_1$ *satisfy*

$$\sup_{t \in [0,T]} \mathbb{E} \, |\mathbf{X}(t, \mathbf{x}_0) - \mathbf{X}(t, \mathbf{x}_1)|_H^2 \le K \, |\mathbf{x}_0 - \mathbf{x}_1|_H^2.$$

Finally, (13.2) defines a Feller family on H.

13.1.2 Cylindrical noise

Consider the problem

$$\frac{\partial^2 X}{\partial t^2}(t, \xi) = \frac{\partial^2 X}{\partial \xi^2}(t, \xi) + f(X(t, \xi)) + g_1(X(t, \xi)) \frac{\partial W}{\partial t}(t, \xi)$$

$$+ g_2(X(t, \xi)) \frac{\partial Z}{\partial t}(t, \xi), \qquad t > 0, \ \xi \in (0, 1),$$

$$X(t, 0) = X(t, 1) = 0, \qquad t > 0,$$

$$X(0, \xi) = x(\xi), \qquad \xi \in (0, 1),$$

$$\frac{\partial X}{\partial t}(0, \xi) = y(\xi), \qquad \xi \in (0, 1).$$

(13.4)

Here W is a cylindrical Wiener process on L^2 and Z is an impulsive cylindrical process on L^2.

It is easy to show that in this case conditions (G) and (F) formulated in Section 9.2 are not satisfied for the space H defined in the previous subsection. We will show, however, that the solution exists on a larger state space $\mathbf{H} \hookleftarrow H$. Since the problem is non-linear, X must be function-valued. In fact we are looking for a solution in $\mathbf{H} = (L^2, V)^T$, where V is a space of distributions such that

$L^2 = D((-A)^{1/2})$ for the Laplace operator A considered on V. In order to define V recall that $e_n(\xi) = \sqrt{2}\,\sin(\pi n\xi)$, $n \in \mathbb{N}$, $\xi \in (0, 1)$, is an orthonormal basis of L^2 consisting of eigenvectors of the Laplace operator. Moreover, $\{-\gamma_n\}$, where $\gamma_n = \pi^2 n^2$, is the corresponding sequence of eigenvalues. Define

$$V := \left\{ \sum_n x_n e_n : \ \sum_n \gamma_n^{-1} x_n^2 < \infty \right\}.$$

Then V, with the scalar product

$$\left\langle \sum_n x_n e_n, \ \sum_n y_n e_n \right\rangle_V := \sum_n \gamma_n^{-1} x_n y_n,$$

is a real separable Hilbert space. On V we define A by the formula

$$A \sum_n x_n e_n = \sum_n x_n A e_n = -\sum_n \gamma_n x_n e_n.$$

Clearly

$$D(A) = \left\{ \sum_n x_n e_n : \ -A \sum_n x_n e_n \in V \right\}$$

$$= \left\{ \sum_n x_n e_n : \ \sum_n \gamma_n x_n e_n \in V \right\} = \left\{ \sum_n x_n e_n : \ \sum_n \gamma_n x_n^2 < \infty \right\}.$$

As A is given in diagonal form, it is easy to see that A is a self-adjoint and negative-definite operator. Since $(-A)^{1/2} e_n = \gamma_n^{1/2} e_n$, $n \in \mathbb{N}$, we have

$$\left((-A)^{1/2}\right) = \left\{ \sum_n x_n e_n : \ (-A)^{1/2} \sum_n x_n e_n \in V \right\}$$

$$= \left\{ \sum_n x_n e_n : \ \sum_n x_n (-A)^{1/2} e_n \in V \right\}$$

$$= \left\{ \sum_n x_n e_n : \ \sum_n x_n^2 < \infty \right\} = L^2.$$

As before we assume that $f, g \colon \mathbb{R} \mapsto \mathbb{R}$ are Lipschitz continuous. We treat (13.3) as (13.2) on $\mathbf{H}$. We will show that the mappings $\mathbb{F} \colon \mathbf{H} \mapsto \mathbf{H}$ and $\mathbb{G} \colon \mathbf{H} \mapsto L_{(HS)}(\mathcal{H}, \mathbf{H})$ corresponding to f and g are Lipschitz continuous. The Lipschitz continuity of $\mathbb{F}$ follows from the Lipschitz continuity of $F(u)(\xi) = f(u(\xi))$ from L^2 to L^2. To show the Lipschitz continuity of $\mathbb{G}$, it is suffcient to show the Lipschitz continuity of the mapping

$$G(u)[\psi](\xi) = g(u(\xi))\psi(\xi), \qquad u, \psi \in L^2, \ \xi \in (0, 1),$$

from L^2 to $L_{(HS)}(L^2, V)$. Let $u \in L^2$, and let $\{e_n\}$ be an orthonormal basis of L^2 consisting of eigenvectors of A. Then

$$\sum_n |G(u)e_n|_V^2 = \sum_n \left| \sum_k \langle G(u)e_n, e_k \rangle_{L^2} e_k \right|_V^2$$

$$= \sum_n \sum_k \gamma_k^{-1} \langle G(u)e_n, e_k \rangle_{L^2}^2 = \sum_n \sum_k \gamma_k^{-1} \langle e_n, G(u)e_k \rangle_{L^2}^2$$

$$= \sum_k \gamma_k^{-1} |G(u)e_k|_{L^2}^2 \le 2 \sum_k \gamma_k^{-1} \int_0^1 |g(u(\xi))|^2 \, d\xi,$$

since $|e_k(\xi)| \le \sqrt{2}$, $k \in \mathbb{N}$, $\xi \in (0, 1)$. In the same way one can show that, for $u, v \in L^2$,

$$\sum_n |(G(u) - G(v))e_n|_V^2 \le 2 \sum_k \gamma_k^{-1} \int_0^1 |g(u(\xi)) - g(v(\xi))|^2 \, d\xi.$$

Thus the Lipschitz continuity of G follows from the Lipschitz continuity of g and the summability of the series $\sum \gamma_n^{-1}$. Hence we have the following consequence of the general existence theorem.

Theorem 13.2   *Assume that f and g are Lipschitz continuous. Then for any initial value $(x, y)^T \in \mathbf{H}$ there is a unique solution to (13.4) taking values in $\mathbf{H}$. Moreover, (13.4) defines a Feller family on $\mathbf{H}$.*

13.2 Stochastic wave equation on $\mathbb{R}^d$ driven by impulsive noise

In this section we are concerned with the stochastic wave problem

$$\begin{aligned}
\frac{\partial^2 X}{\partial t^2}(t, \xi) &= \Delta X(t, \xi) + f(X(t, \xi)) \\
&\quad + g_1(\sigma, X(t, \xi))\widehat{\pi}(dt, d\xi, d\sigma), \qquad t > 0, \ \xi \in \mathbb{R}^d, \\
X(0, \xi) &= x(\xi), \qquad \xi \in \mathbb{R}^d, \\
\frac{\partial X}{\partial t}(0, \xi) &= y(\xi), \qquad \xi \in \mathbb{R}^d,
\end{aligned} \tag{13.5}$$

where $\Delta X(t, \xi) = \sum_{j=1}^d \partial^2 X(t, \xi)/\partial \xi^2$ is the Laplace operator with respect to the space variable ξ and π is a Poisson random measure on $[0, \infty) \times \mathbb{R}^d \times S$ with intensity measure $dt \, d\xi \, \nu(d\sigma)$. We will consider (13.5) in dimensions $d = 1, 2$. We are looking for a solution in weighted L^p-spaces, $p \in [1, 2)$. Recall that the spaces $L^p_\rho, p \ge 1, \rho \in \mathbb{R}$, were introduced in Section 2.3. By definition the solution will be a process satisfying a certain integral equation involving the fundamental solution to the wave equation; see Definition 13.8 below. A similar definition of a solution to the wave equation driven by a Wiener process is considered in e.g. Dalang and Frangos (1998), Mueller (1998) and Peszat and Zabczyk (2004). We cannot show the existence of a Markov solution. In fact (see e.g. Da Prato and Giusti 1967 or Littman 1963) the linear deterministic problem for $d \neq 1$ and $p \neq 2$ does not have Markov solutions. The L^2-theory of (Markov) solutions to the stochastic wave equation on $\mathbb{R}^d$ will be considered in Chapter 14.

13.2.1 Linear deterministic wave equation

Consider the linear wave problem on $\mathbb{R}^d$

$$
\begin{aligned}
\frac{\partial^2 u}{\partial t^2}(t, \xi) &= \Delta u(t, \xi), && t > 0, \ \xi \in \mathbb{R}^d, \\
u(0, \xi) &= u_0(\xi), && \xi \in \mathbb{R}^d, \\
\frac{\partial u}{\partial t}(0, \xi) &= v_0(\xi), && \xi \in \mathbb{R}^d,
\end{aligned}
\tag{13.6}
$$

where u_0 and v_0 belong to the space $\mathcal{S}'(\mathbb{R}^d; \mathbb{C})$ of tempered distributions. Denote by $\mathfrak{F}u$ the Fourier transform of u. Then we arrive at the problem

$$
\frac{d^2}{dt^2}\mathfrak{F}u(t)(\xi) = -4\pi^2 |\xi|^2 \mathfrak{F}u(t)(\xi), \qquad t > 0,
$$

with initial conditions $\mathfrak{F}u(0) = \mathfrak{F}u_0$ and $(d\mathfrak{F}u/dt)(0) = \mathfrak{F}v_0$. Hence, by direct computation we get

$$
\mathfrak{F}u(t)(\xi) = \cos(2\pi |\xi| t)\, \mathfrak{F}u_0(\xi) + \frac{\sin(2\pi |\xi| t)}{2\pi |\xi|}\, \mathfrak{F}v_0(\xi).
$$

In this formula, we have multiplied the distributions u_0 and v_0 by functions $\cos(2\pi |\xi| t)$ and $(2\pi |\xi|)^{-1} \sin(2\pi |\xi| t)$ of the ξ-variable. These products are well defined because the functions are infinitely differentiable, all derivatives being of polynomial growth. Note that $(\mathfrak{F}u(t), \ t \geq 0)$ is an $\mathcal{S}'(\mathbb{R}^d; \mathbb{C})$-valued mapping infinitely differentiable in t.

Let $\mathcal{G}$ be the solution to (13.6) with $\mathcal{G}(0) \equiv 0$ and $\partial \mathcal{G}(0)/\partial t$ equal to the Dirac δ_0-distribution. Then

$$
\mathfrak{F}\left(\frac{\partial}{\partial t}\mathcal{G}(t)\right)(\xi) = \cos(2\pi |\xi| t) \quad \text{and} \quad \mathfrak{F}\mathcal{G}(t)(\xi) = \frac{\sin(2\pi |\xi| t)}{2\pi |\xi|}.
$$

Consequently, the unique solution to (13.6) is given by

$$
u(t) = \frac{\partial}{\partial t}\mathcal{G}(t) * u_0 + \mathcal{G}(t) * v_0,
$$

where $*$ denotes the convolution operator.

We call $\mathcal{G}$ the *fundamental solution* to the wave equation $\partial^2 u / \partial t^2 - \Delta u = 0$. Explicit formulae for $(\mathcal{G}(t), t \geq 0)$ are well known (see e.g. Mizohata 1973, pp. 279–80). Namely, if $d = 1$ then $\mathcal{G}(t)(\xi) = \chi_{\{|\xi| < t\}}/2$. If $d = 2k + 1$ for some integer $k \geq 1$ then

$$\mathcal{G}(t) = \frac{1}{2(2\pi)^k} \left(\frac{1}{t} \frac{\partial}{\partial t} \right)^{k-1} \frac{\sigma_t^d}{t},$$

where σ_t^d is the surface measure on the sphere in $\mathbb{R}^d$ with its center at 0 and with radius t. If $d = 2k$ then

$$\mathcal{G}(t)(\xi) = \frac{1}{(2\pi)^k} \left(\frac{1}{t} \frac{\partial}{\partial t} \right)^{k-1} \frac{1}{(t^2 - |\xi|^2)^{1/2}} \chi_{\{|\xi| < t\}}.$$

Recall that in any dimension $\mathcal{G}(0) \equiv 0$. Clearly, if $d = 1, 2, 3$ then each $\mathcal{G}(t)$ is a finite non-negative measure with support contained in the closed ball $\overline{B(0, t)}$ in $\mathbb{R}^d$ with its center at 0 and radius t.

Recall that the weight ϑ_ρ is a symmetric strictly positive function equal to $e^{-\rho|\xi|}$ for $|\xi| \geq 1$. Let us denote by $W_\rho^{1,p}$ the Sobolev space of order 1 that is the completion of $\mathcal{S}(\mathbb{R}^d)$ with respect to the norm

$$|\psi|_{W_\rho^{1,p}} = \left(\int_{\mathbb{R}^d} \left(|\psi(\xi)|^2 + |\nabla \psi(\xi)|^2 \right)^{p/2} \vartheta_\rho^p(\xi) \, d\xi \right)^{1/p}.$$

Note that for each ρ there is a constant C_ρ such that

$$\vartheta_\rho(\xi - \eta) \leq C_\rho e^{|\rho|t} \vartheta_\rho(\xi) \qquad \text{for } t \geq 0, \ \xi \in \mathbb{R}^d, \ \eta \in \overline{B(0, t)}. \tag{13.7}$$

Lemma 13.3 *Let $d = 1, 2, 3$, $\rho \in \mathbb{R}$, $p \in [1, \infty)$, and let $T < \infty$. Then there is a constant $C < \infty$ such that, for all $t \in [0, T]$ and $\xi \in \mathbb{R}^d$,*

$$\mathcal{G}(t, \cdot) * \vartheta_\rho^p(\xi) = \int_{\mathbb{R}^d} \vartheta_\rho^p(\xi - \eta) \mathcal{G}(t, d\eta) \leq C \vartheta_\rho^p(\xi). \tag{13.8}$$

Proof By direct calculation

$$\int_{\mathbb{R}^d} \mathcal{G}(t)(d\eta) = t, \qquad t \geq 0. \tag{13.9}$$

Hence, by (13.7),

$$\mathcal{G}(t) * \vartheta_\rho^p(\xi) = \int_{\overline{B(0,t)}} \vartheta_\rho^p(\xi - \eta) \mathcal{G}(t)(d\eta) \leq C_\rho e^{|\rho|t} t \vartheta_\rho^p(\xi).$$

$\square$

Lemma 13.4 *Let $\rho \in \mathbb{R}$, $T < \infty$ and $d = 1, 2$. Assume that $p \geq 1$ if $d = 1$ and $p \in [1, 2)$ if $d = 2$. Then there is a constant $C < \infty$ such that, for all $t \in [0, T]$*

and $\xi \in \mathbb{R}^d$,

$$|\mathcal{G}(t)|^p * \vartheta_\rho^p(\xi) = \int_{\mathbb{R}^d} \vartheta_\rho^p(\xi - \eta) |\mathcal{G}(t, \eta)|^p \, d\eta \le C \vartheta_\rho^p(\xi).$$

Proof   If $d = 1$ then $|\mathcal{G}(t, \eta)|^p = 2^{1-p} \mathcal{G}(t, \eta)$, and the desired estimate follows directly from the previous lemma.

If $d = 2$ and $p \in [1, 2)$ then

$$|\mathcal{G}(t)|^p * \vartheta_\rho^p(\xi) = \int_{B(0,t)} \frac{1}{(2\pi)^p} \left(t^2 - |\eta|^2\right)^{-p/2} \vartheta_\rho^p(\xi - \eta) \, d\eta$$

$$\le C \vartheta_\rho^p(\xi) \int_{B(0,t)} \frac{1}{(2\pi)^p} \left(t^2 - |\eta|^2\right)^{-p/2} d\eta.$$

Since

$$\int_{B(0,t)} \frac{1}{2\pi} \left(t^2 - |\eta|^2\right)^{-p/2} d\eta = \int_0^t (t^2 - r^2)^{-p/2} r \, dr < \infty,$$

we have the desired conclusion. $\square$

Lemma 13.5   *Let* $d = 1, 2$, $\rho \in \mathbb{R}$, $p \in [1, 2)$, *and let* $T < \infty$. *Then there is a constant* $C < \infty$ *depending on* ρ, p *and* T *such that, for every* $t \in [0, T]$,

$$|\mathcal{G}(t) * \psi|_{L_\rho^p} \le C |\psi|_{L_\rho^p}, \qquad \forall \, \psi \in L_\rho^p, \tag{13.10}$$

$$\left| \frac{\partial}{\partial t} \mathcal{G}(t) * \psi \right|_{L_\rho^p} \le C \, |\psi|_{W_\rho^{1,p}}, \qquad \forall \, \psi \in W_\rho^{1,p}. \tag{13.11}$$

Proof   Using (13.9), the Jensen inequality and (13.8) we obtain

$$\int_{\mathbb{R}^d} \left| \int_{\mathbb{R}^d} \psi(\xi - \eta) \mathcal{G}(t, d\eta) \right|^p \vartheta_\rho^p(\xi) \, d\xi$$

$$\le t^{p-1} \int_{\mathbb{R}^d} \int_{\mathbb{R}^d} |\psi(\xi - \eta)|^p \mathcal{G}(t, d\eta) \vartheta_\rho^p(\xi) \, d\xi$$

$$\le t^{p-1} \int_{\mathbb{R}^d} \int_{\mathbb{R}^d} |\psi(z)|^p \vartheta_\rho^p(z + \eta) \mathcal{G}(t, d\eta) \, dz$$

$$\le C_1 \int_{\mathbb{R}^d} |\psi(z)|^p \vartheta_\rho^p(z) \, dz,$$

which proves (13.10).

We now prove (13.11). To this end note that if $d = 1$ then

$$\frac{\partial}{\partial t} \mathcal{G}(t) * \psi(\xi) = \frac{1}{2} \frac{\partial}{\partial t} \int_{-t}^{t} \psi(\xi + \eta) \, d\eta = \frac{1}{2} \left(\psi(\xi + t) - \psi(\xi - t) \right).$$

Consequently, using (13.7) we obtain easily

$$\left| \frac{\partial}{\partial t} \mathcal{G}(t) * \psi \right|_{L_\rho^p} \le C |\psi|_{L_\rho^p}.$$

Now, for $d = 2$,

$$\left| \frac{\partial}{\partial t} \mathcal{G}(t) * \psi(\xi) \right|$$

$$= \frac{1}{2\pi} \left| \frac{\partial}{\partial t} \int_{\{|\eta| < t\}} \frac{\psi(\xi - \eta)}{(t^2 - |\eta|^2)^{1/2}} \, d\eta \right|$$

$$= \frac{1}{2\pi} \left| \frac{\partial}{\partial t} \left(t \int_{\{|z| < 1\}} \frac{\psi(\xi - tz)}{(1 - |z|^2)^{1/2}} \, dz \right) \right| \le \frac{1}{t} \mathcal{G}(t) * |\psi|(\xi) + \mathcal{G}(t) * |\nabla \psi|(\xi).$$

By Jensen's inequality and (13.9),

$$\int_{\mathbb{R}^2} \left| \frac{1}{t} \mathcal{G}(t) * |\psi|(\xi) \right|^p \vartheta_\rho^p(\xi) \, d\xi \le \int_{\mathbb{R}^2} \int_{\mathbb{R}^2} \frac{1}{t} \mathcal{G}(t, d\eta) |\psi(\xi - \eta)|^p \vartheta_\rho^p(\xi) \, d\xi$$

$$\le \int_{\mathbb{R}^2} \int_{\mathbb{R}^2} \frac{1}{t} \mathcal{G}(t, d\eta) |\psi(z)|^p \vartheta_\rho^p(z + \eta) \, dz.$$

Since (see the proof of (13.8)) we can choose a constant $C < \infty$ such that, for all $t \in [0, T]$,

$$\frac{1}{t} \int_{\mathbb{R}^2} \vartheta_\rho^p(z + \eta) \mathcal{G}(t, d\eta) \le C \, \vartheta_\rho^p(z), \qquad z \in \mathbb{R}^2,$$

we have

$$\int_{\mathbb{R}^2} \left| \frac{1}{t} \mathcal{G}(t) * |\psi|(\xi) \right|^p \vartheta_\rho^p(\xi) \, d\xi \le C \int_{\mathbb{R}^2} |\psi(\xi)|^p \vartheta_\rho^p(\xi) \, d\xi.$$

By (13.10), $\left| \mathcal{G} * |\nabla \psi| \right|_{L_\rho^p} \le C \, |\psi|_{W_\rho^{1,p}}$, which completes the proof. $\square$

As a direct consequence of the lemma we have the following corollary.

Corollary 13.6 *Let $\rho \in \mathbb{R}$, $p \in [1, 2)$, and let $d = 1, 2$. Then for any $t \ge 0$ there are unique operators $\mathcal{R}(t) \in L(L_\rho^p, L_\rho^p)$ and $\dot{\mathcal{R}}(t) \in L(W_\rho^{1,p}, L_\rho^p)$ such that, for every $\psi \in \mathcal{S}(\mathbb{R}^d)$, $\mathcal{R}(t)\psi = \mathcal{G}(t) * \psi$ and $\dot{\mathcal{R}}(t) = \partial \mathcal{G}(t)/\partial t * \psi$. Moreover, for every $T > 0$ there is a constant C such that*

$$\|\mathcal{R}(t)\|_{L(L_\rho^p, L_\rho^p)} + \|\dot{\mathcal{R}}(t)\|_{L(W_\rho^{1,p}, L_\rho^p)} \le C, \qquad \forall \, t \in (0, T].$$

13.2.2 Stochastic convolution

Let ν be a σ-finite measure on $(S, \mathcal{S})$, and let π be a Poisson random measure on $[0, \infty) \times \mathbb{R}^d \times S$ with intensity $dt \, d\xi \, \nu(d\sigma)$. Assume that π is defined on a filtered probability space $\mathfrak{U} = (\Omega, \mathcal{F}, (\mathcal{F}_t), \mathbb{R})$.

Let

$$\mathcal{L}_\rho^p := L^p\big(\mathbb{R}^d \times S, \ \mathcal{B}(\mathbb{R}^d) \times \mathcal{S}, \ \vartheta_\rho^p(\xi)\,d\xi\,\nu(d\sigma)\big).$$

Given $\rho \in \mathbb{R}$ and $p \in [1,2]$ we denote by $\mathcal{X}_{\rho,T}^p$ and $\mathcal{Y}_{\rho,T}^p$ the spaces of all pre-dictable fields $u\colon \Omega \times [0,T] \mapsto L_\rho^p$ and $v\colon \Omega \times [0,T] \mapsto \mathcal{L}_\rho^p$ such that

$$\|u\|_{\mathcal{X}_{\rho,T}^p} := \sup_{t \in [0,T]} \Big(\mathbb{E}\,|u(t)|_{L_\rho^p}^p\Big)^{1/p} < \infty,$$

$$\|v\|_{\mathcal{Y}_{\rho,T}^p} := \Big(\mathbb{E}\int_0^T |v(t)|_{\mathcal{L}_\rho^p}^p\,dt\Big)^{1/p} < \infty.$$

The main goal of this section is to show that, given $v \in \mathcal{Y}_{\rho,T}^p$, the stochastic integral

$$I_v(t)(\xi) := \int_0^t \int_{\mathbb{R}^d} \int_S \mathcal{G}(t-s, \xi-\eta)v(s)(\eta,\sigma)\widehat{\pi}(ds, d\eta, d\sigma)$$

is a well-defined predictable L_ρ^p-valued process. Note that in dimensions 1, 2 the fundamental solution $\mathcal{G}$ is a function, whereas if $d = 3$ it is a measure. Thus the case of $d = 3$ will need special treatment.

Lemma 13.7 *Let* $\rho \in \mathbb{R}$, $T \in [0,\infty)$ *and* $d = 1, 2$. *Assume that* $p \in [1,2]$ *if* $d = 1$ *and* $p \in [1,2)$ *if* $d = 2$. *Then* $v \mapsto I_v$ *is a bounded linear operator from* $\mathcal{Y}_{\rho,T}^p$ *to* $\mathcal{X}_{\rho,T}^p$.

Proof By Theorem 8.23 there is a constant C_1, depending only on p, such that

$$\mathbb{E}\bigg|\int_0^t \int_{\mathbb{R}^d} \int_S \mathcal{G}(t-s, \xi-\eta)v(s)(\eta,\sigma)\widehat{\pi}(ds, d\eta, d\sigma)\bigg|^p$$

$$\leq C_1 \int_0^t \int_{\mathbb{R}^d} \int_S |\mathcal{G}(t-s, \xi-\eta)|^p\,\mathbb{E}\,|v(s)(\eta,\sigma)|^p\,ds\,d\eta\,\nu(d\sigma).$$

Taking into account Lemma 13.4 one can find a constant C_2 such that

$$\int_{\mathbb{R}^d} \mathbb{E}\bigg|\int_0^t \int_{\mathbb{R}^d} \int_S \mathcal{G}(t-s, \xi-\eta)v(s)(\eta,\sigma)\widehat{\pi}(ds, d\eta, d\sigma)\bigg|^p \vartheta_\rho^p(\xi)\,d\xi$$

$$\leq C_2 \int_0^t \int_{\mathbb{R}^d} \int_S \mathbb{E}\,|v(s)(\eta,\sigma)|^p\,ds\,\nu(d\sigma)\vartheta_\rho^p(\eta)\,d\eta \leq C_2\,\|v\|_{\mathcal{Y}_{\rho,T}^p}^p.$$

What is left is the proof of the predictability of I_v. Clearly I_v is adapted. Thus we need to show that it is stochastically continuous. To do this, given $0 \leq \tilde{t} < t \leq T$ we write $\mathcal{G}_{t,\tilde{t}}(s,z) := \mathcal{G}(t-s,z) - \mathcal{G}(\tilde{t}-s,z)$. We have

$$\mathbb{E}\,\big|I_v(t) - I_v(\tilde{t})\big|_{L_\rho^p}^p$$

$$\leq C \int_{\tilde{t}}^t \int_S \mathbb{E}\,|v(s)(\cdot,\sigma)|_{L_\rho^p}^p\,\nu(d\sigma)\,ds + \int_{\mathbb{R}^d} |I_{t,\tilde{t}}(\xi)|\vartheta_\rho^p(\xi)\,d\xi,$$

where

$$I_{t,\tilde{t}}(\xi) = \mathbb{E}\left|\int_0^{\tilde{t}} \int_{\mathbb{R}^d} \int_S \mathcal{G}_{t,\tilde{t}}(s, \xi - \eta) v(s)(\eta, \sigma)\widehat{\pi}(ds, d\eta, d\sigma)\right|^p$$

$$\leq C_1 \int_0^{\tilde{t}} \int_{\mathbb{R}^d} \int_S |\mathcal{G}_{t,\tilde{t}}(s, \xi - \eta)|^p |v(s)(\eta, \sigma)|^p \, ds \, d\eta \, \nu(d\sigma).$$

Now, there is a constant C independent of $t, \tilde{t} \in [0, T]$ such that

$$\int_{\mathbb{R}^d} |\mathcal{G}_{t,\tilde{t}}(s, \xi - \eta)|^p \vartheta_\rho^p(\xi) \, d\xi \leq C \vartheta_\rho(\xi) \int_{\mathbb{R}^d} |\mathcal{G}_{t,\tilde{t}}(s, \eta)|^p \, d\eta,$$

and the stochastic continuity follows from the identity

$$\lim_{\tilde{t} \to t} \int_{\mathbb{R}^d} |\mathcal{G}_{t,\tilde{t}}(s, \eta)|^p \, d\eta = 0.$$

$\square$

13.2.3 Main result

Let $\rho \in \mathbb{R}$, $p \in [1, 2]$ and $d = 1, 2$. Assume that $u_0 \in W_\rho^{1,p}$, $v_0 \in L_\rho^p$ and that $F(u)(\xi) := f(u(\xi), \xi), \xi \in \mathbb{R}^d$, and $G(u)(\xi, \sigma) := g_1(u(\xi), \xi, \sigma), \xi \in \mathbb{R}^d, \sigma \in S$, are composition operators corresponding to the functions $f \colon \mathbb{R} \times \mathbb{R}^d \mapsto \mathbb{R}$ and $g_1 \colon \mathbb{R} \times \mathbb{R}^d \times S \mapsto \mathbb{R}$.

Definition 13.8 We say that u is an $\mathcal{X}_{\rho,T}^p$-valued solution to (13.5) if $u \in \mathcal{X}_{\rho,T}^p$ and u satisfies the integral equation

$$u(t, \xi) = \frac{\partial}{\partial t} \mathcal{G}(t) * u_0(\xi) + \mathcal{G}(t) * v_0(\xi) + \int_0^t \mathcal{G}(t - s) * F(u(s))(\xi) \, ds$$

$$+ \int_0^t \int_{\mathbb{R}^d} \int_S \mathcal{G}(t - s, \xi - \eta) G(u(s))(\eta, \sigma)\widehat{\pi}(ds, d\eta, d\sigma),$$

where the equality is considered in $\mathcal{X}_{\rho,T}^p$.

Theorem 13.9 *Let $\rho \in \mathbb{R}$, $p \in [1, 2]$, and let $d = 1, 2$. Assume that*

(i) $1 \leq p < 2/d + 1$,
(ii) *$f \colon \mathbb{R} \times \mathbb{R}^d \mapsto \mathbb{R}$ and $g_1 \colon \mathbb{R} \times \mathbb{R}^d \times S \mapsto \mathbb{R}$ are measurable and there are functions $l_1 \in L^p(S, \mathcal{S}, \nu)$, $l_2 \in L_\rho^p$ and a constant L such that, for all $u, v \in \mathbb{R}, \xi \in \mathbb{R}^d$ and $\sigma \in S$,*

$$|f(u, \xi)| \leq L(l_2(\xi) + |u|),$$
$$|f(u, \xi) - f(v, \xi)| \leq L |u - v|$$
$$|g_1(u, \xi, \sigma)| \leq l_1(\sigma)(l_2(\xi) + |u|),$$
$$|g_1(u, \xi, \sigma) - g_1(v, \xi, \sigma)| \leq l_1(\sigma) |u - v|.$$

Then for all $u_0 \in W_\rho^{1,p}$ and $v_0 \in L_\rho^p$ there exists a unique $\mathcal{X}_{\rho,T}^p$-valued solution to (13.5).

Proof   Let us fix u_0, v_0 and $T < \infty$. By (ii), $F \colon \mathcal{X}_{\rho,T}^p \mapsto \mathcal{X}_{\rho,T}^p$ and $G \colon \mathcal{X}_{\rho,T}^p \mapsto \mathcal{Y}_{\rho,T}^p$ are Lipschitz continuous. Write

$$I_1(t) := \frac{\partial}{\partial t} \mathcal{G}(t) * u_0(\xi) + \mathcal{G}(t) * v_0(\xi), \qquad t \in [0,T],$$

$$I_2(u)(t) := \int_0^t \mathcal{G}(t-s) * F(u(s)) \, ds, \qquad t \in [0,T], \ u \in \mathcal{X}_{\rho,T}^p.$$

Recall that I_v was introduced in subsection 13.2.2. Consider on $\mathcal{X}_{\rho,T}^p$ the family of norms

$$\|u\|_\beta := \left(\sup_{t \in [0,T]} e^{-\beta t} \, \mathbb{E} \, |u(t)|_{L_\rho^p}^p \right)^{1/p}, \qquad \beta > 0.$$

The proof will be complete as soon as we can show that for β large enough the mapping $J(u) := I_1 + I_2(u) + I_{G(u)}$, $u \in \mathcal{X}_{\rho,T}^p$, is a contraction on $\mathcal{X}_{\rho,T}^p$. We leave to the reader the proof that if $u \in \mathcal{X}_{\rho,T}^p$ then $I_1, I_2(u), I_{G(u)} \in \mathcal{X}_{\rho,T}^p$. We shall concentrate on the estimate of the Lipschitz constant. To do this we note that, by (13.10) and the Lipschitz continuity of F, there is a constant C such that

$$\|I_2(u) - I_2(v)\|_\beta^p$$

$$\leq \sup_{t \in [0,T]} e^{-\beta t} \, \mathbb{E} \int_0^t \left| \mathcal{G}(t-s) * \big(F(u(s)) - F(v(s)) \big) \right|_{L_\rho^p}^p \, ds$$

$$\leq \sup_{t \in [0,T]} \mathbb{E} \int_0^t e^{-\beta(t-s)} e^{-\beta s} \, C \, |u(s) - v(s)|_{L_\rho^p}^p \, ds$$

$$\leq C \, \|u\|_\beta^p \int_0^T e^{-\beta s} \, ds \leq \frac{C}{\beta} \|u\|_\beta^p.$$

Using Lemma 13.7 we can find a constant C such that

$$\left\| I_{G(u)} - I_{G(v)} \right\|_\beta^p \leq C \sup_{t \in [0,T]} e^{-\beta t} \, \mathbb{E} \int_0^t |u(s) - v(s)|_{L_\rho^p}^p \, ds$$

$$\leq \sup_{t \in [0,T]} \mathbb{E} \int_0^t e^{-\beta(t-s)} e^{-\beta s} \, C \, |u(s) - v(s)|_{L_\rho^p}^p \, ds$$

$$\leq C \, \|u\|_\beta^p \int_0^T e^{-\beta s} \, ds \leq \frac{C}{\beta} \|u\|_\beta^p.$$

$\square$

13.3 Stochastic delay equations

We are concerned with the delay problem

$$dx(t) = \left(\int_{-h}^{0} \mu(d\xi)x(t+\xi) + f(x(t)) \right) dt + g(x(t))\,dW(t) + g_1(x(t))\,dZ(t),$$

$$x(\xi) = \psi(\xi), \xi \in (-h, 0], \tag{13.12}$$

where $h > 0$, μ is an $M(n \times n)$-valued measure on $(-h, 0]$ with finite variation, W is a standard d-dimensional Wiener process, Z is a square integrable mean-zero Lévy process in $\mathbb{R}^m$ and f, g, g_1 are Lipschitz continuous mappings from $\mathbb{R}^n$ to $\mathbb{R}^n$, $M(n \times d)$ and $M(n \times m)$, respectively.

Let

$$L^2(-h, 0; \mathbb{R}^n) := L^2\big((-h, 0], \mathcal{B}((-h, 0]), d\xi; \mathbb{R}^n\big),$$

$$W^{1,2}(-h, 0; \mathbb{R}^n) := W^{1,2}\big((-h, 0], \mathcal{B}((-h, 0]), d\xi; \mathbb{R}^n\big).$$

We interpret (13.12) as (9.4) on the state space

$$H := \begin{pmatrix} L^2(-h, 0; \mathbb{R}^n) \\ \mathbb{R}^n \end{pmatrix},$$

the operator A from (9.4) being given by

$$A \begin{pmatrix} \psi \\ v \end{pmatrix} := \begin{pmatrix} \int_{h}^{0} \mu(d\xi)\psi(\xi) \\ \psi(0) \end{pmatrix},$$

where the domain of A is equal to

$$D(A) = \left\{ \begin{pmatrix} \psi \\ v \end{pmatrix} : \psi \in W^{2,2}(-h, 0; \mathbb{R}^n) \text{ and } v = \psi(0) \right\}.$$

Note that the RKHS of $M = (W, Z)$ is a subspace of $\mathbb{R}^d \times \mathbb{R}^m$ and that

$$F \begin{pmatrix} \psi \\ v \end{pmatrix} := \begin{pmatrix} N_f(\psi) \\ v \end{pmatrix} \quad \text{and} \quad G \begin{pmatrix} \psi \\ v \end{pmatrix} u := \begin{pmatrix} N_g(\psi)u_1 + N_{g_1}(\psi)u_2 \\ v \end{pmatrix}.$$

Then (see e.g. Chojnowska-Michalik 1978 or Da Prato and Zabczyk 1992a) the operator $(A, D(A))$ generates a C_0-semigroup on H.

With the process $x(\cdot)$ on $(-h, +\infty)$ having values in $\mathbb{R}^n$ we associate a process $X(\cdot)$ defined on $[0, +\infty)$ having values in H, whose second coordinate, at time t, is the vector $x(t)$ and whose first coordinate is the *segment*

$$x_t(\xi) := x(t+\xi), \qquad \xi \in (-h, 0],$$

of x. It is easy to show that the Lipschitz continuity of f, g and g_1 guarantees the Lipschitz continuity of F and G, and consequently conditions (F) and (G). Thus we have the following consequence of Theorem 9.29.

Theorem 13.10   *Assume that f, g and g_1 are Lipschitz continuous mappings from $\mathbb{R}^n$ to $\mathbb{R}^n$, $M(n \times d)$ and $M(n \times m)$, respectively. Let W be a standard d-dimensional Wiener process, and let L be a square integrable mean-zero Lévy process in $\mathbb{R}^m$. Then the following hold.*

(i) *For any initial value $x_0 \in H$ there is a unique solution $X(\cdot, x_0)$, $t \geq 0$, of (13.12).*

(ii) *For any $T < \infty$ there is a constant L such that for all $x_0, x_1 \in H$ we have*

$$\sup_{t \in [0,T]} \mathbb{E} \, |X(t, x_0) - X(t, x_1)|_H^2 \leq L \, |x_0 - x_1|_H^2.$$

(iii) *Equation (13.12) defines a Feller family on H.*

14

Equations driven by a spatially homogeneous noise

In this chapter, Lévy processes on the space of tempered distributions are introduced and their reproducing kernel spaces are determined. Non-linear heat and wave equations with spatially homogeneous noise are investigated.

14.1 Tempered distributions

We denote by $S(\mathbb{R}^d)$ and $S(\mathbb{R}^d; \mathbb{C})$ the spaces of infinitely differentiable real- and complex-valued functions ψ such that

$$p_n(\psi) := \sup_{\xi \in \mathbb{R}^d} \sup_{|\alpha| \leq n} \left(1 + |\xi|^2\right)^n \left| \frac{\partial^{|\alpha|} \psi}{\partial \xi^\alpha}(\xi) \right| < \infty, \qquad \forall n \in \mathbb{N}.$$

On $S(\mathbb{R}^d)$ and $S(\mathbb{R}^d; \mathbb{C})$ we consider the topology defined by the family of semi-norms $(p_n, n \in \mathbb{N})$. The *spaces of tempered distributions* $S'(\mathbb{R}^d)$ and $S'(\mathbb{R}^d; \mathbb{C})$ are the dual spaces to $S(\mathbb{R}^d)$ and $S(\mathbb{R}^d; \mathbb{C})$. We denote by $\langle u, \psi \rangle$ the value of the distribution u on a test function ψ. Every complex or real tempered Borel measure λ on $\mathbb{R}^d$, that is, a measure whose variation $\|\lambda\|_{\mathrm{Var}}$ satisfies

$$\int_{\mathbb{R}^d} \left(1 + |\xi|^2\right)^{-n} \|\lambda\|_{\mathrm{Var}}(\mathrm{d}\xi) < \infty \qquad \text{for some } n \in \mathbb{N},$$

can be identified with the distribution $\langle \lambda, \psi \rangle = \int_{\mathbb{R}^d} \psi(\xi) \lambda(\mathrm{d}\xi)$. Sometimes in this book we will denote by (ψ, λ) the integral $\int_E \psi(\xi) \lambda(\mathrm{d}\xi)$.

We denote by $\mathfrak{F}\psi$ the Fourier transform of a test function ψ:

$$\mathfrak{F}\psi(\eta) := \frac{1}{(2\pi)^{d/2}} \int_{\mathbb{R}^d} e^{-2i\pi \eta \xi} \psi(\xi) \, \mathrm{d}\xi,$$

where $\eta \xi = \sum_{i=1}^d \eta_i \xi_i$ and $\xi, \eta \in \mathbb{R}^d$. Next we define $\mathfrak{F}u$ for $u \in S'(\mathbb{R}^d; \mathbb{C})$ by the formula $\langle \mathfrak{F}u, \psi \rangle = \langle u, \mathfrak{F}^{-1}\psi \rangle$, $\psi \in S(\mathbb{R}^d; \mathbb{C})$, where

$$\mathfrak{F}^{-1}\psi(\eta) := \frac{1}{(2\pi)^{d/2}} \int_{\mathbb{R}^d} e^{2i\pi \eta \xi} \psi(\xi) \, d\xi$$

is the inverse Fourier transform. Then $\mathfrak{F}$ is a bijection on $S(\mathbb{R}^d; \mathbb{C})$ and an isometry on $L^2(\mathbb{R}^d; \mathbb{C}) := L^2(\mathbb{R}^d, \mathcal{B}(\mathbb{R}^d), \ell_d; \mathbb{C})$.

Definition 14.1 We say that a topological vector space E is *nuclear* if the topology on E is given by the family of norms $(|\cdot|_{H_n})$, $|u|_{H_n} = \sqrt{\langle u, u \rangle_{H_n}}$, $n \in \mathbb{N}$, $u \in E$, where $((H_n, \langle \cdot, \cdot \rangle_{H_n}))$ is a decreasing sequence of Hilbert spaces such that $E \subset H_n$ and the embedding $H_{n+1} \hookrightarrow H_n$ is nuclear for every n.

Theorem 14.2 *The space* $S(\mathbb{R}^d)$ *is nuclear.*

Proof Let H_n be the completion of $S(\mathbb{R}^d)$ with respect to the norm $|\psi|_{H_n} = \sqrt{\langle \psi, \psi \rangle_{H_n}}$, where

$$\langle \psi, \varphi \rangle_{H_n} = \int_{\mathbb{R}^d} \left(1 + |\xi|^2\right)^{a_n} \sum_{|\alpha| \le a_n} \frac{\partial^{|\alpha|}\psi}{\partial \xi^\alpha}(\xi) \frac{\partial^{|\alpha|}\varphi}{\partial \xi^\alpha}(\xi) \, d\xi.$$

Then choosing the a_n such that they go to ∞ sufficiently fast, one gets the desired conclusion. For more details see e.g. Yosida (1965). $\square$

Let H_n be as above. Then $H_n \hookrightarrow L^2 := L^2(\mathbb{R}^d)$. Identifying $(L^2)^*$ with L^2 we obtain $H_n \hookrightarrow L^2 = (L^2)^* \hookrightarrow H_n^*$, $n \in \mathbb{N}$. Thus the following result holds.

Theorem 14.3 *We have* $S'(\mathbb{R}^d) = \bigcup_{n=1}^{\infty} H_n^*$.

14.2 Lévy processes in $S'(\mathbb{R}^d)$

We now consider equations driven by Lévy processes taking values in $S'(\mathbb{R}^d)$. Since the space of distributions is not metrizable we have to adapt Definition 4.1 to our purposes. Let us note that a Lévy–Khinchin-type description of Lévy processes on $S'(\mathbb{R}^d)$, more generally on co-nuclear spaces, was developed in Ustunel (1984).

Definition 14.4 A stochastic process L taking values in $S'(\mathbb{R}^d)$ is called a *Lévy process in* $S'(\mathbb{R}^d)$ if it is càdlàg, satisfies $L(0) = 0$ and has stationary independent increments.

Throughout the chapter we assume that $(\Omega, \mathcal{F}, (\mathcal{F}_t), \mathbb{P})$ is a filtered probability space satisfying the usual conditions and that L is a Lévy process defined on Ω and adapted to the filtration $(\mathcal{F}_t)$. Moreover, we assume that, for all $t \ge s \ge 0$, $L(t) - L(s)$ is independent of $\mathcal{F}_s$.

For our further purposes, it will be convenient to introduce the class of square integrable Lévy processes.

Definition 14.5 A Lévy process L in $S'(\mathbb{R}^d)$ is *square integrable* (*has mean zero*) if, for every test function $\psi \in S(\mathbb{R}^d)$, the real-valued process $(\langle L(t), \psi \rangle, t \geq 0)$ is square integrable (has mean zero).

Using arguments from the proof of Theorem 4.32 one can easily obtain the following result.

Lemma 14.6 *Assume that L is a square integrable Lévy process in $S'(\mathbb{R}^d)$. Then there are a unique $m_L \in S'(\mathbb{R}^d)$ called the* mean *of L and a unique symmetric positive-definite bilinear form $K : S'(\mathbb{R}^d) \times S'(\mathbb{R}^d) \mapsto \mathbb{R}$ called the* covariance *form of L such that, for all $t, s \geq 0$ and $\psi, \varphi \in S(\mathbb{R}^d)$,*

$$\mathbb{E} \langle L(t), \psi \rangle = t \langle m_L, \psi \rangle,$$

$$\mathbb{E} \langle L(t) - m_L, \psi \rangle \langle L(s) - m_L, \varphi \rangle = t \wedge s \, K(\psi, \varphi).$$

14.2.1 Wiener processes

Definition 14.7 A continuous (in time) mean-zero Lévy process W taking values in $S'(\mathbb{R}^d)$ is called a *Wiener process*.

It follows from this definition that for any finite set of test functions $\{\psi_1, \ldots, \psi_n\} \subset S(\mathbb{R}^d)$ the $\mathbb{R}^n$-valued process

$$(\langle W(t), \psi_1 \rangle, \ldots, \langle W(t), \psi_n \rangle), \qquad t \geq 0,$$

is a continuous mean-zero Lévy process in $\mathbb{R}^n$. Thus (see Theorem 4.12) it is a Wiener process in $\mathbb{R}^n$. In particular it is Gaussian, and hence W is square integrable. Let K be the covariance of W. Then for all non-negative $t_1, \ldots, t_n$ the random vector $(\langle W(t_1), \psi_1 \rangle, \ldots, \langle W(t_n), \psi_n \rangle)$ has the normal distribution $\mathcal{N}(0, [q_{i,j}])$, where $q_{i,j} = t_i \wedge t_j \, K(\psi_i, \psi_j), i, j = 1, \ldots, n$.

14.3 RKHS of a square integrable Lévy process in $S'(\mathbb{R}^d)$

The aims of this section are to show that any square integrable mean-zero Lévy process L in $S'(\mathbb{R}^d)$ can be regarded as a square integrable Lévy process on a properly chosen Hilbert space and to compute the RKHS of such a process L in terms of its covariance form K. To this end, define $\mathcal{H}$ as the set of all $u \in S'(\mathbb{R}^d)$ such that

$$|\langle u, \psi \rangle| \leq L \sqrt{K(\psi, \psi)}, \qquad \forall \psi \in S(\mathbb{R}^d),$$

with a constant $L < \infty$ independent of ψ. Let

$$|u|_{\mathcal{H}} = \sup_{\psi \in S(\mathbb{R}^d):\, K(\psi,\psi)\neq 0} \frac{|\langle u, \psi \rangle|}{\sqrt{K(\psi,\psi)}}, \qquad u \in \mathcal{H}, \tag{14.1}$$

$$\langle u, v \rangle_{\mathcal{H}} = \tfrac{1}{4}\left(|u+v|_{\mathcal{H}}^2 - |u-v|_{\mathcal{H}}^2\right), \qquad u, v \in \mathcal{H}. \tag{14.2}$$

We have the following result.

Theorem 14.8

(i) *The quantity $\langle \cdot, \cdot \rangle_{\mathcal{H}}$ is a scalar product and $|\cdot|_{\mathcal{H}}$ is the corresponding norm. Moreover, $(\mathcal{H}, \langle \cdot, \cdot \rangle_{\mathcal{H}})$ is a real separable Hilbert space.*

(ii) *For any orthonormal basis $\{e_k\}$ of $\mathcal{H}$,*

$$K(\psi, \varphi) = \sum_k \langle e_k, \psi \rangle \langle e_k, \varphi \rangle, \qquad \forall \, \psi, \varphi \in S(\mathbb{R}^d).$$

(iii) *There is a Hilbert space $(U, \langle \cdot, \cdot \rangle_U)$ embedded into $S'(\mathbb{R}^d)$ such that L is a square integrable Lévy process in U. Moreover,[1] the RKHS of L, considered as a U-valued process, is equal to $\mathcal{H}$.*

The proof will be preceded by a few lemmas. Let $((H_n, \langle \cdot, \cdot \rangle_{H_n}))$ be a sequence of separable Hilbert spaces such that $H_{n+1} \hookrightarrow H_n$ with a nuclear embedding and the topology on $S(\mathbb{R}^d)$ is given by a sequence of norms $(|\cdot|_{H_n})$; see Section 14.1. Taking, if necessary, an equivalent scalar product we can assume that $|\psi|_{H_n} \leq |\psi|_{H_{n+1}}$ for all $n \in \mathbb{N}$ and $\psi \in S(\mathbb{R}^d)$. Clearly, we have $S'(\mathbb{R}^d) = \bigcup_{n \in \mathbb{N}} H_n^*$.

Lemma 14.9   *There exist n and C such that $|K(\psi, \varphi)| \leq C\, |\psi|_{H_n} |\varphi|_{H_n}$ for all $\psi, \varphi \in S(\mathbb{R}^d)$.*

Proof   Since

$$K(\psi, \varphi) = \tfrac{1}{4}\left(K(\psi + \varphi, \psi + \varphi) - K(\psi - \varphi, \psi - \varphi)\right)$$

it suffices to show that there exist n and L such that $K(\psi, \psi) \leq C\, |\psi|_{H_n}^2$ for every $\psi \in S(\mathbb{R}^d)$. Suppose, contrary to our claim, that there is a sequence $(\psi_n) \subset S(\mathbb{R}^d)$ such that $K(\psi_n, \psi_n) \geq n\, |\psi_n|_{H_n}^2$ for every $n \in \mathbb{N}$. Set

$$\varphi_n = \frac{\sqrt{n}\, \psi_n}{\sqrt{K(\psi_n, \psi_n)}}.$$

Then, for $n \geq m$,

$$|\varphi_n|_{H_m} = \frac{\sqrt{n}\, |\psi_n|_{H_m}}{\sqrt{K(\psi_n, \psi_n)}} \leq \frac{1}{\sqrt{n}}.$$

---

[1] See Proposition 7.1.

Thus $\varphi_n \to 0$ in $S(\mathbb{R}^d)$ and, since K is continuous on $S(\mathbb{R}^d)$, $K(\varphi_n, \varphi_n) \to 0$ as $n \uparrow \infty$. But $K(\varphi_n, \varphi_n) = n$, which leads to a contradiction. □

Corollary 14.10 *For n sufficiently large, K has a unique extension to a symmetric non-negative continuous bilinear form on $H_n \times H_n$.*

We will denote the extension of K to $H_n \times H_n$ also by K. Since K is symmetric and continuous we have the following (see also e.g. Mlak 1991).

Corollary 14.11 *For n sufficiently large there is a symmetric bounded operator $A_n \colon H_n \mapsto H_n$ such that*

$$K(\psi, \varphi) = \langle A_n \psi, A_n \varphi \rangle_{H_n}, \qquad \psi, \varphi \in H_n. \tag{14.3}$$

Lemma 14.12 *Assume that n is such that $K \colon H_n \times H_n \mapsto \mathbb{R}$ is given by (14.3) with a symmetric and continuous A_n. Then $\mathcal{H} = A_n^* H_n^*$. Moreover, $|v|_{\mathcal{H}} = \left| (A_n^*)^{-1} v \right|_{H_n^*}$ for $v \in A_n^* H_n^*$.*

Proof Let $u \in H_n^*$. Then, for any $\psi \in S(\mathbb{R}^d)$,

$$|\langle A_n^* u, \psi \rangle| = |\langle u, A_n \psi \rangle| \le |u|_{H_n^*} |A_n \psi|_{H_n} = |u|_{H_n^*} \sqrt{K(\psi, \psi)}.$$

Thus $A_n^* u \in \mathcal{H}$. Moreover, for $u \in H_n^*$ orthogonal to Ker A_n^*,

$$|A_n^* u|_{\mathcal{H}} = \sup_{\psi \colon 0 \ne |A_n \psi|_{H_n}} \frac{|\langle u, A_n \psi \rangle|}{|A_n \psi|_{H_n}} = |u|_{H_n^*}.$$

We now show that $\mathcal{H} \subset A_n^* H_n^*$. To this end take a vector $u \in \mathcal{H}$. Then by the definition of $\mathcal{H}$ there is a constant $C < \infty$ such that

$$|\langle u, \psi \rangle| \le C \sqrt{K(\psi, \psi)} = C |A_n \psi|_{H_n}, \qquad \forall \psi \in S(\mathbb{R}^d).$$

Since $S(\mathbb{R}^d)$ is dense in H_n, we have $|\langle u, \psi \rangle| \le C |A_n \psi|_{H_n}$ for every $\psi \in H_n$. □

Proof of Theorem 14.8(i) Let $(A_n^*)^{-1}$ be the operator pseudo-inverse to A_n^*. Then, by Lemma 14.12, $|u|_{\mathcal{H}} = \left| (A_n^*)^{-1} u \right|_{H_n^*}$, $u \in \mathcal{H}$. Thus $| \cdot |_{\mathcal{H}}$ is the norm corresponding to the scalar product

$$\begin{aligned}
\langle u, v \rangle_{\mathcal{H}} &= \left\langle (A_n^*)^{-1} u, (A_n^*)^{-1} v \right\rangle_{H_n^*} \\
&= \tfrac{1}{4} \left(\left| (A_n^*)^{-1} (u + v) \right|_{H_n^*} - \left| (A_n^*)^{-1} (u - v) \right|_{H_n^*} \right) \\
&= \tfrac{1}{4} \left(|u + v|_{\mathcal{H}} - |u - v|_{\mathcal{H}} \right).
\end{aligned}$$

Consequently, $(\mathcal{H}, \langle \cdot, \cdot \rangle_{\mathcal{H}})$ is a real separable Hilbert space, since $A_n^{-1} \colon \mathcal{H} \mapsto H_n^*$ is a linear isometry. □

Proof of Theorem 14.8(ii) Let $\{e_k\}$ be an orthonormal basis of $\mathcal{H}$. Let n be such that $(A_n^*)^{-1} \colon \mathcal{H} \mapsto H_n^*$ is a linear isometry. Set $f_k = (A_n^*)^{-1} e_k$. Then $\{f_k\}$ is an orthonormal (but not necessarily complete) system in H_n^*, and $e_k = A_n^* f_k$. Thus for any $\psi \in S(\mathbb{R}^d)$, in fact for any $\psi \in H_n$,

$$\sum_k \langle e_k, \psi \rangle^2 = \sum_k \langle A_n^* f_k, \psi \rangle^2 = \sum_k \langle f_k, A_n \psi \rangle^2 = |A_n \psi|_{H_n}^2 < \infty.$$

Since $K(\psi, \varphi) = \langle A_n \psi, A_n \varphi \rangle_{H_n}$, we have the desired identity. $\qquad\square$

Proof of Theorem 14.8(iii) Assume that n is such that $K : H_n \times H_n \mapsto \mathbb{R}$ is given by (14.3) with A_n symmetric and continuous. Since the embedding $i : H_{n+1} \hookrightarrow H_n$ is nuclear, it is compact. Then the symmetric operator $i^* i : H_{n+1} \mapsto H_{n+1}$ is also compact. Assume that $\{f_k\}$ is an orthonormal basis of H_{n+1} consisting of eigenvectors of $\kappa := \sqrt{i^* i}$ with corresponding eigenvalues $\{\gamma_k\}$. Since κ is Hilbert–Schmidt,

$$\sum_k \gamma_k^2 < \infty. \tag{14.4}$$

Note that as H_{n+1} is dense in H_n, $\gamma_k \neq 0$ for every k. Next,

$$\langle f_j, f_k \rangle_{H_n} = \langle i f_j, i f_k \rangle_{H_n} = \langle f_j, i^* i f_k \rangle_{H_{n+1}} = \left\langle f_j, \kappa^2 f_k \right\rangle_{H_{n+1}} = \gamma_k^2 \delta_{k,j}.$$

Let $\tilde{f}_k = \gamma_k^{-1} f_k$. Then $\{\tilde{f}_k\}$ is an orthonormal basis of H_n such that

$$\left\langle \tilde{f}_k, \tilde{f}_j \right\rangle_{H_{n+1}} = \gamma_k^{-2} \delta_{k,j}, \qquad k, j \in \mathbb{N}. \tag{14.5}$$

Let $\{\tilde{f}_k^*\}$ be the dual basis, that is, the orthonormal basis of H_n^* such that $\langle \tilde{f}_k^*, \tilde{f}_j \rangle = \delta_{k,j}$ for all k, j. Note that

$$\left\langle \tilde{f}_k^*, \tilde{f}_j^* \right\rangle_{H_{n+1}^*} = \sum_l \langle \tilde{f}_k^*, f_l \rangle \langle \tilde{f}_j^*, f_l \rangle = \delta_{j,k} \left\langle \tilde{f}_k^*, f_k \right\rangle^2$$

$$= \gamma_k^2 \delta_{j,k} \langle \tilde{f}_k^*, \tilde{f}_k \rangle^2 = \gamma_k^2 \delta_{k,j}, \qquad k, j \in \mathbb{N}. \tag{14.6}$$

We will show that

$$L(t) = \sum_k \left\langle L(t), \tilde{f}_k \right\rangle \tilde{f}_k^*, \qquad t \geq 0, \tag{14.7}$$

where the series converges in $L^2(\Omega, \mathcal{F}, \mathbb{P}; H_{n+1}^*)$. First we will show the convergence. By (14.6),

$$\mathbb{E} \left| \sum_{k=N}^M \left\langle L(t), \tilde{f}_k \right\rangle \tilde{f}_k^* \right|_{H_{n+1}^*}^2 = \sum_{k,l=N}^M \mathbb{E} \left\langle L(t), \tilde{f}_k \right\rangle \langle L(t), \tilde{f}_l \rangle \langle \tilde{f}_k^*, \tilde{f}_j^* \rangle_{H_{n+1}}$$

$$= \sum_{k,l=N}^M t \, K \left(\tilde{f}_k, \tilde{f}_l \right) \langle \tilde{f}_k^*, \tilde{f}_j^* \rangle_{H_{n+1}}$$

$$= \sum_{k=N}^M t \, K \left(\tilde{f}_k, \tilde{f}_k \right) \lambda_k^2.$$

Since K is continuous on $H_n \times H_n$ there is a constant C such that

$$\left| K\left(\tilde{f}_k, \tilde{f}_k\right) \right| \leq C \left| \tilde{f}_k \right|_{H_n} \leq C, \qquad \forall k.$$

Consequently, by (14.4),

$$\mathbb{E} \left| \sum_{k=N}^{M} \langle L(t), \tilde{f}_k \rangle \tilde{f}_k^* \right|_{H_{n+1}^*}^2 \leq Ct \sum_{k=N}^{M} \gamma_k^2 \to 0 \qquad \text{as } N, M \to \infty,$$

which proves the convergence of the series. To see the identity we fix a test function $\psi \in S(\mathbb{R}^d)$. We have

$$I_N := \mathbb{E} \left| \langle L(t), \psi \rangle - \sum_{k=1}^{N} \langle L(t), \tilde{f}_k \rangle \langle \tilde{f}_k^*, \psi \rangle \right|^2$$

$$= t \left(K(\psi, \psi) - 2 \sum_{k=1}^{N} K\left(\tilde{f}_k, \psi\right) \langle \tilde{f}_k^*, \psi \rangle + \sum_{k,j=1}^{N} K\left(\tilde{f}_k, \tilde{f}_j\right) \langle \tilde{f}_k^*, \psi \rangle \langle \tilde{f}_j^*, \psi \rangle \right).$$

Since $\psi = \sum_k \langle \tilde{f}_k^*, \psi \rangle \tilde{f}_k$, the sum being convergent in H_n, and since K is continuous on $H_n \times H_n$, we have

$$K(\psi, \psi) = \sum_k K\left(\tilde{f}_k, \psi\right) \langle \tilde{f}_k^*, \psi \rangle$$

and

$$\sum_k K\left(\tilde{f}_k, \psi\right) \langle \tilde{f}_k^*, \psi \rangle = \sum_{k,j} K\left(\tilde{f}_k, \tilde{f}_j\right) \langle f_k^*, \psi \rangle \langle f_l^*, \psi \rangle.$$

Thus $I_N \to 0$, which proves the desired identity. Clearly, L is a process with stationary and independent increments in H_{n+1}^*. It is stochastically continuous since it is mean-square continuous. Indeed,

$$\mathbb{E} |L(t) - L(s)|_{H_{n+1}^*}^2 = \sum_k \mathbb{E} \langle L(t) - L(s), \ \tilde{f}_k \rangle^2 \left| \tilde{f}_k^* \right|_{H_{n+1}^*}^2$$

$$\leq (t-s) \sum_k K\left(\tilde{f}_k, \tilde{f}_k\right) \lambda_k^2 \leq C(t-s),$$

where $C < \infty$. Thus L can be treated as a square integrable mean-zero Lévy process in H_{n+1}^*. By Proposition 7.12 and the first part of the theorem, its RKHS is equal to $\mathcal{H}$. $\qquad\square$

14.4 Spatially homogeneous Lévy processes

Define the *group of translations* $\{\tau_\xi\}_{\xi \in \mathbb{R}^d}$ on $S(\mathbb{R}^d)$ by the formula

$$\tau_\xi \psi(\eta) = \psi(\xi + \eta), \qquad \psi \in S(\mathbb{R}^d), \ \xi, \eta \in \mathbb{R}^d.$$

Next we extend $\{\tau_\xi\}$ on $S'(\mathbb{R}^d)$ by putting $\langle \tau_\xi u, \psi \rangle = \langle u, \tau_{-\xi} \psi \rangle$, $\psi \in S(\mathbb{R}^d)$, $u \in S'(\mathbb{R}^d)$.

Definition 14.13 A Lévy process L taking values in $S'(\mathbb{R}^d)$ is called *spatially homogeneous* or *stationary* if for every $t \geq 0$ the law $\mathcal{L}(L(t))$ of $L(t)$ in $S'(\mathbb{R}^d)$ is invariant with respect to the group of translations $\{\tau_\xi\}$, that is, if for every Borel set $\mathcal{X} \subset S'(\mathbb{R}^d)$

$$\mathcal{L}(L(t))(\mathcal{X}) := \mathbb{P}(L(t) \in \mathcal{X}) = \mathbb{P}\left(L(t) \in \tau_\xi^{-1}\mathcal{X}\right) = \mathcal{L}(L(t))\left(\tau_\xi^{-1}\mathcal{X}\right)$$
$$=: \tau_\xi \circ \mathcal{L}(L(t))(\mathcal{X}), \qquad \xi \in \mathbb{R}^d.$$

Lemma 14.14 *Assume that L is a spatially homogeneous square integrable Lévy process in $S'(\mathbb{R}^d)$. Then its mean and covariance form are invariant with respect to translations, that is, $m_L = \tau_\xi m_L$ for every $\xi \in \mathbb{R}^d$ and*

$$K(\psi, \varphi) = K(\tau_\xi \psi, \tau_\xi \varphi), \qquad \forall \xi \in \mathbb{R}^d, \ \forall \psi, \varphi \in S(\mathbb{R}^d).$$

Proof Let $\psi \in S(\mathbb{R}^d)$, and let $\xi \in \mathbb{R}^d$. Then

$$\langle m_L, \psi \rangle = \mathbb{E} \langle L(1), \psi \rangle = \int_{S'(\mathbb{R}^d)} \langle u, \psi \rangle \mathcal{L}(L(1))(du)$$

$$= \int_{S'(\mathbb{R}^d)} \langle u, \psi \rangle \tau_{-\xi} \circ \mathcal{L}(L(1))(du) = \int_{S'(\mathbb{R}^d)} \langle \tau_\xi u, \psi \rangle \mathcal{L}(L(1))(du)$$

$$= \int_{S'(\mathbb{R}^d)} \langle u, \tau_{-\xi} \psi \rangle \mathcal{L}(L(1))(du) = \mathbb{E} \langle L(1), \tau_{-\xi} \psi \rangle = \langle m_L, \tau_{-\xi} \psi \rangle$$

$$= \langle \tau_\xi m_L, \psi \rangle.$$

In the same way one can show that desired property of the covariance form. □

Remark 14.15 Clearly if L is spatially homogeneous then its Lévy measure

$$\nu(\Gamma) := \mathbb{E} \, \# \, \{t \in [0, 1] \colon L(t) - L(t-) \in \Gamma\}$$

is translation invariant.

Definition 14.16 We say that a square integrable Lévy process L in $S'(\mathbb{R}^d)$ is *weakly spatially homogeneous* if its mean m_L and covariance form are invariant with respect to the group of translations.

Remark 14.17 A Wiener process W is spatially homogeneous if and only if it is weakly spatially homogeneous, since its law $\mathcal{L}(W(t))$ is uniquely determined by the laws of all pairs $\langle W(t), \psi \rangle \langle W(t), \varphi \rangle$, where $\psi, \varphi \in S(\mathbb{R}^d)$, and consequently by its mean and covariance form.

A proof of the following theorem can be found in Gelfand and Vilenkin (1964), p. 169.

Theorem 14.18 *Let $K : S(\mathbb{R}^d) \times S(\mathbb{R}^d) \mapsto \mathbb{R}$. Then K is a continuous symmetric non-negative-definite bilinear form invariant with respect to the group of translations if and only if there exists a non-negative tempered and* *symmetric*[1] *measure μ on $(\mathbb{R}^d, \mathcal{B}(\mathbb{R}^d))$ such that*

$$K(\psi, \varphi) = \langle \mathfrak{F}\mu, \psi * \varphi_{(s)} \rangle, \qquad \psi, \varphi \in S(\mathbb{R}^d), \tag{14.8}$$

where $\varphi_{(s)}(\xi) := \varphi(-\xi)$, $\varphi \in S(\mathbb{R}^d)$ $\xi \in \mathbb{R}^d$ and the convolution operator $*$ *is defined by*

$$\psi * \varphi(\xi) = \int_{\mathbb{R}^d} \psi(\xi - \eta)\varphi(\eta)\,\mathrm{d}\eta, \qquad \xi \in \mathbb{R}^d, \ \psi, \varphi \in S(\mathbb{R}^d).$$

As a direct consequence of Lemma 14.14 and Theorem 14.18 we have the following result.

Corollary 14.19 *Let L be a square integrable weakly homogeneous Lévy process in $S'(\mathbb{R}^d)$. Then its covariance form K satisfies (14.8) with a non-negative measure μ.*

Definition 14.20 We call the quantity μ appearing in (14.8) the *spectral measure* of L and $\Gamma := \mathfrak{F}\mu$ its *covariance*.

14.5 Examples

14.5.1 Wiener case

In this subsection we assume that W is a Wiener process in $S'(\mathbb{R}^d)$.

Example 14.21 Assume that the spectral measure of W is equal to δ_0. Then $\mathfrak{F}\delta_0 = (2\pi)^{-d/2}$, which should be treated as the density of $\mathfrak{F}\delta_0$ with respect to Lebesgue measure $\mathrm{d}\xi$. Thus, for any test functions ψ and φ,

$$K(\psi, \varphi) = (2\pi)^{-d/2} \int_{\mathbb{R}^d} \int_{\mathbb{R}^d} \psi(\eta - \xi)\varphi_{(s)}(\eta)\,\mathrm{d}\eta\,\mathrm{d}\xi$$

$$= (2\pi)^{-d/2} \int_{\mathbb{R}^d} \psi(\xi)\,\mathrm{d}\xi \int_{\mathbb{R}^d} \varphi(\xi)\,\mathrm{d}\xi.$$

Therefore the process W has the form $W(t, \xi) = (2\pi)^{-d/4}\beta(t)$, $t \geq 0$, $\xi \in \mathbb{R}^d$, where β is a one-dimensional (that is, real-valued) Wiener process.

Example 14.22 Assume now that $\mu \equiv \ell_d$. Then $\mathfrak{F}\mu = (2\pi)^{d/2}\delta_0$. Thus

$$K(\psi, \varphi) = (2\pi)^{d/2}\psi * \varphi_{(s)}(0) = (2\pi)^{d/2} \int_{\mathbb{R}^d} \psi(0 - \eta)\varphi_{(s)}(\eta)\,\mathrm{d}\eta$$

$$= (2\pi)^{d/2} \int_{\mathbb{R}^d} \psi(-\eta)\varphi(-\eta)\,\mathrm{d}\eta = (2\pi)^{d/2} \int_{\mathbb{R}^d} \psi(\eta)\varphi(\eta)\,\mathrm{d}\eta.$$

---

[1] For a symmetric measure μ, $\mu(-A) = \mu(A)$ for $A \in \mathcal{B}(\mathbb{R}^d)$.

Consequently, $(2\pi)^{-d/4}W$ is a cylindrical Wiener process on $L^2(\mathbb{R}^d)$.

Example 14.23 Assume that the spectral measure μ of W is finite. Then its co-variance $\Gamma = \mathfrak{F}\mu$ is a bounded uniformly continuous function on $\mathbb{R}^d$. Moreover,

$$K(\psi, \varphi) = \int_{\mathbb{R}^d} \int_{\mathbb{R}^d} \Gamma(\xi)\psi(\xi - \eta)\varphi(-\eta)\,\mathrm{d}\xi\,\mathrm{d}\eta$$

$$= \int_{\mathbb{R}^d} \int_{\mathbb{R}^d} \Gamma(\xi - \eta)\psi(\xi)\varphi(\eta)\,\mathrm{d}\xi\,\mathrm{d}\eta.$$

From the theorem below it follows that W can be treated as a spatially stationary random field on $[0, \infty) \times \mathbb{R}^d$.

Theorem 14.24 *If the spectral measure μ of W is finite then W can be identified with a random field $W(t, \xi)$, $t \geq 0$, $\xi \in \mathbb{R}^d$, that is,*

$$\langle W(t), \psi \rangle = \int_{\mathbb{R}^d} W(t, \xi)\psi(\xi)\,\mathrm{d}\xi, \qquad t \geq 0, \ \psi \in S(\mathbb{R}^d).$$

Moreover, the field W has the following properties.

(i) *W is Gaussian, that is, for any $N \in \mathbb{N}$, $t_1, \ldots, t_n \in [0, \infty)$ and $\xi_1, \ldots, \xi_n \in \mathbb{R}^d$ the random vector $(W(t_1, \xi_1), \ldots, W(t_n, \xi_n))$ is Gaussian.*
(ii) *For any $\xi \in \mathbb{R}^d$, $t \mapsto W(t, \xi)$ is a Wiener process.*
(iii) *Let $\Gamma = \mathfrak{F}\mu$. Then, for all $t, s \geq 0$, $\xi, \eta \in \mathbb{R}^d$,*

$$\mathbb{E}\, W(t, \xi)W(s, \eta) = t \wedge s\, \Gamma(\xi - \eta).$$

Proof Consider an approximation of δ_0 by a sequence (ρ_n) of test functions satisfying

$$\rho_n(\xi) = \rho_n(-\xi), \quad \xi \in \mathbb{R}^d, \rho_n \geq 0, \qquad \int_{\mathbb{R}^d} \rho_n(\xi)\,\mathrm{d}\xi = 1, \qquad \forall n \in \mathbb{N},$$

$$\lim_{n,m\to\infty} (\Gamma * \rho_n) * \rho_m(0) = \Gamma(0) \tag{14.9}$$

and

$$\lim_{n\to\infty} \int_{\mathbb{R}^d} |\psi(\xi) - \rho_n * \psi(\xi)|\,\mathrm{d}\xi = 0, \qquad \forall \psi \in S(\mathbb{R}^d). \tag{14.10}$$

Let $W_n(t, \xi) := \langle W(t), \ \tau_{-\xi}\rho_n \rangle$. Since W is Gaussian, the random fields W_n on $[0, \infty) \times \mathbb{R}^d$ are Gaussian. Moreover,

$$\mathbb{E}\, W_n(t, \xi)W_m(t, \xi) = \mathbb{E}\,\langle W(t), \ \tau_{-\xi}\rho_n \rangle\langle W(t), \ \tau_{-\xi}\rho_m \rangle$$

$$= \langle \Gamma, \ (\tau_{-\xi}\rho_n) * (\tau_{-\xi}\rho_m)_{(s)} \rangle.$$

Thus

$$
\begin{aligned}
\mathbb{E}\,|W_n(t,\xi) - W_m(t,\xi)|^2 \\
= \mathbb{E}\,|W_n(t,\xi)|^2 - 2\,\mathbb{E}\,W_n(t,\xi)W_m(t,\xi) + \mathbb{E}\,|W_m(t,\xi)|^2 \\
= \big\langle \Gamma,\ (\tau_{-\xi}\rho_n) * (\tau_{-\xi}\rho_n)_{(s)} \big\rangle + \big\langle \Gamma,\ (\tau_{-\xi}\rho_m) * (\tau_{-\xi}\rho_m)_{(s)} \big\rangle \\
- 2\big\langle \Gamma,\ (\tau_{-\xi}\rho_n) * (\tau_{-\xi}\rho_m)_{(s)} \big\rangle.
\end{aligned}
$$

Since

$$
\begin{aligned}
\big\langle \Gamma,\ (\tau_{-\xi}\rho_n) * (\tau_{-\xi}\rho_m)_{(s)} \big\rangle &= \int_{\mathbb{R}^d}\int_{\mathbb{R}^d} \Gamma(\eta - \zeta)(\tau_{-\xi}\rho_n)(\eta)(\tau_{-\xi}\rho_m)(\zeta)\,\mathrm{d}\eta\,\mathrm{d}\zeta \\
&= \int_{\mathbb{R}^d}\int_{\mathbb{R}^d} \Gamma(\eta - \zeta)\rho_n(-\xi + \eta)\rho_m(-\xi + \zeta)\,\mathrm{d}\eta\,\mathrm{d}\zeta \\
&= \int_{\mathbb{R}^d}\int_{\mathbb{R}^d} \Gamma(\eta - \zeta)\rho_n(\eta)\rho_m(\zeta)\,\mathrm{d}\eta\,\mathrm{d}\zeta \\
&= (\Gamma * \rho_n) * \rho_m(0),
\end{aligned}
$$

we obtain

$$
\begin{aligned}
\mathbb{E}\,|W_n(t,\xi) - W_m(t,\xi)|^2 \\
= (\Gamma * \rho_n) * \rho_n(0) + (\Gamma * \rho_m) * \rho_m(0) - 2(\Gamma * \rho_n) * \rho_m(0).
\end{aligned}
$$

Consequently, it follows by (14.9) that $(W_n(t,\xi))$ is a Cauchy sequence. We have

$$
\begin{aligned}
\mathbb{E}\left| \langle W(t), \psi \rangle - \int_{\mathbb{R}^d} W_n(t,\xi)\psi(\xi)\,\mathrm{d}\xi \right|^2 \\
= \mathbb{E}\left\langle W(t),\ \psi - \int_{\mathbb{R}^d} \psi(\xi)\tau_{-\xi}\rho_n\,\mathrm{d}\xi \right\rangle^2.
\end{aligned}
$$

Therefore, as

$$
\begin{aligned}
\mathbb{E}\left| \langle W(t), \psi \rangle - \int_{\mathbb{R}^d} W_n(t,\xi)\psi(\xi)\,\mathrm{d}\xi \right|^2 \\
= \int_{\mathbb{R}^d}\int_{\mathbb{R}^d} \Gamma(\eta - \zeta)\left(\psi(\eta) - \int_{\mathbb{R}^d} \rho_n(\eta - \xi)\psi(\xi)\,\mathrm{d}\xi \right) \\
\times \left[\psi(\zeta) - \int_{\mathbb{R}^d} \rho_n(\zeta - \xi)\psi(\xi)\,\mathrm{d}\xi \right]\mathrm{d}\eta\,\mathrm{d}\zeta \\
\leq |\Gamma|_{L^\infty(\mathbb{R}^d)}\left(\int_{\mathbb{R}^d} \left| \psi(\eta) - \int_{\mathbb{R}^d} \rho_n(\eta - \xi)\psi(\xi)\,\mathrm{d}\xi \right|\mathrm{d}\eta \right)^2 \\
\leq |\Gamma|_{L^\infty(\mathbb{R}^d)}\left(\int_{\mathbb{R}^d} |\psi(\eta) - \rho_n * \psi(\eta)|\,\mathrm{d}\eta \right)^2,
\end{aligned}
$$

from (14.10) the limit $W(t, \xi)$ of the sequence $(W_n(t, \xi))$ satisfies

$$\mathbb{E} \left| \langle W(t), \psi \rangle - \int_{\mathbb{R}^d} W(t, \xi) \psi(\xi) \, d\xi \right|^2$$

$$= \lim_{n \to \infty} \mathbb{E} \left| \langle W(t), \psi \rangle - \int_{\mathbb{R}^d} W_n(t, \xi) \psi(\xi) \, d\xi \right|^2 = 0.$$

The field W is Gaussian since, for fixed N, $t_1, \ldots, t_N \in [0, \infty)$ and $\xi_1, \ldots, \xi_N \in \mathbb{R}^d$, the Gaussian vectors $(W_n(t_1, \xi_1), \ldots, W_n(t_N, \xi_N))$, $n \in \mathbb{N}$, converge in $L^2(\Omega, \mathcal{F}, \mathbb{P})$ to $(W(t_1, \xi_1), \ldots, W(t_N, \xi_N))$. Moreover, for a fixed $\xi \in \mathbb{R}^d$, $t \mapsto W(t, \xi)$ is a square integrable martingale with angle bracket

$$\langle W(\cdot, \xi), W(\cdot, \xi) \rangle(t) = \lim_{n \to \infty} \langle W_n(\cdot, \xi), W(\cdot, \xi) \rangle(t) = \lim_{n \to \infty} \mathbb{E} |W_n(t, \xi)|^2$$

$$= t \lim_{n \to \infty} (\Gamma * \rho_n) * \rho_n(0) = t \, \Gamma(0).$$

Thus, for every $\xi \in \mathbb{R}^d$, $W(\cdot, \xi)$ is a Wiener process. The proof will be complete as soon as we show (iii). To this end, let us fix $t, s \in [0, \infty)$ and $\xi, \eta \in \mathbb{R}^d$. Then

$$\mathbb{E} \, W(t, \xi) W(s, \eta) = \lim_{n \to \infty} \mathbb{E} \, W_n(t, \xi) W_n(s, \eta)$$

$$= t \wedge s \lim_{n \to \infty} \mathbb{E} \langle W(1), \tau_{-\xi} \rho_n \rangle \langle W(1), \tau_{-\eta} \rho_n \rangle$$

$$= t \wedge s \lim_{n \to \infty} \langle \Gamma, \tau_{-\xi} \rho_n * (\tau_{-\eta} \rho_n)_{(s)} \rangle$$

$$= t \wedge s \lim_{n \to \infty} \int_{\mathbb{R}^d} \int_{\mathbb{R}^d} \Gamma(\gamma - \zeta) \rho_n(\gamma - \xi) \rho_n(\zeta - \eta) \, d\gamma \, d\zeta$$

$$= t \wedge s \, \Gamma(\xi - \eta).$$

$\square$

14.5.2 Impulsive case

Another important example of a homogeneous Lévy noise is the impulsive cylindrical process on $L^2(\mathbb{R}^d)$ introduced in Section 7.2 (see the first example below). In Example 14.26 we introduce a "colored impulsive" noise.

Example 14.25 Let π be a Poisson measure on $[0, \infty) \times \mathbb{R}^d \times \mathbb{R}$ with intensity measure $dt \, d\xi \, \nu(d\sigma)$, where ν is a measure on $(\mathbb{R}, \mathcal{B}(\mathbb{R}))$ satisfying $\nu(\{0\}) = 0$ and

$$\int_{\mathbb{R}} \sigma^2 \nu(d\sigma) < \infty. \tag{14.11}$$

Let $\widehat{\pi}$ be a compensated Poisson measure, and let

$$Z(t) = \int_0^t \int_{\mathbb{R}} \sigma \widehat{\pi}(ds, d\xi, d\sigma).$$

For more details see Section 7.2. Then Z is a square integrable mean-zero Lévy process taking values in any Hilbert space U such that the embedding $L^2(\mathbb{R}^d) \hookrightarrow U$ is Hilbert–Schmidt. In particular, as U one can take any H_n^*, $n \geq 1$. Moreover, $L^2(\mathbb{R}^d)$ is the RKHS of L, for any U. Since there is a Hilbert space U such that $L^2(\mathbb{R}^d) \hookrightarrow U \hookrightarrow S'(\mathbb{R}^d)$ and the first inclusion is Hilbert–Schmidt, Z is a Lévy process in $S'(\mathbb{R}^d)$. It is spatially homogeneous since the jump position intensity ℓ_d is invariant with respect to the group of translations. Note that

$$K(\psi, \varphi) = \mathbb{E} \langle Z(1), \psi \rangle \langle Z(1), \varphi \rangle = \int_{\mathbb{R}^d} \psi(\xi) \varphi(\xi) \, \mathrm{d}\xi, \qquad \psi, \varphi \in S(\mathbb{R}^d).$$

Hence the spectral measure μ of Z is, up to a constant, equal to ℓ_d. Consequently, again up to a constant, its covariance Γ is equal to the Dirac delta function.

Let $\{U_j\}$ be the partition of $\mathbb{R}^d \times \mathbb{R}$ into disjoint measurable sets such that $a_j := \ell_d \times \nu(U_j) < \infty$, $j \in \mathbb{N}$. It is convenient to write Z in the form

$$Z(t) = \sum_{\tau_j^n \leq t} \sigma_j^n \delta_{\xi_j^n} - t\sigma \nu(\mathrm{d}\sigma) \, \mathrm{d}\xi,$$

where (ξ_j^n, σ_j^n) is a sequence of mutually independent random variables with distributions

$$\mathbb{P}\left((\xi_j^n, \sigma_j^n) \in A \right) = \frac{\ell_d \times \nu(A \cap U_j)}{a_j}, \qquad n, j \in \mathbb{N};$$

$\tau_j^n = r_j^1 + r_j^2 + \cdots + r_j^n$ and the r_j^k are mutually independent random variables (which are also independent of (ξ_j^n, σ_j^n)) with exponential distribution $\mathbb{P}(r_j^n \geq t) = \mathrm{e}^{-a_j t}$, $t \geq 0$.

We interpret σ_j^n as the amount of energy introduced into the system at time τ_j^n. The random variables ξ_j^n represent the sources of energy.

Example 14.26   Let P be a tempered measure, and let Z be an impulsive cylindrical process on $L^2(\mathbb{R}^d)$ with jump size intensity ν. We assume that $P * \ell_d$ and $P * P_{(s)}$ are tempered measures.

Then the convolution $P * Z(t)$ is a well-defined $S'(\mathbb{R}^d)$-valued Lévy process. In fact it is a measure-valued process. It is spatially homogeneous, since

$$\tau_\xi(P * Z(t)) = P * (\tau_\xi Z(t)), \qquad t \geq 0, \ \xi \in \mathbb{R}^d.$$

Indeed, for any test function ψ,

$$\langle \tau_\xi(P * Z(t)), \ \psi \rangle = \langle P * Z(t), \ \tau_{-\xi}\psi \rangle = \int_{\mathbb{R}^d} P_{(s)} * \tau_{-\xi}\psi(\eta) Z(t)(\mathrm{d}\eta)$$

$$= \int_{\mathbb{R}^d} \left(P_{(s)} * \psi \right)(\eta)(\tau_\xi Z(t))(\mathrm{d}\eta) = \langle P * (\tau_\xi Z(t)), \psi \rangle.$$

Note that $P * Z$ is a square integrable mean-zero process with covariance form

$$K(\psi, \varphi) = \mathbb{E} \langle P * Z(1), \ \psi \rangle \langle P * Z(1), \ \varphi \rangle$$

$$= \mathbb{E} \langle Z(1), \ P_{(s)} * \psi \rangle \langle Z(1), \ P_{(s)} * \varphi \rangle = \int_{\mathbb{R}^d} P_{(s)} * \psi(x) P_{(s)} * \varphi(x) \, \mathrm{d}x$$

$$= \langle (P * P_{(s)}) * \psi, \ \varphi \rangle = \langle P * P_{(s)}, \ \varphi * \psi_{(s)} \rangle.$$

Thus the covariance Γ and the spectral measure of $P * Z$ are given by $\Gamma = P * P_{(s)}$, $\mu = \mathfrak{F} P \overline{\mathfrak{F}} P$.

14.6 RKHS of a homogeneous noise

Our goal is to find the RKHS of a spatially homogeneous square integrable mean-zero Lévy process L. Let us recall that its covariance form K is given by $K(\psi, \varphi) = \langle \mathfrak{F} \mu, \ \psi * \varphi_{(s)} \rangle$ for $\psi, \varphi \in S(\mathbb{R}^d)$, where μ is a non-negative symmetric tempered measure on $\mathbb{R}^d$ and $\varphi_{(s)}(\xi) = \varphi(-\xi)$, $\xi \in \mathbb{R}^d$, $\psi \in S(\mathbb{R}^d)$. For $\varphi \in S(\mathbb{R}^d; \mathbb{C})$ we set $\varphi_{(s)}(\xi) := \overline{\varphi(-\xi)}$, $\xi \in \mathbb{R}^d$. Let

$$L_{(s)}^2(\mathbb{R}^d, \mu; \mathbb{C}) := \left\{ \psi \in L^2 \left(\mathbb{R}^d, \mathcal{B}(\mathbb{R}^d), \mu; \mathbb{C} \right) : \psi = \psi_{(s)} \right\}.$$

Note that $L_{(s)}^2(\mathbb{R}^d, \mu; \mathbb{C})$ is a closed subspace of $L^2(\mathbb{R}^d, \mu; \mathbb{C})$.

Remark 14.27 For any $\psi \in L_{(s)}^2(\mathbb{R}^d, \mu; \mathbb{C})$, $\psi \mu \in S'(\mathbb{R}^d; \mathbb{C})$. Moreover $\mathfrak{F}(\psi \mu) \in S'(\mathbb{R}^d)$. For, as μ is a tempered measure there is an $n \in \mathbb{N}$ such that $\int_{\mathbb{R}^d} \left(1 + |\xi|^2 \right)^{-n} \mu(\mathrm{d}\xi) < \infty$. Now $\psi \mu$ is a tempered measure and belongs to $S'(\mathbb{R}^d; \mathbb{C})$ since

$$\int_{\mathbb{R}^d} \left(1 + |\xi|^2 \right)^{-n/2} |\psi(\xi)| \mu(\mathrm{d}\xi)$$

$$\leq \left(\int_{\mathbb{R}^d} |\psi(\xi)|^2 \mu(\mathrm{d}\xi) \right)^{1/2} \left(\int_{\mathbb{R}^d} \left(1 + |\xi|^2 \right)^{-n} \mu(\mathrm{d}\xi) \right)^{1/2} < \infty.$$

In order to show that $\mathfrak{F}(\psi \mu)$ is real-valued we have to show that $\overline{\mathfrak{F}(\psi \mu)} = \mathfrak{F}(\psi \mu)$. This follows from the equalities $\overline{\psi(\eta)} = \psi(-\eta)$ and $\mu(\mathrm{d}\eta) = \mu(-\mathrm{d}\eta)$ in the following way:

$$\overline{\mathfrak{F}(\psi \mu)(\xi)} = (2\pi)^{-d/2} \int_{\mathbb{R}^d} \mathrm{e}^{-\mathrm{i}\xi \eta} \psi(\eta) \mu(\mathrm{d}\eta)$$

$$= (2\pi)^{-d/2} \int_{\mathbb{R}^d} \mathrm{e}^{\mathrm{i}\xi \eta} \overline{\psi(\eta)} \mu(\mathrm{d}\eta) = \mathfrak{F}(\psi \mu)(\xi).$$

Theorem 14.28 *Let L be a spatially homogeneous square integrable mean-zero Lévy process in $S'(\mathbb{R}^d)$ with spectral measure μ. Then its RKHS is given by*

$$\mathcal{H} = \left\{ \mathfrak{F}(\psi \mu) : \psi \in L_{(s)}^2(\mathbb{R}^d, \mu; \mathbb{C}) \right\} \tag{14.12}$$

and the scalar product is equal to

$$\langle \mathfrak{F}(\psi\mu), \mathfrak{F}(\varphi\mu) \rangle_{\mathcal{H}} = \langle \psi, \varphi \rangle_{L^2(\mathbb{R}^d,\mu;\mathbb{C})} = \int_{\mathbb{R}^d} \psi(\xi)\overline{\varphi(\xi)}\mu(\mathrm{d}\xi). \tag{14.13}$$

Proof By Theorem 14.8,

$$\mathcal{H} = \left\{ u \in S'(\mathbb{R}^d) \colon \exists c < \infty \colon \forall \varphi \in S(\mathbb{R}^d), \ |\langle u, \varphi \rangle|^2 \le c \left\langle \mathfrak{F}\mu, \varphi * \varphi_{(s)} \right\rangle \right\}.$$

Assume that $u \in \mathcal{H}$. Then $\langle u, \varphi \rangle = \left\langle \mathfrak{F}^{-1}u, \mathfrak{F}^{-1}\varphi \right\rangle$ and

$$\left\langle \mathfrak{F}\mu, \varphi * \varphi_{(s)} \right\rangle = \left\langle \mu, \mathfrak{F}^{-1}(\varphi * \varphi_{(s)}) \right\rangle = \left\langle \mu, \mathfrak{F}^{-1}\varphi \overline{\mathfrak{F}^{-1}\varphi} \right\rangle$$

because $\mathfrak{F}^{-1}\varphi_{(s)} = \overline{\mathfrak{F}^{-1}\varphi}$. Therefore

$$\left| \left\langle \mathfrak{F}^{-1}u, \mathfrak{F}^{-1}\varphi \right\rangle \right|^2 \le c \int_{\mathbb{R}^d} \mathfrak{F}^{-1}\varphi(\xi)\overline{\mathfrak{F}^{-1}\varphi(\xi)}\mu(\mathrm{d}\xi).$$

Since

$$\mathfrak{F}^{-1} \colon S(\mathbb{R}^d) \mapsto S_{(s)}(\mathbb{R}^d; \mathbb{C}) = \{\varphi \in S(\mathbb{R}^d; \mathbb{C}) \colon \varphi_{(s)} = \varphi\}$$

is a bijection,

$$\left| \left\langle \mathfrak{F}^{-1}u, \varphi \right\rangle \right|^2 \le c \int_{\mathbb{R}^d} |\varphi(\xi)|^2 \mu(\mathrm{d}\xi), \qquad \forall \varphi \in S_{(s)}(\mathbb{R}^d; \mathbb{C}).$$

Since $S_{(s)}(\mathbb{R}^d; \mathbb{C})$ is dense in $L^2_{(s)}(\mathbb{R}^d, \mu; \mathbb{C})$, $\mathfrak{F}^{-1}u$ can be extended to the whole space $L^2_{(s)}(\mathbb{R}^d, \mu; \mathbb{C})$. Thus there is a $\psi \in L^2_{(s)}(\mathbb{R}^d, \mu; \mathbb{C})$ such that

$$\forall \varphi \in S_{(s)}(\mathbb{R}^d; \mathbb{C}), \qquad \left\langle \mathfrak{F}^{-1}u, \varphi \right\rangle = \langle \psi, \varphi \rangle_{L^2(\mathbb{R}^d,\mu;\mathbb{C})} = \left\langle \overline{\psi}\mu, \varphi \right\rangle.$$

Hence $u = \mathfrak{F}\left(\overline{\psi}\mu\right)$. As $\psi \mapsto \overline{\psi}$ is bijective on $L^2_{(s)}(\mathbb{R}^d, \mu; \mathbb{C})$ we have the following fact:

$$\forall u \in \mathcal{H} \ \exists \psi \in L^2_{(s)}(\mathbb{R}^d, \mu; \mathbb{C}) \qquad \text{such that } u = \mathfrak{F}(\psi\mu).$$

To show (14.12) it is sufficient to prove that

$$\forall \psi \in L^2_{(s)}(\mathbb{R}^d, \mu; \mathbb{C}), \qquad u = \mathfrak{F}(\psi\mu) \in \mathcal{H}. \tag{14.14}$$

Let us fix a $\psi \in L^2_{(s)}(\mathbb{R}^d, \mu; \mathbb{C})$. Then for any $\varphi \in S(\mathbb{R}^d)$,

$$
\begin{aligned}
|\langle \mathfrak{F}(\psi\mu), \varphi \rangle|^2 &= |\langle \psi\mu, \mathfrak{F}\varphi \rangle|^2 \\
&\le \int_{\mathbb{R}^d} |\psi(\xi)|^2 \mu(\mathrm{d}\xi) \int_{\mathbb{R}^d} |\mathfrak{F}\varphi|^2 \mu(\mathrm{d}\xi) \\
&\le \int_{\mathbb{R}^d} |\psi(\xi)|^2 \mu(\mathrm{d}\xi) \left\langle \Gamma, \varphi * \varphi_{(s)} \right\rangle,
\end{aligned}
$$

which gives (14.14). What is left is to show that

$$\forall \psi \in L^2_{(s)}(\mathbb{R}^d, \mu; \mathbb{C}), \qquad |\mathfrak{F}(\psi\mu)|^2_{\mathcal{H}} = \int_{\mathbb{R}^d} |\psi(\xi)|^2 \mu(d\xi).$$

To this end we fix a $\psi \in L^2_{(s)}(\mathbb{R}^d, \mu; \mathbb{C})$. Then, by (14.1),

$$|\mathfrak{F}(\psi\mu)|^2_{\mathcal{H}} = \sup_{\varphi \in S(\mathbb{R}^d): \, K(\varphi,\varphi) \neq 0} \frac{|\langle \mathfrak{F}(\psi\mu), \varphi \rangle|^2}{K(\varphi, \varphi)}$$

$$= \sup_{\varphi \in S(\mathbb{R}^d): \, |\mathfrak{F}\varphi|^2_{L^2(\mathbb{R}^d, \mu; \mathbb{C})} \neq 0} \frac{|\langle \mathfrak{F}(\psi\mu), \varphi \rangle|^2}{|\mathfrak{F}\varphi|^2_{L^2(\mathbb{R}^d, \mu; \mathbb{C})}}$$

$$= \sup_{\varphi \in S(\mathbb{R}^d): \, |\mathfrak{F}\varphi|^2_{L^2(\mathbb{R}^d, \mu; \mathbb{C})} \neq 0} \frac{|\langle \mu, \psi\mathfrak{F}^{-1}\varphi \rangle|^2}{|\mathfrak{F}\varphi|^2_{L^2(\mathbb{R}^d, \mu; \mathbb{C})}}.$$

Since

$$\langle \mu, \psi\mathfrak{F}^{-1}\varphi \rangle = \langle \psi, \overline{\mathfrak{F}^{-1}\varphi} \rangle_{L^2(\mathbb{R}^d, \mu; \mathbb{C})} = \langle \psi, \mathfrak{F}\overline{\varphi} \rangle_{L^2(\mathbb{R}^d, \mu; \mathbb{C})} = \langle \psi, \mathfrak{F}\varphi \rangle_{L^2(\mathbb{R}^d, \mu; \mathbb{C})},$$

for real-valued φ, we have

$$|\mathfrak{F}(\psi\mu)|^2_{\mathcal{H}} = \sup_{\varphi \in S(\mathbb{R}^d); \, |\mathfrak{F}\varphi|^2_{L^2(\mathbb{R}^d, \mu; \mathbb{C})} \neq 0} \frac{|\langle \psi, \mathfrak{F}\varphi \rangle_{L^2(\mathbb{R}^d, \mu; \mathbb{C})}|^2}{|\mathfrak{F}\varphi|^2_{L^2(\mathbb{R}^d, \mu; \mathbb{C})}} = |\psi|^2_{L^2(\mathbb{R}^d, \mu; \mathbb{C})},$$

which completes the proof. $\qquad\square$

14.7 Stochastic equations on $\mathbb{R}^d$

In the following sections we shall consider stochastic heat and wave equations driven by spatially homogeneous Lévy process, as follows:

$$\frac{\partial X}{\partial t}(t, \xi) = \Delta X(t, \xi) + f(X(t, \xi)) + g(X(t, \xi))\frac{\partial L}{\partial t}(t, \xi),$$

$$t > 0, \ \xi \in \mathbb{R}^d, \qquad (14.15)$$

$$X(0, \xi) = x(\xi), \qquad \xi \in \mathbb{R}^d.$$

$$\frac{\partial X}{\partial t^2}(t, \xi) = \Delta X(t, \xi) + f(X(t, \xi)) + g(X(t, \xi))\frac{\partial L}{\partial t}(t, \xi),$$

$$t > 0, \ \xi \in \mathbb{R}^d, \qquad (14.16)$$

$$X(0, \xi) = x(\xi), \qquad \frac{\partial X}{\partial t}(0, \xi) = y(\xi), \qquad \xi \in \mathbb{R}^d.$$

We assume that $f, g: \mathbb{R} \mapsto \mathbb{R}$ are Lipschitz continuous and that L is a spatially homogeneous square integrable mean-zero Lévy process with spectral measure μ and covariance Γ.

The results on heat equations are based on Peszat and Zabczyk (1997, 2000) and Peszat and Tindel (2007). Conditions for the existence of function-valued solutions, formulated in the text, appeared for the first time in Dalang and Frangos (1998) in the two-dimensional case and for all dimensions in Karczewska and Zabczyk (2000a, 2001). More information on parabolic equations driven by a spatially homogeneous Wiener process can be found in Brzeźniak and Peszat (1999), Carmona and Molchanow (1994), Da Prato and Zabczyk (1996), Dawson and Salehi (1980), Kifer (1997), Manthey and Zausinger (1999), Peszat (2001, 2002), Karczewska and Zabczyk (2000a, 2001), Peszat and Zabczyk (1997, 2000). The section on wave equations is based on Peszat (2002) and Peszat and Zabczyk (2000). More information on stochastic wave equations driven by a spatially homogeneous Wiener process can be found in Dalang (1999), Dalang and Frangos (1998), Dalang and Sanz-Solé (2006), Karczewska and Zabczyk (2000a, 2001), Millet and Morien (2001), Millet and Sanz-Solé (1999, 2000), Peszat (2001, 2002), Peszat and Tindel (2007), Peszat and Zabczyk (2000) and Quer-Sardanyons and Sanz-Solé (2004b).

First we construct a state space. We require that in the case of the heat equation the Laplace operator, and in the case of the wave equation the operator

$$\begin{pmatrix} 0 & I \\ \Delta & 0 \end{pmatrix},$$

generates a C_0-semigroup on the state space. We are looking for a solution to the heat equation in the spaces L_ρ^2 and $\mathcal{L}_\rho^2$ introduced in Section 2.3. In the case of the wave equation the state space is of the form $\left(L_\rho^2, H_\rho^{-1}\right)^T$, where $H_\rho^{-1} := H^{-1}(\mathbb{R}^d, \vartheta_\rho^2(\xi)\, d\xi)$ is the Sobolev space of order -1 with weight ϑ_ρ. Thus H_ρ^{-1} is a subspace of the space of distributions.

14.8 Stochastic heat equation

This section is devoted to (14.15). Note that (see Example 10.10) the semigroup S is given by $S(t)\psi(\xi) = \mathfrak{g}(t) * \psi(\xi)$, $t > 0$, $\xi \in \mathbb{R}^d$, where $\mathfrak{g}(t)(\xi) := (4\pi t)^{-d/2}\, e^{-|\xi|^2/(4t)}$ for $t > 0$ and $\xi \in \mathbb{R}^d$. Recall that L is a square integrable mean-zero spatially homogeneous Lévy process on $\mathbb{R}^d$ with spectral measure μ. We interpret (14.15) as the stochastic evolution equation

$$dX = (AX + F(X))\, dt + G(X)\, dL$$

on a Hilbert space $H = L_\rho^2$, where A is the generator of the heat semigroup S on L_ρ^2, $F = N_f$, and $G(\psi)\varphi[\xi] = N_g(\psi)(\xi)\varphi(\xi)$ where $\psi \in L_\rho^2$ and φ belongs to RKHS of L.

Theorem 14.29 *Assume that* $\Gamma = \mathfrak{F}\mu$ *is a measure separate from* $-\infty$, *that is, there is a constant* $C < \infty$ *such that* $\Gamma(d\xi) + C\,d\xi$ *is a non-negative measure and*

$$\int_{\{|\xi|\leq 1\}} \log\left(|\xi|^{-1}\right) \Gamma(d\xi) < \infty \qquad \text{if } d = 2,$$

$$\int_{\{|\xi|\leq 1\}} |\xi|^{-d+2}\Gamma(d\xi) < \infty \qquad \text{if } d \neq 2. \tag{14.17}$$

Assume that f *and* g *are Lipschitz continuous. Let* $\rho \geq 0$. *Then for any* $x \in L_\rho^2$ *there is a unique solution to (14.15) in* L_ρ^2. *If* $\rho \leq 0$ *and* $f(0) = 0 = g(0)$ *then for each* $x \in L_\rho^2$ *there is a a unique solution to (14.15) in* L_ρ^2. *Moreover, (14.15) defines a Feller family.*

Proof By Theorem B.9, the heat semigroup is C_0 on L_ρ^2. By Lemma 2.5 the composition operators N_f and N_g are Lipschitz with linear growth on L_ρ^2. In particular condition (F) of the general existence theorem is satisfied. We need to verify condition (G) for the diffusion term. We have to show that if ψ is from a dense set $H_0 \subset L_\rho^2$ then $S(t)G(\psi) \in L_{(HS)}\left(\mathcal{H}, L_\rho^2\right)$ and, moreover, that there is a locally square integrable function $a \colon (0, \infty) \mapsto \mathbb{R}$ such that

$$\|S(t)G(\psi)\|_{L_{(HS)}(\mathcal{H},L_\rho^2)} \leq a(t)\left(1 + |\psi|_{L_\rho^2}\right),$$

$$\|S(t)(G(\psi) - G(\phi))\|_{L_{(HS)}(\mathcal{H},L_\rho^2)} \leq a(t)\,|\psi - \phi|_{L_\rho^2}$$

for $t \in (0, \infty)$ and $\psi, \phi \in H_0$. For H_0 we take $S(\mathbb{R}^d)$. Let $M_t(\psi)\varphi = S(t)(\psi\varphi)$. As $N_g \colon L_\rho^2 \mapsto L_\rho^2$ is Lipschitz with linear growth and $S(t)G(\psi)\varphi = M_t(N_g(\psi))\varphi$, it is sufficient to show that, for every $t > 0$, M_t is a continuous linear operator from L_ρ^2 to $L_{(HS)}\left(\mathcal{H}, L_\rho^2\right)$ and

$$\int_0^T \|M_t\|_{L(L_\rho^2, L_{(HS)}(\mathcal{H};L_\rho^2))}^2 \, dt < \infty, \qquad \forall\, T < \infty. \tag{14.18}$$

Then for the function a we can take $\|M_t\|_{L(L_\rho^2, L_{(HS)}(\mathcal{H};L_\rho^2))}$ multiplied by a Lipschitz constant of the mapping N_g.

In order to show (14.18) we take $\psi \in S(\mathbb{R}^d)$, $t > 0$ and an orthonormal basis $\{e_n\}$ of $\mathcal{H}$. It follows from the form of $\mathcal{H}$ that one can choose $\{e_n\} \subset C_b(\mathbb{R}^d)$. Indeed, $e_n = \mathfrak{F}(u_n\mu)$ where $\{u_n\}$ is an orthonormal basis in $L_{(s)}^2(\mathbb{R}^d, \mu; \mathbb{C})$. Taking $\{u_n\} \subset S(\mathbb{R}^d; \mathbb{C})$ we obtain $\{e_n\} \subset C_b(\mathbb{R}^d)$. Thus $\psi e_n \in L_\rho^2$ for all $\psi \in S(\mathbb{R}^d)$ and n. In particular $S(t)(\psi e_n)$ is well defined and belongs to L_ρ^2 for all ψ, t and n. We have

$$\sum_n |M_t(\psi)e_n|_{L_\rho^2}^2 = \sum_n |S(t)(\psi e_n)|_{L_\rho^2}^2$$

$$= \sum_n \int_{\mathbb{R}^d} \left|\int_{\mathbb{R}^d} \mathfrak{g}(t)(\xi - \eta)\psi(\eta)e_n(\eta)\,d\eta\right|^2 \vartheta_\rho^2(\xi)\,d\xi.$$

Recall (see Theorem 14.8(ii)) that any square integrable mean-zero Lévy process L in $S'(\mathbb{R}^d)$ has covariance form

$$K(\psi, \psi) = \sum_n \langle e_n, \psi \rangle^2, \qquad \psi \in S(\mathbb{R}^d),$$

where $\{e_n\}$ is an orthonormal basis of RKHS of L. Hence

$$\sum_n \left| \int_{\mathbb{R}^d} \mathfrak{g}(t)(\xi - \eta)\psi(\eta)e_n(\eta)\,d\eta \right|^2$$

$$= \sum_n \langle e_n, \mathfrak{g}(t)(\xi - \cdot)\psi(\cdot) \rangle^2$$

$$= K\big(\mathfrak{g}(t)(\xi - \cdot)\psi(\cdot), \mathfrak{g}(t)(\xi - \cdot)\psi(\cdot)\big)$$

$$= \big\langle \Gamma, \mathfrak{g}(t)(\xi - \cdot)\psi(\cdot) * \big(\mathfrak{g}(t)(\xi - \cdot)\psi(\cdot)\big)_{(s)} \big\rangle.$$

Let C be such that $\tilde{\Gamma}(d\eta) =: \Gamma(d\eta) + C\,d\eta$ is a non-negative measure. From the inequality $\langle \ell_d, \phi * \phi_{(s)} \rangle \geq 0$, valid for $\phi \in S(\mathbb{R}^d)$, we have

$$\big\langle \Gamma, \ \mathfrak{g}(t)(\xi - \cdot)\psi(\cdot) * \big(\mathfrak{g}(t)(\xi - \cdot)\psi(\cdot)\big)_{(s)} \big\rangle$$

$$\leq \big\langle \tilde{\Gamma}, \ \mathfrak{g}(t)(\xi - \cdot)\psi(\cdot) * \big(\mathfrak{g}(t)(\xi - \cdot)\psi(\cdot)\big)_{(s)} \big\rangle.$$

Hence

$$\sum_n |M_t(\psi)e_n|^2_{L^2_\rho}$$

$$\leq \int_{\mathbb{R}^d} \big\langle \tilde{\Gamma}, \ \mathfrak{g}(t)(\xi - \cdot)\psi(\cdot) * \big(\mathfrak{g}(t)(\xi - \cdot)\psi(\cdot)\big)_{(s)} \big\rangle \vartheta^2_\rho(\xi)\,d\xi.$$

Since

$$\mathfrak{g}(t)(\xi - \cdot)\psi(\cdot) * \big(\mathfrak{g}(t)(\xi - \cdot)\psi(\cdot)\big)_{(s)}(\eta)$$

$$= \int_{\mathbb{R}^d} \mathfrak{g}(t)(\xi - \eta + \zeta)\psi(\eta - \zeta)\mathfrak{g}(t)(\xi + \zeta)\psi(-\zeta)\,d\zeta,$$

we have

$$\sum_n |M_t(\psi)e_n|^2_{L^2_\rho}$$

$$\leq \int_{\mathbb{R}^d}\int_{\mathbb{R}^d}\int_{\mathbb{R}^d} \mathfrak{g}(t)(\xi - \eta + \zeta)\psi(\eta - \zeta)\mathfrak{g}(t)(\xi + \zeta)\psi(-\zeta)\vartheta^2_\rho(\xi)\tilde{\Gamma}(d\eta)\,d\zeta\,d\xi$$

$$\leq \int_{\mathbb{R}^d}\int_{\mathbb{R}^d}\int_{\mathbb{R}^d} |\psi(\eta + \xi - z)||\psi(\xi - z)|\vartheta^2_\rho(\xi)\,d\xi\, \mathfrak{g}(t)(z - \eta)\mathfrak{g}(t)(z)\tilde{\Gamma}(d\eta)\,dz.$$

Note that there is a constant c_1 depending on ρ such that, for all $\eta, z \in \mathbb{R}^d$, $\vartheta_\rho(\eta - z) \leq \vartheta_\rho(z)c_1 e^{|\rho||\eta|}$. Thus, by the Schwarz inequality,

$$\int_{\mathbb{R}^d} |\psi(\eta + \xi - z)||\psi(\xi - z)|\vartheta_\rho^2(\xi)\,\mathrm{d}\xi$$

$$\leq \left(\int_{\mathbb{R}^d} |\psi(\eta + \xi - z)|^2\vartheta_\rho^2(\xi)\,\mathrm{d}\xi\right)^{1/2}\left(\int_{\mathbb{R}^d} |\psi(\xi - z)|^2\vartheta_\rho^2(\xi)\,\mathrm{d}\xi\right)^{1/2}$$

$$\leq \left(\int_{\mathbb{R}^d} |\psi(\varsigma)|^2\vartheta_\rho^2(-\eta + z + \varsigma)\,\mathrm{d}\varsigma\right)^{1/2}\left(\int_{\mathbb{R}^d} |\psi(\varsigma)|^2\vartheta_\rho^2(\varsigma + z)\,\mathrm{d}\varsigma\right)^{1/2}$$

$$\leq |\psi|_{L_\rho^2}^2 c_1^2\, e^{|\rho||z-\eta|+|\rho||z|} \leq |\psi|_{L_\rho^2}^2 c_1^2\, e^{|\rho|(|z-\eta|+|z|)}.$$

Consequently,

$$\sum_n |M_t(\psi)e_n|_{L_\rho^2}^2 \leq c_1^2|\psi|_{L_\rho^2}^2 \int_{\mathbb{R}^d}\int_{\mathbb{R}^d} e^{|\rho|(|z-\eta|+|z|)}\mathrm{g}(t)(z - \eta)\mathrm{g}(t)(z)\tilde{\Gamma}(\mathrm{d}\eta)\,\mathrm{d}z.$$

Using now the fact that for all t, ς and ρ,

$$-\frac{|\varsigma|^2}{2t} + \frac{|\rho||\varsigma|}{2} = -\frac{|\varsigma|^2}{4t} + \left(-\frac{|\varsigma|^2}{4t} + \frac{|\rho||\varsigma|}{2}\right) \leq -\frac{|\varsigma|^2}{4t} + \frac{t\rho^2}{4}$$

we obtain

$$\int_{\mathbb{R}^d}\int_{\mathbb{R}^d} e^{|\rho|(|z-\eta|+|z|)}\mathrm{g}(t)(z - \eta)\mathrm{g}(t)(z)\tilde{\Gamma}(\mathrm{d}\eta)\,\mathrm{d}z$$

$$\leq 2^d e^{2t\rho^2} \int_{\mathbb{R}^d}\int_{\mathbb{R}^d} \mathrm{g}(2t)(z - \eta)\mathrm{g}(2t)(z)\tilde{\Gamma}(\mathrm{d}\eta)\,\mathrm{d}z.$$

Since S is a semigroup we have $\mathrm{g}(2t) * \mathrm{g}(2t) = \mathrm{g}(4t)$. Hence

$$\sum_n |M_t(\psi)e_n|_{L_\rho^2}^2 \leq 2^d c_1^2 e^{2t\rho^2}\, |\psi|_{L_\rho^2}^2 \int_{\mathbb{R}^d} \mathrm{g}(4t)(\eta)\tilde{\Gamma}(\mathrm{d}\eta).$$

To show (14.18) it is sufficient to prove that (14.17) implies

$$\int_0^T \int_{\mathbb{R}^d} \mathrm{g}(4t)(\eta)\tilde{\Gamma}(\mathrm{d}\eta)\,\mathrm{d}t < \infty.$$

To this end we need to estimate

$$K(T, d, \eta) := \int_0^T \mathrm{g}(4t)(\eta)\,\mathrm{d}t.$$

By the lemma below it follows that $K(T, d, \eta)$ decreases to 0 as η goes to infinity faster than $|\eta|^{-m}$ with any m. Since Γ is a measure and the Fourier transform of a tempered distribution, Γ is a tempered measure. Thus

$$\int_{\{|\eta|\geq 1\}} K(T, d, \eta)\tilde{\Gamma}(\mathrm{d}\eta) < \infty$$

and it suffices to show that (14.17) ensures that

$$\int_{\{|\eta|\leq 1\}} K(T,d,\eta)\tilde{\Gamma}(\mathrm{d}\eta) < \infty. \tag{14.19}$$

If $d=1$ then, by Lemma 14.30 below, $K(T,1,\cdot)$ is bounded on $[-1,1]$. Thus (14.19) holds true. If $d=2$ then, by Lemma 14.30(iii), $K(T,2,\eta) \leq C\log(|\eta|^{-1})$. Thus (14.19) follows from (14.17) and from the fact that

$$\int_{\{|\eta|\leq 1\}} \log\left(|\eta|^{-1}\right)\mathrm{d}\eta < \infty.$$

If $d>2$, then by Lemma 14.30(iv) we have $K(T,2,\eta) \leq C|\eta|^{2-d}$. Thus again (14.19) follows from (14.17) and $\int_{\{|\eta|\leq 1\}} |\eta|^{2-d}\,\mathrm{d}\eta < \infty$. $\qquad\square$

Let $T > 0$, and $\alpha \geq 0$. For $r \geq 0$ we set

$$\kappa(\alpha,T,r) := \int_0^T t^{-\alpha}\mathrm{e}^{-r/t}\,\mathrm{d}t. \tag{14.20}$$

Lemma 14.30

(i) *For $\alpha \geq 0$, $n \in \mathbb{N}$ and $T > 0$, $r^n\kappa(\alpha,T,r) \to 0$ if $r \to \infty$.*

(ii) *If $\alpha \in [0,1)$ then $\kappa(\alpha,T,\cdot)$ is a bounded function.*

(iii) *There are constants $C_1, C_2 \in (0,\infty)$ such that*

$$C_1\log\left(|r|^{-1}\right) \leq \kappa(1,T,r) \leq C_2\log\left(|r|^{-1}\right), \qquad \forall\, r \in (0,1].$$

(iv) *If $\alpha > 1$ then there are constants $C_1, C_2 \in (0,\infty)$ such that*

$$C_1|r|^{1-\alpha} \leq \kappa(\alpha,T,r) \leq C_2|r|^{1-\alpha}, \qquad \forall\, r \in (0,1].$$

Proof After changing variables to $s = t/r$ we obtain

$$\kappa(\alpha,T,r) = r^{1-\alpha}\int_0^{T/r} s^{-\alpha}\mathrm{e}^{-1/s}\,\mathrm{d}s.$$

Thus (i) follows from the fact that

$$\lim_{a\downarrow 0} a^{-m}\int_0^a s^{-\alpha}\mathrm{e}^{-1/s}\,\mathrm{d}s = 0, \qquad \forall\, \alpha > 0,\ \forall\, m \in \mathbb{N}.$$

If $\alpha \in [0,1)$ then

$$\kappa(\alpha,T,r) \leq \int_0^T t^{-\alpha}\,\mathrm{d}t < \infty,$$

which proves (ii). Let $\alpha = 1$. Then (iii) follows from

$$\lim_{a\to+\infty} \frac{\int_0^a t^{-1}\mathrm{e}^{-1/t}\,\mathrm{d}t}{\log a} = \lim_{a\to+\infty} \frac{a^{-1}\mathrm{e}^{-1/a}}{a^{-1}} = \lim_{a\to+\infty} \mathrm{e}^{-1/a} = 1.$$

Finally (iv) follows from

$$\lim_{a \to +\infty} \frac{a^{\alpha-1} \int_0^a t^{-\alpha} e^{-1/t} \, dt}{a^{\alpha-1}} = \int_0^\infty t^{-\alpha} e^{-1/t} \, dt \in (0, \infty).$$

$\square$

Remark 14.31 We leave it as an exercise for the reader (see, however, Karczewska and Zabczyk 2000a, 2001) to show that if Γ is separate from $-\infty$ (see Theorem 14.29) then (14.17) holds if and only if

$$\int_{\mathbb{R}^d} \frac{\mu(d\xi)}{1 + |\xi|^2} < \infty. \tag{14.21}$$

Remark 14.32 From the remark above it follows that the Fourier transform of a measure satisfying (14.21) does not has to be separate from $-\infty$. As an example one can take the measure μ on $\mathbb{R}$ given by

$$\mu = \sum_{n \in \mathbb{N}} n^\alpha (\delta_{-n} + \delta_n),$$

where $\alpha \in (0, 1)$.

Remark 14.33 If Γ is a measure separate from $-\infty$ then it is finite on any ball. Thus (14.17) is always satisfied for dimension $d = 1$.

14.9 Space–time regularity in the Wiener case

In this section we consider the stochastic problem

$$\frac{\partial X}{\partial t}(t, \xi) = \Delta X(t, \xi) + f(X(t, \xi)) + g(X(t, \xi)) \frac{\partial W}{\partial t}(t, \xi), \qquad t > 0, \ \xi \in \mathbb{R}^d,$$

$$X(0, \xi) = x(\xi), \qquad \xi \in \mathbb{R}^d \tag{14.22}$$

where W is a spatially homogeneous Wiener process with spectral measure μ and covariance Γ. We will assume that $\Gamma = \mathfrak{F}\mu$ is a measure separate from $-\infty$ and that f and g are Lipschitz continuous.

Let $p \in [2, \infty)$, and let $\rho \in \mathbb{R}$. Let $\mathcal{C}_\rho := \mathcal{C}_\rho(\mathbb{R}^d)$; see Section 2.3. Given $T < \infty$ we denote by $\mathfrak{Z}_T$ and $\mathfrak{C}_T$ the classes of all adapted processes $Y : \Omega \times [0, T] \mapsto L_\rho^p$ and $Y : \Omega \times [0, T] \mapsto L_\rho^p \cap \mathcal{C}_\rho$ with continuous trajectories in L_ρ^p and in $L_\rho^p \cap \mathcal{C}_\rho$, respectively, and such that

$$\|Y\|_{\mathfrak{Z}_T} := \left(\mathbb{E} \sup_{t \in [0,T]} |Y(t)|_{L_\rho^p}^p \right)^{1/p} < \infty$$

and

$$\|Y\|_{\mathfrak{C}_T} := \left(\mathbb{E} \sup_{t \in [0,T]} \left(|Y(t)|_{L^p_\rho}^p + |Y(t)|_{\mathcal{C}_\rho}^p \right) \right)^{1/p} < \infty.$$

Theorem 14.34 *Let $\rho \in \mathbb{R}$, and $p \in (2, \infty)$. If $\rho \le 0$ assume that $f(0) = 0 = g(0)$.*

(i) *If there exists an $\alpha \in (1/p, 1/2)$ such that*

$$\int_{\{|\xi| \le 1\}} |\xi|^{-d-4\alpha+2} \Gamma(\mathrm{d}\xi) < \infty^2 \tag{14.23}$$

then for every $x \in L^p_\rho$ there is a unique solution X to (14.22) satisfying $X \in \mathfrak{Z}_T$ for every $T > 0$. Moreover, (14.22) defines a homogeneous Markov family on L^p_ρ satisfying the Feller property.

(ii) *If $1/p + 2d/p < 1/2$ and there is an $\alpha \in (1/p + 2d/p, 1/2)$ such that (14.23) holds then for every $x \in L^p_\rho \cap C_\rho$ there is a unique solution X to (14.22) such that $X \in \mathfrak{C}_T$ for every $T > 0$. Moreover, (14.22) defines a homogeneous Markov family on $L^p_\rho \cap C_\rho$ satisfying the Feller property.*

14.9.1 Proof of Theorem 14.34

By Theorem B.9 the Laplace operator generates a C_0-semigroup S on L^p_ρ. Let $\mathcal{H}$ be the RKHS of W, and let $U \subset S'(\mathbb{R}^d)$ be a Hilbert space such that the embedding $\mathcal{H} \hookrightarrow U$ is Hilbert–Schmidt. By Theorem 14.8 such a space exists. We may assume that L^2 is a dense subspace of U, and identifying L^2 with $(L^2)^*$ we obtain $U^* \hookrightarrow (L^2)^* \equiv L^2 \hookrightarrow U$. Then in particular $S(\mathbb{R}^d)$ is a dense subspace of U^*, and the covariance form (introduced in Section 7.1) of W on $U^* \times U^*$ is the unique continuous extension of the bilinear form K appearing in Theorem 14.18. Recall that the space $R(\mathcal{H}, L^p_\rho)$, the norm $\|\cdot\|_{R(\mathcal{H}, L^p_\rho)}$ and the spaces $\mathcal{L}^p_{W,T}(L^p_\rho)$ of integrable processes were introduced in subsection 8.8.2. In our considerations a fundamental role is played by the following result.

Lemma 14.35 *For all $t > 0$ and $\psi \in S(\mathbb{R}^d)$ the operator $M_t(\psi)$ given by $M_t(\psi)\varphi := S(t)(\psi\varphi)$ belongs to $R_{U,0}(\mathcal{H}, L^p_\rho)$. Moreover, there are constants C, c independent of t and ψ such that*

$$\|M_t(\psi)\|_{R_{U,0}(\mathcal{H}, L^p_\rho)} \le Ce^{Ct} |\psi|_{L^p_\rho} \sqrt{\langle \tilde{\Gamma}, \mathfrak{g}(ct) \rangle}.$$

Proof   We will use some arguments from the proof of (14.18). Let us fix t and ψ. We have $M_t(\psi)(\varphi)(\xi) = \langle Q(\xi), \varphi \rangle$, where $Q(\xi) = \mathfrak{g}(t)(\xi - \cdot)\psi(\cdot) \in S(\mathbb{R}^d)$. Then

---

2 Note that as Γ is a tempered measure (14.23) is always satisfied if $-d - 4\alpha + 2 \ge 0$.

(for the definition of $\Delta_p(\mathcal{Q})$ see Section 8.8) we have

$$\Delta_p^p(\mathcal{Q}) = \int_{\mathbb{R}^d} K(\mathcal{Q}(\xi), \mathcal{Q}(\xi))^{p/2} \vartheta_\rho^p(\xi) \, d\xi$$

$$= \int_{\mathbb{R}^d} \left\langle \Gamma, \, \mathfrak{g}(t)(\xi - \cdot)\psi(\cdot) * \left(\mathfrak{g}(t)(\xi - \cdot)\psi(\cdot)\right)_{(s)} \right\rangle^{p/2} \vartheta_\rho^p(\xi) \, d\xi$$

$$\leq \int_{\mathbb{R}^d} \left(\int_{\mathbb{R}^d} \int_{\mathbb{R}^d} |\psi(\xi + \eta - \zeta)||\psi(\xi + \eta)|\mu_t(d\zeta, d\eta) \right)^{p/2} \vartheta_\rho^p(\xi) \, d\xi,$$

where

$$\mu_t(d\zeta, d\eta) := \mathfrak{g}(t)(\eta - \zeta)\mathfrak{g}(t)(\eta)\tilde{\Gamma}(d\zeta) \, d\eta.$$

Thus, by Jensen's inequality,

$$\Delta_p^p(\mathcal{Q}) \leq \left(\int_{\mathbb{R}^d} \int_{\mathbb{R}^d} \mu_t(d\zeta, d\eta) \right)^{p/2-1}$$

$$\times \int_{\mathbb{R}^d} \int_{\mathbb{R}^d} \int_{\mathbb{R}^d} |\psi(\xi + \eta - \zeta)|^{p/2} |\psi(\xi + \eta)|^{p/2} \vartheta_\rho^p(\xi) \, d\xi \, \mu_t(d\zeta, d\eta).$$

We have

$$\int_{\mathbb{R}^d} \int_{\mathbb{R}^d} \mu_t(d\zeta, d\eta) = \langle \tilde{\Gamma}, \mathfrak{g}(t) * \mathfrak{g}(t) \rangle = \langle \tilde{\Gamma}, \mathfrak{g}(2t) \rangle$$

and

$$\int_{\mathbb{R}^d} |\psi(\xi + \eta - \zeta)|^{p/2} |\psi(\xi + \eta)|^{p/2} \vartheta_\rho^p(\xi) \, d\xi$$

$$\leq \left(\int_{\mathbb{R}^d} |\psi(\xi + \eta - \zeta)|^p \vartheta_\rho^p(\xi) \, d\xi \right)^{1/2} \left(\int_{\mathbb{R}^d} |\psi(\xi + \eta)|^p \vartheta_\rho^p(\xi) \, d\xi \right)^{1/2}$$

$$\leq C_1 |\psi|_{L_\rho^p}^p \exp\left\{ \tfrac{1}{2} |p\rho| \left(|\zeta - \eta| + |\eta| \right) \right\}.$$

Thus

$$\Delta_p^p(\mathcal{Q}) \leq C_1 |\psi|_{L_\rho^p}^p \langle \tilde{\Gamma}, \mathfrak{g}(2t) \rangle^{p/2-1} R,$$

where

$$R := \int_{\mathbb{R}^d} \int_{\mathbb{R}^d} \exp\left\{ \tfrac{1}{2} |p\rho| \left(|\zeta - \eta| + |\eta| \right) \right\} \mu_t(d\zeta, d\eta).$$

It is easy to see (compare with the proof of (14.18)) that

$$R \leq C_2 e^{C_2 t} \langle \tilde{\Gamma}, \mathfrak{g}(C_3 t) \rangle.$$

Consequently, there are constants C_4, C_5 such that

$$\|M_t(\psi)\|_{R_{U,0}(\mathcal{H}, L_\rho^p)}^p = \Delta_p^p(\mathcal{Q}) \leq C_4 e^{C_4 t} |\psi|_{L_\rho^p}^p \langle \tilde{\Gamma}, \mathfrak{g}(C_5 t) \rangle^{p/2},$$

as required. □

Since $S(\mathbb{R}^d)$ is dense in L_ρ^p and since $\psi \in L_\rho^p \implies g(\psi) \in L_\rho^p$ we have the following consequence of Lemma 14.35.

Corollary 14.36 *For all $\psi \in L_\rho^p$ and $t > 0$, $S(t)G(\psi) \in R(\mathcal{H}, L_\rho^p)$. Moreover, there are constants C, c such that, for all $\psi, \varphi \in L_\rho^p$ and $t > 0$,*

$$\|S(t)G(\psi)\|_{R(\mathcal{H}, L^p)} \leq Ce^{Ct}\left(1 + |\psi|_{L_\rho^p}\right)\sqrt{\langle\tilde{\Gamma}, g(ct)\rangle},$$

$$\|S(t)(G(\psi) - G(\varphi))\|_{R(\mathcal{H}, L^p)} \leq Ce^{Ct}|\psi - \varphi|_{L_\rho^p}\sqrt{\langle\tilde{\Gamma}, g(ct)\rangle}.$$

Proof of Theorem 14.34(i)   Let us fix a $T < \infty$. In the proof we use the Banach contraction principle on the space $\mathfrak{Z} := \mathfrak{Z}_T$. On $\mathfrak{Z}$ we consider the family of equivalent norms

$$\|Y\|_\beta := \left(\mathbb{E}\sup_{t\in[0,T]} e^{-\beta t}|Y(t)|_{L_\rho^p}^p\right)^{1/p}, \qquad \beta > 0.$$

Let

$$J_1(Y)(t) := S(t)x + \int_0^t S(t - s)F((Y(s))\,\mathrm{d}s,$$

$$J_2(Y)(t) := \int_0^t S(t - s)G((Y(s))\,\mathrm{d}W(s).$$

We should show that $J_1, J_2 \colon \mathfrak{Z} \mapsto \mathfrak{Z}$ and that there is a β such that

$$\|J_i(Y) - J_i(Z)\|_\beta \leq \tfrac{1}{3}\|Y - Z\|_\beta, \qquad i = 1, 2, \ \forall Y, Z \in \mathfrak{Z}.$$

However, we will consider only the stochastic term.

We use the factorization method introduced in Chapter 11. Recall that, for $\alpha > 0$, the infinite-dimensional Liouville–Riemann operator $I_{A,\alpha}$ is given by (11.2). By Theorem 11.3, $I_{A,\alpha}$ maps $L^p(0, T; L_\rho^p)$ into $C([0, T]; L_\rho^p)$ provided that $1/p < \alpha < 1$.

Let $\alpha \in (1/p, 1/2)$ be such that

$$\int_{\{|\xi|\leq 1\}} |\xi|^{-d-4\alpha+2}\Gamma(\mathrm{d}\xi) < \infty.$$

Note that, since $\alpha < 1/2$, $\int_{\{|\xi|\leq 1\}} |\xi|^{-d-4\alpha+2}\,\mathrm{d}\xi < \infty$. Consequently,

$$\int_{\{|\xi|\leq 1\}} |\xi|^{-d-4\alpha+2}\tilde{\Gamma}(\mathrm{d}\xi) < \infty. \tag{14.24}$$

Replacing, if necessary, α by $\alpha' \in (1/p, \alpha)$ we may assume that $d/2 + 2\alpha \neq 1$. As a consequence of (14.24) we have

$$\int_0^T t^{-2\alpha}\langle\tilde{\Gamma}, g(ct)\rangle\,\mathrm{d}t < \infty, \qquad \forall c > 0. \tag{14.25}$$

Indeed,

$$\int_0^T t^{-2\alpha} \langle \tilde{\Gamma}, \mathfrak{g}(ct) \rangle \, dt = c_1 \int_{\mathbb{R}^d} \kappa \left(2\alpha + \tfrac{1}{2}d, c_2 T, |\xi|^2 \right) \tilde{\Gamma}(d\xi),$$

where κ is given by (14.20), and (14.25) then follows from Lemma 14.30(iv).

Let $Y \in \mathfrak{Z}$. Let $t \in [0, T]$. It follows from Corollary 14.36 and (14.25) that

$$\Psi \colon \Omega \times [0, t] \ni (\omega, s) \mapsto (t - s)^{-\alpha} S(t - s) G(Y(s, \omega)) \in R_U \left(\mathcal{H}, L_\rho^p \right)$$

belongs to the space $\mathcal{L}_{W,t}^p \left(L_\rho^p \right)$ of integrable processes and that

$$\mathbb{E} \left(\int_0^t (t - s)^{-2\alpha} \| S(t - s) G(Y(s)) \|_{R(\mathcal{H}, L_\rho^p)}^2 \, ds \right)^{p/2}$$

$$\leq C_1 \left(1 + \|Y\|_0^p \right) \left(\int_0^t s^{-2\alpha} \langle \tilde{\Gamma}, \mathfrak{g}(ct) \rangle \, dt \right)^{p/2} \leq C_2 \left(1 + \|Y\|_0 \right)^p,$$

where C_2 is a constant independent of Z and $t \in [0, T]$ and $\| \cdot \|_0$ is the norm on $\mathfrak{Z}$ corresponding to $\beta = 0$. Let

$$I(Y)(t) = \frac{1}{\Gamma(1 - \alpha)} \int_0^t (t - s)^{-\alpha} S(t - s) G(Y(s)) \, dW(s), \qquad t \in [0, T].$$

By the estimate above and Theorem 8.23, $I(Y)$ is a well-defined process satisfying

$$\sup_{t \in [0,T]} \mathbb{E} \, |I(Y)(t)|_{L_\rho^p}^p < \infty.$$

Thus $I(Y)$ has trajectories in $L^p \left(0, T; L_\rho^p \right)$. Using arguments from the proof of Theorem 11.5, we can show that $J_2(Y) = \Gamma(1) I_{A,\alpha}(I(Y))$ and consequently $J_2(Y) \in \mathfrak{Z}$. We now show the contractivity of J_2. In what follows, the C_j are constants independent of $t \in [0, T]$, $Y, Z \in \mathfrak{Z}$ and β. We have

$$\| J_2(Y) - J_2(Z) \|_\beta^p \leq C_1 \mathbb{E} \sup_{t \in [0,T]} e^{-\beta t} | J_2(Y)(t) - J_2(Z)(t) |_{L_\rho^p}^p$$

$$\leq C_2 \mathbb{E} \sup_{t \in [0,T]} e^{-\beta t} | I_{A,\alpha}(I(Y) - I(Z))(t) |_{L_\rho^p}^p.$$

Let $\beta' = \beta/p$ and $q = p/(p - 1)$. Since

$$\left| \int_0^t (t - s)^{\alpha - 1} e^{-\beta'(t-s)} S(t - s) e^{-\beta' s} \left(I(Y)(s) - I(Z)(s) \right) ds \right|_{L_\rho^p}$$

$$\leq C_3 \left(\int_0^T t^{q(\alpha - 1)} e^{-q\beta' t} dt \right)^{1/q} \left(\int_0^T e^{-\beta s} | I(Y)(s) - I(Z)(s) |_{L_\rho^p}^p \, ds \right)^{1/p},$$

we have

$$\| J_2(Y) - J_2(Z) \|_\beta \leq c(\beta) C_4 \left(\int_0^T e^{-\beta t} \mathbb{E} \, | I(Y)(t) - I(Z)(t) |_{L_\rho^p}^p \, dt \right)^{1/p},$$

where

$$c(\beta) := \left(\int_0^T t^{q(\alpha-1)} e^{-q\beta' t} \, dt \right)^{p/q} \to 0 \qquad \text{as } \beta \uparrow \infty. \tag{14.26}$$

Now, by Corollary 14.36 and Theorem 8.23,

$$\mathbb{E} \, |I(Y)(t) - I(Z)(t)|_{L_\rho^p}^p \le C_5 \, \mathbb{E} \sup_{s \in [0,t]} |Y(s) - Z(s)|_{L_\rho^p}^p.$$

Thus

$$e^{-\beta t} \, \mathbb{E} \, |I(Y)(t) - I(Z)(t)|_{L_\rho^p}^p \le C_5 \, \mathbb{E} \sup_{s \in [0,t]} e^{-\beta s} |Y(s) - Z(s)|_{L_\rho^p}^p$$

$$\le C_5 \|Y - Z\|_\beta^p.$$

Summing up, we have

$$\|J_2(Y) - J_2(Z)\|_\beta \le C_6 c(\beta) \, \|Y - Z\|_\beta,$$

and the desired estimate follows from (14.26). ☐

In the proof of the second part of the theorem we need the following analytical result.

Lemma 14.37 *Let $p > 2$ and $\alpha \in (1/p + d/2p, 1/2)$. Then $I_{A,\alpha}$ is a bounded linear operator from $L^p\big(0, T; L_\rho^p\big)$ to $C\big([0, T]; C_\rho \cap L_\rho^p\big)$.*

Proof By Theorem B.7, $\|S(t)\|_{L(L_\rho^p, C_\rho)} \le C t^{-d/2p}$ for $t \in [0, T]$. Let q satisfy $1/q + 1/p = 1$. By the assumption, $(\alpha - 1)q - dq/2p > -1$ and hence

$$\int_0^t s^{(\alpha-1)q - dq/2p} \, ds < \infty, \qquad \forall t > 0.$$

Thus for all $\psi \in L^p\big(0, T; L_\rho^p\big)$ and $t > 0$, $I_{A,\alpha}(\psi)(t) \in C_\rho$ and

$$|I_{A,\alpha}(\psi)(t)|_{C_\rho} \le C_1 \left(\int_0^t s^{(\alpha-1)q - dq/2p} \, ds \right)^{1/q} |\psi|_{L^p(0,T;L_\rho^p)}.$$

Since $I_{A,\alpha}(\psi) \in C([0, T]; C_\rho)$ for $\psi \in C([0, T]; C_\rho)$ (see the proof of Theorem 11.3) and $C([0, T]; C_\rho)$ is dense in $L^p\big(0, T; L_\rho^p\big)$, the desired conclusion follows. ☐

Proof of Theorem 14.34(ii) Let us fix a $T < \infty$. By the first part of the theorem we know that there is a solution $X \in \mathfrak{Z}_T$. Our goal is to show that $X \in \mathfrak{C}_T \subset \mathfrak{Z}_T$. By the definition of a solution we have $X = J_1 + J_2$, where

$$J_1(t) := S(t)x + \int_0^t S(t - s)F((X(s)) \, ds,$$

$$J_2(t) := \int_0^t S(t - s)G((X(s)) \, dW(s).$$

We will show that $J_2 \in \mathfrak{C}_T$; the proof that $J_1 \in \mathfrak{C}_T$, which is simpler, is left to the reader. We use the factorization method. This method can be applied also to establish that $J_1 \in \mathfrak{C}_T$.

Let $\alpha \in (1/p + d/2p, 1/2)$ be such that (14.23) holds. We have $J_2 = \Gamma(1)I_{A,\alpha}(I)$, where

$$I(t) := \frac{1}{\Gamma(1-\alpha)} \int_0^t (t-s)^{-\alpha} S(t-s)G(X(s)) \, \mathrm{d}W(s), \qquad t \in [0, T].$$

By Theorem 8.23 and Corollary 14.36,

$$\mathbb{E}\left(\int_0^t (t-s)^{-2\alpha} \|S(t-s)G(X(s))\|^2_{R_U(\mathcal{H}, L_\rho^p)} \, \mathrm{d}s \right)$$
$$\leq C\left(1 + \|X\|_{3_T}\right)^p \left(\int_0^T s^{-2\alpha} \langle \tilde{\Gamma}, \mathfrak{g}(cs) \rangle \, \mathrm{d}s \right)^{p/2}.$$

Since (see Lemma 14.30) $\int_0^T s^{-2\alpha} \mathfrak{g}(cs)(\xi) \, \mathrm{d}s$ is of order $|\xi|^{2-d-4\alpha}$, we conclude from (14.23) that

$$\left(\int_0^T s^{-2\alpha} \langle \tilde{\Gamma}, \mathfrak{g}(cs) \rangle \, \mathrm{d}s \right)^{p/2} < \infty$$

and, consequently,

$$\sup_{t \in [0,T]} \mathbb{E}\left(\int_0^t (t-s)^{-2\alpha} \|S(t-s)G(X(s))\|^2_{R_U(\mathcal{H}, L_\rho^p)} \, \mathrm{d}s \right)^{p/2} < \infty.$$

Therefore,

$$\mathbb{E} \int_0^T |I(t)|^p_{L_\rho^p} \, \mathrm{d}t < \infty.$$

Since, by Lemma 14.37, $I_{A,\alpha} \in L\left(L^p\left(0, T; L_\rho^p\right), C\left([0, T]; C_\rho \cap L_\rho^p\right)\right)$ the process $J_2 = \Gamma(1)I_{A,\alpha}(I)$ has trajectories in $C\left([0, T]; C_\rho \cap L_\rho^p\right)$ and there is a constant C such that

$$\|J_2\|_{\mathfrak{C}_T} \leq C \left(\mathbb{E} \int_0^T |I(t)|^p_{L_\rho^p} \, \mathrm{d}t \right)^{1/p} < \infty,$$

which completes the proof. $\qquad\square$

14.10 Stochastic wave equation

14.10.1 Preliminaries

As in the case of the heat equation, first of all we need to construct a state space. Recall that (ϑ_ρ) is a family of exponential weights. Given $r \in \mathbb{R}$ let H_ρ^r be the completion of $S(\mathbb{R}^d)$ with respect to the norm

$$|\psi|_{H_\rho^r}^2 := \int_{\mathbb{R}^d} \left(1 + |\xi|^2\right)^r \left|\mathfrak{F}\left(\vartheta_\rho^{1/2}\psi\right)(\xi)\right|^2 \, d\xi.$$

Obviously H_ρ^r is a Hilbert space. Let

$$\mathbb{H}_\rho := \begin{pmatrix} L_\rho^2 \\ H_\rho^{-2} \end{pmatrix},$$

and let

$$A := \begin{pmatrix} 0 & I \\ \Delta & 0 \end{pmatrix}, \qquad D(A) := \begin{pmatrix} H_\rho^1 \\ L_\rho^2 \end{pmatrix}. \qquad (14.27)$$

Lemma 14.38 *Let $\rho \in \mathbb{R}$. Then $(A, D(A))$ generates a C_0-semigroup on $\mathbb{H}_\rho$.*

Proof If $\rho = 0$ then the result follows from the Lumer–Phillips theorem (see Theorem B.4 and Lemma B.3) in the same way as for the wave equation on a bounded interval. Assume that $\rho \neq 0$. Consider the isometric isomorphism of $\mathbb{H}_\rho$ onto $\mathbb{H}_0$ given by the formula

$$j \begin{pmatrix} \psi \\ \varphi \end{pmatrix} = \begin{pmatrix} \vartheta_\rho^{1/2}\psi \\ \vartheta_\rho^{1/2}\varphi \end{pmatrix}.$$

Now let S be the C_0-semigroup generated by A on $\mathbb{H}_0$. Let $\tilde{S}(t) = j^{-1}S(t)j, t \geq 0$. Clearly $\tilde{S}$ is a C_0-semigroup on $\mathbb{H}_\rho$. Its generator is equal to

$$\tilde{A} = j^{-1}Aj = \begin{pmatrix} 0 & I \\ \vartheta_\rho^{-1/2}\Delta\vartheta_\rho^{1/2} & 0 \end{pmatrix}$$

with domain

$$D(\tilde{A}) = \{v \colon jv \in D(A)\} = \begin{pmatrix} H_\rho^1 \\ L_\rho^2 \end{pmatrix}.$$

Thus the domain of $\tilde{A}$ is equal to the domain of A considered on $\mathbb{H}_\rho$. Note that for any function ψ,

$$\vartheta_\rho^{-1/2}\Delta\vartheta_\rho^{1/2}\psi = \Delta\psi + \sum_{j=1}^d \vartheta_\rho^{-1/2}\frac{\partial\vartheta_\rho}{\partial\xi_j}\frac{\partial\psi}{\partial\xi_j} + \left(\vartheta_\rho^{-1/2}\Delta\vartheta_\rho^{1/2}\right)\psi.$$

Since the functions $\vartheta_\rho^{-1/2}\partial\vartheta_\rho/\partial\xi_j$, $j = 1, \ldots, d$, and $\vartheta_\rho^{-1/2}\Delta\vartheta_\rho^{1/2}$ are of class $C_b(\mathbb{R}^d)$, we have

$$\vartheta_\rho^{-1/2}\Delta\vartheta_\rho^{1/2}\psi = \Delta\psi + B\psi,$$

where B is a differential operator of the first order with coefficients of the class $C_b(\mathbb{R}^d)$. Thus $\tilde{A} = A + \tilde{B}$, where

$$\tilde{B} := \begin{pmatrix} 0 & 0 \\ B & 0 \end{pmatrix}.$$

Note that $\tilde{B}$ is a bounded operator on $\mathbb{H}_\rho$. Consequently, A generates a C_0-semigroup on $\mathbb{H}_\rho$ as a bounded perturbation of the generator $\tilde{A}$. $\qquad\square$

For polynomial weights we have the following analogue of Lemma 14.38. Namely let K_ρ^r be the completion of $S(\mathbb{R}^d)$ to a Hilbert space with respect to the norm

$$|\psi|^2_{K_\rho^r} := \int_{\mathbb{R}^d} \left(1 + |\xi|^2\right)^r \left|\mathfrak{F}\left(\theta_\rho^{1/2}\psi\right)(\xi)\right|^2 d\xi.$$

Let

$$\mathbb{K}_\rho =: \begin{pmatrix} \mathcal{L}_\rho^2 \\ K_\rho^{-2} \end{pmatrix},$$

and let

$$A := \begin{pmatrix} 0 & I \\ \Delta & 0 \end{pmatrix}, \qquad D(A) := \begin{pmatrix} K_\rho^1 \\ \mathcal{L}_\rho^2 \end{pmatrix}. \tag{14.28}$$

Lemma 14.39 *Let $\rho \in \mathbb{R}$. Then $(A, D(A))$ generates a C_0-semigroup on $\mathbb{K}_\rho$.*

Since the proof of the lemma is similar to the proof of Lemma 14.38 we leave it to the reader.

Given a function $f: \mathbb{R} \mapsto \mathbb{R}$ write

$$\mathcal{N}_f \begin{pmatrix} \psi \\ \varphi \end{pmatrix} := \begin{pmatrix} 0 \\ N_f(\psi) \end{pmatrix}.$$

The proofs of the lemmas below are standard and are left to the reader.

Lemma 14.40 *Assume that $f: \mathbb{R} \mapsto \mathbb{R}$ is Lipschitz continuous. For $\rho > 0$, $\mathcal{N}_f$ is a Lipschitz mapping with linear growth on $\mathbb{H}_\rho$. If $\rho \le 0$ then $\mathcal{N}_f$ is Lipschitz with linear growth on $\mathbb{H}_\rho$ if and only if $f(0) = 0$.*

A similar result can be stated for polynomial weights.

Lemma 14.41 *Assume that $f: \mathbb{R} \mapsto \mathbb{R}$ is Lipschitz continuous. Then, for $\rho > d/2$, $\mathcal{N}_f$ is Lipschitz with linear growth on $\mathbb{K}_\rho$. If $\rho \le d/2$ then $\mathcal{N}_f$ is Lipschitz with linear growth on $\mathbb{K}_\rho$ if and only if $f(0) = 0$.*

14.10.2 Main result

Let

$$\mathbb{X} := \begin{pmatrix} X \\ \dfrac{\partial X}{\partial t} \end{pmatrix} = \begin{pmatrix} X \\ Y \end{pmatrix}.$$

Then

$$\frac{\partial \mathbb{X}}{\partial t} = \begin{pmatrix} \dfrac{\partial X}{\partial t} \\ \dfrac{\partial Y}{\partial t} \end{pmatrix} = \begin{pmatrix} Y \\ \dfrac{\partial^2 X}{\partial t^2} \end{pmatrix}.$$

Thus we can treat (14.16) as the equation

$$d\mathbb{X} = (A\mathbb{X} + F(\mathbb{X}))\, dt + G(\mathbb{X})\, dL$$

on a Hilbert space $\mathbb{H}_\rho$ with A given by (14.27), $F = \mathcal{N}_f$ and $G(u)\varphi[\xi] = \mathcal{N}_g(u)(\xi)\varphi(\xi)$; $u \in \mathbb{H}_\rho$ and φ belongs to the RKHS of L.

Theorem 14.42   *Assume that the spectral measure μ of L satisfies the following integral condition:*

$$\sup_{\xi \in \mathbb{R}^d} \int_{\mathbb{R}^d} \frac{\mu(d\eta)}{1 + |\xi + \eta|^2} < \infty. \tag{14.29}$$

Assume also that the functions f and g are Lipschitz continuous. Let $\rho \geq 0$. Then for any $z_0 \in \mathbb{H}_\rho$ there is a unique solution to (14.16) in $\mathbb{H}_\rho$. If $\rho \leq 0$ and $f(0) = 0 = g(0)$, then for any $z_0 \in \mathbb{H}_\rho$ there is a unique solution to (14.16) in $\mathbb{H}_\rho$. Moreover, (14.16) defines a Feller family on $\mathbb{H}_\rho$.

Proof   It follows from Lemma 14.38 that A generates a C_0-semigroup on $\mathbb{H}_\rho$. By Lemma 14.40, $\mathcal{N}_f$ and $\mathcal{N}_g$ are Lipschitz continuous. Since $G = \mathcal{M}\mathcal{N}_g$, where

$$\mathcal{M}\begin{pmatrix} \psi \\ \phi \end{pmatrix}\varphi = \begin{pmatrix} 0 \\ \varphi\phi \end{pmatrix},$$

it is sufficient to show that $\mathbb{M}\colon \phi \mapsto M_\phi$ is a bounded linear operator from L_ρ^2 to $L_{(HS)}(\mathcal{H}, H_\rho^{-1})$. To this end fix $\phi \in S(\mathbb{R}^d)$ and an orthonormal basis $\{e_n\}$ of $L_{(s)}^2(\mathbb{R}^d, \mu; \mathbb{C})$. Then $\{\mathfrak{F}(e_k\mu)\}$ is an orthonormal basis of $\mathcal{H}$ and

$$\sum_k |M_\phi \mathfrak{F}(e_k\mu)|_{H_\rho^{-1}}^2 = \sum_k \int_{\mathbb{R}^d} \left(1 + |\xi|^2\right)^{-1} \left|\mathfrak{F}(\vartheta_\rho^{1/2}\phi\, \mathfrak{F}(e_k\mu))(\xi)\right|^2 d\xi$$

$$= \int_{\mathbb{R}^d} \left(1 + |\xi|^2\right)^{-1} \left(\sum_k \left|\mathfrak{F}(\vartheta_\rho^{1/2}\phi\, \mathfrak{F}(e_k\mu))(\xi)\right|^2\right) d\xi.$$

Since $\{e_k\}$ is an orthonormal sequence in $L^2(\mathbb{R}^d, \mu; \mathbb{C})$, we obtain

$$\sum_k \left|\mathfrak{F}(\vartheta_\rho^{1/2}\phi\,\mathfrak{F}(e_k\mu))(\xi)\right|^2 = \sum_k \left|\int_{\mathbb{R}^d} \mathfrak{F}(\vartheta_\rho^{1/2}u)(\xi - \eta)e_k(\eta)\mu(d\eta)\right|^2$$

$$\leq \int_{\mathbb{R}^d} \left|\mathfrak{F}(\vartheta_\rho^{1/2}u)(\xi - \eta)\right|^2\mu(d\eta).$$

Hence

$$\sum_k |M_\phi\mathfrak{F}(e_k\mu)|^2_{H_\rho^{-1}} \leq \int_{\mathbb{R}^d} \left(1 + |\xi|^2\right)^{-1} \int_{\mathbb{R}^d} \left|\mathfrak{F}(\vartheta_\rho^{1/2}\phi)(\xi - \eta)\right|^2\mu(d\eta)\,d\xi$$

$$\leq \int_{\mathbb{R}^d} \int_{\mathbb{R}^d} \left(1 + |\xi + \eta|^2\right)^{-1}\left|\mathfrak{F}(\vartheta_\rho^{1/2}\phi)(\xi)\right|^2\mu(d\eta)\,d\xi$$

$$\leq \left(\sup_{\xi \in \mathbb{R}^d} \int_{\mathbb{R}^d} \frac{\mu(d\eta)}{1 + |\xi + \eta|^2}\right) \int_{\mathbb{R}^d} \left|\mathfrak{F}(\vartheta_\rho^{1/2}\phi)(\xi)\right|^2\,d\xi$$

$$\leq |\phi|^2_{L_\rho^2} \sup_{\xi \in \mathbb{R}^d} \int_{\mathbb{R}^d} \frac{\mu(d\eta)}{1 + |\xi + \eta|^2}.$$

$\square$

Remark 14.43 Under the assumption that $\Gamma = \mathfrak{F}(\mu)$ is a measure separate from $-\infty$, condition (14.29) is equivalent to either of the following conditions:

$$\int_{\mathbb{R}^d} \frac{\mu(d\eta)}{1 + |\eta|^2} < \infty,$$

$$\begin{cases} \displaystyle\int_{\{|\eta|\leq 1\}} \log\left(|\eta|^{-1}\right)\Gamma(d\eta) < \infty & \text{if } d = 2, \\[3mm] \displaystyle\int_{\{|\eta|\leq 1\}} |\eta|^{-d+2}\Gamma(d\eta) < \infty & \text{if } d \neq 2. \end{cases}$$

15

Equations with noise on the boundary

This chapter is devoted to partial differential equations of the second order with non-homogeneous boundary conditions of white-noise type. The chapter is organized as follows. First an integral version of the equation is derived in a slightly informal way. Then in the remaining part of the chapter problems involving the Laplace operator with Dirichlet or Neumann boundary conditions are considered and the equivalence of the concepts of weak and mild solutions is discussed.

15.1 Introduction

Assume that either $\mathcal{O} = (0, 1)$ or $\mathcal{O} \subset \mathbb{R}^d$ is a bounded domain with boundary $\partial\mathcal{O}$ of class C^∞,

$$A := \sum_{|\alpha| \le 2} a_\alpha(\xi) \frac{\partial^{|\alpha|}}{\partial \xi^\alpha}, \qquad \xi \in \overline{\mathcal{O}},$$

is a second-order differential operator on $\mathcal{O}$ and

$$B := \sum_{|\alpha| \le 1} b_\alpha(\xi) \frac{\partial^{|\alpha|}}{\partial \xi^\alpha}, \qquad \xi \in \overline{\mathcal{O}},$$

is a first-order differential operator. Let Z be a Lévy process taking values in a Hilbert space U embedded into the space $\mathcal{D}(\partial\mathcal{O})$ of distributions on the boundary $\partial\mathcal{O}$. We are concerned with the following problem:

$$\frac{\partial X}{\partial t}(t, \xi) = AX(t, \xi) + f(X(t, \xi)), \qquad t > 0, \ \xi \in \mathcal{O},$$

$$X(0, \xi) = x(\xi), \qquad \xi \in \mathcal{O}, \tag{15.1}$$

$$BX(t, \xi) = \dot{Z}(t, \xi) = \frac{\partial Z}{\partial t}(t, \xi), \qquad t > 0, \ \xi \in \partial\mathcal{O},$$

272

where $f: \mathbb{R} \mapsto \mathbb{R}$ is Lipschitz continuous and the initial value x belongs to $L^p(\mathcal{O})$ for some $p \in [1, \infty)$. In this book we are concerned only with solutions that are p-integrable processes taking values in $L^p(\mathcal{O})$. We will impose the following assumptions on A, B and $\mathcal{O}$.

(i) The operator A with domain $D(A) := \{\psi \in W^{2,p}(\mathcal{O}): B\psi = 0\}$ generates a C_0-semigroup S on $L^p(\mathcal{O})$.

(ii) There are $\gamma \in \mathbb{R}$ and $D_{B,\gamma} \in L(L^p(\partial\mathcal{O}), L^p(\mathcal{O}))$ such that $\operatorname{Im} D_{B,\gamma} \subset W^{2,p}(\mathcal{O})$ and, for every $\psi \in L^p(\mathcal{O})$, $v = D_{B,\gamma}\psi$ is a solution of the elliptic problem $Av(\xi) = \gamma v(\xi)$ for $\xi \in \mathcal{O}$ and $Bv(\xi) = \psi(\xi)$ for $\xi \in \partial\mathcal{O}$.

Following Da Prato and Zabczyk (1993) we derive the integral version of (15.1) by replacing the distributional-valued process $\dot{Z}$ by a process V regular in t and ξ. Namely, assume that X is a solution to the problem

$$\frac{\partial X}{\partial t}(t,\xi) = AX(t,\xi) + f(X(t,\xi)), \qquad t > 0, \ \xi \in \mathcal{O},$$
$$X(0,\xi) = x(\xi), \qquad \xi \in \mathcal{O},$$
$$BX(t,\xi) = V(t,\xi), \qquad t > 0, \ \xi \in \partial\mathcal{O}.$$

Then $Y := X - D_{B,\gamma}V$ satisfies

$$BY(t,\xi) = BX(t,\xi) - BD_{B,\gamma}V(t,\xi) = 0, \qquad \xi \in \partial\mathcal{O}, \ t > 0.$$

Moreover

$$Y(0,\xi) = x(\xi) - D_{B,\gamma}V(0,\xi), \qquad \xi \in \mathcal{O},$$

and, for $t > 0$ and $\xi \in \mathcal{O}$, we have

$$\begin{aligned}
\frac{\partial Y}{\partial t}(t,\xi) &= \frac{\partial X}{\partial t}(t,\xi) - \frac{\partial D_{B,\gamma}V}{\partial t}(t,\xi) \\
&= AX(t,\xi) + f(X(t,\xi)) - AD_{B,\gamma}V(t,\xi) + \gamma D_{B,\gamma}V(t,\xi) \\
&\quad - \frac{\partial D_{B,\gamma}V}{\partial t}(t,\xi) \\
&= AY(t,\xi) + f(X(t,\xi)) + \gamma D_{B,\gamma}V(t,\xi) - \frac{\partial D_{B,\gamma}V}{\partial t}(t,\xi).
\end{aligned}$$

Thus Y solves the problem

$$\frac{\partial Y}{\partial t}(t,\xi) = AY(t,\xi) + f(X(t,\xi)) + \gamma D_{B,\gamma}V(t,\xi) - \frac{\partial D_{B,\gamma}V}{\partial t}(t,\xi),$$
$$t > 0, \ \xi \in \mathcal{O},$$
$$Y(0,\xi) = x(\xi) - D_{B,\gamma}V(0,\xi), \qquad \xi \in \mathcal{O},$$
$$BY(t,\xi) = 0, \qquad t > 0, \ \xi \in \partial\mathcal{O}.$$

Hence (see Section 9.1) Y satisfies the integral equation

$$Y(t) = S(t)(x - D_{B,\gamma} V(0))$$
$$+ \int_0^t S(t - s) \left(F(X(s)) + \gamma D_{B,\gamma} V(s) - \frac{\partial D_{B,\gamma} V}{\partial s}(s) \right) ds,$$

where $F(\psi)(\xi) = f(\psi(\xi)) = N_f(\psi)(\xi)$, $\psi \in L^p(\mathcal{O})$ and $\xi \in \mathcal{O}$. Integrating by parts we obtain

$$- \int_0^t S(t - s) \frac{\partial D_{B,\gamma} V}{\partial s}(s) \, ds$$
$$= -D_{B,\gamma} V(t) + S(t) D_{B,\gamma} V(0) - \int_0^t A S(t - s) D_{B,\gamma} V(s) \, ds.$$

Hence X satisfies

$$X(t) = S(t)x + \int_0^t S(t - s) F(X(s)) \, ds + \int_0^t (\gamma - A) S(t - s) D_{B,\gamma} V(s) \, ds.$$

Putting $V = \dot{Z}$ we obtain the integral version of (15.1),

$$X(t) = S(t)x + \int_0^t S(t - s) F(X(s)) \, ds + \int_0^t (\gamma - A) S(t - s) D_{B,\gamma} \, dZ(s).$$
$$(15.2)$$

Note that, somewhat informally, X solves the stochastic evolution equation

$$dX = (AX + F(X)) \, dt + (\gamma - A) D_{B,\gamma} \, dZ$$

obtained in Da Prato and Zabczyk (1993).

Let us denote by $\mathcal{X}_T^p$, $T < \infty$, the space of all predictable processes $X : [0, T] \times \Omega \mapsto L^p(\mathcal{O})$ satisfying

$$\|X\|_{\mathcal{X}_T^p} := \sup_{t \in [0,T]} \left(\mathbb{E} \, |X(t)|_{L^p(\mathcal{O})}^p \right)^{1/p} < \infty.$$

Let $\mathcal{X}^p$ be the class of all processes $X : [0, \infty) \times \Omega \mapsto L^p(\mathcal{O})$ whose restriction to each finite interval $[0, T]$ belongs to $\mathcal{X}_T^p$.

Theorem 15.1 *Assume that the stochastic integral*

$$I(t) = \int_0^t (\gamma - A) S(t - s) D_{B,\gamma} \, dZ(s), \qquad t \in [0, \infty)$$

is a well-defined process from $\mathcal{X}^p$ and that $f : \mathbb{R} \mapsto \mathbb{R}$ is Lipschitz continuous. Then for any $x \in L^p(\mathcal{O})$ there is a unique $X \in \mathcal{X}^p$ satisfying (15.2).

The theorem can easily be shown by using the Banach contraction principle in the space $\mathcal{X}_T^p$, $T < \infty$.

In the important case $p = 2$ we have the following more explicit result.

Theorem 15.2 *Assume that Z is a square integrable mean zero Lévy process with RKHS $\mathcal{H}$ and that $f : \mathbb{R} \mapsto \mathbb{R}$ is Lipschitz continuous. If*

$$\int_0^t \left\| (\gamma - A) S(s) D_{B,\gamma} \right\|^2_{L_{(HS)}(\mathcal{H}, L^2(\mathcal{O}))} \, ds < \infty \qquad \text{for } t > 0 \tag{15.3}$$

then for any $x \in L^2(\mathcal{O})$ there is a unique solution $X \in \mathcal{X}^2$ to (15.1).

By Corollary 8.17, (15.3) is a necessary and sufficient condition for the stochastic integral

$$I(t) = \int_0^t (\gamma - A) S(t - s) D_{B,\gamma} \, dZ(s), \qquad t \ge 0,$$

to be a well-defined square integrable process taking values in $L^2(\mathcal{O})$. Moreover, under this assumption $I \in \mathcal{X}^2$. Thus the theorem follows from our previous result.

15.2 Weak and mild solutions

This section is based on Peszat and Russo (2007). Here we assume that A is the Laplace operator Δ. Since, by Theorems 15.1 and 15.2, the existence of a solution to the linear equation implies the existence of a solution to the non-linear problem with Lipschitz f, we restrict our attention to the case of $f \equiv 0$.

We consider (15.1) with Dirichlet or Neumann boundary conditions. In the Dirichlet case the boundary operator $B = B_D$ is given by $B_D \psi(\xi) = \psi(\xi)$, $\xi \in \partial \mathcal{O}$, whereas in the Neumann case $B_N \psi(\xi) = \partial \psi(\xi)/\partial \mathbf{n}$, $\xi \in \partial \mathcal{O}$, where $\mathbf{n}$ is the interior normal to the boundary. Thus we are concerned with the following problems:

$$\frac{\partial X}{\partial t} = \Delta X \qquad \text{on } (0, \infty) \times \mathcal{O},$$

$$X = \frac{dZ}{dt} \qquad \text{on } (0, \infty) \times \partial \mathcal{O}, \tag{15.4}$$

$$X(0, \xi) = x(\xi), \qquad \xi \in \mathcal{O}$$

and

$$\frac{\partial X}{\partial t} = \Delta X \qquad \text{on } (0, \infty) \times \mathcal{O},$$

$$\frac{\partial X}{\partial \mathbf{n}} = \frac{dZ}{dt} \qquad \text{on } (0, \infty) \times \partial \mathcal{O}, \tag{15.5}$$

$$X(0, \xi) = x(\xi).$$

Note that Δ considered with homogeneous Dirichlet or Neumann boundary conditions is a self-adjoint operator on $L^2(\mathcal{O})$. Taking into account Green's formula

$$\int_{\mathcal{O}} \Delta u v \, d\ell_d = \int_{\partial \mathcal{O}} \left(-\frac{\partial u}{\partial \mathbf{n}} v + \frac{\partial v}{\partial \mathbf{n}} u \right) d\sigma + \int_{\mathcal{O}} u \Delta v \, d\ell_d$$

we arrive at the following definition of a weak solution. Below, $(\cdot, \cdot)$ stands for the canonical bilinear form on either $\mathcal{D}(\mathcal{O}) \times C^{\infty}(\mathcal{O})$ or $\mathcal{D}(\partial \mathcal{O}) \times C^{\infty}(\partial \mathcal{O})$.

Definition 15.3 We say that an $L^p(\mathcal{O})$-valued process X is a *weak solution* to (15.4) if

$$(X(t), \psi) = (x, \psi) + \int_0^t (X(s), \Delta \psi) \, ds + \left(Z(t), \frac{\partial \psi}{\partial \mathbf{n}} \right)$$

for all $t > 0$ and $\psi \in C^{\infty}(\overline{\mathcal{O}})$ satisfying $\psi = 0$ on $\partial \mathcal{O}$.

We call an $L^p(\mathcal{O})$-valued process X a *weak solution* to (15.5) if

$$(X(t), \psi) = (x, \psi) + \int_0^t (X(s), \Delta \psi) \, ds - (Z(t), \psi)$$

for all $t > 0$ and $\psi \in C^{\infty}(\overline{\mathcal{O}})$ satisfying $\partial \psi / \partial \mathbf{n} = 0$ on $\partial \mathcal{O}$.

Let S_D and S_N be the semigroups generated by the Laplace operators Δ_D and Δ_N with homogeneous Dirichlet and Neumann boundary conditions, respectively. Let D_D and D_N be the boundary operators. In the Dirichlet case we set the parameter γ appearing in the definition of the operator $D_{\tau, \gamma}$ equal to 0. Then $D_D \psi$ is the unique solution to the Laplace problem $\Delta u = 0$ on $\mathcal{O}$ and $u = \psi$ on $\partial \mathcal{O}$. Since the problem $\Delta u = 0$ on $\mathcal{O}$ and $\partial u / \partial \mathbf{n} = \psi$ on $\partial \mathcal{O}$ is not well posed, we have to take $\gamma \neq 0$ in the Neumann case.

Let

$$I_D(t) := -\int_0^t \Delta_D S_D(t-s) D_D \, dZ(s),$$

$$I_N(t) := \int_0^t (\gamma - \Delta_N) S_N(t-s) D_N \, dZ(s).$$

Theorem 15.4

(i) *Assume that $I_D \in \mathcal{X}_T^p$ for every $T > 0$. Then there is a unique weak solution X to (15.4) such that $X \in \mathcal{X}_T^p$ for every $T > 0$. Moreover,*

$$X(t) = S_D(t)x + I_D(t). \tag{15.6}$$

(ii) *Assume that $I_N \in \mathcal{X}_T^p$ for every $T > 0$. Then there is a unique weak solution X to (15.5) such that $X \in \mathcal{X}_T^p$ for every $T > 0$. Moreover,*

$$X(t) = S_N(t)x + I_N(t). \tag{15.7}$$

Proof   Assume that $I_D \in \mathcal{X}_T^p$ for every $T > 0$. First we show that every weak solution X of (15.4) is given by the mild formula (15.6). To this end we adopt the strategy from the proof of our general equivalence theorem 9.15. Namely, we fix $t > 0$ and $\psi \in C^\infty(\overline{\mathcal{O}})$ satisfying $\psi = 0$ on $\partial\mathcal{O}$. Then ψ belongs to the domain $D(\Delta_D)$ of Δ_D. Define $z(s) := S_D(t - s)\psi$. Then

$$d(X(s), z(s)) = \big(X(s), \ \Delta_D z(s) + \dot{z}(s)\big)\, ds + \left(dZ(s), \frac{\partial z}{\partial \mathbf{n}}(s)\right)$$

$$= \big(X(s), \ \Delta_D z(s) - \Delta_D z(s)\big)\, ds + \left(dZ(s), \frac{\partial z}{\partial \mathbf{n}}(s)\right)$$

$$= -\big(D_D\, dZ(s), \ \Delta_D z(s)\big).$$

Thus

$$(X(t), \psi) = (x, S_D(t)\psi) - \int_0^t \big(D_D\, dZ(s), \ \Delta_D S_D(t - s)\psi\big)$$

$$= (S_D(t)x, \psi) - \int_0^t \big(\Delta_D S_D(t - s) D_D dZ(s), \ \psi\big).$$

Since the class of all $\psi \in C^\infty(\overline{\mathcal{O}})$ satisfying $\psi = 0$ on $\partial\mathcal{O}$ is dense in $L^2(\mathcal{O})$ we obtain $X(t) = S_D(t)x + I_D(t)$.

In order to show that the mild solution X is weak note that, by the stochastic Fubini theorem for every $\psi \in C^\infty(\overline{\mathcal{O}})$ satisfying $\psi = 0$ on $\partial\mathcal{O}$,

$$\int_0^t \big(X(s), \ \Delta_D\psi\big)\, ds = \left(\int_0^t \Delta_D S_D(t - s)x\, ds, \ \psi\right)$$

$$- \left(\int_0^t \int_r^t \Delta_D S_D(s - r)\, ds\, D_D\, dZ(r), \ \Delta_D\psi\right)$$

$$= (-x + S_D(t)x, \ \psi) - \left(\int_0^t (S_D(t - r) - I)\, D_D\, dZ(r), \ \Delta_D\psi\right)$$

$$- (\psi, x) + (\psi, X(t)) + \big(D_D Z(t) - D_D Z(0), \ \Delta_D\psi\big)$$

$$= -(x, \psi) + (X(t), \psi) - \left(Z(t), \frac{\partial\psi}{\partial \mathbf{n}}\right).$$

The case of the Neumann boundary condition is left to the reader.   □

15.3 Analytical preliminaries

Let us denote by $D_D(\theta, p)$ and $D_N(\theta, p)$, $\theta \in [0, 1]$, the real-interpolation spaces between $L^p(\mathcal{O})$ and the domains of Δ_D and Δ_N considered as generators on $L^p(\mathcal{O})$. For an exposition of the theory of interpolation spaces we refer the reader to e.g. Lunardi (1995) or Grisvard (1966). The following theorem was proven in Peszat and Russo (2007).

Theorem 15.5 *Let $p \in [1, \infty)$ and $\theta \in [0, 1]$. Then, for every $t > 0$, $\Delta_D S_D(t)$ and $\Delta_N S_N(t)$ are bounded linear operators from $D_{\Delta_D}(\theta, p)$ to $L^p(\mathcal{O})$ and from $D_{\Delta_N}(\theta, p)$ to $L^p(\mathcal{O})$, respectively. Moreover, if $p \geq 1$ and $\theta \in (0, 1]$ are such that $(\theta - 1)p > -1$ then*

$$\int_0^t \|\Delta_D S_D(s)\|_{L(D_{\Delta_D}(\theta,p),L^p(\mathcal{O}))}^p \, ds < \infty, \qquad \forall t > 0,$$

$$\int_0^t \|\Delta_N S_N(s)\|_{L(D_{\Delta_N}(\theta,p),L^p(\mathcal{O}))}^p \, ds < \infty, \qquad \forall t > 0.$$

Proof Since S_D is an analytic semigroup, the operators $\Delta_D S_D(t)$, $t > 0$, are bounded from $L^p(\mathcal{O})$ into $L^p(\mathcal{O})$, and from $D(\Delta_D)$ into $D(\Delta_D)$. Moreover (see Definition 2.4), for every $T > 0$,

$$\sup_{t \in (0,T]} \left\{ t \|\Delta_D S_D(t)\|_{L(L^p(\mathcal{O}),L^p(\mathcal{O}))} + \|\Delta_D S_D(t)\|_{L(D(\Delta_D),L^p(\mathcal{O}))} \right\} < \infty.$$

By definition $D_{\Delta_D}(0, p) = L^p(\mathcal{O})$ and $D_{\Delta_D}(1, p)$ is equal to the domain of Δ_D equipped with the graph norm. Hence

$$\sup_{t \in (0,T]} \left\{ t \|\Delta_D S_D(t)\|_{L(D_{\Delta_D}(0,p),L^p(\mathcal{O}))} + \|\Delta_D S_D(t)\|_{L(D_{\Delta_D}(1,p),L^p(\mathcal{O}))} \right\} < \infty.$$

By the interpolation property,

$$\sup_{t \in (0,T]} t^{1-\theta} \|\Delta_D S_D(t)\|_{L(D_{\Delta_D}(\theta,p),L^p(\mathcal{O}))} < \infty, \qquad \forall T > 0,$$

which gives the desired estimate for the Dirichlet case. The estimates in the Neumann case can be proven using the above arguments. □

For further references we present the following theorem, which provides characterizations of $D_D(\theta, p)$ and $D_N(\theta, p)$ in terms of Sobolev spaces. It follows as a special case from Theorem 3.2.3 of Lunardi (1995); see also Grisvard (1966).

Theorem 15.6 *Let $p \in (1, \infty)$ and $\theta \in (0, 1)$ be such that 2θ and $2\theta - 1/p$ are not integers. Then*

$$D_{\Delta_D}(\theta, p) = \begin{cases} W^{2\theta,p}(\mathcal{O}) & \text{if } 2\theta < 1/p, \\ \{u \in W^{2\theta,p}(\mathcal{O}) : u = 0 \text{ on } \partial\mathcal{O}\} & \text{if } 2\theta > 1/p, \end{cases}$$

$$D_{\Delta_N}(\theta, p) = \begin{cases} W^{2\theta,p}(\mathcal{O}) & \text{if } 2\theta < 1 + 1/p, \\ \{u \in W^{2\theta,p}(\mathcal{O}) : \partial u/\partial\mathbf{n} = 0 \text{ on } \partial\mathcal{O}\} & \text{if } 2\theta > 1 + 1/p. \end{cases}$$

15.4 L^2 **case**

In this section we will show that unless $Z = 0$ the Dirichlet problem (15.4) does not have a solution from the class $\mathcal{X}^2$. However, the Neumann problem (15.5) can be solved in $\mathcal{X}^2$ for some non-trivial random perturbations. We will first consider the problems on $\mathcal{O} = (0, 1)$. This part of the discussion is based on Da Prato and Zabczyk (1993). The multidimensional case will be considered in the last subsection.

15.4.1 Stochastic heat equation on $(0, 1)$

Assume that $\mathcal{O} = (0, 1)$ and $x \in L^2(0, 1)$. Then $\partial \mathcal{O} = \{0, 1\}$. Therefore we can identify $L^2(\partial \mathcal{O})$ with $\mathbb{R}^2$, and any Lévy process Z on $L^2(\partial \mathcal{O})$ is of the form $Z = (Z_0, Z_1)$, where Z_0 and Z_1 are real-valued Lévy processes. Hence problems (15.4) and (15.5) can be written as follows:

$$\frac{\partial X}{\partial t}(t, \xi) = \frac{\partial^2 X}{\partial \xi^2}(t, \xi), \qquad t > 0, \ \xi \in (0, 1),$$
$$X(0, \xi) = x(\xi), \qquad \xi \in (0, 1), \tag{15.8}$$
$$X(t, 0) = \dot{Z}_0(t), \quad X(t, 1) = \dot{Z}_1(t), \qquad t > 0$$

and

$$\frac{\partial X}{\partial t}(t, \xi) = \frac{\partial^2 X}{\partial \xi^2}(t, \xi), \qquad t > 0, \ \xi \in (0, 1),$$
$$X(0, \xi) = x(\xi), \qquad \xi \in (0, 1), \tag{15.9}$$
$$\frac{\partial X}{\partial \xi}(t, 0) = \dot{Z}_0(t), \quad \frac{\partial X}{\partial \xi}(t, 1) = \dot{Z}_1(t), \qquad t > 0.$$

Denote by B_D and B_N the corresponding boundary operators, that is,

$$B_D v = (v(0), v(1)) \quad \text{and} \quad B_N v = \left(\frac{dv}{d\xi}(0), \frac{dv}{d\xi}(1) \right).$$

For the Dirichlet problem we put $\gamma = 0$. Then $D_D = D_{\tau_D, 0}$ can be computed as follows. Given $w = (w_0, w_1) \in \mathbb{R}^2$, $v = D_D w$ solves

$$\frac{d^2 v}{d\xi^2}(\xi) = 0 \quad \text{for } \xi \in (0, 1), \qquad v(0) = w_0, \quad v(1) = w_1.$$

Thus $D_D(w)(\xi) = w_0 + (w_1 - w_0)\xi$ for $\xi \in (0, 1)$ or, equivalently,

$$D_D(w) = w_0 \psi_0 + (w_1 - w_0)\psi_1,$$
$$\psi_0(\xi) := 1, \quad \psi_1(\xi) := \xi, \qquad \xi \in (0, 1). \tag{15.10}$$

For the Neumann problem we take $\gamma = 1$. Then $D_N = D_{B_N, 1}$ can be computed as for the Dirichlet problem. Namely, given $w = (w_0, w_1) \in \mathbb{R}^2$, $v = D_N w$ solves

$$\frac{d^2 v}{d\xi^2}(\xi) = v(\xi) \quad \text{for } \xi \in (0, 1), \qquad \frac{dv}{d\xi}(0) = w_0, \qquad \frac{dv}{d\xi}(1) = w_1.$$

Hence

$$D_N(w) = \frac{e w_0 - w_1}{e^{-1} - e} \varphi_0 + \frac{e^{-1} w_0 - w_1}{e^{-1} - e} \varphi_1, \tag{15.11}$$

$$\varphi_0(\xi) := e^{-\xi}, \quad \varphi_1(\xi) := e^{\xi}, \qquad \xi \in (0, 1).$$

We have the following result.

Theorem 15.7

(i) *For the Dirichlet boundary problem, condition (15.3) is satisfied if and only if $Z = 0$. Thus problem (15.8) does not have a square integrable solution in $L^2(0, 1)$ if $Z \neq 0$.*

(ii) *For the Neumann boundary problem, condition (15.3) is satisfied for any square integrable Lévy process in $\mathbb{R}^2$. Thus if Z is square integrable then for every $x \in L^2(0, 1)$ there is a unique solution to (15.9) from $\mathcal{X}^2$.*

Proof of (i) Assume that $Z \neq 0$. Then its RKHS $\mathcal{H}$ is a non-zero subspace of $\mathbb{R}^2$. Hence there exists $w = (w_0, w_1) \in \mathcal{H}$ such that $|w|_\mathcal{H} = 1$. Since

$$l(t) := \int_0^t |\Delta_D S_D(s) D_D w|_{L^2(0,1)}^2 \, ds$$

$$\leq \int_0^t \|\Delta_D S_D(s) D_D\|_{L_{(HS)}(\mathcal{H}, L^2(0,1))}^2 \, ds$$

it is enough to show that $l(t) = \infty$ for $t > 0$. From (15.10),

$$l(t) = \int_0^t \left| \Delta_D S_D(s) (w_0 \psi_0 + (w_1 - w_0) \psi_1) \right|_{L^2(0,1)}^2 ds.$$

Let $\{e_n\}$, $e_n(\xi) = \sqrt{2} \sin(\pi n \xi)$, be an orthonormal basis consisting of eigenvectors of Δ_D. Then, for $\psi \in L^2(0, 1)$,

$$\int_0^t |\Delta_D S_D(s)\psi|_{L^2(0,1)}^2 \, ds = \sum_n \int_0^t \langle \Delta_D S_D(s)\psi, e_n \rangle_{L^2(0,1)}^2 \, ds$$

$$= \sum_n \int_0^t \langle \psi, \Delta_D S_D(s) e_n \rangle_{L^2(0,1)}^2 \, ds$$

$$= \sum_n \langle \psi, e_n \rangle_{L^2(0,1)}^2 \int_0^t e^{-2\pi^2 n^2 s} \pi^4 n^4 \, ds$$

$$= \sum_n \langle \psi, e_n \rangle_{L^2(0,1)}^2 \pi^2 n^2 \frac{1}{2} \left(1 - e^{-2\pi^2 n^2 t}\right).$$

Thus $l(t) < \infty$ for every (or equivalently for some) $t > 0$ if and only if

$$l := \sum_n \pi^2 n^2 \langle w_0 \psi_0 + (w_1 - w_0) \psi_1, \ e_n \rangle^2_{L^2(0,1)} < \infty.$$

Since $\pi^2 n^2$, $n \in \mathbb{N}$, is the sequence of eigenvalues of Δ_D,

$$l = \left| (-\Delta_D)^{1/2} (w_0 \psi_0 + (w_1 - w_0) \psi_1) \right|^2_{L^2(0,1)}$$
$$= |w_0 \psi_0 + (w_1 - w_0) \psi_1|^2_{D((-\Delta_D)^{1/2})}.$$

Since (see e.g. Grisvard 1966, Lions and Magenes 1972 or Lunardi 1995) $D\big((-\Delta_D)^{1/2}\big)$ is equal to $D_{\Delta_D}(1/2, 2)$, Theorem 15.6 yields

$$D\big((-\Delta_D)^{1/2}\big) = \left\{ u \in W^{1,2}(0,1) \colon u(0) = 0 = u(1) \right\}.$$

Thus it is enough to observe that $w_0 \psi_0 + (w_1 - w_0) \psi_1$ does not vanish on the boundary $\{0, 1\}$ provided that $(w_0, w_1) \neq (0, 0)$.

Proof of (ii) Let $\varepsilon_0 := (1, 0)$ and $\varepsilon_1 := (0, 1)$. Since the RKHS $\mathcal{H}$ of Z is a subspace of $\mathbb{R}^2$ there is a constant C such that

$$J(t) := \int_0^t \|\Delta_N S_N(s) D_N\|^2_{L_{(HS)}(\mathcal{H}, L^2(0,1))} \, ds$$
$$\leq C \sum_{i=0,1} \int_0^t |\Delta_N S_N(s) D_N \varepsilon_i|^2_{L^2(0,1)} \, ds.$$

We have to show that $J(t) < \infty$ for $t > 0$. Taking into account (15.11) it is enough to show that

$$\int_0^t |\Delta_N S_N(s) \varphi_i|^2_{L^2(0,1)} \, ds < \infty, \qquad i = 0, 1, \ t > 0,$$

where φ_i, $i = 0, 1$, are the functions appearing in (15.11). This (see the proof of the first part of the theorem) is equivalent to the following estimates:

$$\sum_n \pi^2 n^2 \langle \varphi_i, f_n \rangle^2_{L^2(0,1)} < \infty, \qquad i = 0, 1, \tag{15.12}$$

where $f_n(\xi) = \sqrt{2} \cos(\pi n \xi)$, $n \in \mathbb{N}$, is an orthonormal basis of $L^2(0, 1)$ consisting of eigenvectors of Δ_N with corresponding eigenvalues $\{-n\pi\}$. By an elementary calculation, the estimates in (15.12) are equivalent to the claim that $\varphi_0, \varphi_1 \in D\big((-\Delta_N)^{1/2}\big)$. Since, by Theorem 15.6,

$$D\big((-\Delta_N)^{1/2}\big) = D_{\Delta_N}(1/2, 2) = W^{1,2}(0, 1)$$

it is enough to observe that $\varphi_0, \varphi_1 \in W^{1,2}(0, 1)$. This follows from the facts that φ_0 and φ_1 are smooth functions and that in $W^{1,2}(0, 1)$ there are no hidden boundary conditions. $\qquad \square$

15.4.2 Multidimensional case

Let us now consider (15.4) and (15.5) on a domain $\mathcal{O} \subset \mathbb{R}^d$. We assume that Z is a Lévy process of the form

$$Z(t) = \sum_k \gamma_k Z_k(t) f_k, \tag{15.13}$$

where the Z_k are uncorrelated square integrable real-valued Lévy processes, (γ_k) is a sequence of real numbers and (f_k) is a sequence of functions defined on $\partial\mathcal{O}$. Let D, equal either to D_D or to D_N, be a suitable boundary operator. Then (see the proof of Theorem 15.7) it is easy to see that condition (15.3) can be written in the form

$$\sum_k \gamma_k^2 |Df_k|^2_{D((-\Delta_D)^{1/2})} = \sum_k \gamma_k^2 |Df_k|^2_{D_{\Delta_D}(1/2, 2)} < \infty.$$

By Theorem 15.5, $D_\Delta(1/2, 2) = \{u \in W^{1,2}(\mathcal{O}) : u = 0 \text{ on } \partial\mathcal{O}\}$ in the Dirichlet case and $D_\Delta(1/2, 2) = W^{1,2}(\mathcal{O})$ in the Neumann case. Clearly $\tau_D Du = 0$ if and only if $u = 0$. Thus we have the following result.

Theorem 15.8   *Assume that Z is given by (15.13).*

 (i) *For the Dirichlet boundary problem (15.4), condition (15.3) is satisfied if and only if $Z = 0$. Thus problem (15.4) does not have a solution in $L^2(\mathcal{O})$ if $Z \neq 0$.*

(ii) *For the Neumann boundary problem (15.5), condition (15.3) if satisfied if and only if*

$$\sum_k \gamma_k^2 |D_N f_k|^2_{W^{1,2}(\mathcal{O})} < \infty. \tag{15.14}$$

Thus, provided (15.14) holds, for every $x \in L^2(\mathcal{O})$ there is a unique solution to (15.5) in $\mathcal{X}^2$.

In order to study the Dirichlet problem (15.4) with Gaussian white noise one should introduce a weighted L^2-space, namely, $L^2(\mathcal{O}, \kappa(\xi) \, d\xi)$ where κ vanishes on the boundary $\partial\mathcal{O}$. For more details we refer the reader to Da Prato and Zabczyk (1993), Sowers (1994), Alòs and Bonaccorsi (2002a, b) or Peszat and Russo (2007).

15.5 Poisson perturbation

Let π be a Poisson random measure on $[0, \infty) \times \partial\mathcal{O} \times S$ with intensity $ds \, \sigma^d(d\xi)\nu(d\kappa)$. Following Peszat and Russo (2007) we assume that

$$Z(t)(\xi) = \int_0^t \int_{\partial\mathcal{O}} \int_S \phi(\xi, \eta, \kappa)\widehat{\pi}(ds, d\eta, d\kappa),$$

where $\phi\colon [0, \infty) \times \partial\mathcal{O} \times S \mapsto \overline{\mathbb{R}}$ is a certain (regularization) kernel. We first deal with the following Dirichlet boundary problem:

$$
\begin{aligned}
\frac{\partial X}{\partial t}(t, \xi) &= \Delta X(t, \xi) + f(X(t, \xi)), & t > 0, \ \xi \in \mathcal{O}, \\
X(0, \xi) &= x(\xi), & \xi \in \mathcal{O}, \\
X(t, \xi) &= \dot{Z}(t)(\xi), & t > 0, \ \xi \in \partial\mathcal{O},
\end{aligned}
\tag{15.15}
$$

where $f\colon \mathbb{R} \mapsto \mathbb{R}$ is Lipschitz continuous.

Theorem 15.9 *Let $p \in (1, 2]$. Assume that there is a parameter $\theta \in (0, 1]$, satisfying*

$$
1 - \frac{1}{p} < \theta < \frac{1}{2p},
\tag{15.16}
$$

such that $D_D\phi(\cdot, \eta, \kappa)$ is well defined, belongs to $W^{2\theta, p}(\mathcal{O})$ for almost all η and κ and satisfies

$$
b_\phi := \int_{\partial\mathcal{O}} \int_S |D_D\phi(\cdot, \eta, \kappa)|_{W^{2\theta,p}(\mathcal{O})}^p \sigma^d(d\eta)\nu(d\kappa) < \infty.
\tag{15.17}
$$

Then for every $x \in L^p(\mathcal{O})$ there is a unique solution $X \in \mathcal{X}^p$ to (15.15). Moreover, (15.15) defines a Feller family on $L^p(\mathcal{O})$.

Proof It is enough to show that

$$
I(t)(\xi) := \int_0^t \int_{\partial\mathcal{O}} \int_S \Delta_D S_D(t - s)D_D\phi(\xi, \eta, \kappa)\widehat{\pi}(ds, d\eta, d\kappa), \qquad t > 0, \ \xi \in \mathcal{O},
$$

belongs to $\mathcal{X}^p$. To this end note that, by Theorem 8.26, there is a constant c_p such that for all $t > 0$,

$$
\begin{aligned}
\mathbb{E}\int_{\mathcal{O}} &\left| \int_0^t \int_{\partial\mathcal{O}} \int_S (\Delta_D S_D(t - s)D_D\phi)(\cdot, \eta, \kappa)(\xi)\widehat{\pi}(ds, d\eta, d\kappa) \right|^p d\xi \\
&\leq c_p \int_0^t \int_{\partial\mathcal{O}} \int_S |(\Delta_D S_D(s)D_D\phi)(\cdot, \eta, \kappa)|_{L^p(\mathcal{O})}^p \, ds \, \sigma^d(d\eta)\nu(d\kappa).
\end{aligned}
$$

We need to estimate $|\Delta_D S_D(s)D_D\phi(\cdot, \eta, \kappa)|_{L^p(\mathcal{O})}^p$. We have

$$
\begin{aligned}
\int_0^t &|\Delta_D S_D(s)D_D\phi(\cdot, \eta, \kappa)|_{L^p(\mathcal{O})}^p \, ds \\
&\leq \int_0^t \|\Delta_D S_D(s)\|_{L(W^{2\theta,p}(\mathcal{O}), L^p(\mathcal{O}))}^p \, ds \, |D_D\phi(\cdot, \eta, \kappa)|_{W^{2\theta,p}(\mathcal{O})}^p.
\end{aligned}
$$

Write

$$
a(t) := \int_0^t \|\Delta_D S_D(s)\|_{L(W^{2\theta,p}(\mathcal{O}), L^p(\mathcal{O}))}^p \, ds.
$$

By Theorem 15.6 and the assumption that $2\theta < 1/p$ we have $W^{2\theta,p}(\mathcal{O}) = D_{\Delta_D}(\theta, p)$. Hence, by Theorem 15.5 and the assumption that $(\theta - 1)p > -1$, a is a continuous function. Thus, for $t \in [0, T]$,

$$\mathbb{E} \left| \int_0^t \int_{\partial\mathcal{O}} \int_S \Delta_D S_D(t - s) D_D\phi(\cdot, \eta, \kappa)(\xi)\widehat{\pi}(ds, d\eta, d\kappa) \right|_{L^p(\mathcal{O})}^p$$

$$\leq c_p a(T) b_f < \infty.$$

$\square$

Remark 15.10   It follows from the proof of the theorem that it is enough to assume that, with a certain θ satisfying $(\theta - 1)p > -1$,

$$\int_{\partial\mathcal{O}} \int_S |D_D\phi(\cdot, \eta, \kappa)|_{D_{\Delta_D}(\theta,p)}^p \sigma^d(d\eta)\nu(d\kappa) < \infty.$$

However, unless $2\theta < 1/p$, it follows from $D_D\phi \in D_{\Delta_D}(\theta, p)$ that $D_D\phi(\xi, \eta, \kappa)$ must equal 0 for $\xi \in \partial\mathcal{O}$. Since $D_D\phi(\xi, \eta, \kappa) = \phi(\xi, \eta, \kappa)$ for $\xi \in \partial\mathcal{O}$, the function ϕ must equal 0.

Remark 15.11   If (15.16) holds then necessarily $p \in (1, 3/2)$. Since (see e.g. Lions and Magenes 1972) D_D is a linear operator from $W^{s+3/2,2}(\partial\mathcal{O})$ to $W^{s+2,2}(\mathcal{O}) \subset W^{s+2,p}(\mathcal{O})$, the hypotheses of the theorem are satisfied providing that

$$\int_{\partial\mathcal{O}} \int_S |\phi(\cdot, \eta, \kappa)|_{W^{3/2,2}(\partial\mathcal{O})}^p \sigma^d(d\eta)\nu(d\kappa) < \infty.$$

A similar result holds true for the Neumann boundary problem

$$\frac{\partial X}{\partial t}(t, \xi) = \Delta X(t, \xi) + f(X(t, \xi)), \qquad t > 0, \ \xi \in \mathcal{O},$$

$$X(0, \xi) = x(\xi), \qquad \xi \in \mathcal{O}, \qquad\qquad\qquad (15.18)$$

$$\frac{\partial X}{\partial \mathbf{n}}(t, \xi) = \dot{Z}(t)(\xi), \qquad t > 0, \ \xi \in \partial\mathcal{O}.$$

Taking into account that $D_{\Delta_N}(\theta, p) = W^{2\theta,p}(\mathcal{O})$ for $2\theta < 1 + 1/p$ we obtain the following result.

Theorem 15.12   *Let $p \in (1, 2]$. Assume that there is a parameter $\theta \in (0, 1]$ satisfying $1 - 1/p < \theta < 1/2 + 1/(2p)$ such that $D_N\phi(\cdot, \eta, \kappa)$ is well defined, belongs to $W^{2\theta,p}(\mathcal{O})$ for almost all η and κ and satisfies*

$$b_\phi := \int_{\partial\mathcal{O}} \int_S |D_N\phi(\cdot, \eta, \kappa)|_{W^{2\theta,p}(\mathcal{O})}^p \sigma^d(d\eta)\nu(d\kappa) < \infty.$$

Then for every $x \in L^p(\mathcal{O})$ there is a unique solution $X \in \mathcal{X}^p$ to (15.18). Moreover, (15.18) defines a Feller family on $L^p(\mathcal{O})$.

Part III

Applications

16

Invariant measures

One of the most important problems in applying stochastic analysis is the question of the existence and uniqueness of invariant measures for specific processes. This chapter is devoted to this question for a class of so-called dissipative systems of great physical significance. The results obtained will be applied, in the following chapters, to lattice systems and equations of financial mathematics.

16.1 Basic definitions

Let (P_t) be the transition function of a Markov process $X = (X(t), t \geq 0)$ on a Polish space E.

Definition 16.1 A probability measure μ is *invariant with respect to the transition function* (P_t) or *invariant for* X if, for any Borel set $\Gamma \subset E$ and any $t \geq 0$,

$$\mu(\Gamma) = \int \mu(\mathrm{d}x) P_t(x, \Gamma).$$

If the initial position $X(0)$ of X is a random variable with distribution μ then the distribution of $X(t)$ is equal to μ for any $t \geq 0$. Thus one can expect that processes with invariant measures exhibit some kind of stability.

Let $\mathcal{M}_1(E)$ denote the space of all Borel probability measures on E, and let $(\psi, \lambda) = \int_E \psi(y)\lambda(\mathrm{d}y)$, $\psi \in B_b(E)$, $\lambda \in \mathcal{M}_1(E)$.

By definition μ is an invariant measure for (P_t) if and only if $P_t^* \mu = \mu$, $t \geq 0$, where (P_t^*) denotes the adjoint transition semigroup defined on $\mathcal{M}_1(E)$ by

$$P_t^* \lambda(\Gamma) = \int_E P_t(x, \Gamma) \lambda(\mathrm{d}x).$$

Clearly, $(P_t \psi, \lambda) = (\psi, P_t^* \lambda)$ for $t \geq 0$, $\psi \in B_b(E)$ and $\lambda \in \mathcal{M}_1(E)$.

287

The following classical result provides a method of proving the existence of invariant measures for Feller semigroups.

Theorem 16.2 (Krylov–Bogolyubov) *Assume that (P_t) is a Feller transition semigroup on E and there is an $x \in E$ such that $P_t(x, \cdot)$ converges weakly to a probability measure μ. Then μ is an invariant measure.*

Proof Let $t > 0$. Then $P_t^* \mu = \mu$ if and only if, for any $\psi \in C_b(E)$, $(\psi, P_t^* \mu) = (\psi, \mu)$. Let $\psi \in C_b(E)$. Since (P_t) is Feller, we have $P_t \psi \in C_b(E)$. Thus

$$(\psi, P_t^* \mu) = (P_t \psi, \mu) = \lim_{s \uparrow \infty} (P_t \psi, P_s(x, \cdot)) = \lim_{s \uparrow \infty} (\psi, P_{t+s}(x, \cdot)) = (\psi, \mu).$$

$\square$

Assume now that E is a Banach space. Let $\mathrm{Lip}\,(E)$ be the space of all Lipschitz continuous functions $\psi \colon E \mapsto \mathbb{R}$; see Section 2.4. In the following sections we will show the existence, uniqueness and exponential mixing of an invariant measure for so-called dissipative systems.

Definition 16.3 We say that an invariant measure μ for (P_t) is *exponentially mixing* with *exponent* $\omega > 0$ and *function* $c \colon E \mapsto (0, +\infty)$ if

$$|P_t \psi(x) - (\psi, \mu)| \le c(x) \mathrm{e}^{-\omega t} \|\psi\|_{\mathrm{Lip}}, \qquad \forall x \in E, \ \forall t > 0.$$

Let us equip the space $\mathcal{M}_1(E)$ with the so-called *Fortet–Mourier norm*

$$\|\rho\|_{FM} = \sup \left\{ |(\psi, \rho)| \colon \psi \in \mathrm{Lip}\,(E), \ \|\psi\|_\infty \le 1, \ \|\psi\|_{\mathrm{Lip}} \le 1 \right\}.$$

Then μ is exponentially mixing with exponent ω and function c if and only if

$$\|P_t(x, \cdot) - \mu\|_{FM} \le c(x) \mathrm{e}^{-\omega t}, \qquad \forall t > 0, \ \forall x \in E.$$

It is known (see e.g. Lasota and Yorke 1994) that weak convergence on $\mathcal{M}_1(E)$ is equivalent to convergence in the Fortet–Mourier norm. Thus, if μ is exponentially mixing then, for any $x \in E$, $P_t(x, \cdot)$ converges weakly to μ as $t \uparrow \infty$. In fact one can show (see e.g. Da Prato and Zabczyk 1996) that if μ is exponentially mixing then $P_t(x, \Gamma) \to \mu(\Gamma)$ for any $\Gamma \in \mathcal{B}(E)$.

The following result provides useful conditions for the existence, uniqueness and exponential mixing of an invariant measure. For a different approach see e.g. Lasota and Szarek (2006).

Proposition 16.4 *Assume that (P_t) is a Feller transition semigroup satisfying the following conditions.*

(i) *There is an $x_0 \in E$ such that $P_t(x_0, \cdot)$ converges weakly to a probability measure μ.*

(ii) *There are functions* $c\colon E \mapsto (0, +\infty)$ *and* $\tilde{c}\colon E \times E \mapsto (0, +\infty)$ *and a constant* $\omega > 0$ *such that, for all* $s \geq t \geq 0$, $\psi \in \mathrm{Lip}\,(E)$ *and* $x, \tilde{x} \in E$,

$$|P_t\psi(x) - P_s\psi(x)| \leq c(x)\mathrm{e}^{-\omega t}\,\|\psi\|_{\mathrm{Lip}},$$
$$|P_t\psi(x) - P_t\psi(\tilde{x})| \leq \tilde{c}(x, \tilde{x})\mathrm{e}^{-\omega t}\,\|\psi\|_{\mathrm{Lip}}.$$

Then μ *is a unique invariant measure for* (P_t) *and it is exponentially mixing with exponent* ω *and function* c.

Proof By the Krylov–Bogolyubov theorem, μ is an invariant measure. The fact that it is exponentially mixing follows from (ii). Indeed, let $x \in E$, $t \geq 0$, and $\psi \in \mathrm{Lip}\,(E)$. Then

$$|P_t\psi(x) - (\psi, \mu)| = \lim_{s\uparrow\infty} |P_t\psi(x) - P_s\psi(x_0)|$$
$$\leq \limsup_{s\uparrow\infty} |P_t\psi(x) - P_s\psi(x)| + \limsup_{s\uparrow\infty} |P_s\psi(x) - P_s\psi(x_0)|$$
$$\leq c(x)\mathrm{e}^{-\omega t}\|\psi\|_{\mathrm{Lip}}.$$

The uniqueness of the invariant measure follows from the exponential mixing of μ. Indeed, assume that $\tilde{\mu}$ is an invariant measure. Since μ is exponentially mixing, for every $\psi \in \mathrm{Lip}\,(E)$ and every $x \in E$ we have $P_t\psi(x) \to (\psi, \mu)$ as $t \uparrow \infty$. Hence

$$(\psi, \tilde{\mu}) = (P_t\psi, \tilde{\mu}) \to \int_E (\psi, \mu)\tilde{\mu}(\mathrm{d}x) = (\psi, \mu).$$

$\square$

16.2 Existence results

The next two subsections are concerned with the exponential mixing and uniqueness of an invariant measure for the Markov family defined by the equation

$$\mathrm{d}X = (AX + F(X))\,\mathrm{d}t + G(X)\,\mathrm{d}Z \tag{16.1}$$

driven by a square integrable Lévy process Z. We outline the strategy of the proofs. Let $X(t, x)$ be the value at time t of a solution $X(t, x)$ starting at time 0 from x. Clearly, we have imposed conditions that guarantee the existence and uniqueness of the solution. Taking into account the Krylov–Bogolyubov theorem we would like to show the weak convergence of $P_t(x, \cdot) = \mathcal{L}(X(t, x))$. To do this one can ask whether $X(t, x)$ converges in probability (or in L^2) to a random variable. This is however not true even in the simplest case, that of one-dimensional Ornstein–Uhlenbeck diffusion, for which

$$\mathrm{d}X = -\tfrac{1}{2}X\,\mathrm{d}t + \mathrm{d}W, \qquad X(0) = 0.$$

Indeed, in this case

$$\mathcal{L}(X(t,0)) = \mathcal{L}\left(\int_0^t e^{-(t-s)/2}\,dW(s)\right)$$

$$= \mathcal{L}\left(\int_0^t e^{-s/2}\,dW(s)\right) \Rightarrow \mathcal{L}\left(\int_0^\infty e^{-s/2}\,dW(s)\right) = \mathcal{N}(0,1),$$

but $X(t,0)$ does not converge in probability to any random variable.

To overcome this difficulty we consider a *double-sided Lévy process* $\overline{Z}$, that is, a process defined on $\mathbb{R}$ such that $\overline{Z}(t) = Z(t)$, $t \geq 0$, and $\overline{Z}(-t)$, $t \geq 0$, are independent, identically distributed, Lévy processes. Given $-\infty < t_0 \leq t < +\infty$ and x, let $X(t, t_0, x)$ be the value at time t of the (mild) solution to the equation

$$dX = (AX + F(X))\,dt + G(X)\,d\overline{Z}, \qquad X(t_0) = x. \tag{16.2}$$

From the uniqueness of the solution, $\mathcal{L}(X(t,x)) = \mathcal{L}(X(t_0 + t, t_0, x))$. We will show that, under certain conditions on A, F and G, $X(t_0, 0, x)$ converges in probability as $t_0 \downarrow -\infty$.[1] In this way we obtain the existence of an invariant measure. To show its exponential mixing we will use Proposition 16.4.

16.2.1 Regular case

We are concerned with (16.1), where Z is a square integrable mean-zero Lévy process with RKHS $\mathcal{H}$, $(A, D(A))$ is the generator of a C_0-semigroup S on a Hilbert space H and $F\colon H \mapsto H, G\colon H \mapsto L_{(HS)}(\mathcal{H}, H)$ are Lipschitz continuous mappings. Below, A_n stands for the Yosida approximation of A; see appendix section B.1. The following result is taken from Rusinek (2006a); see also Da Prato and Zabczyk (1996).

Theorem 16.5   *Assume that there is a constant $\omega > 0$ such that, for all $x, y \in H$ and $n \in \mathbb{N}$,*

$$2\langle A_n(x-y) + F(x) - F(y),\ x-y\rangle_H + \|G(x) - G(y)\|^2_{L_{(HS)}(\mathcal{H},H)}$$
$$\leq -\omega\,|x-y|^2_H. \tag{16.3}$$

Then there exists exactly one invariant measure μ for (16.1) and it is exponentially mixing with exponent $\omega/2$ and function c of linear growth.

Proof   Let $\overline{Z}$ be the double-sided Lévy process corresponding to Z. Given $n \in \mathbb{N}$, $t_0 \in \mathbb{R}$ and $x \in H$ we consider the regularized problem

$$dX(t) = \big(A_n X(t) + F(X(t))\big)dt + G(X(t))\,d\overline{Z}(t), \qquad t \geq t_0, \qquad X(t_0) = x,$$

---

[1] For the Ornstein–Uhlenbeck equation this is very simple.

with a straightforward generalization of the stochastic integral. It is easy to show that the regularization equation has a unique solution $X_n(t) = X_n(t, t_0, x)$ and that, since A_n is a bounded linear operator, X_n is a strong solution, that is,

$$X_n(t) = x + \int_{t_0}^t \big(A_n X_n(s) + F(X_n(s))\big)ds + \int_{t_0}^t G(X_n(s)) \, d\overline{Z}(s).$$

Moreover, for each $t \ge t_0$, $X_n(t)$ converges in $L^2(\Omega, \mathcal{F}, \mathbb{P}; H)$ to the unique (mild) solution $X(t, t_0, x)$ of (16.1).

We divide the proof of Theorem 16.5 into three steps.

Step 1   This step consists of the proof of the following estimate:

$$\mathbb{E} \, |X(t, t_0, x)|_H^2 \le C\big(1 + |x|_H^2\big), \qquad \forall t > t_0, t_0 \in \mathbb{R}, \, \forall x \in H. \tag{16.4}$$

To show (16.4) we apply the Itô formula to the semimartingale X_n and the function $\psi(x) = |x|_H^2$. By Lemma D.3,

$$\mathbb{E} \, |X_n(t)|_H^2 = |x|_H^2 + \mathbb{E} \int_{t_0}^t \Big\{ 2\langle X_n(s), \; A_n X_n(s) + F(X_n(s))\rangle_H$$

$$+ \|G(X_n)\|_{L_{(HS)}(U,H)}^2 \Big\} \, ds.$$

Note that

$$2\langle A_n x + F(x), \; x\rangle_H + \|G(x)\|_{L_{(HS)}(\mathcal{H},H)}^2$$

$$\le 2\langle A_n x + F(x) - F(0), \; x\rangle_H + \|G(x) - G(0)\|_{L_{(HS)}(\mathcal{H},H)}^2 + I(x),$$

where

$$I(x) := 2|F(0)|_H \, |x|_H + 2\|G(x) - G(0)\|_{L_{(HS)}(\mathcal{H},H)} \|G(0)\|_{L_{(HS)}(\mathcal{H},H)}$$

$$+ \|G(0)\|_{L_{(HS)}(\mathcal{H},H)}^2.$$

Clearly, for any $\varepsilon > 0$ there is a constant C_ε such that

$$I(x) \le \varepsilon \left(\|G(x) - G(0)\|_{L_{(HS)}(\mathcal{H},H)}^2 + |x|_H^2 \right) + C_\varepsilon.$$

Thus, since G is Lipschitz continuous, there is a constant C_1 such that, for all n and x,

$$2\langle A_n x + F(x), \; x\rangle_H + \|G(x)\|_{L_{(HS)}(\mathcal{H},H)}^2 \le -\frac{\omega}{2}|x|_H^2 + C_1.$$

Hence

$$\mathbb{E} \, |X_n(t)|_H^2 \le |x|_H^2 - \frac{\omega}{2} \int_{t_0}^t \mathbb{E} \, |X_n(s)|_H^2 \, ds + C_1(t - t_0),$$

and consequently, by Gronwall's inequality,

$$\mathbb{E} |X_n(t)|_H^2 \le e^{-\omega(t-t_0)/2} \left(|x|_H^2 + C_1(t-t_0)\right).$$

Letting $n \to \infty$ we obtain (16.4).

Step 2   Recall that $X(t, t_0, x)$ is the value at t of the solution to (16.2). We will show that there is a constant K such that, for all $x \in H$, $t_0 < 0$ and $h > 0$,

$$\mathbb{E} |X(0, t_0, x) - X(0, t_0 - h, x)|_H^2 \le K e^{\omega t_0} \left(1 + |x|_H^2\right), \qquad (16.5)$$

$$\mathbb{E} |X(0, t_0, x) - X(0, t_0, \tilde{x})|_H^2 \le K e^{\omega t_0} |x - \tilde{x}|_H^2. \qquad (16.6)$$

To do this, observe that $X_n(t, t_0 - h, x) = X_n (t, t_0, X_n(t_0, t_0 - h, x))$. Thus, by Lemma D.3,

$$\Delta_n(t, t_0, h, x) := \mathbb{E} |X_n(t, t_0, x) - X_n(t, t_0 - h, x)|_H^2$$

satisfies

$$\Delta_n(t, t_0, h, x) \le \Delta_n(t_0, t_0, h, x) - \omega \int_{t_0}^t \Delta_n(s, t_0, h, x)\, ds$$

and hence, by Gronwall's inequality,

$$\Delta_n(t, t_0, h, x) \le e^{-\omega(t-t_0)} \mathbb{E} |X_n(t_0, t_0 - h, x) - x|_H^2.$$

Since (see Step 1) there is a constant C such that

$$\mathbb{E} |X_n(t_0, t_0 - h, x)|_H^2 \le C\left(1 + |x|_H^2\right), \qquad \forall t_0, h, \ \forall x,$$

we have

$$\Delta_n(t, t_0, h, x) \le e^{-\omega(t-t_0)} 2C\left(1 + |x|_H^2\right).$$

Letting $n \uparrow \infty$ and $t = t_0$ we obtain (16.5). In the same way one can show (16.6).

Step 3   We will show that the assumptions of Proposition 16.4 are satisfied. It follows from (16.5) that $X(0, t_0, x)$ converges in $L^2(\Omega, \mathcal{F}, \mathbb{P}; H)$ as $t_0 \downarrow -\infty$ to a random variable $\tilde{X}$. Therefore $\mathcal{L}(X(-t_0, x)) = \mathcal{L}(X(0, t_0, x))$ converges weakly to $\mu := \mathcal{L}(\tilde{X})$. Now, for $\psi \in \mathrm{Lip}(H)$, $s \ge t \ge 0$ and $x, \tilde{x} \in H$,

$$|P_t\psi(x) - P_s\psi(x)|_H^2 = \left|\mathbb{E}\big(\psi(X(t, x)) - \psi(X(s, x))\big)\right|_H^2$$
$$\le \|\psi\|_{\mathrm{Lip}}^2 \mathbb{E}\big|X(0, -t, x) - X(0, -s, x)\big|^2$$

and

$$|P_t\psi(x) - P_t\psi(\tilde{x})|_H^2 = \left|\mathbb{E}\big(\psi(X(t, x)) - \psi(X(t, \tilde{x}))\big)\right|_H^2$$
$$\le \|\psi\|_{\mathrm{Lip}}^2 \mathbb{E}\big|X(0, -t, x) - X(0, -t, \tilde{x})\big|^2.$$

Hence, by (16.5),

$$|P_t\psi(x) - P_s\psi(x)|_H^2 \le K\mathrm{e}^{-\omega t}\left(1 + |x|_H^2\right)\|\psi\|_{\mathrm{Lip}}^2$$

and, by (16.6),

$$|P_t\psi(x) - P_t\psi(\tilde{x})|_H^2 \le K\mathrm{e}^{-\omega t}\,|x - \tilde{x}|_H^2\,\|\psi\|_{\mathrm{Lip}}^2.$$

$\square$

The assumption that G is a Hilbert–Schmidt operator-valued mapping from the RKHS of Z is rather restrictive. It is not satisfied by composition operators and cylindrical processes on L^2-spaces. However, it has some natural applications in models of mathematical finance (see Theorems 20.19 and 20.20).

16.2.2 Non-Lipschitz case

We now consider a problem with additive noise,

$$\mathrm{d}X = (AX + F(X))\,\mathrm{d}t + \mathrm{d}Z, \qquad X(0) = X_0, \qquad (16.7)$$

driven by a square integrable mean-zero Lévy process taking values in a Hilbert space U. However, as in Chapter 10 we are dealing here with a general non-Lipschitz drift F.

Let Z_A be a solution to the linear problem

$$\mathrm{d}Y = AY\,\mathrm{d}t + \mathrm{d}Z, \qquad Y(0) = 0. \qquad (16.8)$$

Theorem 16.6 *If the assumptions of Theorem 10.1 are satisfied and in addition*

(i) *the mappings $A + \omega_1$ and $F + \omega_2$ are dissipative with $\omega_1 + \omega_2 = \omega > 0$,*
(ii) $\sup\limits_{t \ge 0} \mathbb{E}\left(|Z_A(t)|_H + |F(Z_A(t))|_H\right) < \infty$,

then there exists a unique invariant measure μ for (16.7) and it is exponentially mixing with exponent ω and function c of linear growth.

Proof Given a càdlàg $f\colon [0, \infty) \mapsto B$ and $x \in H$, consider the following deterministic equation:

$$y(t) = S(t)x + \int_0^t S(t - s)F(y(s) + f(s-))\,\mathrm{d}s, \qquad t \ge 0. \qquad (16.9)$$

As in the proof of Theorem 10.14 we consider the approximate problem

$$\frac{d^-}{dt} y_{\alpha\beta}(t) = (A + \omega_1)_\beta y_{\alpha\beta} + (F + \omega_2)_\alpha(y_{\alpha\beta}(t) + f(t-))$$
$$- (\omega_1 + \omega_2) y_{\alpha\beta}(t) - \omega_2 f(t-), \qquad t \geq 0,$$
$$y_{\alpha\beta}(0) = x.$$

Step 1   The first step consists of proving the estimate

$$|y(t, x)|_H \leq e^{-\omega t}|x|_H + \int_0^t e^{-\omega(t-s)}\big(|F(f(s-))|_H + \gamma|f(s-)|_H\big)\, ds, \quad (16.10)$$

for the solution y to (16.9), where $\gamma = \omega_2 + |\omega_2|$.

To show (16.10) note that by Proposition 10.1 and the elementary properties of dissipative mappings (see Section 10.1),

$$\frac{d^-}{dt}|y_{\alpha\beta}(t)|_H$$

$$= \left\langle \frac{d^-}{dt} y_{\alpha\beta}(t), \frac{y_{\alpha\beta}(t)}{|y_{\alpha\beta}(t)|}\right\rangle_H$$

$$= \frac{1}{|y_{\alpha\beta}(t)|_H}\big\langle (A + \omega_1)_\beta y_{\alpha\beta}(t) + (F + \omega_2)_\alpha(y_{\alpha\beta}(t) + f(t-)),\ y_{\alpha\beta}(t)\big\rangle_H$$

$$- \frac{\omega_1 + \omega_2}{|y_{\alpha\beta}(t)|_H}\langle y_{\alpha\beta}(t), y_{\alpha\beta}(t)\rangle_H - \frac{\omega_2}{|y_{\alpha\beta}(t)|_H}\langle f(t-), y_{\alpha\beta}(t)\rangle_H.$$

Thus

$$\frac{d^-}{dt}|y_{\alpha\beta}(t)|_H$$

$$\leq \frac{1}{|y_{\alpha\beta}(t)|_H}\big\langle (F + \omega_2)_\alpha(y_{\alpha\beta}(t) + f(t-)) - (F + \omega_2)_\alpha(f(t-)),\ y_{\alpha\beta}(t)\big\rangle_H$$

$$+ \frac{1}{|y_{\alpha\beta}(t)|_H}\big\langle (F + \omega_2)_\alpha(f(t-)),\ y_{\alpha\beta}(t)\big\rangle_H$$

$$- (\omega_1 + \omega_2)|y_{\alpha\beta}(t)|_H - \frac{\omega_2}{|y_{\alpha\beta}(t)|_H}\langle f(t-), y_{\alpha\beta}(t)\rangle_H$$

$$\leq -(\omega_1 + \omega_2)|y_{\alpha\beta}(t)|_H + |(F + \omega_2)_\alpha(f(t-))|_H + \omega_2|f(t-)|_H$$

$$\leq -(\omega_1 + \omega_2)|y_{\alpha\beta}(t)|_H + |(F + \omega_2)(f(t-))|_H + \omega_2|f(t-)|_H$$

$$\leq -(\omega_1 + \omega_2)|y_{\alpha\beta}(t)|_H + |F(f(t-)|_H + (\omega_2 + |\omega_2|)|f(t-)|_H.$$

Letting $\beta \to 0$ and then $\alpha \to 0$ we arrive at (16.10).

Step 2   We show that if $y(t, x)$, $y(t, \tilde{x})$ are solutions to (16.9) with initial conditions $x, \tilde{x} \in H$ then

$$|y(t, x) - y(t, \tilde{x})|_H \leq e^{-\omega t}|x - \tilde{x}|_H, \qquad t \geq 0. \qquad (16.11)$$

We can assume that $x, \tilde{x} \in B$, where the space B appears in the formulation of Theorem 10.14. By calculations similar to those in the Step 1 we obtain

$$\frac{d^-}{dt}\left|y_{\alpha\beta}(t, x) - y_{\alpha\beta}(t, \tilde{x})\right|_H \leq -\omega\left|y_{\alpha\beta}(t, x) - y_{\alpha\beta}(t, \tilde{x})\right|_H,$$

where $y_{\alpha\beta}(\cdot, x)$ and $y_{\alpha\beta}(\cdot, \tilde{x})$ are solutions to the corresponding approximating problems. Passing to the limits we obtain (16.11).

Step 3 Let $\overline{Z}$ be a double-sided Lévy process corresponding to Z. Fix an $x \in H$ and, for each $a > 0$, denote by $X^a(t, x)$, $t \geq -a$, the unique generalized solution of the equation

$$dX = (AX + F(X))\,dt + d\overline{Z}, \qquad X(-a) = x.$$

Let

$$\overline{Z}_A^a(t) := \int_{-a}^{t} S(t - s)\,d\overline{Z}(s), \qquad t \geq -a.$$

For $t \geq -a \geq -b$ we have $X^b(t, x) = X^a\big(t, X^b(-a, x)\big)$. But

$$X^a\big(t, X^b(-a, x)\big) = \overline{Z}_A^a(t) + y(t + a),$$

where $y(s)$, $s \geq 0$, is a generalized solution to

$$\frac{d^-}{ds}y(s) = Ay(s) + F(y(s) + f(s-)), \qquad y(0) = X^b(-a, x),$$

with

$$f(s) = \int_{-a}^{-a+s} S(-a + s - \sigma)\,d\overline{Z}(\sigma), \qquad s \geq 0.$$

By Step 2

$$\left|X^a(t, x) - X^b(t, x)\right|_H = \left|X^a(t, x) - X^a\big(t, X^b(-a, x)\big)\right|_H$$
$$\leq e^{-\omega(t+a)}\big(|x|_H + \left|X^b(-a, x)\right|_H\big)$$

and by Step 1

$$|X^b(-a, x)|_H$$
$$\leq \left|\overline{Z}_A^b(-a)\right|_H + e^{-\omega(b+a)}|x|_H$$
$$+ \int_0^{a+b} e^{-\omega(a+b-\sigma)}\left|F\left(\int_{-b}^{-b+\sigma} S(-b + \sigma - u)\,d\overline{Z}(u)\right)\right|_H d\sigma$$
$$+ \gamma \int_0^{a+b} e^{-\omega(a+b-\sigma)}\left|\int_{-b}^{-b+\sigma} S(-b + \sigma - u)\,d\overline{Z}(u)\right|_H d\sigma.$$

Consequently,

$$\mathbb{E}\left|X^a(0, x) - X^b(0, x)\right|_H$$

$$\leq e^{-\omega a}\left(2|x|_H + \mathbb{E}\left|Z_A(b+a)\right|_H\right.$$

$$\left. + \int_0^{b+a} e^{-\omega(b+a-\sigma)}\,\mathbb{E}\big(|F(Z_A(\sigma))|_H + \gamma\,|Z_A(\sigma)|_H\big)\mathrm{d}\sigma\right)$$

$$\leq e^{-\omega a}\left[2|x|_H + \sup_{t\geq 0}\mathbb{E}\left(\frac{\omega+\gamma}{\omega}|Z_A(t)|_H + \frac{1}{\omega}|F(Z_A(t))|_H\right)\right]$$

$$\leq e^{-\omega a}\left(2|x|_H + C\right),$$

where

$$C = \sup_{t\geq 0}\mathbb{E}\left(\frac{\omega+\gamma}{\omega}|Z_A(t)|_H + \frac{1}{\omega}|F(Z_A(t))|_H\right).$$

From the estimate above it is clear that $X^a(0, x)$ converges in $L^1(\Omega, \mathcal{F}, \mathbb{P}; H)$ as $a \to +\infty$ to some $\tilde{X} \in L^1(\Omega, \mathcal{F}, \mathbb{P}; H)$. Let us denote by $\mathcal{L}(X(a, x))$, $\mathcal{L}(X^a(0, x))$ and $\mathcal{L}(\tilde{X})$ the laws of the corresponding random elements. Then, in particular, $\mathcal{L}(X(a, x)) = \mathcal{L}\big(X^a(0, x)\big) \Rightarrow \mathcal{L}(\tilde{X})$ as $a \to +\infty$. From this it follows easily that $\mu = \mathcal{L}(\tilde{X})$ is an invariant measure for (16.7).

In order to show the uniqueness of the invariant measure and the exponential convergence, note that $P_t(x, \cdot) = \mathcal{L}(X^t(0, x)) = \mathcal{L}(X(t, x))$. Moreover, for $s \geq t \geq 0$,

$$|P_t\phi(x) - P_s\phi(x)| = \left|\mathbb{E}\big(\phi(X(t, x)) - \phi(X(s, x))\big)\right|$$

$$= \left|\mathbb{E}\left(\phi\left(X^t(0, x)\right) - \phi\left(X^s(0, x)\right)\right)\right|$$

$$\leq \|\phi\|_{\mathrm{Lip}}\,\mathbb{E}\left|X^t(0, x) - X^s(0, x)\right|_H$$

$$\leq \|\phi\|_{\mathrm{Lip}}\,e^{-\omega t}\left(2|x|_H + C\right).$$

In a similar way we can show that

$$|P_t\phi(x) - P_t\phi(\tilde{x})| = |\mathbb{E}\,\phi(X(t, x)) - \mathbb{E}\,\phi(X(t, \tilde{x}))|$$

$$= \left|\mathbb{E}\left(\phi\left(X^t(0, x)\right) - \phi\left(X^t(0, \tilde{x})\right)\right)\right|$$

$$\leq \|\phi\|_{\mathrm{Lip}}\,\mathbb{E}\left|X^t(0, x) - X^t(0, \tilde{x})\right|_H$$

$$\leq \|\phi\|_{\mathrm{Lip}}\,e^{-\omega t}|x - \tilde{x}|_H.$$

This, by means of Proposition 16.4, proves the desired result. $\qquad\square$

16.3 Invariant measures for the reaction–diffusion equation

As an application of Theorem 16.6 we will show the existence and uniqueness of an invariant measure for the stochastic reaction–diffusion problem

$$dX(t, \xi) = \big(\Delta X(t, \xi) - X^3(t, \xi) + \gamma X(t, \xi))\big) \, dt + dZ(t, \xi),$$
$$X(t, \xi) = 0, \qquad \xi \in \partial \mathcal{O}, \tag{16.12}$$
$$X(0, \xi) = x(\xi), \qquad \xi \in \mathcal{O},$$

where $\mathcal{O}$ is an open bounded subset of $\mathbb{R}^d$, $\gamma \in \mathbb{R}$ and Z is a Lévy process in $L^2 = L^2(\mathcal{O})$. We consider (16.12) as an evolution equation on $H = L^2(\mathcal{O})$ with $F(x)(\xi) = -x^3(\xi) + \gamma x(\xi)$ and A equal to the Laplace operator Δ on $L^2(\mathcal{O})$ with Dirichlet boundary conditions. Note that

$$\langle \Delta \psi, \psi \rangle_{L^2(\mathcal{O})} \leq -C_{\mathcal{O}} |\psi|_{L^2}, \qquad \psi \in D(\Delta), \tag{16.13}$$
$$\langle F(\psi), \psi \rangle_{L^2(\mathcal{O})} \leq \gamma |\psi|_{L^2}, \qquad \psi \in D(F), \tag{16.14}$$

where $C_{\mathcal{O}} = \big(\omega_d^{-1} \ell_d(\mathcal{O})\big)^{2/d}$ is the constant appearing in the Poincaré inequality with $p = 2$; see Theorem 2.2 and Example B.26. We denote by Δ_p the generator of the heat semigroup $S = (S(t), \, t \geq 0)$, with Dirichlet boundary conditions, on $L^p := L^p(\mathcal{O})$.

Theorem 16.7   *Let $Z = W + L$, where W is a Wiener process and L is a Lévy martingale. Assume that*

(i) *the process $W_A(t) = \int_0^t S(t - s) \, dW(s)$, $t \geq 0$, has continuous trajectories in L^{18} and $\sup_{t \geq 0} \mathbb{E} \, |W_A(t)|_{L^{18}}^3 < \infty$,*
(ii) *L is a Lévy process in L^2 with càdlàg trajectories in the domain of $(\Delta_{18} - \kappa I)^\alpha$ for some $\alpha > 0$ and representation $L(t) = \int_0^t \int_{L^2} x \widehat{\pi}(dt, dx)$, where the intensity measure ν of π satisfies*

$$\int_{L^2(\mathcal{O})} \big(|x|_{L^{18}}^3 + |x|_{L^{18}}\big) \, \nu(dx) < \infty,$$

(iii) *$\omega := C_{\mathcal{O}}^{-1} - \gamma > 0$.*

Then there is a unique invariant measure μ for (16.12) and it is exponentially ergodic with exponent ω and function c of linear growth.

Proof   We have to verify the hypotheses of Theorems 10.14 and 16.6 with $B = L^6(\mathcal{O})$ and $F(x)(\xi) = -x^3(\xi) + \gamma x(\xi)$. As in the proof of Theorem 10.15 we verify the assumptions of Theorem 10.14. Condition (i) of Theorem 16.6 follows from (16.13), (16.14), Example 10.8, Proposition 10.11 and Lemma 10.17. We show below that part (ii) is a consequence of Proposition 6.7. To prove that

$\sup_{t\geq 0} \mathbb{E}|F(L_A(t))|_{L^6} < \infty$ we need to show that $\sup_{t\geq 0} I(t) < \infty$, where

$$I(t) := \mathbb{E}\,|L_A(t)|^3_{L^{18}} = \mathbb{E}\left|\int_0^t \int_{L^2} S(t-s)x\widehat{\pi}(ds, dx)\right|^3_{L^{18}}.$$

By Proposition 6.7, there is a polynomial R such that

$$I(t) = R(I_1(t), I_2(t), I_3(t)),$$

where

$$I_j(t) =: \int_0^t \int_{L^2} |S(t-s)x|^j_{L^{18}}\, ds\, \nu(dx), \qquad j = 1, 2, 3.$$

For $t > 1$,

$$I_j(t) = I_j(1) + \int_1^t \int_{L^2} |S(s)x|^j_{L^{18}}\, ds\, \nu(dx)$$

$$\leq I_j(1) + \int_1^t \int_{L^2} \|S(1)\|^j_{L(L^2, L^{18})}|x|^j_{L^2}\|S(s-1)\|^j_{L(L^2, L^2)}\, ds\, \nu(dx).$$

From (16.13), $\|S(s)\|_{L(L^2, L^2)} \leq e^{-C_0 s}$, $s \geq 0$. By Theorem B.7, $S(1)$ is a bounded operator from L^2 to L^{18}. Therefore $\sup_{t\geq 0} I^j(t) < \infty$, $j = 1, 2, 3$, as required. □

Remark 16.8 If the jump measure of π satisfies $\int_{L^2} |x|_{W^{\alpha,18}(\mathcal{O})}\nu(dx) < \infty$ then, by Proposition 6.7, L is a Lévy process (and consequently càdlàg) in $D(\Delta_{18}-\kappa I)^\alpha$.

Example 16.9 Let $\mathcal{O} = [a, b]$ be a bounded interval in $\mathbb{R}$, let W be a cylindrical Wiener process on L^2, or any Wiener process with RKHS contained in $L^2(a, b)$. Assume that $L(t) = \sum_{k=1}^N f_k L_k$, where the L_k are real-valued square integrable Lévy martingales and $\{f_k\} \subset C^1_c(a, b)$. Then the assumptions of Theorem 16.7 are satisfied.

17

Lattice systems

In this chapter, the existence and the long-time behavior of solutions to systems of equations on the lattice $\mathbb{Z}^d$ are investigated. The connection with Gibbs measures is shown.

17.1 Introduction

Consider the following system of equations on the lattice $\mathbb{Z}^d$:

$$dX_k = \left(\sum_{j \in \mathbb{Z}^d} a_{k,j} X_j + f(X_k) \right) dt + g(X_k) dL_k + r(X) dP_k, \qquad k \in \mathbb{Z}^d,$$

$$(17.1)$$

where $a_{k,l} \in \mathbb{R}$ for $k, j \in \mathbb{Z}^d$, $f \colon \mathbb{R} \mapsto \mathbb{R}$, $g \colon \mathbb{R} \mapsto M(2 \times 1) \equiv L(\mathbb{R}^2, \mathbb{R})$, $r \colon \mathbb{R} \mapsto \mathbb{R}$ and L_k, P_k are Lévy processes in $\mathbb{R}^2$ and $\mathbb{R}$, respectively. We assume that, for each $k \in \mathbb{Z}^d$, $L_k = (W_k, Z_k)$ where W_k, Z_k and P_k are independent, W_k is a real-valued Wiener processes with covariance $q_k^2 = \mathbb{E} \, W_k^2(1)$ and

$$\mathbb{E} \, e^{iZ_k(t)z} = e^{-t\psi_k(z)}, \qquad \mathbb{E} \, e^{iP_k(t)z} = e^{-t\psi_k^0(z)}.$$

Here

$$\psi_k(z) = \int_{\mathbb{R}} \left(1 - e^{izy} - izy \right) \nu_k(dy),$$

$$\psi_k^0(z) = \int_{\mathbb{R}} \left(1 - e^{izy} \right) \mu_k(dy)$$

and $\int_{\mathbb{R}} |y|^2 \nu_k(dy) < \infty$, $\mu_k(\mathbb{R}) < \infty$. We will call (q_k, ν_k, μ_k) the *characteristics* of (L_k, P_k). Elements of the matrix $(a_{k,j})$ describe global interactions between the sites of the lattice, and a primitive $\nu(x)$ of f is the so-called *potential of local interactions*.

299

We are looking for solutions to (17.1) taking values in weighted l^p-spaces. Namely, given a sequence $\rho = (\rho_k)$ of positive numbers, we denote by l^p_ρ the space of sequences $x = (x_k)$ of real numbers such that

$$|x|_{l^p_\rho} = \left(\sum_{k \in \mathbb{Z}^d} \rho_k x_k^p \right)^{1/p} < \infty.$$

Note that if $\rho \equiv 1$ then $l^2_\rho = l^2$ is the classical Hilbert space of sequences.

Denote by A the linear operator on l^p_ρ defined by the matrix $(a_{k,j})$, that is,

$$(Ax)_k := \sum_{j=1}^{\infty} a_{k,j} x_j, \qquad k \in \mathbb{Z}^d, \; x \in l^p_\rho. \tag{17.2}$$

Let F be given by $F(x)_k := f(x_k)$, $k \in \mathbb{Z}^d$, $x \in l^p_\rho$, let

$$G(x)[\mathbf{z}]_k := g_k(x_k)\big(z^1_k, z^2_k\big), \qquad x \in l^p_\rho, \; \mathbf{z} = \big(z^1_k, z^2_k\big) \in \big(\mathbb{R}^2\big)^{\mathbb{Z}^d}, \; k \in \mathbb{Z}^d,$$

and let $R(x)[z]_k := r_k(x_k)z$, $x \in l^p_\rho$, $z \in \mathbb{R}$. Finally, let $L := (L_k)$, $P := (P_k)$. We will treat (17.1) as a stochastic evolution equation

$$dX = (AX + F(X)) \, dt + G(X) \, dL + R(X) \, dP, \tag{17.3}$$

of the type investigated in Section 9.7.

The chapter is organized as follows. First we study the properties of the operator A. Then we formulate existence results for f and g Lipschitz and also for f monotone and g constant. Finally, we show the existence and exponential mixing of invariant measures for (17.1) and discuss physical applications.

17.2 Global interactions

We will now show that, under rather general conditions on the matrix $(a_{k,j})$, the corresponding operator A is bounded on an appropriate weighted space.

Theorem 17.1 *Let $\rho = (\rho_k)$ be a sequence of positive numbers. Assume that $\sup_{k \in \mathbb{Z}^d} \sum_{j \in \mathbb{Z}^d} |a_{k,j}| = \alpha < \infty$ and that there is a constant $\beta > 0$ such that*

$$\sum_k |a_{k,j}| \rho_k \le \beta \rho_j, \qquad \forall \, j \in \mathbb{Z}^d.$$

Then, for any $p \in (1, \infty)$, $A \in L\big(l^p_\rho, l^p_\rho\big)$ with operator norm less than or equal to $\alpha^{1/q} \beta^{1/p}$, where $1/p + 1/q = 1$. In particular $A \in L\big(l^2_\rho, l^2_\rho\big)$, with operator norm less than or equal to $\sqrt{\alpha\beta}$.

Proof By the Hölder inequality,

$$\left(\sum_j |a_{k,j}||x_j|\right)^p \le \left(\sum_j \left(|a_{k,j}|^{1/p}|x_j|\right)|a_{k,j}|^{1/q}\right)^p \le \alpha^{p/q}\sum_j |a_{k,j}||x_j|^p$$

and, consequently,

$$\sum_k \rho_k \left|\sum_{j\in\mathbb{Z}^d} a_{k,j}x_j\right|^p \le \alpha^{p/q}\sum_{k,j}\rho_k|a_{k,j}||x_j|^p \le \alpha^{p/q}\beta\sum_j \rho_j|x_j|^p.$$

$\square$

Corollary 17.2 *If, for some $\alpha < \infty$,*

$$\sup_k \sum_j |a_{k,l}| \le \alpha \qquad and \qquad \sup_j \sum_k |a_{k,j}| \le \alpha \qquad (17.4)$$

then $A \in L(l^p, l^p)$ with norm $\le \alpha$.

Definition 17.3 The interactions $a_{k,j}$ have a *bounded range* if there are constants $R, M > 0$ such that

$$a_{k,j} = 0 \quad \text{if } |k-j| > R, \qquad |a_{k,j}| \le M \text{ for all } k, j \in \mathbb{Z}^d. \qquad (17.5)$$

Lemma 17.4 *Assume (17.5) and that*

$$\frac{\rho_k}{\rho_j} \le M \quad \text{if } |k-j| \le R, \qquad \sum_{k\in\mathbb{Z}^d}\rho_k < \infty. \qquad (17.6)$$

Then $A \in L\left(l_\rho^p, l_\rho^p\right)$ for every $p \ge 1$.

Proof We show that the assumptions of Theorem 17.1 are satisfied. Note that

$$\sum_k |a_{k,j}|\rho_k = \sum_k |a_{k,j}|\frac{\rho_k}{\rho_j}\rho_j \le M\left(\sum_k |a_{k,j}|\right)\rho_j \le \beta\rho_j,$$

where $\beta = M\sup_j \sum_k |a_{k,j}|$. $\square$

Lemma 17.5 *Condition (17.6) is satisfied for weights as follows:*

$$either \quad \rho_k := \rho_k^\kappa := e^{-\kappa|k|} \quad or \quad \rho_k = \rho_k^{\kappa,r} = \frac{1}{1+\kappa|k|^r},$$

$$\kappa > 0, \ r > d. \qquad (17.7)$$

Proof For the weight ρ^κ we have $\rho_k/\rho_j = e^{-\kappa|k|+\kappa|j|}$. Since $\kappa(|j|-|k|) \le$

$\kappa|j - k|$, $\rho_k/\rho_j \leq e^{\kappa|j-k|} \leq e^{\kappa R}$ for $|k - j| \leq R$. For the weight $\rho^{\kappa,r}$,

$$\frac{\rho_k}{\rho_j} = \frac{1 + \kappa|j|^r}{1 + \kappa|k|^r} = \frac{1 + \kappa(|k + j - k|)^r}{1 + \kappa|k|^r} \leq \frac{1 + \kappa(|k| + |j - k|)^r}{1 + \kappa|k|^r}$$

$$\leq \frac{1 + \kappa 2^{r-1}(|k|^r + |j - k|^r)}{1 + \kappa|k|^r} \leq \frac{1 + \kappa 2^{r-1}|k|^r}{1 + \kappa|k|^r} + \frac{1 + \kappa 2^{r-1}R^r}{1 + \kappa|k|^r}$$

$$\leq 2^{r-1}(1 + \kappa R^r) + 1.$$

$\square$

An operator A with important physical applications is

$$A = P - (1 + \alpha)I, \tag{17.8}$$

where α is a positive constant and P is a symmetric transition matrix. In particular, of special interest are the operators $A = \Delta_d - \alpha I$, where Δ_d is the so-called *discrete Laplacian*. To define Δ_d consider the transition matrix P_d of a symmetric random walk on a d-dimensional lattice $\mathbb{Z}^d$ with entries

$$P_{(k_1,...,k_d),(j_1,...,j_d)} := \begin{cases} (2d)^{-1} & \text{if } \sum_{i=1}^d |k_i - j_i| = 1, \\ 0 & \text{if } \sum_{i=1}^d |k_i - j_i| > 1. \end{cases}$$

Then $\Delta_d := P_d - I$. It follows from the proposition below that if the matrix A is of the form (17.8) then the operator $A + \alpha I$ is dissipative on l^2, which is an essential property required for the ergodicity of system (17.1).

Proposition 17.6   *Let P be a symmetric transition matrix acting as a linear operator on l^2. Then $\langle Px, x \rangle_{l^2} \leq |x|_{l^2}^2$ for all $x \in l^2$.*

Proof   We will use the Frobenius–Perron theorem (see e.g. *Encyclopedia of Mathematics* 1987). It says that an arbitrary non-zero quadratic finite matrix P, with non-negative entries, has a non-negative eigenvalue $\hat{\lambda}$ greater than or equal to the moduli of any of the remaining eigenvalues. Moreover, one can choose an eigenvector $\hat{x}$, corresponding to $\hat{\lambda}$, with non-negative coordinates.

Assume first that $P = p_{i,j} \in M(n \times n)$ is a symmetric transition matrix. We show that $\sup_{|x|_{\mathbb{R}^n} \leq 1} \langle Px, x \rangle_{\mathbb{R}^n} = 1$. Let $P\hat{x} = \hat{\lambda}\hat{x}$. Hence $\sum_j p_{ij}\hat{x}_j = \hat{\lambda}\hat{x}_i$ and consequently, since P is stochastic,

$$\sum_{i,j} p_{i,j}\hat{x}_j = \sum_j \hat{x}_j = \hat{\lambda} \sum_i \hat{x}_i.$$

Thus $\hat{\lambda} = 1$. But $\sup_{|x|_{\mathbb{R}^n} \leq 1} \langle Px, x \rangle_{\mathbb{R}^n} = \hat{\lambda}$, and the desired identity follows.

If P is an infinite matrix, fix a finite set Γ of coordinates and let x^Γ and P^Γ be the restrictions of x and P to the set Γ. Then

$$\langle P^\Gamma x^\Gamma, x^\Gamma \rangle_{l^2} = \sum_{k,j \in \Gamma} p_{k,j} x_k x_j.$$

Thus $\langle Px^\Gamma, x^\Gamma \rangle_{l^2} = \langle P^\Gamma x^\Gamma, x^\Gamma \rangle_{l^2}$ and consequently $\langle P^\Gamma x^\Gamma, x^\Gamma \rangle_{l^2} \le |x^\Gamma|_{l^2}^2$. An elementary approximation gives the result. □

Proposition 17.7   *If A is given by (17.8) then $\langle Ax, x \rangle_{l^2} \le -\alpha |x|_{l^2}^2$.*

Proof   By Proposition 17.6,

$$\sup_{|x|_{l^2} \le 1} \langle Ax, x \rangle_{l^2} = \sup_{|x|_{l^2} \le 1} \langle (P - (1 + \alpha)I)x, \ x \rangle_{l^2} \le 1 - (\alpha + 1).$$

□

17.3 Regular case

In this section we consider the Lipschitz coefficients f, g and b.

17.3.1 Existence of solution to (17.1)

Theorem 17.8   *Let $\rho = (\rho_k)$ be as in (17.7). Assume that*

  (i) *the matrix $A = (a_{k,j})$ of local interactions has a bounded range,*
 (ii) *f, g and r are Lipschitz functions,*
(iii) *Z_k and P_k are independent, with characteristics (q_k, v_k, μ_k) satisfying*

$$\sum_{k \in \mathbb{Z}^d} \left(q_k^2 \rho_k + \int_\mathbb{R} |y|^2 v_k^\rho(dy) + \mu_k(\mathbb{R}) \right) < \infty, \qquad (17.9)$$

where $v_k^\rho(\Gamma) = v_k(\Gamma / \rho_k^{1/2})$, $\Gamma \in \mathcal{B}(\mathbb{R})$, $k \in \mathbb{Z}^d$.

Then for any $x \in l_\rho^2 := l_\rho^2(\mathbb{Z}^d)$ there is a unique solution to (17.1) in l_ρ^2 satisfying $X(0) = x$. Moreover, (17.1) defines a Feller family on l_ρ^2.

Proof   Note that $W = (W_k)$, $Z = (Z_k)$ and $P = (P_k)$ are independent Lévy processes in l_ρ^2. Moreover, Z is a square integrable martingale and P is a compound Lévy process. These properties follow from Theorem 4.40 and (17.9) since $\mathbb{E}\,|W(t)|_{l_\rho^2}^2 = t \sum_k q_k^2 \rho_k < \infty$. Next, let $\{\delta_k\}$ be the canonical basis of l^2. Then $\{e_k\}$, $e_k = \delta_k \rho_k^{-1/2}$, is an orthonormal basis of l_ρ^2. We have $Z(t) = \sum_k Z_k^\rho(t)e_k$ and $P(t) = \sum_k P_k^\rho(t)e_k$, where $Z_k^\rho(t) = Z_k(t)\rho_k^{1/2}$ and $P_k^\rho(t) = P_k(t)\rho_k^{1/2}$. Then $\mathbb{E}\,e^{iZ_k^\rho(t)z} = e^{-t\psi_k^\rho(z)}$ and $\mathbb{E}\,e^{iP_k^\rho(t)z} = e^{-t\phi_k^0(z)}$, where

$$\psi_k^\rho(z) = \int_\mathbb{R} \left(1 - e^{izy} - izy \right) v_k^\rho(dy),$$

$$\phi_k^0(z) = \int_\mathbb{R} \left(1 - e^{izy} \right) \mu_k^\rho(dy), \qquad \mu_k^\rho(\Gamma) = \mu_k(\Gamma / \rho_k^{1/2}).$$

By Lemmas 17.4 and 17.5, A is a bounded operator on l_ρ^2 and hence generates a C_0-semigroup on l_ρ^2 Thus the result follows from the (easy to prove) fact that $F: l_\rho^2 \mapsto l_\rho^2$ and $G, R: l_\rho^2 \mapsto L(l_\rho^2, l_\rho^2)$ are Lipschitz continuous.   □

Using the same arguments one can show the following result.

Theorem 17.9   *Assume that*

(i) $A = (a_{k,j})$ is given by (17.8),
(ii) f, g and r are Lipschitz functions and $f(0) = 0$,
(iii) L_k and P_k are independent with characteristics (q_k, ν_k, μ_k) satisfying

$$\sum_{k \in \mathbb{Z}^d} \left(q_k^2 + \int_\mathbb{R} |y|^2 \nu_k(dy) + \mu_k(\mathbb{R}) \right) < \infty. \qquad (17.10)$$

Then for any $x \in l^2 := l^2(\mathbb{Z}^d)$ there is a unique solution to (17.1) in l^2 satisfying $X(0) = x$. Moreover, (17.1) defines a Feller family on l^2.

17.3.2 Ergodicity

Assume that $r \equiv 0$ and (17.9) holds. Then the RKHS of (W, L) is equal to $\mathcal{H}_W \times \mathcal{H}_Z$, where

$$\mathcal{H}_W := \left\{ (x_k) \in l_\rho^2 : \sum_k x_k^2 q_k^{-2} < \infty \right\}, \qquad \langle (x_k), (y_k) \rangle_{\mathcal{H}_W} = \sum_k x_k y_k q_k^{-2},$$

$$\mathcal{H}_Z := \left\{ (x_k) \in l_\rho^2 : \sum_k x_k^2 p_k^{-2} < \infty \right\}, \qquad \langle (x_k), (y_k) \rangle_{\mathcal{H}_W} = \sum_k x_k y_k p_k^{-2}$$

and $p_k^2 := \int_{\{|y| \le 1\}} |y|^2 \nu_k(dy)$. Then

$$\langle F(x) - F(y), \ x - y \rangle_{l^2} + \|G(x) - G(y)\|_{L_{(HS)}(\mathcal{H}, l_\rho^2)}^2$$
$$= \sum_k \rho_k \left((f(x_k) - f(y_k))(x_k - y_k) + \langle g(x_k) - g(y_k), (q_k, p_k) \rangle_{\mathbb{R}^2}^2 \right).$$

By Theorem 16.5 we have the following.

Theorem 17.10

(i) *Assume that the hypotheses of Theorem 17.8 are satisfied and that there is a constant $\omega > 0$ such that, for all $x, y \in l_\rho^2$,*

$$\langle A(x - y) - x - y \rangle_{l_\rho^2} + \langle F(x) - F(y), x - y \rangle_{L_\rho^2}$$
$$+ \|G(x) - G(y)\|_{L_{(HS)}(\mathcal{H}, l_\rho^2)}^2 \le -\omega |x - y|_{l_\rho^2}^2.$$

Then there is a unique invariant measure μ for (17.1) considered on L_ρ^2 and
it is exponentially mixing with exponent $\omega/2$.

(ii) Assume that the hypotheses of Theorem 17.9 are satisfied and that

$$\omega = \alpha - \|f\|_{\text{Lip}} - \|g\|_{\text{Lip}}^2 \sum_k \left(q_k^2 + p_k^2\right) > 0.$$

Then there is a unique invariant measure μ for (17.1) considered on L^2 and
it is exponentially mixing with exponent $\omega/2$.

17.4 Non-Lipschitz case

Assume that $r \equiv 0$ and $g(x)(z_1, z_2) = z_1 + z_2$.

Theorem 17.11 *Assume that $f(z) = f_0(z) + f_1(z)$, $z \in \mathbb{R}$, where f_0 is a Lipschitz continuous function and f_1 is a decreasing function such that, for some
$C > 0$ and $N \in \mathbb{N}$,*

$$|f_1(z)| \le C\left(1 + |z|^N\right), \qquad \forall z \in \mathbb{R}.$$

*Assume that the matrix $A = (a_{k,j})$ of local interactions has a bounded range, that
the characteristics (q_k, v_k) satisfy $(q_k) \in l^\infty$ and that*

$$\sum_{k \in \mathbb{Z}^d} \int_{\mathbb{R}} \left(|y| + |y|^{N^2}\right) v_k^\rho(\mathrm{d}y) < \infty, \qquad (17.11)$$

*where $v_k^\rho(\Gamma) = v_k\left(\Gamma/\rho_k^{1/2}\right)$, $\Gamma \in \mathcal{B}(\mathbb{R})$, $k \in \mathbb{Z}^d$. Then for any $x \in l_\rho^2$ there is a
unique generalized solution to (17.1) in l_ρ^2 satisfying $X(0) = x$, and for any $x \in l_\rho^N$
there is a unique càdlàg solution to (17.1) in l_ρ^N. Moreover, (17.1) defines a Feller
family on l_ρ^2 and on l_ρ^N.*

Proof We have to check the assumptions of Theorem 10.14 with $H = l_\rho^2$ and $B = l_\rho^N$. By Example 10.7, the composition operator corresponding to $f = f_0 + f_1$ is
almost m-dissipative on H and B. By Lemmas 17.4 and 17.5, A is a bounded
operator on H and B. Since $(q_k) \in l^\infty$, W takes values in any l_ρ^p-space, $p < \infty$.
By (17.11) and Proposition 6.9, L is a Lévy process in $l_\rho^{N^2}$ and, as A is a bounded
operator,

$$Z_A(t) = \int_0^t e^{A(t-s)} \, \mathrm{d}(W(s) + L(s)), \qquad t \ge 0,$$

is càdlàg in $l_\rho^{N^2}$ and $\mathbb{E} \int_0^T |F(Z_A(t))|_B \, \mathrm{d}t < \infty$. $\qquad\qquad \square$

As a consequence of Theorem 16.6 we have the following result.

Theorem 17.12 *Let the assumptions of Theorem 17.11 be satisfied and moreover let the mappings $A + \omega_1$ and $F + \omega_2$ be dissipative on l_ρ^2 and l_ρ^N with $\omega = \omega_1 + \omega_2 > 0$. Then there is a unique invariant measure for (17.1) and it is exponentially mixing with exponent ω.*

17.5 Kolmogorov's formula

The following fact was first observed by A. N. Kolmogorov (1937).

Theorem 17.13 *Let $U \in C_b^2(\mathbb{R}^d)$ be a function such that $c := \int_{\mathbb{R}^d} e^{U(x)} \, dx < \infty$. Then the unique invariant probability measure μ for the equation*

$$dX = \tfrac{1}{2} DU(X) \, dt + dW \tag{17.12}$$

is of the form $\mu(dx) = c^{-1} e^{U(x)} \, dx$.

Proof   We give only a sketch of the proof. The equation

$$dX = F(X) \, dt + dW, \qquad X(0) = x \in \mathbb{R}^d,$$

determines the following differential operator:

$$L\varphi(x) = \tfrac{1}{2} \mathrm{Tr} \, Q D^2 \varphi(x) + \langle D\varphi(x), F(x) \rangle \,.$$

It is not difficult to show (see e.g. Da Prato and Zabczyk 1996) that if ϕ is the density of an invariant measure of the solution X to (17.12) on $\mathbb{R}^d$ then

$$L^* \phi(x) = 0, \qquad x \in \mathbb{R}^d. \tag{17.13}$$

Conversely, if $\phi \in C_b^2(\mathbb{R}^d)$ is a non-negative and integrable solution to the *Focker–Planck equation (17.13)* then, after normalization, it is the density of an invariant measure for X. It is therefore enough to check that the function $\phi := e^U/c$ satisfies (17.13). In fact,

$$\int_{\mathbb{R}^d} e^U \tfrac{1}{2} \sum_{j=1}^d \left(\frac{\partial^2 \phi}{\partial x_j^2} + \tfrac{1}{2} \frac{\partial \phi}{\partial x_j} \frac{\partial U}{\partial x_j} \right) \, dx$$

$$= -\tfrac{1}{2} \int_{\mathbb{R}^d} \left[\sum_{j=1}^d \left(\frac{\partial}{\partial x_j} e^U \right) \frac{\partial \phi}{\partial x_j} \right] \, dx + \tfrac{1}{2} \int_{\mathbb{R}^d} e^U \sum_{j=1}^d \frac{\partial \phi}{\partial x_j} \frac{\partial U}{\partial x_j} \, dx$$

$$= -\tfrac{1}{2} \int_{\mathbb{R}^d} \left(\sum_{j=1}^d e^U \frac{\partial U}{\partial x_j} \frac{\partial \phi}{\partial x_j} \right) \, dx + \tfrac{1}{2} \int_{\mathbb{R}^d} e^U \left(\sum_{j=1}^d \frac{\partial \phi}{\partial x_j} \frac{\partial U}{\partial x_j} \right) \, dx = 0,$$

as required. $\square$

Often (17.12) is called a *gradient equation*. Unfortunately the theorem is not true for a larger class of (non-Gaussian) noise processes.

We need to generalize Kolmogorov's result to infinite dimensions. There are however some obstacles to doing so. In the infinite-dimensional case an analogue of Lebesgue measure does not exist, and the concept of density loses its natural meaning. Nevertheless, for some systems a generalization can be established. Consider the equation

$$dX = \left(AX + \tfrac{1}{2} DU(X) \right) dt + dW, \tag{17.14}$$

in which A is a non-negative-definite unbounded operator. Formally, the operator A is the gradient of a negative-definite form $Ax = \tfrac{1}{2} D \langle Ax, x \rangle_H$. Moreover, if $\mathfrak{H}(x) := \langle Ax, x \rangle_H + U(x)$ then $Ax + \tfrac{1}{2} DU(x) = \tfrac{1}{2} D\mathfrak{H}(x)$, and (17.14) becomes

$$dX = \tfrac{1}{2} D\mathfrak{H}(X) dt + dW.$$

Therefore, if dx denotes a (non-existing) infinite-dimensional analogue of Lebesgue measure, the formula

$$\mu(dx) = \frac{1}{c} \exp\{\mathfrak{H}(x)\} dx = \frac{1}{c} \exp\{\langle Lx, x \rangle_H + U(x)\} dx \tag{17.15}$$

suggests how an invariant measure for (17.14) could be defined. To deal with the dx term, denote by μ_0 an invariant measure for the Ornstein–Uhlenbeck process

$$dY = AY dt + dW, \qquad Y(0) = x.$$

Since the elements of the spectrum of A have strictly negative real parts, such a measure exists and is unique. Then, again formally, $\mu_0(dx) = \exp\{\langle Ax, x \rangle_H\} dx$. Consequently the measure (17.15) can be written as

$$\mu(dx) = \frac{1}{C} \exp\{U(x)\} \mu_0(dx),$$

with all terms well defined. This is a generalization of the Kolmogorov formula to an infinite-dimensional situation. The fact that it defines an invariant measure for (17.14) can be shown in several important cases. For more details we refer the reader to Da Prato and Zabczyk (1996) and Zabczyk (1989). The formula also provides a candidate for the Gibbs measure for some spin systems discussed in the next chapter. In physics the procedure described above is sometimes called *renormalization*.

17.6 Gibbs measures

This section is concerned with an application of the theory of stochastic evolution equations to statistical mechanics. On the basis of work presented in Da Prato and Zabczyk (1996) we sketch the proof of the existence of a Gibbs measure for a certain class of spin systems. We start with some motivation leading to the abstract definition of a Gibbs measure.

Let $\mathbb{Z}^d$ be a lattice, and let M be a metric space. In applications M is either a finite set, a compact manifold, the space $\mathbb{R}^d$ or a Hilbert space. We denote by $\mathcal{F}$ the family of all finite subsets of $\mathbb{Z}^d$. A *configuration* is an arbitrary function x defined on $\mathbb{Z}^d$, with values in M. The set of all configurations is denoted by $\mathcal{M}$. Thus $x = (x_\gamma, \gamma \in \mathbb{Z}^d)$ and $\mathcal{M} = M^{\mathbb{Z}^d}$. A *potential* $(J_\Delta, \Delta \in \mathcal{F})$ is a family of real-valued functions, indexed by elements of $\Delta \in \mathcal{F}$, such that if $\Delta = \{\gamma^1, \ldots, \gamma^k\}$ then $J_\Delta(x) = f_\Delta(x_{\gamma^1}, \ldots, x_{\gamma^k})$, $x \in \mathcal{M}$, for some function $f_\Delta \colon \mathbb{R}^k \mapsto \mathbb{R}$. A *Hamiltonian* $\mathfrak{H}$ is a sum of potentials

$$\mathfrak{H}(x) = \sum_{\Delta \in \mathcal{F}} J_\Delta(x), \qquad x \in \mathcal{M}.$$

Let ν_0 be a probability measure on M, and let $\nu(\mathrm{d}x) = \prod_{\gamma \in \mathbb{Z}^d} \nu_0(\mathrm{d}x_\gamma)$ be a measure on $\mathcal{M}$. By the *Gibbs measure* we mean the probability measure $\mu(\mathrm{d}x) = C^{-1} \mathrm{e}^{\mathfrak{H}(x)} \nu(\mathrm{d}x)$. The function $\mathrm{e}^{\mathfrak{H}}$ is called the *Gibbs density* and C is called a *normalizing constant*. Often the density has the form $\mathrm{e}^{\beta \mathfrak{H}}$, where the parameter β has a physical interpretation as the inverse temperature.

In the so-called *Ising model*, $M = \{-1, 1\}$,

$$J_\Delta(x) = \begin{cases} x_k x_l, & \text{for } \Delta = \{k, l\} \text{ and } |k - l| = 1, \\ 0, & \text{for the remaining configurations.} \end{cases}$$

In this case $\nu_0\{-1\} = \nu_0\{1\} = 1/2$. Note that the Hamiltonian is not well defined since for any x the sum $\sum_{\Delta \in \mathcal{F}} J_\Delta(x)$ has an infinite number of components of modulus 1. Its meaning can be given by the *Dobrushin–Lanford–Ruelle (DLR) equation*; see Definition 17.15 below. Before we formulate this equation we will recall without proof a fact from the theory of conditional distributions. Let ν_1 and ν_2 be non-negative, not necessarily finite, measures defined on E_1 and E_2, and let g be an arbitrary density with respect to $\nu_1 \times \nu_2$ on $E_1 \times E_2$. Define the *conditional density*

$$g(x|y) := \frac{g(x, y)}{\int_{E_1} g(z, y) \nu_1(\mathrm{d}z)}, \qquad (x, y) \in E_1 \times E_2.$$

Theorem 17.14   *If (X, Y) is a random variable in $E_1 \times E_2$ and $\mu = \mathcal{L}(X, Y) = g(x, y)\nu_1(\mathrm{d}x)\nu_2(\mathrm{d}y)$ then*

$$\mathbb{E}\left(\phi(X)|\sigma(Y)\right) = \int_{E_1} \phi(x) g(x|Y) \nu_1(\mathrm{d}x).$$

In particular,

$$\int_E \phi(x)\mu(\mathrm{d}x, \mathrm{d}y) = \int_E \left(\int_{E_1} \phi(z) g(z|y) \nu_1(\mathrm{d}z) \right) \mu(\mathrm{d}x, \mathrm{d}y).$$

For $\Gamma \in \mathcal{F}$ and any $x \in \mathcal{M}$ write $x = (x^\Gamma, y^{\Gamma^c})$. Then for the Gibbs density we have

$$g_{\mathfrak{H}}(x) = g_{\mathfrak{H}}(x^\Gamma, y^{\Gamma^c}) := \exp\left\{\mathfrak{H}(x^\Gamma, y^{\Gamma^c})\right\}$$

$$= \exp\left\{\sum_{\Delta \subset \Gamma^c} J_\Delta(x^\Gamma, y^{\Gamma^c}) + \sum_{\Delta \cap \Gamma \neq \emptyset} J_\Delta(x^\Gamma, y^{\Gamma^c})\right\}$$

and, for the conditional density,

$$g_{\mathfrak{H}}(x^\Gamma | y^{\Gamma^c}) = \frac{\exp\left\{\sum_{\Delta \subset \Gamma^c} J_\Delta(x^\Gamma, y^{\Gamma^c}) + \sum_{\Delta \cap \Gamma \neq \emptyset} J_\Delta(x^\Gamma, y^{\Gamma^c})\right\}}{\int \exp\left\{\sum_{\Delta \subset \Gamma^c} J_\Delta(z^\Gamma, y^{\Gamma^c}) + \sum_{\Delta \cap \Gamma \neq \emptyset} J_\Delta(z^\Gamma, y^{\Gamma^c})\right\} \nu_1(\mathrm{d}z\Gamma)},$$

where $\nu_1(\mathrm{d}z^\Gamma) = \prod_{\gamma \in \Gamma} \nu_0(\mathrm{d}z_\gamma)$. But if $\Delta \subset \Gamma^c$ then for $x^\Gamma, y^{\Gamma^c}, z^{\Gamma^c}$ we have $J_\Delta(x^\Gamma, y^{\Gamma^c}) = J_\Delta(z^\Gamma, y^{\Gamma^c})$ and consequently

$$g_{\mathfrak{H}}(x^\Gamma | y^{\Gamma^c}) = \frac{\exp\left\{\sum_{\Delta \cap \Gamma \neq \emptyset} J_\Delta(x^\Gamma, y^{\Gamma^c})\right\}}{\int \exp\left\{\sum_{\Delta \cap \Gamma \neq \emptyset} J_\Delta(z^\Gamma, y^{\Gamma^c})\right\} \nu_1(\mathrm{d}z^\Gamma)}. \tag{17.16}$$

The sums appearing above often have only a finite number of non-zero components. Thus the conditional densities can be well defined even in the case when the density of (X, Y) is not well defined. The quantity appearing in (17.16) can be interpreted as the conditional probability density for the occurrence in Γ of the configuration x^Γ, if outside Γ one observes y^{Γ^c}. This leads to the following rigorous definition.

Definition 17.15 A *Gibbs measure* on the space $\mathcal{M}$ of configurations corresponding to the potential $(J_\Delta, \Delta \in \mathcal{F})$ is any probability measure μ such that, for any $\Gamma = \{\gamma_1, \ldots, \gamma_k\} \in \mathcal{F}$ and any bounded function ϕ of k variables, we have

$$\int_\mathcal{M} \phi(x_{\gamma_1}, \ldots, x_{\gamma_k}) \mu(\mathrm{d}x^\Gamma, \mathrm{d}y^{\Gamma^c})$$

$$= \int_\mathcal{M} \left(\int_{M^\Gamma} \phi(z_{\gamma_1}, \ldots, z_{\gamma_k}) g_{\mathfrak{H}}(z_{\gamma_1}, \ldots, z_{\gamma_k} | y^{\Gamma^c}) \nu_1(\mathrm{d}z^\Gamma)\right) \mu(\mathrm{d}x^\Gamma, \mathrm{d}y^{\Gamma^c}).$$

The equality above is called the *DLR equation*.

17.6.1 Gibbs measure for continuous spin systems

Assume that the configuration space $\mathcal{M}$ consists of all real-valued functions defined on $\mathbb{Z}^d$, and that the Hamiltonian operator is of the form

$$\mathfrak{H}(x) = \sum_{k,j} a_{k,j} x_k x_j + \sum_k v(x_k). \tag{17.17}$$

The function v is the so-called *potential of local interactions*.

Assume that the interactions $a_{k,j}$ have a bounded range. The Gibbs measure has the form

$$\mathrm{e}^{\mathfrak{H}(x)} \prod_{\gamma \in \mathbb{Z}^d} \mathrm{d}x_\gamma = \mathrm{e}^{\mathfrak{H}(x)} \, \mathrm{d}\mathbf{x},$$

where $\mathrm{d}\mathbf{x}$ is a product of an infinite number of Lebesgue measures on $\mathbb{R}$. The series in $\mathfrak{H}$ usually do not converge. However, the density appearing in the DLR equation is well defined. Indeed, set

$$J_\Delta(x) = \begin{cases} v(x_\gamma), & \text{for } \Delta = \{\gamma\}, \\ a_{kj} x_k x_j, & \text{for } \Delta = \{k, j\}, \\ 0, & \text{in the remaining cases.} \end{cases}$$

If $\Gamma = \{\gamma_1, \ldots, \gamma_k\}$ is finite then in the sum $\sum_{\Delta \cap \Gamma \neq \emptyset} J_\Delta(x)$ there are only a finite number of non-zero functions. Also, the measure ν_1 is a product of a finite number of Lebesgue measures.

Given a Hamiltonian (17.17) we look for Gibbs measures among the measures on the weighted l^2-space. To do this, consider the following infinite sequence of equations:

$$\mathrm{d}X_k(t) = \left(\sum_{k \in \mathbb{Z}^d} a_{k,j} X_j(t) + \tfrac{1}{2} v'(X_j(t)) \right) \mathrm{d}t + \mathrm{d}W_k(t),$$

$$X_k(0) = x_k, \tag{17.18}$$

where v' is the derivative of the real function v and $\left(W_k, \ k \in \mathbb{Z}^d \right)$ is a sequence of independent real-valued standard Wiener processes.

Defining $X(t) := \left(X_k(t), \ k \in \mathbb{Z}^d \right)$, $W(t) := \left(W_k(t), \ k \in \mathbb{Z}^d \right)$ and taking into account the formula for $\mathfrak{H}$, system (17.18) can be written formally as follows:

$$\mathrm{d}X = \tfrac{1}{2} D\mathfrak{H}(X) \, \mathrm{d}t + \mathrm{d}W.$$

According to our discussion on invariant measures for gradient systems in the previous section, the invariant measure for (17.18) should be the required Gibbs measure. This can be proved rigorously by considering an increasing sequence of finite sets $M_n \subset \mathbb{Z}^d$, covering $\mathbb{Z}^d$, and exploiting the form of the Kolmogorov density together with the DLR equation. The following existence result from Da Prato and Zabczyk (1996) follows from Theorem 17.12.

Theorem 17.16 *Assume that the interactions $a_{k,j}$ have a finite range and that the derivative v' of the potential function can be decomposed as $v' = f_0 + f_1$, where f_0 is a decreasing function such that, for some $c_0 > 0$ and $N \geq 1$, $|f_0(z)| \leq c_0 \left(1 + |z|^N \right)$ for $z \in \mathbb{R}$ and f_1 satisfies the Lipschitz condition. Assume also that the operator $A + \eta I$ is dissipative on l^2 for some $\eta > \| f_1 \|_{\mathrm{Lip}}$. Then there is a constant $\kappa_0 > 0$ such that if $\kappa \in (0, \kappa_0)$ then in each space $l^2_{\rho^\kappa}$ and $l^2_{\rho^{\kappa,r}}$, $\kappa \in (0, \kappa_0)$, there is a unique invariant measure for (17.18) and it is exponentially mixing.*

18

Stochastic Burgers equation

In this chapter, the existence of a solution to the so-called stochastic Burgers system driven by a Wiener process is investigated. The existence of a solution to the stochastic Burgers equation with an additive Lévy noise is shown as well.

18.1 Burgers system

We consider a stochastic variant of a model of turbulence introduced by Burgers (1939). The general theory developed in this book cannot be applied here without some modification. In fact our exposition is based on Twardowska and Zabczyk (2004); see also Twardowska and Zabczyk (2006). However, we will limit our considerations to the existence of local solutions only.

Let $U = U(t)$ be the primary velocity of a fluid parallel to the walls of the channel, let $v = v(t, \xi)$, $t \geq 0$, $\xi \in (0, 1)$, be the secondary velocity, of turbulent motion, and let P be the exterior force. We denote the density of the fluid by ρ and its viscosity by μ. We will write $v := \mu/\rho$.

The stochastic version of the Burgers system looks as follows:

$$\frac{dU(t)}{dt} = P - vU(t) - \int_0^1 v^2(t, \xi) \, d\xi, \qquad \text{for } t > 0, \qquad (18.1)$$

and, for $\{(t, \xi): t > 0, \ \xi \in (0, 1)\}$,

$$\frac{\partial v}{\partial t}(t, \xi) = v \frac{\partial^2 v}{\partial \xi^2}(t, \xi) + U(t)v(t, \xi)$$

$$- \frac{\partial}{\partial \xi}\left(v^2(t, \xi)\right) + g\left(v(t, \xi)\right)\frac{\partial^2 \mathcal{W}(t, \xi)}{\partial t \, \partial \xi}. \qquad (18.2)$$

The Burgers system is considered with the following initial and Dirichlet boundary conditions:

$$U(0) = 0, \quad v(0, \xi) = v_0(\xi), \qquad \xi \in (0, 1), \tag{18.3}$$

and

$$v(t, 0) = v(t, 1) = 0, \qquad t > 0. \tag{18.4}$$

In (18.2), $\mathcal{W}(t, \xi)$ is a Brownian sheet (see Remark 7.19), and thus

$$W(t, \xi) = \frac{\partial \mathcal{W}(t, \xi)}{\partial \xi}$$

is a cylindrical Wiener process on $L^2 := L^2(0, 1)$. Note that the system does not satisfy the conditions of the general existence theorem 9.29. As in Chapter 9 we have two solution concepts.

Definition 18.1 A *weak solution* of the system (18.1)–(18.4) is a pair $(U(t), v(t))$ of continuous adapted processes with values in $\mathbb{R}$ and L^2, respectively, such that (18.1) holds and, for any $\varphi \in C_c^\infty(0, 1)$,

$$\langle v(t), \varphi \rangle_{L^2} = \langle v_0, \varphi \rangle_{L^2} + \int_0^t \langle \varphi g\left(v(s)\right), dW(s) \rangle_{L^2}$$
$$+ \int_0^t \left(U(s)\langle v(s), \varphi \rangle_{L^2} + v\left\langle v(s), \frac{d^2\varphi}{d\xi^2} \right\rangle_{L^2} + \left\langle v^2(s), \frac{d\varphi}{d\xi} \right\rangle_{L^2} \right) ds.$$

To define the so-called *mild solution* we denote by $(S(t), t \geq 0)$ the classical heat semigroup, given by

$$S(t)v = \sum_{k=1}^\infty e^{-v\pi^2 k^2 t} \langle v, e_k \rangle_{L^2} e_k, \qquad v \in L^2, \; t > 0,$$

where $e_k(\xi) = \sqrt{2/\pi} \, \sin(k\pi\xi)$. For more details see Example B.12. We need the following lemma.

Lemma 18.2 *The operators $S(t)$, $t > 0$, can be extended linearly to the space of all distributions of the form $dv/d\xi$, $v \in L^1 := L^1(0, 1)$, in such a way that $S(t)(dv/d\xi) \in L^2$ for all $v \in L^1$ and*

$$\left| S(t) \frac{dv}{d\xi} \right|_{L^2} \leq |v|_{L^1} \sqrt{2\pi} \left(k^2 \sum_k e^{-2v\pi^2 k^2 t} \right)^{1/2}, \qquad \forall v \in L^1.$$

Proof By Parseval's identity,

$$|S(t)u|_{L^2}^2 = \sum_{k=1}^\infty e^{-2v\pi^2 k^2 t} \langle u, e_k \rangle_{L^2}^2.$$

Let $v \in L^2$ be an absolutely continuous function such that $dv/d\xi \in L^2$. Then

$$\left| S(t)\frac{dv}{d\xi} \right|_{L^2}^2 = \frac{2}{\pi} \sum_{k=1}^{\infty} e^{-2v\pi^2 k^2 t} \left(\int_0^1 \frac{dv}{d\xi} \sin(k\pi\xi)\,d\xi \right)^2$$

$$= \frac{2}{\pi} \sum_{k=1}^{\infty} e^{-2v\pi^2 k^2 t}(k\pi)^2 \left(\int_0^1 v(\xi)\cos(k\pi\xi)\,d\xi \right)^2$$

$$\leq 2\pi \left(\sum_{k=1}^{\infty} k^2 e^{-2v\pi^2 k^2 t} \right) \left(\int_0^1 |v(\xi)|\,d\xi \right)^2 .$$

□

Definition 18.3 A *mild solution* to the Burgers system (18.1)–(18.4) is a pair (U, v) of continuous adapted processes with values in $\mathbb{R}$ and L^2, respectively, such that (18.1) holds and

$$v(t) = S(t)v_0 + \int_0^t U(s)S(t-s)v(s)\,ds$$

$$+ \int_0^t S(t-s)\frac{\partial}{\partial\xi}v^2(s)\,ds + \int_0^t S(t-s)g(v(s))\,dW(s).$$

Owing to Lemma 18.2, all terms in the equation above have a well-defined meaning. The following result can be proved using arguments from Section 9.3.

Proposition 18.4 *A process*

$$X = \begin{pmatrix} U \\ v \end{pmatrix}$$

is a weak solution of (18.1)–(18.4) if and only if it is a mild solution.

18.2 Uniqueness and local existence of solutions

The theorem below ensures the uniqueness of a weak solution to the Burgers system. However (see Proposition 18.6), only the local existence of a solution will be established.

Theorem 18.5 *System (18.1)–(18.4) has a unique weak solution.*

We will divide the proof into that of two propositions.

Proposition 18.6   *For any* $T > 0$ *there exists a constant* c *such that, for every* $t \leq T$ *and for every measurable bounded* L^1-*valued mapping* $(v(s), \ s \in (0, t))$,

$$\int_0^t \left| S(\sigma) \frac{\partial}{\partial \xi} \, v(\sigma) \right|_{L^2} \, d\sigma \leq c \, t^{1/4} \sup_{s \leq t} |v(s)|_{L^1}.$$

Proof   Taking into account Lemma 18.2, we have to show that

$$\int_0^t \left(\sum_k k^2 e^{-sk^2} \right)^{1/2} ds = O(t^{1/4}) \qquad \text{as } t \downarrow 0.$$

To this end note that

$$h(t) = \sum_k k^2 e^{-tk^2} = \int_0^\infty e^{-tr} \mu(dr),$$

where $\mu(dr) = \sum_k k^2 \delta_{\{k^2\}}(dr)$. Define

$$U(\sigma) = \mu((0, \sigma]) = \sum_{k^2 \leq \sigma} k^2.$$

Since $1 + 2^2 + \cdots + k^2 = \frac{1}{3} k(k+1)\left(k + \frac{1}{2}\right)$, we have that $U(\sigma y)/U(\sigma) \to y^{3/2}$ as $\sigma \to \infty$. By the Tauberian theorem (see e.g. Feller 1971),

$$\lim_{t \to 0} \frac{h(t)}{U(1/t)} = \Gamma(5/2).$$

But $U(1/t) \, 3t^{3/2} \to 1$ as $t \to 0$ and therefore

$$\frac{h(t)}{2\Gamma(5/2)t^{3/2}} \to 1 \qquad \text{as } t \to 0.$$

Since $\int_0^T t^{-3/4} \, dt = 4T^{1/4}$, the required inequality holds. $\square$

Given $p \geq 1$, $T > 0$, we denote by $\mathcal{H}_T^p$ the space of all continuous adapted processes $X = (U, v)^T$ with values in $(\mathbb{R}, L^2)^T$ such that

$$\|X\|_{\mathcal{H}_T^p} = \left(\mathbb{E} \sup_{t \in [0,T]} |U(t)|^p \right)^{1/p} + \left(\mathbb{E} \sup_{t \in [0,T]} |v(t)|_{L^2}^p \right)^{1/p}.$$

Next, for $n \in \mathbb{N}$ define

$$B_1(0, n) := [-n, n], \quad B_2(0, n) := \{v \in L^2 : |v|_{L^2} \leq n\}, \qquad n = 1, 2, \ldots,$$

$$\pi_{n1}(U) := \begin{cases} U & \text{if } |U| \leq n, \\ \dfrac{nU}{|U|} & \text{if } |U| > n, \end{cases} \qquad \pi_{n2}(v) := \begin{cases} v & \text{if } |v|_{L^2} \leq n, \\ \dfrac{nv}{|v|_{L^2}} & \text{if } |v|_{L^2} > n. \end{cases}$$

The existence and uniqueness of local solutions to (18.1)–(18.4) are consequences of the following proposition. For the existence of global solutions we refer the reader to Twardowska and Zabczyk (2004).

Proposition 18.7   *For any $p > 4$ and $n = 1, 2, \ldots,$ the system of equations*

$$U(t) = e^{-\nu t} U_0 + \int_0^t e^{-\nu(t-s)} \left(P - |\pi_{n2} v(s)|_{L^2}^2 \right) ds,$$

$$v(t) = S(t) v_0 + \int_0^t \pi_{n1} \big(U(s) \big) S(t-s) \pi_{n2} v(s) \, ds$$

$$+ \int_0^t S(t-s) \frac{\partial}{\partial \xi} \big(\pi_{n2} v(s) \big)^2 ds + \int_0^t S(t-s) g \big(v(s) \big) \, dW(s)$$

has a unique solution in $\mathcal{H}_T^p$.

Proof   Let us fix n and introduce the operators

$$I_1(U, v) := \int_0^t e^{-\nu(t-s)} \left(P - |\pi_{n2} v(s)|_{L^2}^2 \right) ds,$$

$$I_2(U, v) := \int_0^t \pi_{n1} \big(U(s) \big) S(t-s) \pi_{n2} v(s) \, ds,$$

$$I_3(v) := \int_0^t S(t-s) \frac{\partial}{\partial \xi} \big(\pi_{n2} v(s) \big)^2 ds,$$

$$I_4(v) := \int_0^t S(t-s) g \big(v(s) \big) \, dW(s).$$

The result follows from the contraction principle in the space $\mathcal{H}_T^p$. We show only how to estimate the Lipschitz constant of the mapping $I \colon \mathcal{H}_T^p \mapsto \mathcal{H}_T^p$ given by

$$I(Y)(t) = \int_0^t S(t-s) \frac{\partial}{\partial \xi} \big(\pi_{n2} Y(s) \big)^2 ds,$$

To do this, take two processes Y and $\tilde{Y}$ from $\mathcal{H}_T^p$ such that $Y(0) = \tilde{Y}(0) = v_0$. We have

$$\left| (\pi_{n2} Y(\sigma))^2 - (\pi_{n2} \tilde{Y}(\sigma))^2 \right|_{L^1}$$

$$= \left| (\pi_{n2} Y(\sigma) + \pi_{n2} \tilde{Y}(\sigma)) (\pi_{n2} Y(\sigma) - \pi_{n2} \tilde{Y}(\sigma)) \right|_{L^1}$$

$$\leq \left| \pi_{n2} Y(\sigma) + \pi_{n2} \tilde{Y}(\sigma) \right|_{L^2} \left| \pi_{n2} Y(\sigma) - \pi_{n2} \tilde{Y}(\sigma) \right|_{L^2}$$

$$\leq 2n \left| Y(\sigma) - \tilde{Y}(\sigma) \right|_{L^2}.$$

Consequently, by Proposition 18.6,

$$\sup_{t \leq T} \left| I(Y)(t) - I(\tilde{Y})(t) \right|_{L^2}$$

$$= \sup_{t \leq T} \left| \int_0^t S(t - \sigma) \frac{\partial}{\partial \xi} \left((\pi_{n2} Y(\sigma))^2 - (\pi_{n2} \tilde{Y}(\sigma))^2 \right) d\sigma \right|_{L^2}$$

$$\leq C \sup_{t \in T} t^{1/4} \sup_{\sigma \leq t} \left| (\pi_{n2} Y(\sigma))^2 - (\pi_{n2} \tilde{Y}(\sigma))^2 \right|_{L^1}$$

$$\leq 2nC \sup_{t \leq T} t^{1/4} \sup_{\sigma \leq t} |Y(s) - \tilde{Y}(s)|_{L^2}$$

$$\leq 2nC T^{1/4} \sup_{s \leq T} |Y(s) - \tilde{Y}(s)|_{L^2},$$

which gives the required estimate. $\qquad\square$

Remark 18.8   As noticed by Z. Brzeźniak (private communication), the proofs of Lemma 18.2 and Proposition 18.6 are valid for operators more general than second-order differential operators, for instance, for some fractional powers of the Laplace operator.

18.3 Stochastic Burgers equation with additive noise

Given $p \in [1, +\infty]$ set $L^p := L^p([0, 1], \mathcal{B}([0, 1]), \ell_1)$. Let us fix a finite time interval $[0, T]$ and let $Z = (Z(t), \ t \in [0, T])$ be a Lévy process taking values in a Hilbert space U containing L^2. We assume that Z is defined on a filtered probability space $(\Omega, \mathcal{F}, (\mathcal{F}_t), \mathbb{P})$.

Given $v_0 \in L^2$ we are concerned with the one-dimensional stochastic Burgers problem

$$\frac{\partial v}{\partial t}(t, \xi) = v \frac{\partial^2 v}{\partial \xi^2}(t, \xi) + \frac{1}{2} \frac{\partial v^2}{\partial \xi}(t, \xi) + \frac{\partial Z}{\partial t}, \qquad \xi \in (0, 1), \ t \in [0, T],$$

$$v(t, 0) = v(t, 1), \qquad t \in (0, T],$$

$$v(0, \xi) = v_0(\xi), \qquad \xi \in [0, 1].$$

Our first task is to present a rigorous definition of a weak solution. To this end, denote by A the Laplace operator on L^2 with Dirichlet boundary conditions (see subsection 12.4.1). Recall that $D(A) = W^{2,2}(0, 1) \cap W_0^{1,2}(0, 1)$. The space $D(A)$ is considered with a graph norm. Let $V := W_0^{1,2}(0, 1)$. Then $D(A) \hookrightarrow V \hookrightarrow L^2$, where the embeddings are continuous and dense. Let $(D(A))^*$ and V^* be the dual spaces corresponding to $D(A)$ and V, respectively, where the duality form is given by the scalar product in L^2. Thus

$$D(A) \hookrightarrow V \hookrightarrow L^2 \equiv (L^2)^* \hookrightarrow V^* \hookrightarrow (D(A))^*.$$

We denote by $\langle \cdot, \cdot \rangle$ the scalar product on L^2 or the duality form on $V^* \times V$ and $(D(A))^* \times D(A)$.

Since for $u \in V$, $u(\xi) = \int_0^\xi (du(\eta)/d\eta) \, d\eta$, the space V is embedded into $C([0, 1]) \subset L^\infty$. For $u, z \in C^1([0, 1])$ write

$$B(u, z)(\xi) := \tfrac{1}{2} \frac{d(uz)}{d\xi}(\xi), \qquad \xi \in [0, 1].$$

Lemma 18.9   *Let $p, q > 1$ be such that $1/p + 1/q = 1/2$. Then B can be uniquely extended to a bounded linear mapping from $L^p \times L^q$ to V^*.*

Proof   Let $\phi \in V$ and $u, z \in C^1([0, 1])$. Then

$$|\langle B(u, z), \phi \rangle| = \tfrac{1}{2} \left| \int_0^1 u(\xi) z(\xi) \frac{d\phi}{d\xi}(\xi) \, d\xi \right| \le \tfrac{1}{2} |uz|_{L^2} \left| \frac{d\phi}{d\xi} \right|_{L^2}$$

$$\le \tfrac{1}{2} |u|_{L^p} |z|_{L^q} |\phi|_V.$$

$\square$

Lemma 18.10   *The Laplace operator A can be uniquely extended to a bounded linear operator[1] from V to V^* and from L^2 into $(D(A))^*$.*

Proof   Let $u \in D(A)$ and $\phi \in V$. Then, integrating by parts, we obtain

$$|\langle Au, \phi \rangle| = \left| \int_0^1 \frac{du}{d\xi}(\xi) \frac{d\phi}{d\xi}(\xi) \, d\xi \right| \le |u|_V |\phi|_V,$$

which proves the first claim. Since for $u, \phi \in D(A)$

$$|\langle Au, \phi \rangle| = |\langle u, A\phi \rangle| \le |u|_{L^2} |A\phi|_{L^2} \le |u|_{L^2} |\phi|_{D(A)},$$

the second claim follows.

$\square$

Given a Banach space U write $L^2(0, T; U) := L^2([0, T], \mathcal{B}([0, T]), \ell_1; U)$. From now on we assume that Z takes values in $(D(A))^*$.

Definition 18.11   A random element $v \colon \Omega \mapsto L^2(0, T; L^4)$ is a solution to the Burgers problem if and only if, for each $\phi \in D(A)$, $\mathbb{P}$-a.s.,

$$\langle v(t), \phi \rangle = \langle v_0, \phi \rangle + \int_0^t \left\langle \nu Av(s) + B(v(s), v(s)), \ \phi \right\rangle ds + \langle Z(t), \phi \rangle \quad (18.5)$$

for ℓ_1-almost all $t \in [0, T]$.

By Lemmas 18.9 and 18.10, for every $u \in L^2(0, T; L^4)$ the mappings $t \mapsto B(u(t), u(t))$ and $t \mapsto Au(t)$ belong to $L^2(0, T; V^*)$. Thus the assumption in the

---

[1]   This operator is also denoted by A.

definition above that X has trajectories in $L^2(0, T; L^4)$ guarantees that all terms on the right-hand side of (18.5) are well defined.

We will formulate a sufficient condition for the existence of a solution to the Burgers problem in terms of the existence of a sufficiently regular solution Z_A to the linear problem

$$
\begin{aligned}
\frac{\partial Z_A}{\partial t}(t, \xi) &= \nu \frac{\partial^2 Z_A}{\partial \xi^2}(t, \xi) + \frac{\partial Z}{\partial t}, && \xi \in (0, 1), \ t \in [0, T], \\
Z_A(t, 0) &= Z(t, 1), && t \in (0, T], \\
Z_A(0, \xi) &= 0, && \xi \in [0, 1].
\end{aligned}
\tag{18.6}
$$

Definition 18.12 A random element $Z_A \colon \Omega \mapsto L^2(0, T; L^2)$ is a *solution to* (18.6) if, for every $\phi \in D(A)$, $\mathbb{P}$-a.s.,

$$
\langle Z_A(t), \phi \rangle = \int_0^t \langle \nu A Z_A(s), \phi \rangle \, \mathrm{d}s + \langle Z(t), \phi \rangle
$$

for ℓ_1-almost all $t \in [0, T]$.

Remark 18.13 Let A denote the Laplace operator on $[0, 1]$ with Dirichlet boundary conditions, and let S be the semigroup generated by νA. It is easy to see that S has a unique extension to a C_0-semigroup of contractions on V^*. Next, by the Lévy–Khinchin decomposition, Z can be seen as a sum of a square integrable martingale and a process with finite variation in V^*. Thus (see subsection 8.6.2, the Kotelenez inequality from subsection 9.4.1) the solution to (18.2) is a càdlàg process in V^* given by

$$
Z_A(t) := \int_0^t S(t - s) \, \mathrm{d}Z(s), \qquad t \geq 0.
$$

We can now formulate our main existence result.

Theorem 18.14 *Assume that there is a solution Z_A to (18.6) with trajectories in $L^4(0, T; L^2)$. Then for each $v_0 \in L^2$ there is a unique solution to the Burgers problem.*

Proof Note that $v \colon \Omega \mapsto L^2(0, T; L^4)$ is a solution to the Burgers problem if and only if $Y := v - Z_A$ solves

$$
\mathrm{d}Y = (\nu A Y + B(Y + Z_A, Y + Z_A)) \, \mathrm{d}t, \qquad Y(0) = v_0,
$$

that is, if and only if for every $\phi \in D(A)$, $\mathbb{P}$-a.s.,

$$
\langle Y(t), \phi \rangle = \langle Y(0), v_0 \rangle + \int_0^t \langle \nu A Y(s) + B(Y(s) + Z_A(s), \ Y(s) + Z_A(s)), \ \phi \rangle \, \mathrm{d}s
$$

for ℓ_1-almost all $t \in [0, T]$. Thus the theorem is a direct consequence of the following deterministic result from Brzeźniak (2006); see Proposition 3.13.   □

Lemma 18.15   *Assume that $z \in L^2(0, T, L^4)$. Then for every $v_0 \in L^2$ there is a unique $y \in L^2(0, T; L^4)$ such that, for any $\phi \in D(A)$, $\mathbb{P}$-a.s.*

$$\langle y(t), \phi \rangle = \langle y(0), v_0 \rangle + \int_0^t \langle \nu Ay(s) + B(y(s) + z(s), \ y(s) + z(s)), \ \phi \rangle \, ds$$

for ℓ_1-almost all $t \in [0, T]$.

For the proof of the lemma we refer the reader to Brzeźniak (2006). It is worth noting that the claim in this paper is much stronger than that presented in our lemma. Namely, it was shown that for every $z \in L^4(0, T; L^2)$, $g \in L^2(0, T; V^*)$ and $v_0 \in L^2$ there is a unique solution $y \in \mathcal{X} := L^\infty(0, T; L^2) \cap L^2(0, T; V) \cap W^{1,2}(0, T; V^*)$ to the problem

$$dy = (\nu Ay + B(y, z) + B(z, y) + B(y, y) + g) \, dt, \qquad y(0) = v_0. \quad (18.7)$$

Moreover, it was shown that the mapping $L^2(0, T; V^*) \times L^2 \ni (g, x_0) \mapsto z \in \mathcal{X}$, where z is the solution to (18.7), is real analytic. This result was then applied to the construction of the so-called random dynamical system associated with the Burgers equation and to the investigation of its random attractors and asymptotic compactness and the existence of invariant measures.

Example 18.16   Assume that $Z = W + L + \tilde{L}$, where W is a Wiener process with RKHS contained in L^2, L is a square integrable martingale in L^2 and $\tilde{L}$ is a compound Poisson process in L^4. Assume, moreover, that $L(t) = \int_0^t \int_{L^2} x\widehat{\pi}(ds, dx)$, where the intensity measure ν of π satisfies

$$\int_{L^2} |x|_{L^2}^j \nu(dx) < \infty, \qquad j = 1, 2.$$

Then $Z_A = W_A + L_A + \tilde{L}_A$. Clearly $\tilde{L}_A$ is càdlàg in L^4. By Theorem 12.17, W_A has continuous trajectories in $C([0, 1])$. To see that L_A has trajectories in $L^2(0, T; L^4)$ we use Proposition 6.7, as we did in the proof of Theorem 16.7. Namely,

$$I(t) := \mathbb{E} \, |L_A(t)|_{L^4}^2 = \mathbb{E} \left| \int_0^t \int_{L^2} S(t - s)x\widehat{\pi}(ds, dx) \right|_{L^4}^2.$$

By Proposition 6.7, there is a polynomial R such that $I(t) = R(I_1(t), I_2(t))$, where

$$I_j(t) =: \int_0^t \int_{L^2} |S(t - s)x|_{L^4}^j \, ds \, \nu(dx), \qquad j = 1, 2.$$

We have

$$I_j(t) = \int_0^t \int_{L^2} |S(s)x|_{L^4}^j \, ds \, \nu(\mathrm{d}x)$$

$$\leq \int_0^t \int_{L^2} \|S(s)\|_{L(L^2,L^4)}^j |x|_{L^2}^j \, ds \, \nu(\mathrm{d}x).$$

By Theorem B.7, $\|S(s)\|_{L(L^2,L^4)} \leq Cs^{-1/8}$, $s \in [0, T]$, and hence

$$\sup_{t\in[0,T]} I^j(t) < \infty, \qquad j = 1, 2.$$

Summing up, Z_A has trajectories in $L^2(0, T; L^4)$ and consequently for any $v_0 \in L^2$ there is a unique solution to the Burgers problem.

Remark 18.17 If Z is an impulsive white noise on L^2 then Z_A has trajectories in $L^2(0, T; L^2)$ but not in $L^2(0, T; L^4)$.

19

Environmental pollution model

In this chapter, following Kallianpur and Xiong (1995) we show how one can model a process of environmental pollution using stochastic parabolic equations driven by a Poisson measure.

19.1 Model

The chemical concentration $X(t, \xi)$ at time t at a location $\xi \in \mathcal{O} \subset \mathbb{R}^d$, in the absence of random deposits, satisfies the following partial differential equation:

$$\frac{\partial X}{\partial t}(t, \xi) = AX(t, \xi) - \alpha X(t, \xi),$$

where $A = \delta\Delta + \sum_i V_i \partial/\partial \xi_i$, δ is the dispersion coefficient, $V = (V_1, \dots, V_d)$ is the drift vector and α is the leakage rate. For physical reasons the equation is considered with Neumann boundary conditions.

Assume now that chemicals are deposited at sites in $\mathcal{O}$ at random times $\tau_1, \tau_2, \dots$ and locations $\xi_1, \xi_2, \dots$ with positive random magnitudes $\sigma_1, \sigma_2, \dots$ Assume that the time intervals $\tau_1, \tau_2 - \tau_1, \tau_3 - \tau_3, \dots$, the random magnitudes $\sigma_1, \sigma_2, \dots$ and the random locations $\xi_1, \xi_2, \dots$ are independent random variables and that τ_1, $\tau_2 - \tau_1, \tau_3 - \tau_3, \dots$ are exponentially distributed with exponent β. Then $Z(t) = \sum_{\tau_k \le t} \sigma_k \delta_{\xi_k}$ is a compound Poisson process with values in the space of discrete measures. The concentration X corresponding to the random input Z satisfies

$$dX(t) = (AX(t) - \alpha X) dt + dZ(t),$$
$$X(0) = x.$$
$$\tag{19.1}$$

Let $\mathcal{G}$ be the Green function for $A - \alpha I$. Then the solution to (19.1) is given by

$$X(t) = S(t)x + \int_0^t S(t - s) \, dZ(s) = S(t)x + \sum_{\tau_k \le t} \sigma_k S(t - \tau_k)\delta_{\xi_k},$$

where $S(t)\psi(\xi) = \int_{\mathcal{O}} \mathcal{G}(t, \xi, \eta)\psi(\eta)\,\mathrm{d}\eta$. This leads to the following function-valued process:

$$X(t, \xi) = \int_{\mathcal{O}} \mathcal{G}(t, \xi, \eta)x(\eta)\,\mathrm{d}\eta + \sum_{\tau_k \leq t} \sigma_k \mathcal{G}(t - \tau_k, \xi, \xi_k).$$

Let $\pi(\mathrm{d}s, \mathrm{d}\xi, \mathrm{d}\sigma)$ be the Poisson random measure on $[0, \infty) \times \mathcal{O} \times [0, \infty)$, with intensity $\mathrm{d}s\,\mathrm{d}\xi\,\nu(\mathrm{d}\sigma)$. Then

$$X(t, \xi) = \int_{\mathcal{O}} \mathcal{G}(t, \xi, \eta)x(\eta)\,\mathrm{d}\eta + \int_0^t \int_{\mathcal{O}} \int_0^\infty \mathcal{G}(t - s, \xi, \eta)\sigma\pi(\mathrm{d}s, \mathrm{d}\eta, \mathrm{d}\sigma)$$
$$= I_0(t, \xi) + I_1(t, \xi).$$

Let $\widehat{\pi}$ be the compensated measure. Assume that $a := \int_0^\infty \sigma\nu(\mathrm{d}\sigma) < \infty$. Then, for $t \geq 0$ and $x \in \mathcal{O}$, we have

$$I_1(t, \xi) = aS(t)\chi_{\mathcal{O}}(\xi) + \int_0^t \int_{\mathcal{O}} \int_0^\infty \mathcal{G}(t - s, \xi, \eta)\sigma\widehat{\pi}(\mathrm{d}s, \mathrm{d}\eta, \mathrm{d}\sigma).$$

We therefore have the following corollary to Theorem 12.11, or, in the case of $\mathcal{O} = \mathbb{R}^d$, to Theorem 12.13.

Proposition 19.1 *Let $T < \infty$. Assume that $1 \leq p < 2/d + 1$ and*

$$\int_0^\infty |\sigma|^p \nu(\mathrm{d}\sigma) < \infty.$$

Then, for each $x \in L^p(\mathcal{O})$ $\left(x \in L_\rho^p\right)$, there exists a unique $\mathcal{X}_{T,p}(\mathcal{X}_{T,p,\rho})$ solution to (19.1).

Different situations (see Kallianpur and Xiong 1995) can be considered with similar results. Suppose, for instance, that factories are located at fixed sites $a_1, a_2, \ldots, a_r \in [0, l]$ and that the interval $[0, l]$ represents a river. The factories deposit chemicals with magnitudes σ_i, $i = 1, 2, \ldots, r$, distributed uniformly in the intervals $(a_i - \varepsilon_i, a_i + \varepsilon_i) \subset [0, l]$. They pollute the river independently according to Poisson processes with parameters $\lambda_1, \ldots, \lambda_r$. The distributions of σ_i, $i = 1, \ldots, r$, are measures $\mu_1, \ldots, \mu_r$ on $\mathbb{R}_+$. The corresponding Lévy process Z is then function-valued:

$$Z(t, \xi) = \sum_{k=1}^r \sum_{\tau_k^j \leq t} \sigma_k^j \chi_{(a_k - \varepsilon_k, a_k + \varepsilon_k)}(\xi).$$

20

Bond market models

Stochastic infinite-dimensional calculus and the theory of stochastic evolution equations find an interesting application in the mathematical finance of bond markets. In this chapter, models proposed by Heath, Jarrow and Morton (1992), but with Lévy noise, are investigated. The so-called HJM condition is derived, and a generalized Heath, Jarrow and Morton equation is considered. The mean reversion property of the solutions is established. The final part of the chapter is devoted to the so-called consistency problem in the Gaussian case.

20.1 Forward curves and the HJM postulate

The basic concept in bond market theory is the forward rate function. Denote by $P(t, \theta)$, $0 \leq t \leq \theta$, the market price at time t of a bond paying the amount 1 at time θ and by $(R(t),\ t \geq 0)$ the short-rate process offered by a bank. Functions $f(t, \theta)$, $0 \leq t \leq \theta$, defined by the relation

$$P(t, \theta) = \exp \left\{ - \int_t^\theta f(t, \sigma)\, d\sigma \right\}, \qquad t \leq \theta,$$

are called *forward rate functions*.

In Heath, Jarrow and Morton (1992) it was assumed that

$$df(t, \theta) = \alpha(t, \theta)\, dt + \langle \sigma(t, \theta), dW(t) \rangle, \tag{20.1}$$

where W is a d-dimensional Wiener process with covariance Q. According to the observed data, the (random) function $f(t, \theta)$ is regular in θ for fixed t and chaotic in t for fixed θ. The latter property is implied by the presence of W in the representation, and the former is implied by the regular dependence of $\alpha(t, \theta)$ and $\sigma(t, \theta)$ on θ for fixed t.

For practical implementation of bond market models it is useful to replace the Wiener process W by a Lévy process Z defined on a filtered probability space $(\Omega, \mathcal{F}, (\mathcal{F}_t), \mathbb{P})$ and taking values in a possibly infinite-dimensional Hilbert space $(U, \langle \cdot, \cdot \rangle_U)$. Thus we assume that the dynamics of the forward rate functions is given by the equation

$$\mathrm{d}f(t, \theta) = \alpha(t, \theta)\,\mathrm{d}t + \langle \sigma(t, \theta), \mathrm{d}Z(t) \rangle_U, \qquad t \le \theta. \tag{20.2}$$

For each $\theta \ge 0$ $\alpha(t, \theta)$ and $\sigma(t, \theta)$ are predictable processes. We may extend α, σ and P, putting

$$\alpha(t, \theta) := 0, \quad \sigma(t, \theta) := 0 \qquad \text{for } t \ge \theta \tag{20.3}$$

and

$$P(t, \theta) := \exp\left\{ \int_\theta^t R(s)\,\mathrm{d}s \right\} \qquad \text{for } t \ge \theta. \tag{20.4}$$

Let us note that if $t \le \theta$ then

$$f(t, \theta) = f(0, \theta) + \int_0^t \alpha(s, \theta)\,\mathrm{d}s + \int_0^t \langle \sigma(s, \theta), \mathrm{d}Z(s) \rangle_U,$$

and if $\theta \le t$ then

$$R(\theta) = f(0, \theta) + \int_0^\theta \alpha(s, \theta)\,\mathrm{d}s + \int_0^\theta \langle \sigma(s, \theta), \mathrm{d}Z(s) \rangle_U = f(t, \theta). \tag{20.5}$$

From now on we assume (20.2)–(20.4) and that the short rate is given by (20.5).

Let x be the initial capital $V(0)$ of an investor and assume that she or he buys $\pi_1(0), \ldots, \pi_N(0)$ bonds with maturities $0 < \theta_1 < \theta_2 < \cdots < \theta_N$ and puts the rest $\pi_0(0)$ into a bank account. The random variables $\pi_1(0), \ldots, \pi_N(0)$ should be $\mathcal{F}_0$-measurable. Assume in addition that the *portfolio*

$$\pi(t) := (\pi_1(t), \ldots, \pi_N(t)) = (\pi_1(0), \ldots, \pi_N(0))$$

is unchanged in the time interval $[0, t_1)$. Write

$$S_0(t) := \exp\left\{ \int_0^t R(s)\,\mathrm{d}s \right\}, \quad S_i(t) := P(t, \theta_i), \qquad t \ge 0, \ i = 1, \ldots, N.$$

Then

$$V(0) := x = \pi_0(0)S_0(0) + \pi_1(0)S_1(0) + \cdots + \pi_N(0)S_N(0)$$

and the total capital $V(t)$ of the investor at time $t \le t_1$ is equal to

$$V(t) := \pi_0(0)S_0(t) + \pi_1(0)S_1(t) + \cdots + \pi_N(0)S_N(t).$$

At time t_1 the investor fixes a new portfolio

$$\pi(t_1) = (\pi_1(t_1), \ldots, \pi_N(t_1))$$

of $\mathcal{F}_{t_1}$-measurable random variables, which is constant for $t \in (t_1, t_2]$. This procedure can be repeated for consecutive intervals and in this way one arrives at the concept of a *simple self-financing strategy* $(\pi(t), t \geq 0)$. We have the following result involving so-called *discounted capitals* $(\hat{V}(t), t \geq 0)$ and *discounted prices* $(\hat{S}(t), t \geq 0)$. This lemma can be extended to a large family of predictable processes π.

Lemma 20.1 *Let*

$$\hat{V}(t) := \frac{V(t)}{S_0(t)}, \qquad \hat{S}(t) := \left(\frac{S_1(t)}{S_0(t)}, \dots, \frac{S_N(t)}{S_0(t)} \right).$$

Then, for any simple self-financing strategy $(\pi_0(t), \dots, \pi_N(t))$,

$$\hat{V}(t) = V(0) + \int_0^t \langle \pi(s), d\hat{S}(s) \rangle. \tag{20.6}$$

Proof It is enough to consider $t \leq t_1$. Then

$$\hat{V}(t) = \pi_0(0) + \pi_1(0)\hat{S}_1(t) + \cdots + \pi_N(0)\hat{S}_N(t).$$

Therefore

$$
\begin{aligned}
\hat{V}(t) - \hat{V}(0) &= \hat{V}(t) - x \\
&= \pi_1(0)\big(\hat{S}_1(t) - \hat{S}_1(0)\big) + \cdots + \pi_N(0)\big(\hat{S}_N(t) - \hat{S}_1(0)\big),
\end{aligned}
$$

and the result follows. $\qquad\square$

Let us identify the strategy $\pi(t)$ with a measure-valued process

$$\Pi(t) := \sum_{j=1}^N \delta_{\{\theta_j\}} \pi_j(t), \qquad t \geq 0.$$

Let

$$\hat{P}(t, \theta) := \exp\left\{ -\int_0^t R(s)\,ds \right\} P(t, \theta), \qquad t \geq 0, \tag{20.7}$$

be the *discounted price* of the bond. Then $\langle \pi(t), \hat{S}(t) \rangle = (\hat{P}(t), \Pi(t))$, where (φ, λ) denotes the integral of the function φ with respect to the (not necessarily positive) measure λ. Then (20.6) can be rewritten with a stochastic integral,

$$\hat{V}(t) = V(0) + \int_0^t (d\hat{P}(s), \Pi(s)), \qquad t \geq 0. \tag{20.8}$$

with respect to the infinite-dimensional process $(\hat{P}(t), t \geq 0)$.

Let us assume that the process $(\hat{P}(t), t \geq 0)$ evolves on the space $H^1 := W^{1,2}(0, \infty, \mathcal{B}((0, \infty)), \nu)$ of absolutely continuous functions whose first derivative is square integrable with respect to a given non-negative measure ν. Then the dual space H^{1*} contains all measures of the form $\sum_{j=1}^{N} \gamma_j \delta_{\{\theta_j\}}$ as a dense subset. We therefore adopt (20.8) as the definition of the discounted capital for a strategy Π that is an H^{1*}-valued process stochastically integrable with respect to $\hat{P}$. For more details we refer the reader to Filipović (2001) and Zabczyk (2001b).

Definition 20.2   A strategy Π is called an *arbitrage opportunity* if $V(0) = 0$ and for some $t > 0$ one has $\mathbb{P}(V(t) \geq 0) = 1$ and $\mathbb{P}(V(t) > 0) > 0$.

It should be clear that in the above definition one can replace the capital V by the discounted capital.

The fundamental theorem of asset-pricing theory from Delbaen and Schachermayer (1994) states that under fairly general conditions there are no arbitrage-opportunity strategies if and only if there exists a probability measure $\hat{\mathbb{P}}$, equivalent to the initial one, $\mathbb{P}$, such that $(\hat{S}(t),\ t \geq 0)$ is a local martingale on $(\Omega, \mathcal{F}, \hat{\mathbb{P}})$. Measures $\hat{\mathbb{P}}$ with these properties are called *equivalent martingale measures*. In Heath, Jarrow and Morton (1992) the so-called HJM postulate was introduced:

HJM postulate   *For each $\theta \in [0, T]$ the process $\hat{P}(t, \theta)$, $t \leq \theta$, is a local martingale on $(\Omega, \mathcal{F}, (\mathcal{F}_t), \hat{\mathbb{P}})$ for some probability measure $\hat{\mathbb{P}}$ equivalent to $\mathbb{P}$.*

In applications of the theory one usually assumes that the measure $\mathbb{P}$ is a martingale measure. Note that if for each $\theta \geq 0$ the process $(\hat{P}(t, \theta),\ t \geq 0)$ is a local martingale then, under mild conditions on the strategy Π, the corresponding discounted capital is a local martingale (as a stochastic integral with respect to a local martingale) and therefore Π is not an arbitrage opportunity.

20.2 HJM condition

Let us recall (see Theorem 4.27) that $\mathbb{E}\, e^{i\langle x, Z(t)\rangle_U} = e^{-t\psi(x)}$, $x \in U$, $t \geq 0$, where

$$\psi(x) := -i\langle a, x\rangle_U + \tfrac{1}{2}\langle Qx, x\rangle_U + \psi_0(x),$$

$$\psi_0(x) := \int_U \left(1 - e^{i\langle x, y\rangle_U} + i\langle x, y\rangle_U \chi_{\{|y|_U \leq 1\}}(y)\right) \nu(dy), \qquad x \in U.$$

Moreover,

$$Z(t) = at + W(t) + \int_0^t \int_{\{|y|_U \leq 1\}} y\big(\pi(ds, dy) - ds\, \nu(dy)\big)$$

$$+ \int_0^t \int_{\{|y|_U > 1\}} y\, \pi(ds, dy),$$

where π is the Poisson random measure corresponding to Z. Under additional conditions the process Z has exponential moments, its Laplace transform exists and

$$\mathbb{E}\,e^{-\langle x, Z(t)\rangle_U} = e^{-t\tilde{\psi}(x)}, \qquad x \in U,$$

where $\tilde{\psi}$ is given in Theorem 4.30. Then $J := -\tilde{\psi}$ is given by

$$J(x) := -\langle a, x\rangle_U + \tfrac{1}{2}\langle Qx, x\rangle_U + J_0(x),$$
$$J_0(x) := \int_U \left(e^{-\langle x, y\rangle_U} - 1 + \langle x, y\rangle_U \chi_{\{|y|_U \le 1\}}\right) \nu(dy). \tag{20.9}$$

Let b be the Laplace transform of the measure ν restricted to the complement of the ball $\{y \colon |y|_U \le 1\}$, that is,

$$b(x) := \int_{\{|y|_U > 1\}} e^{-\langle x, y\rangle_U} \nu(dy),$$

and let B be the set of those $x \in U$ for which the Laplace transform is finite. Thus $B = \{x \in U \colon b(x) < \infty\}$. We intend now to prove a theorem from Jakubowski and Zabczyk (2004, 2007), which states "if and only if" conditions under which the discounted-price processes (20.7) are local martingales with respect to the probability $\mathbb{P}$. We will regard the coefficients α and σ in (20.2) as, respectively, $H = L^2([0, T])$-valued and $L(U, H)$-valued predictable processes given by

$$\alpha(t)(\theta) = \alpha(t, \theta), \qquad \theta \in [0, T],$$
$$\sigma(t)x(\theta) = \langle \sigma(t, \theta), x\rangle_U, \qquad x \in U,\ \theta \in [0, T].$$

For our purposes it is convenient to introduce the following condition on the jump intensity measure ν:

$$\forall\, r > 0 \text{ the function } b \text{ is bounded on } \{x \in B \colon |x|_U \le r\}. \tag{20.10}$$

In the theorem below $J \colon U \mapsto \mathbb{R}$ is given by (20.9).

Theorem 20.3 *Assume that the predictable processes α and σ have bounded trajectories and that (20.10) is satisfied.*

(i) *If the discounted-price processes given by (20.7) are local martingales then, for all $\theta \le T$,*

$$\int_t^\theta \sigma(t, v)\, dv \in B, \qquad \mathbb{P}\text{-a.s. for almost all } t \in [0, \theta]. \tag{20.11}$$

(ii) *Assume (20.11). Then the discounted-price processes (20.7) are local martingales if and only if*

$$\int_t^\theta \alpha(t, v)\,\mathrm{d}v = J\left(\int_t^\theta \sigma(t, v)\,\mathrm{d}v\right), \tag{20.12}$$

$\forall \theta \le T$, P-*a.s. for almost all* $t \in [0, \theta]$.

Remark 20.4 We call (20.12) the *HJM condition*. Let D be the derivative operator acting on functions defined on U. Note that the theorem says that under very mild assumptions the discounted-price processes are local martingales if and only if (20.12) holds or, equivalently, if and only if

$$\alpha(t, \theta) = \frac{\mathrm{d}}{\mathrm{d}\theta} J\left(\int_t^\theta \sigma(t, v)\,\mathrm{d}v\right) = \left\langle DJ\left(\int_t^\theta \sigma(t, v)\,\mathrm{d}v\right), \ \sigma(t, \theta)\right\rangle_U.$$

Thus the dynamics of the forward-rate functions is given by

$$\mathrm{d}f(t, \theta) = \left\langle DJ\left(\int_t^\theta \sigma(t, v)\,\mathrm{d}v\right), \ \sigma(t, \theta)\right\rangle_U \mathrm{d}t + \langle\sigma(t, \theta), \mathrm{d}Z(t)\rangle_U.$$

Note that the drift term is completely determined by the diffusion term.

Remark 20.5 In the particular case where Z is a Wiener process with covariance Q one arrives at the *classical HJM condition*

$$\int_t^\theta \alpha(t, v)\,\mathrm{d}v = \tfrac{1}{2}\left\langle Q\int_t^\theta \sigma(t, v)\,\mathrm{d}v, \ \int_t^\theta \sigma(t, v)\,\mathrm{d}v\right\rangle_U$$

introduced in Heath, Jarrow and Morton (1992). Clearly the condition above holds if and only if

$$\alpha(t, \theta) = \left\langle Q\sigma(t, \theta), \ \int_t^\theta \sigma(t, v)\,\mathrm{d}v\right\rangle_U$$

for every $\theta \le T$, P-a.s. for almost all $t \in [0, \theta]$.

Remark 20.6 Formulae similar to (20.12) were obtained in Björk *et al.* (1997), Björk, Kabanov and Runggaldier (1997) and Eberlein and Raible (1999).

Proof of Theorem 20.3   Fix a time $\theta \le T$. For $t \in [0, T]$ set

$$A(t, \theta) := \left\langle \chi_{[0,\theta]}, \alpha(t)\right\rangle_H = \int_0^\theta \alpha(t, \eta)\,\mathrm{d}\eta = \int_t^\theta \alpha(t, \eta)\,\mathrm{d}\eta,$$

$$\Sigma(t, \theta) := \sigma^*(t)\chi_{[0,\theta]} = \int_t^\theta \sigma(t, \eta)\,\mathrm{d}\eta.$$

Since θ is fixed, in the following calculations we omit θ and write $A(t)$ and $\Sigma(t)$. Let

$$X(t) := \langle \chi_{[0,\theta]}, f(t) \rangle_H = \int_0^\theta f(t, \eta)\, d\eta.$$

Then

$$
\begin{aligned}
dX(t) &= A(t)\, dt + \langle \Sigma(t), dZ(t) \rangle_U \\
&= A(t)\, dt + \langle \Sigma(t), a\, dt + dW(t) \rangle_U \\
&\quad + \int_U \chi_{\{|y|_U \le 1\}}(y)\langle \Sigma(s), y \rangle_U \left(\pi(dt, dy) - dt\, \nu(dy) \right) \\
&\quad + \int_U \chi_{\{|y|_U > 1\}}(y)\langle \Sigma(s), y \rangle_U \pi(ds, dy).
\end{aligned}
$$

To apply the Itô formula to $\psi(X(t))$ for a given function $\psi \in C^2(\mathbb{R})$, denote by μ_X the jump measure of the semimartingale X. We have $\Delta X(t) = \langle \Sigma(t), \Delta Z(t) \rangle_U$. Therefore

$$\mu_X([0, t], \Gamma) = \sum_{s \le t} \chi_\Gamma\left(\langle \Sigma(s), \Delta Z(s) \rangle_U \right) = \int_0^t \int_U \chi_\Gamma\left(\langle \Sigma(s), y \rangle_U \right) \pi(ds, dy),$$

and, more generally, for a non-negative predictable field $\varphi(s, z)$, $s \ge 0$, $z \in \mathbb{R}$,

$$\int_0^t \int_{\mathbb{R}} \varphi(s, z)\, \mu_X(ds, dz) = \int_0^t \int_U \varphi\left(s, \langle \Sigma(s), y \rangle_U \right) \pi(ds, dy).$$

Moreover, the quadratic-variation process of $\int_0^t \langle \Sigma(s), dW(s) \rangle_U$ is equal to $\int_0^t \langle Q\Sigma(s), \Sigma(s) \rangle_U\, ds$. Consequently, the Itô formula from Appendix D yields

$$
\begin{aligned}
\psi(X(t)) ={}& \psi(X(0)) + \int_0^t \psi'(X(s-))\, dX(s) \\
&+ \frac{1}{2} \int_0^t \psi''(X(s))\langle Q\Sigma(s), \Sigma(s) \rangle_U\, ds \\
&+ \sum_{s \le t} \left(\psi(X(s)) - \psi(X(s-)) - \psi'(X(s-))\Delta X(s) \right) \\
=:{}& \psi(X(0)) + I_1(t) + I_2(t) + I_3(t).
\end{aligned}
$$

We have

$$
\begin{aligned}
I_1(t) ={}& M_1(t) + \int_0^t \psi'(X(s-))\left(A(t) + \langle \Sigma(s), a \rangle_U \right) ds \\
&+ \int_0^t \int_U \chi_{\{|y|_U > 1\}}(y)\psi'(X(s-))\langle \Sigma(s), y \rangle_U\, \pi(ds, dy),
\end{aligned}
$$

where M_1 is a local martingale, since it is the sum of a Wiener integral and a stochastic integral with respect to the compensated jump measure $\widehat{\pi}$. Write

$$\Psi(t)(y) := \psi\big(X(t-) + \langle \Sigma(t), y \rangle_U\big) - \psi(X(t-))$$

and

$$\Phi(t)(y) := \psi'(X(t-))\langle \Sigma(t), y \rangle_U.$$

Then

$$I_3 = \int_0^t \int_{\mathbb{R}} \big(\psi(X(s-) + z) - \psi(X(s-)) - \psi'(X(s-))z\big)\, \mu_X(\mathrm{d}s, \mathrm{d}z)$$

$$= \int_0^t \int_U \big(\Psi(s)(y) - \Phi(s)(y)\big)\pi(\mathrm{d}s, \mathrm{d}y).$$

Consequently,

$$\psi(X(t)) = \psi(X(0)) + M_1(t)$$

$$+ \int_0^t \psi'(X(s-))\big(A(t) + \langle \Sigma(s), a \rangle_U\big)\, \mathrm{d}s$$

$$+ \frac{1}{2} \int_0^t \psi''(X(s))\langle Q\Sigma(s), \Sigma(s) \rangle_U\, \mathrm{d}s$$

$$+ \int_0^t \int_U \big(\Psi(s)(y) - \Phi(s)(y)\big)\pi(\mathrm{d}s, \mathrm{d}y). \qquad (20.13)$$

The discounted prices are local martingales if $(\psi(X(t)),\ t \geq 0)$ is a local martingale for $\psi(x) = e^{-x}, x \in \mathbb{R}$. Thus, there exists an increasing sequence (τ_n) of stopping times such that the integrals

$$\int_0^t \int_U \chi_{[0,\tau_n](s)}\big(\Psi(s)(y) - \Phi(s)(y)\big)\pi(\mathrm{d}s, \mathrm{d}y), \qquad n \in \mathbb{N},$$

are random variables with finite expectation. Let

$$\xi_n(t, y) := \chi_{[0,\tau_n](t)}\big(\Psi(t)(y) - \Phi(t)(y)\big).$$

Then

$$\int_0^t \int_U \xi_n(s, y)^+ \pi(\mathrm{d}s, \mathrm{d}y) \quad \text{and} \quad \int_0^t \int_U \xi_n(s, y)^- \pi(\mathrm{d}s, \mathrm{d}y)$$

are also integrable random variables. Since the random fields ξ_n^+, ξ_n^- are predictable, we have

$$\mathbb{E} \int_0^t \int_U |\xi_n(s, y)|\pi(\mathrm{d}s, \mathrm{d}y) = \mathbb{E} \int_0^t \int_U |\xi_n(s, y)|\, \mathrm{d}s\, \nu(\mathrm{d}y) < \infty.$$

Consequently,

$$\mathbb{E} \int_0^t \int_U \chi_{[0,\tau_n]}(s) \chi_{\{|y|_U > 1\}} \big| \psi \big(X(s-) + \langle \Sigma(s), y \rangle_U \big) - \psi(X(s-)) \big| \, ds \, \nu(dy)$$

is finite, and hence

$$\mathbb{E} \int_0^t \int_U \chi_{[0,\tau_n]}(s) e^{-X(s-)} \int_{\{|y|_U > 1\}} \big| e^{-\langle \Sigma(s), y \rangle_U} - 1 \big| \, \nu(dy) \, ds < \infty.$$

Hence, for each natural n, $\mathbb{P}$-a.s. $\int_0^{\tau_n} b(\Sigma(s)) \, ds < \infty$, and the first part of the theorem follows.

To prove the second part, consider formula (20.13) with $\psi(x) = e^{-x}$. Let Ψ and Φ be as in the proof of the first part. Note that if (20.10) holds then

$$J(t) := \int_0^t \int_U |\Psi(s) - \Phi(s)| \, \nu(dz) \, ds < \infty.$$

Therefore

$$\int_0^t \int_U \big(\Psi(s)(y) - \Phi(s)(y) \big) \pi(ds, dy)$$

$$= \int_0^t \int_U \big(\Psi(s)(y) - \Phi(s)(y) \big) \big(\pi(ds, dy) - ds \, \nu(dy) \big)$$

$$+ \int_0^t \int_U \big(\Psi(s)(y) - \Phi(s)(y) \big) ds \, \nu(dy).$$

Consequently, (20.13) can be rewritten as

$$e^{-X(t)} = e^{-X(0)} + M_2(t) + \int_0^t \psi'(X(s-)) \big(A(s) - J(\Sigma(s)) \big) \, ds,$$

where M_2 is a local martingale. This finishes the proof of the theorem. □

20.3 HJMM equation

An important link between HJM modeling and stochastic partial differential equations is provided by the so-called *Musiela parametrization*. Assume that

$$df(t, \theta) = \alpha(t, \theta) \, dt + \langle \sigma(t, \theta), dZ(t) \rangle_U,$$

and for $t \geq 0, \xi \geq 0$ and $u \in U$ define

$$r(t)(\xi) := f(t, t + \xi), \quad a(t)(\xi) := \alpha(t, t + \xi),$$

$$(b(t)u)(\xi) := \langle \sigma(t, t + \xi), u \rangle_U.$$

We will call r the *forward curve*. Next, let $S(t)\varphi(\xi) = \varphi(\xi + t)$ be the shift semi-group. Then

$$r(t)(\xi) = f(t, t + \xi)$$

$$= f(0, t + \xi) + \int_0^t \alpha(s, t + \xi)\,ds + \int_0^t \langle \sigma(s, t + \xi), dZ(s) \rangle_U$$

$$= r(0)(t + \xi) + \int_0^t a(s)(t - s + \xi)\,ds + \int_0^t b(s)(t - s + \xi)\,dZ(s)$$

$$= S(t)r(0)(\xi) + \int_0^t S(t - s)a(s)(\xi)\,ds + \int_0^t S(t - s)b(s)(\xi)\,dZ(s).$$

Thus

$$r(t) = S(t)r(0) + \int_0^t S(t - s)a(s)\,ds + \int_0^t S(t - s)b(s)\,dZ(s)$$

is a mild solution to the equation

$$dr(t) = \left(\frac{\partial}{\partial \xi} r(t) + a(t) \right) dt + b(t)\,dZ(t),$$

where $\partial/\partial\xi$ denotes the generator of $(S(t), t \geq 0)$. Identifying the $L(U, \mathbb{R})$-valued process $b(\cdot)(\xi)$ with the corresponding U-valued process (denoted also by $b(\cdot)(\xi)$) we note that if the HJM condition is satisfied then

$$dr(t)(\xi) = \left(\frac{\partial}{\partial \xi} r(t)(\xi) + \left\langle b(t)(\xi), \ DJ \left(\int_0^\xi b(t)(\eta)\,d\eta \right) \right\rangle_U \right) dt$$

$$+ b(t)(\xi)\,dZ(t)$$

$$= \frac{\partial}{\partial \xi} \left(r(t)(\xi) + J \left(\int_0^\xi b(t)(\eta)\,d\eta \right) \right) dt + b(t)(\xi)\,dZ(t). \quad (20.14)$$

Let the *volatility* b depend on the forward curve r according to, say, $b(t)(\xi) = G(t, r(t))(\xi)$, and let

$$F(t, r)(\xi) := \left\langle G(t, r(t))(\xi), \ DJ \left(\int_0^\xi G(t, r(t))(\eta)\,d\eta \right) \right\rangle_U$$

$$= \frac{\partial}{\partial \xi} J \left(\int_0^\xi G(t, r(t))(\eta)\,d\eta \right). \quad (20.15)$$

Then the forward-curve process becomes a solution of the so-called *Heath–Jarrow–Morton–Musiela (HJMM) equation*

$$dr(t)(\xi) = \left(\frac{\partial}{\partial \xi} r(t)(\xi) + F(t, r(t))(\xi) \right) dt + G(t, r(t))(\xi)\,dZ(t). \quad (20.16)$$

20.3.1 Existence of solutions

Following Peszat, Rusinek and Zabczyk (2007), we derive the existence of a solution to (20.16) from the general theorem 9.29. We assume that the driving noise Z is a finite-dimensional, say $\mathbb{R}^d$-valued, square integrable martingale. Thus $U = \mathbb{R}^d$ and Z is a sum of a Wiener process and a compensated jump process, and therefore

$$J(z) = \tfrac{1}{2}\langle Qz, z\rangle + \int_{\mathbb{R}^d} \left(e^{-\langle z, y\rangle} - 1 + \langle z, y\rangle \right) \nu(\mathrm{d}y), \qquad (20.17)$$

where Q is a symmetric non-negative-definite matrix and the jump measure ν satisfies $\int_{\mathbb{R}^d} |y|^2 \nu(\mathrm{d}y) < \infty$. Here we denote by $\langle \cdot, \cdot \rangle$ the scalar product on $\mathbb{R}^d$ and by $|\cdot|$ the corresponding norm. We have the following elementary fact.

Lemma 20.7

(i) *If $z \in \mathbb{R}^d$ is such that $\int_{\{|y|\geq 1\}} |y| e^{|z||y|} \nu(\mathrm{d}y) < \infty$ then J is differentiable at z and*

$$DJ(z) = Qz + \int_{\mathbb{R}^d} y \left(1 - e^{-\langle z, y\rangle} \right) \nu(\mathrm{d}y).$$

(ii) *If $z \in \mathbb{R}^d$ is such that $\int_{\mathbb{R}^d} |y|^2 e^{|z||y|} \nu(\mathrm{d}y) < \infty$ then J is twice differentiable at z and*

$$D^2 J(z) = Q + \int_{\mathbb{R}^d} y \otimes y\, e^{-\langle z, y\rangle} \nu(\mathrm{d}y),$$

where $y \otimes y[v] = \langle y, v\rangle y$, $v \in \mathbb{R}^d$.

We assume that G is of composition type, that is,

$$G(t, r(t))(\xi)[z] = \langle g(t, \xi, r(t)(\xi)),\ z\rangle, \qquad t, \xi \in [0, +\infty),\ z \in \mathbb{R}^d, \quad (20.18)$$

where $g \colon [0, +\infty) \times [0, +\infty) \times \mathbb{R} \mapsto \mathbb{R}^d$. We identify $G(t, \psi)(\xi)$ with the vector $g(t, \xi, \psi(\xi))$ in $\mathbb{R}^d$.

Given $\gamma > 0$ we consider the equation on the state space $\mathbf{H}_\gamma := H_\gamma \oplus \{\text{constant functions}\}$, where $H_\gamma := L^2 \left([0, +\infty), \mathcal{B}([0, +\infty)), e^{\gamma \xi} \mathrm{d}\xi \right)$. Note that $\mathbf{H}_\gamma$, equipped with the scalar product $\langle \psi + u, \varphi + v\rangle_{\mathbf{H}_\gamma} := \langle \psi, \varphi\rangle_{H_\gamma} + uv$, $\psi, \varphi \in H_\gamma$, $u, v \in \mathbb{R}$, is a real separable Hilbert space.

Let S be the shift semigroup. Then, for $\psi \in H_\gamma$,

$$|S(t)\psi|^2_{H_\gamma} = \int_0^{+\infty} |\psi(\xi + t)|^2 e^{\gamma \xi}\, \mathrm{d}\xi = \int_t^{+\infty} |\psi(\eta)|^2 e^{\gamma(\eta - t)}\, \mathrm{d}\eta \leq e^{-\gamma t} |\psi|^2_{H_\gamma},$$

and hence the following lemma holds.

Lemma 20.8   *S is a C_0-semigroup on H_γ and $\mathbf{H}_\gamma$. Moreover,*

$$\|S(t)\|_{L(H_\gamma, H_\gamma)} \leq e^{-\gamma t/2}, \quad \|S(t)\|_{L(\mathbf{H}_\gamma, \mathbf{H}_\gamma)} = 1, \qquad t \geq 0.$$

By the Hölder inequality, for every $\gamma > 0$ the space H_γ is continuously embedded into $L^1 := L^1([0, +\infty), \mathcal{B}([0, +\infty)), \mathrm{d}\xi)$ and $|\psi|_{L^1} \leq \gamma^{-1/2}|\psi|_{H_\gamma}$ for all $\psi \in H_\gamma$. We can formulate our first existence theorem.

Theorem 20.9 *Let Z be an $\mathbb{R}^d$-valued square integrable mean-zero Lévy process with jump measure v, and let G be given by (20.18). Assume that there are functions $\overline{g} \in H_\gamma$ and $\overline{h} \in H_\gamma \cap L^\infty$ such that*

(i) $\int_{\mathbb{R}} y^2 \exp\{|\overline{g}|_{L^1}|y|\} v(\mathrm{d}y) < \infty,$
(ii) *for all $t, \xi \in [0, +\infty)$ and $u, v \in \mathbb{R}$*

$$|g(t, \xi, u)| \leq \overline{g}(\xi), \qquad |g(t, \xi, u) - g(t, \xi, v)| \leq \overline{h}(\xi)|u - v|.$$

Then, for each $r_0 \in \mathbf{H}_\gamma$ (for each $r_0 \in H_\gamma$) there is a unique solution r to (20.16) in $\mathbf{H}_\gamma$ (in H_γ) satisfying $r(0) = r_0$. Moreover, if the coefficient g does not depend on t then (20.16) defines (time-homogeneous) Feller families on $\mathbf{H}_\gamma$ and on H_γ.

We note that this book is indeed concerned mainly with time-homogeneous equations. Clearly (20.16) can be written as a time-homogeneous equation of variables $X = (r, t)$ on the state space $\mathbf{H}_\gamma \times \mathbb{R}$. Therefore the theorem is a direct consequence of Lemmas 20.7 and 20.8, Theorem 9.29 and the following lemma, which will be useful elsewhere also (see the proof of Theorem 20.19 below). It will be convenient to introduce the following notation:

$$K_1(J, \overline{g}) := \sup_{z:\, |z| \leq |\overline{g}|_{L^1}} |DJ(z)|, \qquad K_2(J, \overline{g}) := \sup_{z:\, |z| \leq |\overline{g}|_{L^1}} \left\| D^2 J(z) \right\|_{L(\mathbb{R}^d, \mathbb{R}^d)}.$$

Clearly, if assumption (i) of Theorem 20.9 is satisfied then, by Lemma 20.7, J is twice differentiable at an arbitrary z with $|z| \leq |\overline{g}|_{L^1}$ and $K_i(J, \overline{g}) < \infty, i = 1, 2$.

Lemma 20.10 *Under the assumptions of Theorem 20.9, for every $t \geq 0$ we have $G(t, \cdot): \mathbf{H}_\gamma \mapsto L_{(HS)}(\mathbb{R}^d, H_\gamma)$ and $F(t, \cdot): \mathbf{H}_\gamma \mapsto H_\gamma$. Moreover, the following estimates hold.*

(i) *For all $t \geq 0$ and $\psi \in \mathbf{H}_\gamma$,*

$$|F(t, \psi)|^2_{H_\gamma} + \|G(t, \psi)\|^2_{L_{(HS)}(\mathbb{R}^d, H_\gamma)} \leq \left(K_1^2(J, \overline{g}) + 1\right)|\overline{g}|^2_{H_\gamma}.$$

(ii) *For all $t \leq 0$ and $\psi, \varphi \in H_\gamma$,*

$$\|G(t, \psi) - G(t, \varphi)\|_{L_{(HS)}(\mathbb{R}^d, H_\gamma)} \leq |\overline{h}|_{L^\infty} |\psi - \varphi|_{H_\gamma},$$
$$|F(t, \psi) - F(t, \varphi)|_{H_\gamma} \leq K|\psi - \varphi|_{H_\gamma},$$

where $K := |\overline{h}|_{L^\infty}\left(2K_2(J, \overline{g})|\overline{g}|^2_{H_\gamma} + 2K_1(J, \overline{g})\right)^{1/2}.$

(iii) *For all $t \geq 0$ and $\psi, \varphi \in \mathbf{H}_\gamma$,*

$$|F(t, \psi) - F(t, \varphi)|_{H_\gamma}^2 + \|G(t, \psi) - G(t, \varphi)\|_{L_{(HS)}(\mathbb{R}^d, H_\gamma)}^2 \leq \tilde{K} \, |\psi - \varphi|_{\mathbf{H}_\gamma}^2,$$

where $\tilde{K} := 2\big(|\overline{h}|_{L^\infty}^2 + |\overline{h}|_{H_\gamma}^2\big)\big(1 + 2K_2(J, \overline{g})|\overline{g}|_{H_\gamma}^2 + 2K_1(J, \overline{g})\big).$

Proof Take $t \geq 0$ and $\psi \in \mathbf{H}_\gamma$. Then

$$\|G(t, \psi)\|_{L_{(HS)}(\mathbb{R}^d, H_\gamma)}^2 = \int_0^\infty |g(t, \xi, \psi(\xi))|^2 e^{\gamma\xi} \, d\xi \leq |\overline{g}|_{H_\gamma}^2.$$

Next, treating $G(t, \psi)(\eta)$ as a vector in $\mathbb{R}^d$,

$$\left| \int_0^\xi G(t, \psi)(\eta) \, d\eta \right| \leq \int_0^\infty |\overline{g}(\eta)| \, d\eta = |\overline{g}|_{L^1}. \tag{20.19}$$

Hence, by the first assumption of the theorem and Lemma 20.7, for every $\xi > 0$ we have that $\int_0^\xi G(t, \psi)(\eta) \, d\eta$ belongs to the domain of the derivative of J,

$$\left| DJ \left(\int_0^\xi G(t, \psi)(\eta) \, d\eta \right) \right| \leq K_1(J, \overline{g}) < \infty,$$

and (i) follows.

To see the Lipschitz continuity note that, for $\psi, \varphi \in \mathbf{H}_\gamma$ and $t \geq 0$,

$$\|G(t, \psi) - G(t, \varphi)\|_{L_{(HS)}(\mathbb{R}^d, H_\gamma)}^2 \leq \int_0^\infty |\overline{h}(\xi)|^2 |\psi(\xi) - \varphi(\xi)|^2 e^{\gamma\xi} \, d\xi.$$

Thus $C(\psi, \varphi) := \|G(t, \psi) - G(t, \varphi)\|_{L_{(HS)}(\mathbb{R}^d, H_\gamma)}^2$ is estimated as

$$\begin{cases} |\overline{h}|_{L^\infty}^2 |\psi - \varphi|_{H_\gamma}^2 & \text{if } \psi, \varphi \in H_\gamma, \\ \overline{h}|_{H_\gamma}^2 |\psi - \varphi|^2 & \text{if } \psi, \varphi \in \mathbb{R}, \\ 2\big(|\overline{h}|_{H_\gamma}^2 + |\overline{h}|_{L^\infty}^2\big)\big(|\psi|_{H_\gamma}^2 + |\varphi|^2\big) & \text{if } \psi \in H_\gamma, \varphi \in \mathbb{R}. \end{cases}$$

Since for $\psi \in H_\gamma$ and $\varphi \in \mathbb{R}$ we have $|\psi - \varphi|_{\mathbf{H}_\gamma}^2 = |\psi|_{H_\gamma}^2 + |\varphi|^2$,

$$C(\psi, \varphi) \leq \begin{cases} |\overline{h}|_{L^\infty}^2 |\psi - \varphi|_{H_\gamma}^2, & \text{if } \psi, \varphi \in H_\gamma, \\ 2\big(|\overline{h}|_{L^\infty}^2 + |\overline{h}|_{H_\gamma}^2\big)|\psi - \varphi|_{\mathbf{H}_\gamma}^2, & \text{if } \psi, \varphi \in \mathbf{H}_\gamma. \end{cases}$$

We will now show the Lipschitz continuity of F. Let $\psi, \varphi \in \mathbf{H}_\gamma$ and $t \geq 0$. Clearly

$$|F(t, \psi) - F(t, \varphi)|_{H_\gamma}^2 \leq 2(I_1 + I_2),$$

where

$$I_1 := \int_0^\infty |G(t, \psi)(\xi)|^2 \left| DJ \left(\int_0^\xi G(t, \psi)(\eta) \, d\eta \right) \right.$$

$$\left. - DJ \left(\int_0^\xi G(t, \varphi)(\eta) \, d\eta \right) \right|^2 e^{\gamma\xi} \, d\xi$$

and

$$I_2 := \int_0^\infty |G(t,\psi)(\xi) - G(t,\varphi)(\xi)|^2 \left| DJ\left(\int_0^\xi G(t,\varphi)(\eta)\,d\eta\right)\right|^2 e^{\gamma\xi}\,d\xi.$$

By (20.19),

$$I_1 \le K_2(J,\overline{g}) \int_0^\infty |\overline{g}(\xi)|^2 \left(\int_0^\xi |G(t,\psi)(\eta) - G(t,\varphi)(\eta)|\,d\eta\right)^2 e^{\gamma\xi}\,d\xi$$

$$\le K_2(J,\overline{g})\,|\overline{g}|_{H_\gamma}^2 \int_0^\infty |G(t,\psi)(\eta) - G(t,\varphi)(\eta)|^2\,d\eta$$

$$\le K_2(J,\overline{g})\,|\overline{g}|_{H_\gamma}^2\,\|G(t,\psi) - G(t,\varphi)\|_{L_{(HS)}(\mathbb{R}^d,H_\gamma)}^2$$

and

$$I_2 \le K_1(J,\overline{g})\,\|G(t,\psi) - G(t,\varphi)\|_{L_{(HS)}(\mathbb{R}^d,H_\gamma)}^2.$$

□

Remark 20.11 Note that for any initial value $r_0 \in \mathbf{H}_\gamma$ the solution to (20.16) is a càdlàg process in $\mathbf{H}_\gamma$. Indeed, the stochastic term $\int_0^t S(t-s)G(s,r(s))\,dZ(s)$ can be written as $\int_0^t S(t-s)\,dM(s)$, where $M(t) = \int_0^t G(s,r(s))\,dZ(s)$ is a square integrable martingale in $\mathbf{H}_\gamma$. Since, by Lemma 20.8, S is a semigroup of contractions on $\mathbf{H}_\gamma$ we use inference from the Kotelenez theorem; see Theorem 9.20. Next, if Z is a Wiener process then by factorization one obtains the continuity of r in $\mathbf{H}_\gamma$; see Theorem 11.8.

20.3.2 A special case

We restrict our attention to the special case of (20.16) for which Z is two-dimensional and G is, as in subsection 20.3.1, of composition type. In fact we assume that $Z = (W,L)$ where W is a standard real-valued Wiener process and L is an independent square integrable real-valued Lévy martingale with Laplace exponent

$$\log \mathbb{E}\,e^{-zL(1)} = \int_{\mathbb{R}} (e^{-zy} - 1 + zy)\nu(dy), \qquad \int_{\mathbb{R}} y^2\nu(dy) < \infty. \tag{20.20}$$

Note that, using the notation from the previous subsection, $d = 2$ and

$$J(z_1,z_2) = \tfrac{1}{2}z_1^2 + \int_{\mathbb{R}} \left(e^{-z_2y} - 1 + z_2y\right)\nu(dy). \tag{20.21}$$

Therefore we are concerned with the following equation:

$$dr(t)(\xi) = \left(\frac{\partial}{\partial\xi}r(t)(\xi) + F(r)(\xi)\right)dt + g_1(t,\xi,r(t)(\xi))\,dW(t)$$
$$+ g_2(t,\xi,r(t)(\xi))\,dL(t), \tag{20.22}$$

where $g_i : [0, +\infty) \times [0, +\infty) \times \mathbb{R} \mapsto \mathbb{R}$, $i = 1, 2$ and

$$F(t, \psi)(\xi) = g_1(t, \xi, \psi(\xi)) \int_0^\xi g_1(t, \eta, \psi(\eta)) \, \mathrm{d}\eta$$

$$+ g_2(t, \xi, \psi(\xi)) \int_\mathbb{R} y \left(1 - \exp \left\{ -y \int_0^\xi g_2(t, \eta, \psi(\eta)) \, \mathrm{d}\eta \right\} \right) \nu(\mathrm{d}y).$$

Note that $G(t, \psi)[z_1, z_2](\xi) = g_1(t, \xi, \psi(\xi)) z_1 + g_2(t, \xi, \psi(\xi)) z_2$.

Theorem 20.12 *Assume that ν is supported in $[-m, +\infty)$ for some $m \geq 0$ and that $g_2(t, \xi, u) \geq 0$ for all $t, \xi \geq 0$ and $u \in \mathbb{R}$. Moreover, we assume that there are functions $\overline{g} \in \mathbf{H}_\gamma$ and $\overline{h} \in H_\gamma \cap L^\infty$ such that, for all $t, \xi \in [0, +\infty)$ and $u, v \in \mathbb{R}$,*

$$|g_i(t, \xi, u)| \leq \overline{g}(\xi), \quad |g_i(t, \xi, u) - g_i(t, \xi, v)| \leq \overline{h}(\xi) |u - v|, \qquad i = 1, 2.$$

Then for each $r_0 \in \mathbf{H}_\gamma$ (each $r_0 \in H_\gamma$) there is a unique solution r to (20.22) in $\mathbf{H}_\gamma$ (in H_γ) satisfying $r(0) = r_0$. Moreover, if the coefficients g_i do not depend on t then (20.22) defines (time-homogeneous) Feller families on $\mathbf{H}_\gamma$ (on H_γ).

Proof The proof follows the ideas of the proof of Theorem 20.9. Only the fact that we do not assume that $\int_\mathbb{R} y^2 \exp\{|\overline{g}|_{L^1} |y|\} \nu(\mathrm{d}y) < \infty$ needs to be explained. This assumption is allowed since g_2 is non-negative and ν has its support in $[-m, +\infty)$. In fact we have the following version of Lemma 20.10. Its proof is left to the reader. $\qquad\square$

To formulate the result we need the following analogues of $K_1(J, \overline{g})$ and $K_2(J, \overline{g})$. Let

$$\tilde{J}(z) = \int_{-m}^{+\infty} \left(\mathrm{e}^{-zy} - 1 + zy \right) \nu(\mathrm{d}y),$$

and let

$$\tilde{K}_1(\tilde{J}, \overline{g}) := \sup_{0 \leq z \leq |\overline{g}|_{L^1}} |\tilde{J}'(z)|, \qquad K_2(\tilde{J}, \overline{g}) := \sup_{0 \leq z \leq |\overline{g}|_{L^1}} |\tilde{J}''(z)|.$$

Note that $\tilde{K}_i(\tilde{J}, \overline{g}_2) < \infty$, $i = 1, 2$.

Lemma 20.13 *Under the assumptions of Theorem 20.12, for every $t \geq 0$, $G(t, \cdot) : \mathbf{H}_\gamma \mapsto L_{(HS)}(\mathbb{R}^2, H_\gamma)$ and $F(t, \cdot) : \mathbf{H}_\gamma \mapsto H_\gamma$. Moreover, we have the following.*

(i) *For all $t \geq 0$ and $\psi \in \mathbf{H}_\gamma$,*

$$|F(t, \psi)|_{H_\gamma}^2 + \|G(t, \psi)\|_{L_{(HS)}(\mathbb{R}^2, H_\gamma)}^2 \leq 2|\overline{g}|_{H_\gamma}^2 \left(1 + |\overline{g}|_{L^1}^2 + \tilde{K}_1^2(\tilde{J}, \overline{g}) \right).$$

(ii) *For all $t \leq 0$ and $\psi, \varphi \in H_\gamma$,*

$$\|G(t, \psi) - G(t, \varphi)\|_{L_{(HS)}(\mathbb{R}^2, H_\gamma)} \leq 2|\overline{h}|_{L^\infty} |\psi - \varphi|_{H_\gamma},$$

$$|F(t, \psi) - F(t, \varphi)|_{H_\gamma} \leq K |\psi - \varphi|_{H_\gamma},$$

where $K := |\overline{h}|_{L^\infty}\big(1 + |\overline{g}|_{H_\gamma} + 2\tilde{K}_2(\tilde{J}, \overline{g})|\overline{g}|^2_{H_\gamma} + 2\tilde{K}_1(\tilde{J}, \overline{g})\big)^{1/2}$.

(iii) *For all $t \geq 0$ and $\psi, \varphi \in \mathbf{H}_\gamma$,*

$$|F(t, \psi) - F(t, \varphi)|^2_{H_\gamma} + \|G(t, \psi) - G(t, \varphi)\|^2_{L_{(HS)}(\mathbb{R}^2, H_\gamma)} \leq \tilde{K} \, |\psi - \varphi|^2_{\mathbf{H}_\gamma},$$

where

$$\tilde{K} := 8\big(|\overline{h}|^2_{L^\infty} + |\overline{h}|^2_{H_\gamma}\big)\big(1 + |\overline{g}|^2_{H_\gamma} + 2\tilde{K}_2(\tilde{J}, \overline{g})|\overline{g}|^2_{H_\gamma} + 2\tilde{K}_1(\tilde{J}, \overline{g})\big).$$

20.3.3 Positivity in the special case

We call a function $\psi \colon [0, +\infty) \mapsto \mathbb{R}$ *non-negative* if $\psi(\xi) \geq 0$ for almost all $\xi \geq 0$. Clearly, in all models, the forward-curve functions need to take non-negative values. We present here sufficient conditions on the coefficient G and the noise Z under which (20.16) preserves positivity, that is, for every non-negative initial value $r(0)$ the functions $r(t)$, $t \geq 0$, are non-negative.

We restrict our attention to the special case of (20.16) considered in subsection 20.3.2. To simplify the exposition we consider only time-independent coefficients g_i, $i = 1, 2$.

Theorem 20.14 *Assume that $g_1(\xi, 0) = 0$ for $\xi \in [0, \infty)$ and that one of the following conditions holds:*

(i) *$g = (g_1, g_2)$ satisfies the assumptions of Theorem 20.9, ν is supported on $[-m, M]$ for some $m, M > 0$ and $|g_2(\xi, u)| \leq uM^{-1} \wedge m^{-1}$ for all $\xi \geq 0$ and $u \geq 0$;*

(ii) *ν is supported on $[-m, +\infty)$ for some m, g_1 and g_2 satisfy the assumptions of Theorem 20.12 and $0 \leq g_2(\xi, u) \leq um^{-1}$ for all $\xi \geq 0$ and $u \geq 0$.*

Then (20.19) preserves positivity, that is, for every non-negative $r(0) \in \mathbf{H}_\gamma$ the functions $r(t)$, $t \geq 0$, are non-negative.

By Theorems 20.9 and 20.12, for any $r(0) \in \mathbf{H}_\gamma$ there is a unique solution $(r(t), t \geq 0)$ starting from $r(0)$. Moreover, see Remark 20.11, r is a càdlàg process in $\mathbf{H}_\gamma$. In the proof we will use the following theorem from Milian (2002) concerning the preserving of positivity by the equation

$$dX = (AX + F(t, X)) \, dt + B(t, X) \, dW \qquad (20.23)$$

driven by a Wiener process W taking values in a Hilbert space U. In its formulation, the state space H is $L^2(\mathcal{O}, \mathcal{B}(\mathcal{O}), \rho(\xi)\,d\xi)$, where $\mathcal{O}$ is an open domain in $\mathbb{R}^d$ and ρ is a non-negative weight, A generates a C_0-semigroup S on H and $F: [0, +\infty) \times H \mapsto H$, $B: [0, +\infty) \times H \mapsto L_{(HS)}(U, H)$.

Theorem 20.15 (Milian) *Assume the following.*

(i) *The semigroup S preserves positivity.*
(ii) *There is a constant C such that, for all $t, s > 0$ and $x, y \in H$,*

$$|F(t, x) - F(s, y)| + \|B(t, x) - B(s, y)\|_{L_{(HS)}(U,H)}$$
$$\leq C\,(|t - s| + |x - y|_H).$$

(iii) *For every $t \geq 0$ and for all non-negative continuous $x, f \in H$ satisfying $\langle f, x\rangle_H = 0$ one has $\langle F(t, x), f\rangle_H \geq 0$ and $\langle B(t, x)v, f\rangle_H = 0$ for every $v \in U$.*

Then (20.23) preserves positivity.

Remark 20.16 The original theorem of Milian is a little more general. In particular, it covers also the case where W is a cylindrical Wiener process. Furthermore, it shows that (iii) is necessary for preserving positivity. The assumption of Lipschitz continuity in t can be easily replaced by that of uniform continuity.

The problem of preserving positivity and the so-called comparison principle have been studied by several authors; see e.g. Aubin and Da Prato (1990), Goncharuk and Kotelenez (1998), Jachimiak (1996) and Kotelenez (1992b).

Proof of Theorem 20.14 First of all note that the semigroup S preserves positivity. Let $D(x)(\xi) = g_2(\xi, x(\xi))$ and let us approximate L by a sequence (L_n) of processes satisfying $|\Delta L_n(t)| \geq 1/n, t \geq 0, n \in \mathbb{N}$. We assume that L_n converges $\mathbb{P}$-a.s. to L uniformly on each compact time interval. The existence of such a sequence follows from the Lévy–Khinchin decomposition. Let r_n be the solution to the problem

$$dr = \left(\frac{\partial}{\partial \xi}r + F(r)\right)dt + B(r)\,dW(t) + D(r)\,dL_n, \qquad r_n(0) = r(0). \quad (20.24)$$

Since r_n converges to r it is sufficient to show that (20.24) preserves positivity. To do this note that L_n has only isolated jumps. Between the jumps positivity is preserved by Theorem 20.15, as the driving process is Wiener. Assume that the solution is positive until the jump at time τ. Then

$$r_n(\tau)(\xi) = r_n(\tau-)(\xi) + g_2(\xi, r_n(\tau-)(\xi))(L_n(\tau) - L_n(\tau-)).$$

Hence if (i) holds then

$$r_n(\tau-)(\xi) + g_2(\xi, r_n(\tau-)(\xi))(L_n(\tau) - L_n(\tau-))$$
$$\geq r_n(\tau-)(\xi) - (m \vee M)|g_2(\xi, r_n(\tau-)(\xi))| \geq 0$$

and if (ii) holds then

$$r_n(\tau-)(\xi) + g_2(\xi, r_n(\tau-)(\xi))(L_n(\tau) - L_n(\tau-))$$
$$\geq r_n(\tau-)(\xi) - mg_2(\xi, r_n(\tau-)(\xi)) \geq 0,$$

and the result follows. □

Example 20.17 Let $Z = (W, L)$ be as in Theorem 20.14, and let the jump measure ν of L be supported on $[-m, +\infty)$ for a certain $m > 0$. Let $g_i(\xi, u) = h_i(\xi)v_i(u)$, $i = 1, 2$. Assume that:

(i) the v_i, $i = 1, 2$, are bounded and Lipschitz continuous and the h_i, $i = 1, 2$, are bounded and belong to H_γ;
(ii) $v_1(0) = 0$, h_2 and v_2 are non-negative and $0 \leq v_2(u) \leq u/(m|h_2|_{L^\infty})$ for $u \geq 0$.

Then the assumptions of Theorem 20.14 are satisfies and (20.22) defines Feller-family-preserving positivity.

An important example of a jump measure supported on $[0, +\infty)$ is given below.

Example 20.18 Given $\alpha > 0$ let $\nu(d\xi) = \chi_{\{\xi > 0\}}\xi^{-1-\beta}e^{-\alpha\xi}$. Then there is a constant $c = c(\alpha, \beta)$ such that, for $z > 0$,

$$J'(z) = \begin{cases} c\left(\alpha^{-1+\beta} - (z+\alpha)^{-1+\beta}\right) & \text{if } 0 < \beta < 1, \\ c\left((z+\alpha)^{-1+\beta} - \alpha^{-1+\beta}\right) & \text{if } 1 < \beta < 2. \end{cases}$$

20.3.4 Invariant measures

We will now consider the long-time behavior of the time-homogeneous HJM equation. Thus we assume that the coefficients G and F in (20.16) are time independent. As a direct consequence of Lemmas 20.8 and 20.10 and Theorem 16.5 we have the following result. For more details, see Rusinek (2006b).

Theorem 20.19 *Let G and F satisfy the assumptions of Theorem 20.9 for functions $\bar{g}$ and $\bar{h}$. Let K be as in Lemma 20.10. If $\omega := \gamma - |\bar{h}|_{L^\infty}^2 - 2K^2 > 0$ then, for any $C \geq 0$, there is a unique invariant measure for (20.16), considered on $H_\gamma + C$, and it is exponentially mixing with exponent $\omega/2$ and function c of linear growth.*

A similar result can be stated for (20.22), as follows.

Theorem 20.20   *Let G and F satisfy the assumptions of Theorem 20.12 for functions $\overline{g}$ and $\overline{h}$. Let K be as in Lemma 20.13. If $\omega := \gamma - 4|\overline{h}|_{L^\infty}^2 - 2K^2 > 0$ then, for any $C \geq 0$, there is a unique invariant measure for (20.22), considered on $H_\gamma + C$, and it is exponentially mixing with exponent $\omega/2$ and function c of linear growth.*

20.4 Linear volatility

As in subsection 20.3.2 concerning a special case, we assume that $Z = (W, L)$, where W is a standard Wiener process in $\mathbb{R}$ and L is a real-valued Lévy martingale with Laplace transform (20.20). Moreover we assume that the volatility G is a linear function of r, that is,

$$G(t, \psi)[z_1, z_2](\xi) = g_1(t)\psi(\xi)z_1 + g_2(t)\psi(\xi)z_2, \qquad z_1, z_2 \in \mathbb{R}, \ \xi, t \geq 0.$$

Here g_1 and g_2 are predictable random processes independent of ξ. Then (20.22) becomes

$$dr(t)(\xi) = \left(\frac{\partial}{\partial \xi} r(t)(\xi) + F(t, r)(\xi)\right) dt + g_1(t)r(t)(\xi)\, dW(t)$$
$$+ g_2(t)r(t)(\xi)\, dL(t), \tag{20.25}$$

where

$$F(t, \psi)(\xi) = g_1^2(t)\psi(\xi) \int_0^\xi \psi(\eta)\, d\eta$$
$$+ g_2(t)\psi(\xi) \int_{\mathbb{R}} y \left(1 - \exp\left\{-yg_2(t) \int_0^\xi \psi(\eta)\, d\eta\right\}\right) v(dy).$$

We will always assume that r_0 is a non-negative function. Let $u(t)(\xi) = \int_0^\xi r(t)(\eta)\, d\eta$ be a primitive of $r(t)$. Then

$$du(t)(\xi) = \left(\frac{\partial}{\partial \xi} u(t)(\xi) + \tfrac{1}{2}(g_1(t)u(t)(\xi)^2 + \tilde{J}\left(g_2(t)u(t)\right)\right) dt$$
$$+ g_1(t)u(t)\, dW(t) + g_2(t)u(t)(\xi)\, dL(t), \tag{20.26}$$

where

$$\tilde{J}(z) := \int_{\mathbb{R}} \left(e^{-zy} - 1 + zu\right) v(dy). \tag{20.27}$$

We assume that the jump measure v of L satisfies $\int_{\mathbb{R}} |y|^2 v(dy) < \infty$.

20.4.1 Pure jump case

Assume that $g_1 \equiv 0$. To simplify, we assume also that $g_2 \equiv 1$. Then

$$du(t)(\xi) = \left(\frac{\partial}{\partial \xi} u(t)(\xi) + \tilde{J}\left(u(t)(\xi) \right) \right) dt + u(t)(\xi)\, dL(t). \tag{20.28}$$

To ensure positivity we assume that ν is supported on $[-1, +\infty)$.

In order to solve (20.28) and consequently the HJMM equation for r we first solve the following auxiliary problem:

$$dv(t) = \tilde{J}(v(t))\, dt + v(t)\, dL(t), \qquad v(0) = \xi. \tag{20.29}$$

Clearly $\tilde{J}: [0, +\infty) \mapsto \mathbb{R}$ is Lipschitz continuous if its derivative is bounded. This is guaranteed by the assumptions that

$$\nu \text{ has support in } [0, +\infty) \quad \text{and} \quad \int_0^\infty y\nu(dy) < \infty. \tag{20.30}$$

Proposition 20.21 *Assume (20.30). Then we have the following.*

(i) *For each non-negative $\xi \geq 0$ there is a unique non-negative solution $(v(t)(\xi),\ t \geq 0)$ to (20.29). Moreover, for each t, $v(t)(\xi)$ depends continuously on ξ, $\mathbb{P}$-a.s.*
(ii) *There exists a version of v differentiable in ξ, and*

$$\frac{\partial}{\partial \xi} v(t)(\xi) = \exp\left\{ \int_0^t \tilde{J}'(v(s)(\xi))\, ds \right\} A(t) \tag{20.31}$$

where A is a positive càdlàg process.

Proof By Protter (2005), Theorem 37, p. 308, the unique solution to (20.29) depends continuously on ξ. Since $\tilde{J}'(z) = \int_{[0,\infty)} y^2 e^{-zy}\nu(dy)$, $z \geq 0$, is locally Lipschitz, by Protter (2005), Theorem 39, the solution v is differentiable in ξ and $v_\xi(t)(\xi) = (\partial/\partial\xi)v(t)(\xi)$ satisfies

$$dv_\xi(t)(\xi) = \tilde{J}'\left(v(t-)(\xi) \right) v_\xi(t-)(\xi)\, dt + v_\xi(t-)(\xi)\, dL(t) = v_\xi(t-)(\xi)\, dX(t),$$

where

$$X(t) = L(t) + \int_0^t \tilde{J}'\left(v(s-)(\xi) \right) ds.$$

By Doléan's formula (see e.g. Protter 2005),

$$v_\xi(t)(\xi) = e^{X(t)} \prod_{s \leq t} (1 + \Delta X(s))\, e^{-\Delta X(s)}.$$

Since $\Delta X(s) = \Delta L(s)$ and $\int_0^t \tilde{J}'\left(v_\xi(s-)(\xi)\right) ds = \int_0^t \tilde{J}'\left(v_\xi(s)(\xi)\right) ds$,

$$v_\xi(t)(\xi) = \exp\left\{\int_0^t \tilde{J}'\left(v(s)(\xi)\right) ds\right\} e^{L(t)} \prod_{s \leq t} (1 + \Delta L(s)) e^{-\Delta L(s)}.$$

$\square$

Proposition 20.22   *Assume (20.30). Let $v(t)(\xi)$, $t \geq 0$, $\xi \geq 0$, be the solution to (20.29). Then*

$$u(t)(\xi) = v(t)(u_0(t + \xi)), \qquad t \geq 0, \ \xi \geq 0,$$

is the unique solution to (20.28).

Proof   We claim that

$$dv(t)(u_0(t + \xi)) = \left(\frac{\partial}{\partial \xi} v(t)(u_0(t + \xi)) + \tilde{J}\left(v(t)(u_0(t + \xi))\right)\right) dt$$
$$+ v(t)(u_0(t + \xi)) \, dL(t).$$

Let $\psi \colon [0, +\infty) \times [0, +\infty) \mapsto \mathbb{R}$ be a differentiable function of both variables. Given $\xi > 0$ consider the process $(v(t)(\psi(t, \xi)), \ t \geq 0)$. Then, for any partition $0 = t_0 < t_1 < \cdots < t_N = t$,

$$v(t)(\psi(t, \xi)) - v(0)(\psi(0, \xi)) = I_1 + I_2,$$

where

$$I_1 := \sum_n \left(v(t_{n+1})(\psi(t_{n+1}, \xi)) - v(t_{n+1})(\psi(t_n, \xi))\right),$$

$$I_2 := \sum_n \left(v(t_{n+1})(\psi(t_n, \xi)) - v(t_n)(\psi(t_n, \xi))\right).$$

But

$$I_1 = \sum_n \left[v_\xi(t_{n+1})\left(\psi(t_{n+1}, \xi)\right.\right.$$
$$+ \left.\left.\tilde{\varepsilon}(\psi(t_n, \xi) - \psi(t_{n+1}, \xi))(\psi(t_{n+1}, \xi) - \psi(t_n, \xi))\right)\right]$$
$$= \sum_n \left(v_\xi(t_{n+1})(\psi(t_{n+1}, \xi) + \tilde{\varepsilon}(\psi(t_n, \xi) - \psi(t_{n+1}, \xi)))\right)$$
$$\times \psi'(t_n + \tilde{\eta}(t_{n+1} - t_n), \xi)(t_{n+1} - t_n),$$

where $\tilde{\varepsilon}$ and $\tilde{\eta}$ are such that $|\tilde{\varepsilon}| \leq 1$ and $|\tilde{\eta}| \leq 1$. Taking into account the continuous dependence of v_ξ on the second variable (see Proposition 20.21) we obtain $I_1 \to \int_0^t v_\xi(s)(\psi(s, \xi))\psi^1(s, \xi) ds$ as $n \uparrow \infty$. Since v satisfies (20.29),

$$I_2 = \sum_n \left(\int_{t_n}^{t_{n+1}} J\left(v(s-)(\psi(t_{n+1}, \xi))\right) ds + \int_{t_n}^{t_{n+1}} v(s-)(\psi(t_n, \xi)) dL(s)\right)$$

and therefore

$$I_2 \to \int_0^t J\left(v(s-)(\psi(s,\xi))\,\mathrm{d}s + \int_0^t v(s-)(\psi(s,\xi))\,\mathrm{d}L(s).\right.$$

$\square$

Corollary 20.23 *For any non-negative r_0, the unique solution r to (20.28) is given by*

$$r(t)(\xi) = \frac{\partial}{\partial\xi}u(t)(\xi) = \frac{\partial}{\partial\xi}v(t)(u_0(t+\xi)) = v_\xi(t)(u_0(t+\xi))r_0(t+\xi),$$

where v_ξ has the representation (20.31).

20.4.2 Gaussian case

This subsection is concerned with the Gaussian case. Namely, we assume that the jump in (20.25) vanishes, and consequently $L = W$ is a standard Brownian motion. To shorten the notation we write $g_1(t) = h(t)$. We assume that h is a predictable process satisfying $\mathbb{E}\int_0^T h^2(t)\,\mathrm{d}t < \infty, \forall\, T > 0$. Therefore (20.25) becomes

$$\mathrm{d}r(t)(\xi) = \left(\frac{\partial}{\partial\xi}r(t)(\xi) + h^2(t)r(t)(\xi)\int_0^\xi r(t)(\eta)\,\mathrm{d}\eta\right)\mathrm{d}t$$
$$+ h(t)r(t)(\xi)\,\mathrm{d}W(t), \tag{20.32}$$
$$r(0)(\xi) = r_0(\xi).$$

We assume that r_0 is a non-negative function. Set

$$M_h(t) := \exp\left\{-\tfrac{1}{2}\int_0^t h^2(s)\,\mathrm{d}s + \int_0^t h(s)\,\mathrm{d}W(s)\right\}.$$

The theorem below shows that (20.32) can be solved explicitly. However, the solution may blow up in a finite time.

Theorem 20.24 *The unique solution to (20.32) is given by*

$$r(t)(\xi) = \frac{\partial}{\partial\xi}\left[\left(\int_0^{t+\xi} r_0(\eta)\,\mathrm{d}\eta\right)^{-1} - \tfrac{1}{2}\int_0^t h^2(s)M_h(s)\,\mathrm{d}s\right]^{-1} M_h(t). \tag{20.33}$$

Proof Let us denote by $u(t)$ the primitive of $r(t)$ given by $u(t)(\xi) := \int_0^\xi r(t)(\eta)\,\mathrm{d}\eta$. Then $r(t)(\xi) = \partial u(t)(\xi)/\partial\xi$ and consequently we have the following equation for u:

$$\mathrm{d}\frac{\partial u}{\partial\xi} = \left(\frac{\partial^2 u}{\partial\xi^2} + h^2\frac{\partial u}{\partial\xi}u\right)\mathrm{d}t + h\frac{\partial u}{\partial\xi}\,\mathrm{d}W$$
$$= \frac{\partial}{\partial\xi}\left[\left(\frac{\partial u}{\partial\xi} + \tfrac{1}{2}h^2u^2\right)\mathrm{d}t + hu\,\mathrm{d}W\right].$$

Hence,

$$du(t)(\xi) = \left(\frac{\partial u}{\partial \xi}(t)(\xi) + \tfrac{1}{2} h^2(t) u^2(t)(\xi) \right) dt + h(t) u(t)(\xi) \, dW(t),$$

$$u(0)(\xi) = \int_0^\xi r_0(\eta) \, d\eta.$$

In order to solve this equation we use ideas from the previous subsection. Namely, $u(t)(\xi) = v(t)(u(0)(t + \xi))$, where $v(t)(\xi)$ is the unique solution to the stochastic Bernoulli problem

$$dv(t)(\xi) = \tfrac{1}{2} h^2(t) v^2(t)(\xi) \, dt + h(t) v(t)(\xi) \, dW(t), \qquad v(0)(\xi) = \xi.$$

Let us fix ξ. We can solve the equation using the substitution $z(s) = 1/v(s)(\xi)$. By Itô's formula,

$$dz(s) = \left(-\frac{1}{v^2(s)} \tfrac{1}{2} h^2(s) v^2(s) + \tfrac{1}{2} 2 \frac{1}{v^3(s)} h(s) v^2(s) \right) ds - \frac{1}{v^2(s)} h(s) v(s) \, dW(s)$$

$$= h^2(s) \left(-\tfrac{1}{2} + z(s) \right) ds - h(s) z(s) \, dW(s).$$

To solve the equation we use the variation-of-constants formula. Thus, we look for a solution in the form $z(s) = c(s) z_0(s)$, where z_0 is a solution to the homogeneous equation

$$dz_0(s) = h^2(s) z_0(s) \, ds - h(s) z_0(s) \, dW(s).$$

Then

$$dz = d(c z_0) = c' z_0 \, ds + c \, dz_0 = c' z_0 \, ds + h^2 z \, ds - h z \, dW.$$

This leads to the condition

$$c'(s) = -\tfrac{1}{2} h^2(s) \frac{1}{z_0(s)}.$$

Clearly, for z_0 we can take

$$z_0(s) = \exp \left\{ \tfrac{1}{2} \int_0^s h^2(\eta) \, d\eta - \int_0^s h(\eta) \, dW(\eta) \right\} = \frac{1}{M_h(s)}.$$

Taking into account the initial-value condition $z(0) = 1/\xi$ we obtain

$$z(s) = \left(\xi^{-1} - \tfrac{1}{2} \int_0^s h^2(\eta) M_h(\eta) \, d\eta \right) \frac{1}{M_h(s)},$$

which gives the desired formula. □

Corollary 20.25 *As a direct consequence of (20.33), the set $\mathcal{D}$ of all (t, ξ) for which the solution blows up is given by*

$$\mathcal{D} = \left\{ (t, \xi) \colon \tfrac{1}{2} \int_0^t h^2(s) M_h(s) \, \mathrm{d}s = \left(\int_0^{t+\xi} r_0(\eta) \, \mathrm{d}\eta \right)^{-1} \right\}.$$

Since $r_0 \geq 0$, the processes $t \mapsto \left(\int_0^{t+\xi} r_0(\eta) \, \mathrm{d}\eta \right)^{-1}$, where $\xi \geq 0$ is fixed, and $\xi \mapsto \left(\int_0^{t+\xi} r_0(\eta) \, \mathrm{d}\eta \right)^{-1}$, where $t \geq 0$ is fixed, have decreasing trajectories. Clearly, the process $\tfrac{1}{2} \int_0^t h^2(s) M_h(s) \, \mathrm{d}s$, $t \geq 0$, is increasing and starts from 0. Therefore, if $\int_0^\infty r_0(\eta) \, \mathrm{d}\eta = +\infty$ and $\mathbb{P}\left(\int_0^\infty h^2(s) \, \mathrm{d}s > 0 \right) = 1$ then, with probability 1, for any ξ there is a time t such that $(t, \xi) \in \mathcal{D}$.

Let $\phi \colon \mathbb{R} \mapsto \mathbb{R}$ be a bounded non-negative Lipschitz function satisfying $\phi(1) \neq 0$, $\phi(0) = 0$ and

$$m(\phi) := \sup_{x \neq 0} \frac{|\phi(x)|^2}{|x|} < \infty. \tag{20.34}$$

Let X be a solution to the equation

$$\mathrm{d}X = \phi(X) \, \mathrm{d}W, \qquad X(0) = 1.$$

Note that X is not identically 0 as $\phi(1) \neq 0$. Clearly, $X = M_h$ for a non-negative process

$$h(t) = \frac{\phi(X(t))}{X(t)}, \qquad t \geq 0.$$

From (20.34), $\mathbb{P}$-a.s. $h^2(s) M_h(s) \leq |\phi(X(s))|^2 / X(s) \leq m(\phi)$, $t \geq 0$. Hence,

$$\left| \tfrac{1}{2} \int_0^t h^2(s) M_h(s) \, \mathrm{d}s \right| \leq \tfrac{1}{2} t m(\phi)$$

and we have the following consequence of Theorem 20.24.

Theorem 20.26 *Let $T > 0$. Then, for each non-negative initial value $r_0 \in L^1 := L^1([0, +\infty), \mathcal{B}([0, +\infty), \mathrm{d}\xi)$ satisfying $|\psi|_{L^1} < 2/T m(\phi)$, the process r given by (20.33) has trajectories $\mathbb{P}$-a.s. in $C([0, T); L^1)$ and is a unique strong solution to (20.32) on the open interval $[0, T)$. Additionally, if $r_0 \in C_b^n([0, +\infty))$ then $r \in C\left([0, T); C_b^n([0, +\infty)) \right)$, $\mathbb{P}$-a.s.*

20.5 BGM equation

The Brace–Gątarek–Musiela (BGM) equation, which we will now deduce, is also of importance. For derivation and analysis it is convenient to introduce the *integrated volatility*

$$\tilde{\sigma}(t,\xi) := \int_0^\xi b(t,u)\,du\,.$$

In our exposition we follow Jakubowski, Niewęgłowski and Zabczyk (2006). In the new notation one can rewrite (20.14) in the following transparent way:

$$dr(t,\xi) = \frac{\partial}{\partial\xi}\big[(r(t,\xi)+J(\tilde{\sigma}(t,\xi)))\,dt + \langle\tilde{\sigma}(t,\xi),dZ\rangle_U\big]. \qquad (20.35)$$

Instead of starting from (20.35), however, Brace, Gątarek and Musiela proposed to base modeling on the so-called *LIBOR rates* $L(t,s,\theta), 0 \le t \le s \le \theta$, defined in terms of bond prices P and forward curves f by the relation

$$1+(\theta-s)L(t,s,\theta) = \frac{P(t,s)}{P(t,\theta)} = \exp\left\{\int_s^\theta f(t,u)\,du\right\}.$$

Recall that $r(t,u) = f(t,t+u)$. For fixed $\delta > 0$, (in practice δ is equal to three months), we set

$$L(t,\xi) := L(t,\ t+\xi,\ t+\xi+\delta) = \frac{1}{\delta}\left(\exp\left\{\int_\xi^{\xi+\delta} r(t,u)\,du\right\}-1\right), \quad \xi > 0.$$

We can deduce from (20.35) a stochastic equation for $L(t,\xi), t \ge 0, \xi \ge 0$. To do this, we introduce the process

$$Y(t,\xi) := \int_\xi^{\xi+\delta} r(t,u)\,du, \qquad t \ge 0,\ x \ge 0.$$

In the following calculations we use the Itô formula as in the proof of Theorem 20.3. For more details see Jakubowski and Zabczyk (2004). To shorten the notation, given a function g we write

$$\nabla_\delta g(t-,\xi) := g(t-,\xi+\delta) - g(t-,\xi).$$

From (20.35),

$$dY(t,\xi) = \int_\xi^{\xi+\delta} dr(t,u)\,du$$
$$= \big(\nabla_\delta r(t-,\xi)+\nabla_\delta J(\tilde{\sigma}(t-,\xi))\big)\,dt + \big\langle\nabla_\delta\tilde{\sigma}(t-,\xi),\,dZ(t)\big\rangle_U. \qquad (20.36)$$

It is also clear that

$$\frac{\partial}{\partial\xi}L(t,\xi) = \delta^{-1}e^{Y(t,\xi)}\nabla_\delta r(t,\xi). \qquad (20.37)$$

By the Itô formula,

$$de^{Y(t,\xi)} = e^{Y(t-,\xi)}\,dY(t,\xi) + \tfrac{1}{2}e^{Y(t-,\xi)}\big\langle Q\nabla_\delta\tilde{\sigma}(t-,\xi),\,\nabla_\delta\tilde{\sigma}(t-,\xi)\big\rangle_U\,dt$$
$$+ \Delta e^{Y(t,\xi)} - e^{Y(t-,\xi)}\Delta Y(t-,\xi).$$

Inserting (20.37) we obtain

$$
\begin{aligned}
e^{-Y(t-,\xi)}\,dY(t,\xi) \\
= \big(\nabla_\delta r(t-,\xi) + \nabla_\delta J(\tilde\sigma(t-,\xi))\big)\,dt + \big\langle \nabla_\delta \tilde\sigma(t-,\xi),\,dZ(t)\big\rangle_U \\
+ \int_U \big(\exp\{\langle \nabla_\delta \tilde\sigma(t-,\xi),\eta\rangle_U\} - 1 - \langle \nabla_\delta \tilde\sigma(t-,\xi),\,\eta\rangle_U\big)\pi(dt,d\eta),
\end{aligned}
$$

and hence, by (20.37),

$$
\begin{aligned}
dL(t,\xi) = \frac{\partial}{\partial \xi}L(t-,\xi) + \tfrac{1}{2}\delta(1 + \delta L(t-,\xi)) \\
\times \Bigg[\big(\nabla_\delta J(\tilde\sigma(t-,\xi)) + \tfrac{1}{2}\langle Q\nabla_\delta \tilde\sigma(t-,\xi),\nabla_\delta\tilde\sigma(t-,\xi)\rangle_U\big)dt \\
+ \langle \nabla_\delta \tilde\sigma(t-,\xi),\,dZ(t)\rangle_U \\
+ \int_U \big(\exp\{\langle \nabla_\delta \tilde\sigma(t-,\xi),\eta\rangle_U\} - 1 - \langle \nabla_\delta\tilde\sigma(t-,\xi),\eta\rangle\big)\pi(dt,d\eta)\Bigg].
\end{aligned}
$$

Postulating some dependence of the volatility $\tilde\sigma$ on the LIBOR rate L, we arrive at the *BGM equation* in a general form, which at the moment is not tractable, with the exception of some particular cases.

In the original paper Brace, Gątarek and Musiela (1997), the authors assumed that $a = 0$, $\mu \equiv 0$. Moreover, the Wiener process W was one-dimensional. The authors also imposed a log-normality assumption, that there exists a function $\gamma(\xi)$, $\xi \geq 0$, such that

$$
\frac{1}{\delta}\,(1 + \delta L(t,\xi))\,\nabla_\delta\tilde\sigma(\xi) = L(t,\xi)\gamma(\xi),
$$

and arrived at the *classical BGM equation*

$$
\begin{aligned}
dL(t,\xi) = \Bigg(\frac{\partial}{\partial \xi}\,L(t,\xi) + \langle Q\tilde\sigma(L(t),\xi),\ \gamma(\xi)\rangle L(t,\xi) \\
+ \frac{\delta L^2(t,\xi)}{1 + \delta L(t,\xi)}\langle Q\gamma(\xi),\gamma(\xi)\rangle\Bigg)dt + L(t,\xi)\langle\gamma(\xi),dW(t)\rangle \quad (20.38)
\end{aligned}
$$

where

$$
\tilde\sigma(L,\xi) = 0, \qquad\qquad\qquad\qquad \xi \in [0,\delta],
$$

$$
\tilde\sigma(L,\xi) = \sum_{k=0}^{[\xi/\delta]} \frac{\delta L(\xi - k\delta)}{1 + \delta L(x - k\delta)}\gamma(\xi - k\delta), \qquad \xi \geq 0.
$$

In Brace, Gątarek and Musiela (1997) the following theorem was shown.

Theorem 20.27 *Assume that*

$$\int_0^\infty \left(\xi |\gamma'(\xi)|^2 + |\gamma(\xi)| \right) d\xi < \infty, \qquad \sup_{0 \le \xi \le \delta} \sum_{k=0}^\infty |\gamma'(\xi + k\delta)| < \infty.$$

Then we have the following.

(i) *Equation (20.38) has a unique solution in the space of continuous functions having a finite limit at infinity. If the initial function is non-negative then the solution is also non-negative.*

(ii) *There exist invariant measures for (20.38) concentrated on functions having a fixed limit at infinity.*

The existence of an invariant measure is a mathematical equivalent of the so-called *mean-reversion* property. Interest rates tend to drop when they are too high and tend to rise when they are too low.

20.6 Consistency problem

Let K be a parameterized family of functions $\{z(\lambda, \xi) \colon \lambda \in \Lambda, \xi \ge 0\}$.

Definition 20.28 The model (20.16) is said to be *consistent with K* if and only if the set K is invariant for (20.16); that is, for any initial condition $r_0 \in K$, the solution $r(t)$, $t \ge 0$, evolves in $K \colon r(t) \in K$ for all $t \ge 0$.

Assume that (20.16) is used as a model of the time evolution of the forward curve and that for a curve fitting one applies a statistical procedure that uses functions from a given set K. Then the minimal requirement that the set and equation should satisfy is that of consistency. The following families K are used in practice:

(i) the *Nelson–Siegel family*

$$z(\lambda, \xi) = \lambda_1 + \lambda_2 e^{-\lambda_4 \xi} + \lambda_3 \xi e^{-\lambda_4 \xi}, \qquad (\lambda_1, \lambda_2, \lambda_3, \lambda_4) \in \Lambda \subset \mathbb{R}^4;$$

(ii) the *degenerate Nelson–Siegel family*

$$z(\lambda, \xi) = \lambda_1 + \lambda_3 \xi, \qquad (\lambda_1, \lambda_3) \in \Lambda \subset \mathbb{R}^2;$$

(iii) the *augmented Nelson–Siegel family*

$$z(\lambda, \xi) = \lambda_1 + \lambda_2 e^{-a\xi} + \lambda_3 \xi e^{-a\xi} + \lambda_4 e^{-2a\xi}, \qquad (\lambda_1, \lambda_2, \lambda_3, \lambda_4) \in \Lambda \subset \mathbb{R}^4.$$

To solve the *consistency problem* one has to find conditions on the set K and (20.16) which imply the consistency of the model. The question whether for a given equation there exists a set K of smooth functions $z(\lambda, \cdot)$ parameterized by λ from a finite-dimensional set Λ for which consistency holds is of equal importance. Problems of this nature were introduced in Björk and Christensen (1999) and intensively studied by Björk and his collaborators. Important contributions were made by Filipović (2001). These authors adapted an approach based on differential geometry. It is clear that the consistency problem can be formulated equivalently as an *invariance problem*, of finding conditions under which a given set K is invariant for the flow defined by the equation. In fact, following Jachimiak (1996), Zabczyk (2000) and Nakayama (2004a, b), we present below a solution of the consistency problem based on the so-called support theorem.

Consider the evolution equation

$$dX = \big(AX + F(X)\big)\,dt + B(X)\,dW(t), \qquad X(0) = x \in H, \qquad (20.39)$$

driven by a cylindrical Wiener process in U. Denote by $X^x(t)$ the value at t of the solution to (20.39) starting at time 0 from x.

Define the *Wong–Zakai correction term*

$$\tilde{B}(x) := \tfrac{1}{2} \sum_j Db_j(x)\big[b_j(x)\big], \qquad x \in H, \qquad (20.40)$$

where D is the gradient operator, $b_j(x) := B(x)e_j$ and $\{e_j\}$ is an orthonormal basis of U. Given a locally square integrable function $u \colon [0, \infty) \mapsto U$, denote by $y^{x,u}$ the solution to the controlled equation

$$\frac{dy}{dt}(t) = Ay(t) + F(y(t)) - \tilde{B}(y(t)) + B(y(t))u(t), \qquad y(0) = x \in H.$$
$$(20.41)$$

Definition 20.29 A closed set $K \subset H$ is said to be *invariant* for (20.41) if $y^{x,u}(t) \in K$ for all $t \geq 0, x \in K$ and a locally square integrable U-valued function u.

20.6.1 Equations with additive noise

We consider first the invariance of a closed set K with respect to (20.39) with a state-independent linear operator $B \in L(U, H)$. Thus we are dealing with the equation

$$dX = (AX + F(X))\,dt + B\,dW(t), \qquad X(0) = x \in H. \qquad (20.42)$$

Since the correction term vanishes, the system (20.41) becomes

$$\frac{dy}{dt}(t) = Ay(t) + F(y(t)) + Bu(t), \qquad y(0) = x \in H. \qquad (20.43)$$

To ensure the existence of a continuous solution we will impose (see Theorem 11.5) the following condition:

$$\exists \alpha \in (0, 1/2): \int_0^T t^{-\alpha} \|S(t)B\|_{L_{(HS)}(U,H)}^2 \, dt < \infty, \qquad \forall\, T > 0. \qquad (20.44)$$

The following result comes from Zabczyk (2000).

Theorem 20.30 *Assume that A generates a C_0-semigroup on a Hilbert space H, F is a Lipschitz transformation from H into H, and (20.44) holds. Then a closed set $K \subset H$ is invariant for (20.42) if and only if it is invariant for (20.43).*

Proof Write

$$Z(t) := \int_0^t S(t - r)B \, dW(r), \qquad t \geq 0.$$

Condition (20.44) guarantees that Z has continuous paths. Fix a time $T > 0$. Clearly Z is a Gaussian random element in $C([0, T]; H)$. It follows from Da Prato and Zabczyk (1996), p. 141, that the support of its law $\mathcal{L}(Z)$ on $C([0, T]; H)$ is identical with the closure in $C([0, T]; H)$ of the set S_T of all functions

$$f(t) = \int_0^t S(t - r)Bu(r) \, dr \qquad t \in [0, T], u \in L^2(U),$$

where $L^2(U) := L^2([0, T], \mathcal{B}([0, T]), dt; U)$. Let c be the Lipschitz constant of F and let $M = \sup_{t \in [0,T]} \|S(t)\|_{L(H,H)}$. For an arbitrary $u \in L^2(U)$,

$$X^x(t) - y^{x,u}(t) = \int_0^t S(t - r)\left(F(X^x(r)) - F(y^{x,u}(r))\right) dr$$

$$+ Z(t) - \int_0^t S(t - r)Bu(r) \, dr.$$

Hence

$$\left|X^x(t) - y^{x,u}(t)\right| \leq cM \int_0^t \left|X^x(r) - y^{x,u}(r)\right|_U dr$$

$$+ \sup_{t \in [0,T]} \left|Z(t) - \int_0^t S(t - r)Bu(r) \, dr\right|_H.$$

By Gronwall's lemma, for all $t \in [0, T]$,

$$\left|X^x(t) - y^{x,u}(t)\right|_H \leq e^{cMT} \sup_{t \in [0,T]} \left|Z(t) - \int_0^t S(t - r)Bu(r) \, dr\right|_H. \qquad (20.45)$$

Assume that K is invariant for (20.42) and let $u \in L^2(U)$. It follows from the description of the support of the process Z that, for an arbitrary $\delta > 0$,

$$\mathbb{P}\left(\sup_{t \in [0,T]} \left|Z(t) - \int_0^t S(t - r)Bu(r) \, dr\right|_U < \delta\right) > 0.$$

From $\mathbb{P}(X^x(t) \in K) = 1$ for all $t \in [0, T]$ and from (20.45), we obtain dist $(y^{x,u}(t), K) \le e^{cMT}\delta$. Since δ is an arbitrary positive number, we have the desired conclusion, $y^{x,u}(t) \in K$ for all $t \in [0, T]$.

Conversely, assume that K is invariant for (20.43). Since the law $\mathcal{L}(Z)$ is tight in $C([0, T]; H)$ for arbitrary $\delta > 0$ and $\varepsilon > 0$ there exists a finite number of functions $u_1, \ldots, u_N \in L^2(U)$ such that

$$\mathbb{P}\left(\exists n \in \{1, \ldots, N\} : \sup_{t \in [0,T]} \left| Z(t) - \int_0^t S(t-r)Bu_n(r)\,dr \right|_H < \delta\right) > 1 - \varepsilon.$$

Consequently, by (20.45),

$$\mathbb{P}\left(\exists n \in \{1, \ldots, N\} : \sup_{t \in [0,T]} \left| X^x(t) - y^{x,u_n}(t) \right|_H \le e^{TcM}\delta\right) > 1 - \varepsilon.$$

This easily implies the result. $\square$

Remark 20.31 Instead of (20.44) one can require that the semigroup S is a generalized contraction; see Theorem 9.20.

As a corollary one can obtain the following proposition. For a complete proof we refer the reader to Zabczyk (2000), where some applications to non-linear systems with state dependent diffusion operator B are also given.

Proposition 20.32 *There exists a finite-dimensional linear space $K \subset H$ invariant for*

$$dX = (AX + \sigma_0)\,dt + \sum_{j=1}^d \sigma_j\,dW_j,$$

where $\sigma_0, \sigma_1, \ldots, \sigma_d \in H$ if and only if $\sigma_0, \sigma_1, \ldots, \sigma_d \in D(A)$ and there exist $\sigma_j \in D(A)$, $j = d + 1, \ldots, D < +\infty$, such that, for some real numbers γ_{kj}, $k, j = 0, 1, \ldots, D$,

$$A\sigma_k = \sum_{j=0}^D \gamma_{kj}\sigma_j, \qquad k = 0, 1, \ldots, D.$$

In the particular case of the bond market, we have $A = d/d\xi$ and the functions σ_k should satisfy the system of linear differential equations

$$\frac{d\sigma_k}{d\xi}(\xi) = \sum_{j=0}^D \gamma_{kj}\sigma_j(\xi), \qquad k = 0, \ldots, D, \ \xi \ge 0.$$

Therefore, by Jordan decomposition, the functions σ_k, $k = 0, 1, \ldots, d$, are linear combinations of products of polynomials and exponentials,

$$\xi^k e^{\alpha\xi} \cos\beta\xi, \quad \xi^k e^{\alpha\xi} \sin\beta\xi, \qquad k = 0, 1, \ldots, \beta, \alpha \in \mathbb{R}.$$

Note that the Nelson–Siegel families are of that form.

20.6.2 Equations with a general diffusion operator

The following result of T. Nakayama (2004a, b) contains as special cases results of Jachimiak (1996) and Zabczyk (2000). The factorization techniques used by Nakayama are similar to those in Tessitore and Zabczyk (2001a, 2006). Recall that $\tilde{B}$ is the correction term given by (20.40).

Theorem 20.33 *Let $K \subset H$ be a closed subset and assume that the transformations F and $\tilde{B}$ from H into H, and the transformation B from H into the space of Hilbert-Schmidt operators from U to H, are Lipschitz continuous. The following three conditions are equivalent.*

(i) *K is invariant for (20.39).*
(ii) *K is invariant for (20.41).*
(iii) *For every $x \in K$ and $u \in U$,*

$$\liminf_{t \downarrow 0} \frac{1}{t} \, \text{dist} \left(S(t)x + t(F(x) - \tilde{B}(x) + B(x)u), \ K \right) = 0, \qquad (20.46)$$

where $\text{dist}\,(a, K)$ denotes the distance from a to K.

Remark 20.34 One can show that if $K \subset D(A)$ then condition (20.46) is equivalent to

$$\liminf_{t \downarrow 0} \frac{1}{t} \, \text{dist} \left(x + t(A(x) + F(x) - \tilde{B}(x) + B(x)u), \ K \right) = 0.$$

The following important result established by Filipović (2001) contains Proposition 20.32 as a special case. We denote by $T_x K$ the tangent space to K at x.

Theorem 20.35 *Let $K \subset H$ be an n-dimensional closed C^1-submanifold. Then K is invariant for (20.42) if and only if $K \subset D(A)$ and*

$$Ax + F(x) - \tilde{B}(x) \in T_x K, \quad B(x)U \subset T_x K, \qquad \forall\, x \in K.$$

Appendix A   Operators on Hilbert spaces

In this appendix, basic facts on linear operators on Hilbert spaces are gathered. Some details on nuclear and Hilbert–Schmidt operators are given.

A.1 Bounded operators

We start from some elementary facts about Hilbert spaces. Let $\{e_n\}$ be an orthonormal basis of a Hilbert space U with scalar product $\langle \cdot, \cdot \rangle_U$ and norm $|\cdot|_U$. Thus $|e_n|_U = 1$ and $\langle e_n, e_m \rangle_U = 0$ for $n, m = 1, 2, \ldots, n \neq m$, and, for all $x \in U$,

$$x = \sum_n \langle x, e_n \rangle_U e_n, \qquad x \in U,$$

where the series converges in U. The *Parseval identity* holds:

$$\langle x, y \rangle_U = \sum_n \langle x, e_n \rangle_U \langle y, e_n \rangle_U, \qquad x, y \in U.$$

In this appendix U, H, V are real separable Hilbert spaces. We denote by $L(U, H)$ the space of all bounded (i.e. continuous) linear operators from U to H. The space $L(U, H)$ is equipped with the *operator norm*

$$\|R\|_{L(U,H)} := \sup_{x \in U,\, x \neq 0} \frac{|Rx|_H}{|x|_U}, \qquad R \in L(U, H);$$

we say that $R \in L(U, H)$ is a *compact operator* if it transforms any bounded subset of U into a set whose closure is compact, that is, into a *relatively compact set*. We denote by $K(U, H)$ the space of all compact operators from U to H. It can be shown that $K(U, H)$ is a closed subspace of $L(U, H)$.

Given $R \in L(U, H)$ we denote by $R^* \in L(H, U)$ the *adjoint operator*. It is uniquely determined by the following property:

$$\langle R^*h, x \rangle_U = \langle h, Rx \rangle_H, \qquad \forall\, h \in H,\, \forall\, x \in U.$$

We write $L(U)$ instead of $L(U, U)$. The operator $R \in L(U)$ is *symmetric* or *self-adjoint* if $R^* = R$; a symmetric operator $R \in L(U)$ is *non-negative-definite* if $\langle Rx, x \rangle_U \geq 0$ for every $x \in U$.

In this book we use the fact that for any non-negative-definite symmetric compact operator $R \in L(U)$ there exists an orthonormal basis $\{e_n\}$ consisting of eigenvectors of R, i.e. $Re_n = \gamma_n e_n$. Moreover, $\gamma_n \geq 0$, $n \in \mathbb{N}$, and $\gamma_n \to 0$ as $n \to \infty$. For the proof see e.g. Yosida (1965).

A.2 Nuclear and Hilbert–Schmidt operators

Definition A.1   A linear operator $R \in L(U, H)$ is called *nuclear* or *trace class* if it can be represented in the form

$$Ru = \sum_k b_k \langle u, a_k \rangle_U, \qquad u \in U,$$

where $\{a_k\} \subset U$ and $\{b_k\} \subset H$ are such that $\sum_k |a_k|_U \, |b_k|_H < \infty$.

We denote by $L_1(U, H)$ the space of all nuclear operators from U to H. It is a separable Banach space with the *nuclear norm*

$$\|R\|_1 := \inf \left\{ \sum_k |a_k|_U |b_k|_H : Ru = \sum_k b_k \langle u, a_k \rangle_U \right\}.$$

We write $L_1(H)$ instead of $L_1(H, H)$.

Definition A.2   A linear operator $R \in L(U, H)$ is called *Hilbert–Schmidt* if

$$\sum_k |Re_k|_H^2 < \infty$$

for any (or equivalently for a certain) orthonormal basis $\{e_k\}$.

The space of all Hilbert–Schmidt operators $L_{(HS)}(U, H)$ acting from U into H is a separable Hilbert space with scalar product

$$\langle S, R \rangle_{L_{(HS)}(U,H)} := \sum_k \langle Se_k, Re_k \rangle_H.$$

Denote by $\|\cdot\|_{(HS)}$ or $\|\cdot\|_{L_{(HS)}(U,H)}$ the corresponding (Hilbert–Schmidt) norm. We write $L_{(HS)}(H)$ instead of $L_{(HS)}(H, H)$. Let $S \in L_{(HS)}(U, H)$, and let $\{e_k\}$ be an orthonormal basis of U and $\{f_j\}$ an orthonormal basis of H. Then

$$\sum_k |Se_k|_H^2 = \sum_{k,j} \langle Se_k, f_j \rangle_H^2 = \sum_{k,j} \langle e_k, S^* f_j \rangle_U^2 = \sum_j |S^* f_j|_U^2.$$

Hence we have the following result.

Proposition A.3   *The Hilbert–Schmidt norm does not depend on the choice of orthonormal basis for U. Moreover, $S \in L_{(HS)}(U, H)$ if and only if $S^* \in L_{(HS)}(H, U)$ and $\|S\|_{L_{(HS)}(U,H)} = \|S^*\|_{L_{(HS)}(H,U)}$.*

The following proposition gathers basic properties of the spaces of nuclear operators.

Proposition A.4

(i) *If $S \in L_1(U, H)$ and $T \in L(H, V)$ then $TS \in L_1(U, V)$ and*

$$\|TS\|_1 \leq \|S\|_1 \|T\|_{L(H,V)}. \tag{A.1}$$

(ii) *If $S \in L(U, H)$ and $T \in L_1(H, V)$ then $TS \in L_1(U, V)$ and*

$$\|TS\|_1 \leq \|S\|_{L(U,H)} \|T\|_1.$$

Proof   To show (i) assume that $Su = \sum_k b_k \langle u, a_k \rangle_U$, where

$$\sum_k |b_k|_H \, |a_k|_U < \infty.$$

Hence

$$TSu = \sum_k T b_k \langle u, a_k \rangle_U,$$

where

$$\sum_k |T b_k|_V \, |a_k|_U \leq \|T\|_{L(H,V)} \sum_k |a_k|_U \, |b_k|_V < \infty.$$

This proves (A.1). In order to show (ii) assume that $Th = \sum_k b_k \langle h, a_k \rangle_H$. Thus

$$TSu = \sum_k b_k \langle Su, a_k \rangle_H = \sum_k b_k \langle u, S^* a_k \rangle_U$$

and, consequently,

$$\|TS\|_1 \leq \sum_k |b_k|_V \, |S^* a_k|_U \leq \|S^*\| \sum_k |b_k|_V \, |a_k|_H$$

$$\leq \|S\|_{L(U,H)} \sum_k |b_k|_V \, |a_k|_H < \infty.$$

$\square$

If $R \in L_1(H)$ and $\{e_j\}$ is an orthonormal basis of H, then we define the *trace* Tr R of R by the formula

$$\mathrm{Tr}\, R := \sum_j \langle Re_j, e_j \rangle_H.$$

If $Rh = \sum_k \langle h, a_k \rangle_H b_k$ then

$$\sum_j |\langle Re_j, e_j \rangle_H| = \sum_j \left| \sum_k \langle e_j, a_k \rangle_H \langle b_k, e_j \rangle_H \right|$$

$$\leq \sum_k \left(\sum_j |\langle e_j, a_k \rangle_H| \, |\langle b_k, e_j \rangle_H| \right)$$

$$\leq \sum_k \left(\sum_j \langle e_j, a_k \rangle_H^2 \right)^{1/2} \left(\sum_j \langle b_k, e_j \rangle_H^2 \right)^{1/2}$$

$$\leq \sum_k |a_k|_H \, |b_k|_H \leq \|R\|_1.$$

Hence the series defining Tr R is absolutely convergent. Moreover,

$$\text{Tr } R = \sum_j \sum_k \langle e_j, a_k \rangle_H \langle e_j, b_k \rangle_H$$

$$= \sum_k \sum_j \langle e_j, a_k \rangle_H \langle e_j, b_k \rangle_H = \sum_k \langle a_k, b_k \rangle_H,$$

and consequently Tr R does not depend on the choice of basis.

Proposition A.5

(i) *If $S \colon U \mapsto H$ and $T \colon H \mapsto U$ are bounded linear operators and either S or T is nuclear then $TS \in L_1(U)$ and $\text{Tr } TS = \text{Tr } ST$.*

(ii) *If $S, T \in L_{(HS)}(H)$ then $ST \in L_1(H)$, $\|ST\|_1 \leq \|S\|_{(HS)} \|T\|_{(HS)}$ and $\text{Tr } ST = \text{Tr } TS$.*

Proof   Assume that $T \in L_1(H, U)$. Then

$$Th = \sum_k b_k \langle h, a_k \rangle_H, \qquad b_k \in U, \ a_k \in H.$$

Hence

$$TSu = \sum_k b_k \langle Su, a_k \rangle_H = \sum_k b_k \langle u, S^* a_k \rangle_U.$$

Consequently, for any orthonormal bases $\{e_j\}$ of U and $\{f_l\}$ of H,

$$\sum_j \langle TSe_j, e_j \rangle_U = \sum_j \sum_k \langle b_k, e_j \rangle_U \langle e_j, S^* a_k \rangle_U$$

$$= \sum_k \langle b_k, S^* a_k \rangle_U = \sum_k \langle Sb_k, a_k \rangle_H$$

$$= \sum_k \sum_l \langle Sb_k, f_l \rangle_H \langle a_k, f_l \rangle_H = \sum_l \langle STf_l, f_l \rangle_H.$$

In order to show (ii) note that, for any orthonormal basis $\{f_k\}$ of H,

$$Th = \sum_k \langle Th, f_k \rangle_H f_k = \sum_k \langle h, T^* f_k \rangle_H f_k.$$

Hence

$$STh = \sum_k \langle h, T^* f_k \rangle_H Sf_k \tag{A.2}$$

and, consequently,

$$\|ST\|_1 \leq \sum_k |T^* f_k|_H \, |Sf_k|_H \leq \left(\sum_k |T^* f_k|_H^2 \right)^{1/2} \left(\sum_k |Sf_k|_H^2 \right)^{1/2}$$

$$\leq \|T^*\|_{(HS)} \|S\|_{(HS)} = \|T\|_{(HS)} \|S\|_{(HS)}.$$

Moreover, by (A.2) and the Parseval identity,

$$\text{Tr } ST = \sum_k \langle Sf_k, T^* f_k \rangle_H = \sum_k \langle TSf_k, f_k \rangle = \text{Tr } TS.$$

$\square$

Recall that $K(U, H)$ denotes the space of all compact operators from U to H.

Proposition A.6   *We have* $L_1(U, H) \subset L_{(HS)}(U, H) \subset K(U, H)$.

Proof   Let $R \in L_1(U, H)$; then $R = \sum_k b_k \otimes a_k$, where $\{a_k\} \subset U$ and $\{b_k\} \subset H$ are such that $\sum_k |a_k|_U |b_k|_H < \infty$ and $b \otimes a(u) = b\langle a, u \rangle_U$. Let $\{e_n\}$ be an orthonormal basis of U. Then

$$
\begin{aligned}
\sum_n |Re_n|_H^2 &= \sum_n \left| \sum_k b_k \langle a_k, e_n \rangle_U \right|_H^2 \\
&\leq \sum_n \sum_{k,l} |\langle b_k, b_l \rangle_H| \, \|\langle a_k, e_n \rangle_U\| \, |\langle a_l, e_n \rangle_U| \\
&\leq \sum_{k,j} |b_k|_H |b_l|_H \left(\sum_n \langle a_k, e_n \rangle_U^2 \right)^{1/2} \left(\sum_n \langle a_l, e_n \rangle_U^2 \right)^{1/2} \\
&\leq \left(\sum_k |b_k|_H |a_k|_U \right)^2,
\end{aligned}
$$

which proves the first inclusion. The second follows from the following facts: that the Hilbert–Schmidt norm is stronger than the operator norm, that $K(U, H)$ is a closed set of $L(U, H)$, that each $R \in L_{(HS)}(U, H)$ can be approximated in $L_{(HS)}(U, H)$ by the sequence of finite-rank operators

$$
T_n := \sum_{k \leq n} f_k \otimes T^* f_k, \qquad n \in \mathbb{N},
$$

where $\{f_k\}$ is an orthonormal basis of H, and that finite-rank operators are compact. $\qquad\square$

A.3 Nuclear and Hilbert–Schmidt operators on L^2-spaces

Let μ_i be a σ-finite measure on a measurable space $(E_i, \mathcal{E}_i)$, $i = 1, 2$. In this section we will assume that $U := L^2(E_1, \mathcal{E}_1, \mu_1)$ and $H := L^2(E_2, \mathcal{E}_2, \mu_2)$. We write $E := E_1 \times E_2$, $\mathcal{E} := \mathcal{E}_1 \times \mathcal{E}_2$, $\mu := \mu_1 \times \mu_2$. Our first result characterizes the class of Hilbert–Schmidt operators from U to H.

Proposition A.7   *An operator R belongs to $L_{(HS)}(U, H)$ if and only if there is a $K \in L^2(E, \mathcal{E}, \mu)$ such that*

$$
R\psi(\xi) = \int_{E_1} K(\eta, \xi)\psi(\eta)\mu_1(d\eta), \qquad \psi \in U, \ \xi \in E_2. \tag{A.3}
$$

Moreover,

$$
\|R\|_{L_{(HS)}(U,H)}^2 = \int_{E_1} \int_{E_2} |K(\eta, \xi)|^2 \mu_1(d\eta)\mu_2(d\xi).
$$

Proof   Let $\{e_n\}$ be an orthonormal basis of $L^2(E_1, \mathcal{E}_1, \mu_1)$, and let R be given by (A.3). Then, by the Parseval identity,

$$\sum_n |Re_n|_H^2 = \sum_n \int_{E_2} \left(\int_{E_1} K(\eta, \xi) e_n(\eta) \mu_1(d\eta) \right)^2 \mu_2(d\xi)$$

$$= \int_{E_2} \int_{E_1} |K(\eta, \xi)|^2 \mu_1(d\eta) \mu_2(d\xi).$$

Assume now that $R \in L_{(HS)}(U, H)$. Let $\{e_n\}$ be an orthonormal basis of $L^2(E_1, \mathcal{E}_1, \mu_1)$, and let $\{f_k\}$ be an orthonormal basis of $L^2(E_2, \mathcal{E}_2, \mu_2)$. Define the kernel K by

$$K(\eta, \xi) := \sum_{n,k} \langle Re_n, f_k \rangle_H e_n(\eta) f_k(\xi), \qquad \eta \in E_1, \ \xi \in E_2.$$

Note that the series converges in $L^2(E, \mathcal{E}, \mu)$ and that

$$\int_{E_1} \int_{E_2} |K(\eta, \xi)|^2 \mu_1(d\eta) \mu_2(d\xi) = \sum_{n,k} \langle Re_n, f_k \rangle_H^2 = \|R\|_{L_{(HS)}(U,H)}^2.$$

Finally,

$$\int_{E_1} K(\eta, \xi) \psi(\eta) \mu_1(d\eta) = \sum_{n,k} \langle Re_n, f_k \rangle_H \langle \psi, e_n \rangle f_k(\xi) = R\psi(\xi), \qquad \xi \in E_2.$$

$\square$

We pass now to a result useful in the construction of covariance operators of Lévy processes (see Section 4.9 on square integrable Lévy processes). Recall that a continuous function $q \colon \mathcal{O} \times \mathcal{O} \mapsto \mathbb{R}$ is *non-negative-definite* if, for any $M \in \mathbb{N}$, $v_j \in \mathbb{R}$, $j = 1, \ldots, M$, and $\xi_j \in \mathcal{O}$, $j = 1, \ldots, M$,

$$\sum_{i,j=1}^M q(\xi_i, \xi_j) v_i v_j \geq 0.$$

Moreover, q is called *symmetric* if $q(\eta, \xi) = q(\xi, \eta)$ for all $\xi, \eta \in \mathcal{O}$.

Theorem A.8   *Let $\mathcal{O}$ be a bounded closed subset of $\mathbb{R}^d$, let q be a symmetric non-negative-definite continuous function on $\mathcal{O} \times \mathcal{O}$ and let $U := L^2(\mathcal{O}, \mathcal{B}(\mathcal{O}), \ell_d)$. Then the operator $Q \colon U \mapsto U$ given by*

$$Q\psi(\xi) = \int_{\mathcal{O}} q(\eta, \xi) \psi(\eta) \, d\eta, \qquad \psi \in U, \ \xi \in \mathcal{O},$$

is nuclear, symmetric and non-negative-definite.

Proof   We show first that Q is non-negative-definite. To this end it suffices to prove that

$$\int_{\mathcal{O}} \int_{\mathcal{O}} q(\eta, \xi) \psi(\xi) \psi(\eta) \, d\xi \, d\eta \geq 0 \tag{A.4}$$

for every continuous function ψ. Let $\varphi \in U$. Then there is a sequence $\{\psi_n\}$ of continuous functions such that

$$|\varphi - \psi_n|_U^2 = \int_{\mathcal{O}} |\varphi(\xi) - \psi_n(\xi)|^2 \, d\xi \to 0 \qquad \text{as } n \to \infty.$$

Moreover,

$$
\begin{aligned}
I_n &:= \left| \int_{\mathcal{O}} \int_{\mathcal{O}} q(\eta, \xi) \varphi(\xi) \varphi(\eta) \, d\xi \, d\eta - \int_{\mathcal{O}} \int_{\mathcal{O}} q(\eta, \xi) \psi_n(\xi) \psi_n(\eta) \, d\xi \, d\eta \right| \\
&\le \left(\int_{\mathcal{O}} \int_{\mathcal{O}} q^2(\xi, \eta) \, d\xi \, d\eta \right)^{1/2} \left(\int_{\mathcal{O}} \int_{\mathcal{O}} \big(\varphi(\xi) \varphi(\eta) - \psi_n(\xi) \psi_n(\eta) \big)^2 d\xi \, d\eta \right)^{1/2}.
\end{aligned}
$$

Let $\tilde{U} := L^2(\mathcal{O} \times \mathcal{O}, \ \mathcal{B}(\mathcal{O} \times \mathcal{O}), \ d\xi \, d\eta)$. Since

$$
\varphi(\xi)\varphi(\eta) - \psi_n(\xi)\psi_n(\eta) = \big(\varphi(\xi) - \psi_n(\xi)\big)\varphi(\eta) + \psi_n(\xi)\big(\varphi(\eta) - \psi_n(\eta)\big),
$$

we have

$$
\begin{aligned}
I_n &\le |q|_{\tilde{U}} \left(\int_{\mathcal{O}} \int_{\mathcal{O}} \varphi^2(\eta) \big(\varphi(\xi) - \psi_n(\xi) \big)^2 d\xi \, d\eta \right)^{1/2} \\
&\quad + |q|_{\tilde{U}} \left(\int_{\mathcal{O}} \int_{\mathcal{O}} \psi_n^2(\xi) \big(\varphi(\eta) - \psi_n(\eta) \big)^2 d\xi \, d\eta \right)^{1/2} \\
&\le |q|_{\tilde{U}} \big(|\varphi|_U |\varphi - \psi_n|_U + |\psi_n|_U |\varphi - \psi_n|_U \big) \to 0.
\end{aligned}
$$

Thus if (A.4) holds for continuous functions then it holds for all $\varphi \in U$. Assume that $\psi \in U$ is continuous. Let $\varepsilon > 0$. Then there is a disjoint finite partition $A_1, \ldots, A_M$ of $\mathcal{O}$ and points $\xi_1, \ldots, \xi_M$ such that each A_i is measurable, $\xi_i \in A_i, i = 1, \ldots, M$, and, for $\xi \in A_i, \eta \in A_j$, $i, j = 1, \ldots, M$,

$$
|q(\eta, \xi) - q(\xi_j, \xi_i)| \le \varepsilon, \qquad |\psi(\xi) - \psi(\xi_i)| \le \varepsilon.
$$

Since q is non-negative-definite,

$$
\sum_{i,j=1}^{M} q(\xi_i, \xi_j) \psi(\xi_i) \psi(\xi_j) \ell_d(A_i) \ell_d(A_j) \ge 0.
$$

Note that

$$
\begin{aligned}
J &= \left| \int_{\mathcal{O}} \int_{\mathcal{O}} q(\eta, \xi) \psi(\xi) \psi(\eta) \, d\xi \, d\eta - \sum_{i,j=1}^{M} q(\xi_i, \xi_j) \psi(\xi_i) \psi(\xi_j) \ell_d(A_i) \ell_d(A_j) \right| \\
&= \left| \sum_{i,j=1}^{M} \iint_{A_i \times A_j} \big(q(\eta, \xi) \psi(\xi) \psi(\eta) - q(\xi_i, \xi_j) \psi(\xi_i) \psi(\xi_j) \big) \, d\xi \, d\eta \right| \\
&\le \sum_{i,j=1}^{M} \ell_d(A_i) \ell_d(A_j) \sup_{\substack{\eta \in A_i \\ \xi \in A_j}} \big| q(\eta, \xi) \psi(\xi) \psi(\eta) - q(\xi_i, \xi_j) \psi(\xi_i) \psi(\xi_j) \big|.
\end{aligned}
$$

Let $\|\cdot\|_\infty$ be the supremum norm on $C(\mathcal{O})$ and on $C(\mathcal{O} \times \mathcal{O})$. Then

$$\sup_{\substack{\eta \in A_i \\ \xi \in A_j}} \left| q(\eta, \xi)\psi(\xi)\psi(\eta) - q(\xi_i, \xi_j)\psi(\xi_i)\psi(\xi_j) \right|$$

$$\leq \sup_{\substack{\eta \in A_i \\ \xi \in A_j}} \left| q(\eta, \xi)\psi(\xi_i)\psi(\xi_j) - q(\eta, \xi)\psi(\xi)\psi(\eta) \right|$$

$$+ \sup_{\substack{\eta \in A_i \\ \xi \in A_j}} \left| q(\eta, \xi) - q(\xi_i, \xi_j) \right| \|\psi\|_\infty^2$$

$$\leq \|q\|_\infty \sup_{\substack{\eta \in A_i \\ \xi \in A_j}} \left| \psi(\xi)\psi(\eta) - \psi(\xi_i)\psi(\xi_j) \right| + \varepsilon \|\psi\|_\infty^2$$

$$\leq \|q\|_\infty \left(\sup_{\substack{\xi \in A_i \\ \eta \in A_j}} \left| \psi(\xi) - \psi(\xi_i) \right| |\psi(\eta)| + \sup_{\substack{\xi \in A_i \\ \eta \in A_j}} |\psi(\xi_i)| \left| \psi(\eta) - \psi(\xi_j) \right| \right) + \varepsilon \|\psi\|_\infty^2$$

$$\leq 2\varepsilon \|q\|_\infty \|\psi\|_\infty + \varepsilon \|\psi\|_\infty^2.$$

Therefore

$$J \leq \varepsilon \left(2\|q\|_\infty \|\psi\|_\infty + \|\psi\|_\infty^2 \right) \left(\sum_{i=1}^{M} \ell_d(A_i) \right) \left(\sum_{j=1}^{M} \ell_d(A_j) \right)$$

$$\leq \varepsilon \left(2\|q\|_\infty \|\psi\|_\infty + \|\psi\|_\infty^2 \right) \ell_d(\mathcal{O})^2$$

and since $\varepsilon > 0$ is arbitrary we have (A.4).

Now we show that Q is a compact operator. Note that

$$|Q\psi(\xi)| = \left| \int_\mathcal{O} q(\eta, \xi)\psi(\eta)\, d\eta \right| \leq \left(\int_\mathcal{O} q^2(\eta, \xi)\, d\eta \right)^{1/2} \left(\int_\mathcal{O} \psi^2(\eta)\, d\eta \right)^{1/2}$$

$$\leq \|q\|_\infty \ell_d(\mathcal{O})^{1/2} |\psi|_U, \qquad \xi \in \mathcal{O},$$

and

$$|Q\psi(\xi) - Q\psi(\tilde{\xi})| \leq \int_\mathcal{O} |q(\eta, \xi) - q(\eta, \tilde{\xi})| |\psi(\eta)|\, d\eta$$

$$\leq \left(\sup_{\eta \in \mathcal{O}} |q(\eta, \xi) - q(\eta, \tilde{\xi})| \right) \ell_d(\mathcal{O})^{1/2} |\psi|_U.$$

Thus Q transforms the unit ball in U into a set of uniformly bounded and equicontinuous functions and hence into a relatively compact set in $C(\mathcal{O})$ and, in particular, into a relatively compact set in U.

Since Q is a symmetric compact operator, there is an orthonormal basis $\{e_n\}$ of U consisting of eigenvectors of Q. Let $\{\gamma_n\}$ be the corresponding sequence of eigenvalues. Since Q is non-negative-definite we have $\gamma_j \geq 0$, $j = 1, 2 \ldots$ Since $Qe_n = \gamma_n e_n$, we obtain

$e_n \in C(\mathcal{O})$, $n = 1, 2, \ldots$, provided that $\gamma_n \neq 0$. Obviously, without loss of generality we may assume that $\gamma_n > 0$, $n = 1, 2 \ldots$ Write

$$q_N(\eta, \xi) := \sum_{j=1}^{N} \gamma_j e_j(\eta) e_j(\xi), \qquad \xi, \eta \in \mathcal{O},$$

$$Q_N \psi(\xi) := \int_{\mathcal{O}} q_N(\eta, \xi) \psi(\eta) \, d\eta, \qquad \xi \in \mathcal{O} \ \psi \in U.$$

Clearly Q_N is a symmetric operator with continuous kernel q_N. Moreover,

$$(Q - Q_N)\psi = \sum_{j>N} \gamma_j \langle \psi, e_j \rangle_U e_j, \qquad \psi \in U.$$

Thus the operator $Q - Q_N$ is non-negative-definite, that is,

$$\int_{\mathcal{O}} \int_{\mathcal{O}} (q(\eta, \xi) - q_N(\eta, \xi)) \psi(\eta) \psi(\xi) \, d\xi \, d\eta \geq 0, \qquad \psi \in U.$$

It follows that $k_N(\eta, \eta) := q(\eta, \eta) - q_N(\eta, \eta) \geq 0$ for $\eta \in \mathcal{O}$. Indeed if, for some $\tilde{\eta} \in \mathcal{O}$, $k_N(\tilde{\eta}, \tilde{\eta}) = -\varepsilon < 0$ then there is an $r > 0$ such that $k_N(\eta, \xi) \leq -\varepsilon/2$ for $\eta, \xi \in \mathcal{O}$ with $|\eta - \tilde{\eta}| < r$ and $|\xi - \tilde{\eta}| < r$. Let ψ be a continuous non-negative function such that $\psi(\tilde{\eta}) > 0$ and $\psi(\eta) = 0$ for $\eta \in \mathcal{O} \colon |\tilde{\eta} - \eta| \geq r$. Then

$$\int_{\mathcal{O}} \int_{\mathcal{O}} k_N(\eta, \xi) \psi(\xi) \psi(\eta) \, d\eta \, d\xi < \infty,$$

which leads to a contradiction. Eventually, we have $k_N(\eta, \eta) \geq 0$ for $\eta \in \mathcal{O}$ and, in particular, for every non-negative integer N,

$$\sum_{j=1}^{N} \gamma_j e_j^2(\eta) \leq q(\eta, \eta), \qquad \eta \in \mathcal{O}.$$

Hence

$$\sum_{j=1}^{N} \gamma_j = \sum_{j=1}^{N} \gamma_j \int_{\mathcal{O}} e_j^2(\eta) \, d\eta \leq \int_{\mathcal{O}} q(\eta, \eta) \, d\eta < \infty$$

and, consequently, $\sum_j \gamma_j < \infty$. As a by-product we have shown that the series $\sum \gamma_n e_n(\xi) e_n(\eta)$ converges uniformly and absolutely on the set $\mathcal{O} \times \mathcal{O}$ and that

$$q(\eta, \xi) = \sum_n \gamma_n e_n(\eta) e_n(\xi), \qquad \xi, \eta, \in \mathcal{O}, \qquad \text{Tr } Q = \int_{\mathcal{O}} q(\eta, \eta) \, d\eta.$$

$\square$

A.4 Images of linear operators

We present here an important result due to Douglas (1966) on the comparison of the images of two linear operators. We assume that E_1, E_2, E are Hilbert spaces and that $A_1 \in L(E_1, E)$, $A_2 \in L(E_2, E)$ are bounded linear operators. For the proof of the result we refer the reader to Douglas (1966) or Da Prato and Zabczyk (1992a). For a version in Banach spaces see Zabczyk (1995). We denote by A_i^{-1}, $i = 1, 2$, the pseudo-inverse operators.

Theorem A.9 (Douglas)

(i) $A_1(E_1) \subset A_2(E_2)$ *if and only if there is a constant* $c > 0$ *such that* $\left| A_1^* x \right|_{E_1} \leq c \left| A_2^* x \right|_{E_2}$ *for every* $x \in E$.

(ii) *If* $\left| A_1^* x \right|_{E_1} = \left| A_2^* x \right|_{E_2}$ *for every* $x \in E$ *then* $A_1(E_1) = A_2(E_2)$ *and* $\left| A_1^{-1} x \right|_{E_1} = \left| A_2^{-1} x \right|_{E_2}$ *for every* $x \in E$.

Appendix B C_0-semigroups

In this appendix, basic results on C_0-semigroups, introduced at the beginning of Chapter 9, are gathered. Examples of important specific generators and semigroups are provided.

B.1 Generation theorems

Let A be a closed, densely defined, linear operator on a Banach space B. The *resolvent set of A* is the set $\rho(A)$ of all $\alpha \in \mathbb{C}$ such that $\alpha I - A$ is invertible. The inverse $(\alpha I - A)^{-1}$ is a bounded operator also denoted by $R(\alpha)$. The family of operators $(\alpha I - A)^{-1}, \alpha \in \rho(A)$, is called the *resolvent of A*. The following classical characterization is due to E. Hille and K. Yosida. For its proof we refer the reader to Davies (1980), Engel and Nagel (2000), Pazy (1983) or Yosida (1965).

Theorem B.1 (Hille–Yosida)

(i) *A densely defined closed operator A generates a C_0-semigroup S such that, for some ω and $M > 0$, $|S(t)z|_B \leq e^{\omega t} M |z|_B$ for all $z \in B$ and $t \geq 0$ if and only if $(\omega, \infty) \subset \rho(A)$ and*

$$\|R^m(\alpha)\|_{L(B,B)} \leq \frac{M}{(\alpha - \omega)^m}, \qquad \forall m \in \mathbb{N}, \ \forall \alpha > \omega. \tag{B.1}$$

Moreover, if (B.1) holds then

$$R(\alpha) = \int_0^\infty e^{-\alpha t} S(t) \, dt, \qquad \alpha > \omega,$$

and $S(t)z = \lim_{\alpha \to \infty} e^{t A_\alpha} z$, where $A_\alpha := \alpha (\alpha R(\alpha) - I), \alpha > \omega$.

(ii) *If, for some $z \in B$, $A_\alpha z$ converges as α tends to infinity then $z \in D(A)$ and*

$$\lim_{\alpha \to \infty} A_\alpha z = Az.$$

The operators (A_α) appearing in the theorem above are called the *Yosida approximations of A*.

Let us now recall the classical concept of the *adjoint operator*. Namely if A is a densely defined linear operator A on a Hilbert space V then $y \in D(A^*)$ if and only if the linear

365

functional that associates with $z \in D(A)$ the value $\langle y, Az \rangle_V$ has a continuous extension to the whole V. In this case there exists a unique element w such that $\langle w, z \rangle_V = \langle y, Az \rangle_V$ for all $z \in D(A)$, and we may define $A^* y = w$. The operator A^* with domain $D(A^*)$ is called the *operator adjoint* to A.

The following result is a consequence of the Hille–Yosida theorem.

Theorem B.2   *A closed densely defined linear operator A on a Hilbert space H generates a C_0-semigroup S satisfying $\|S(t)\|_{L(H,H)} \leq e^{\omega t}$ for all $t > 0$ if and only if*

$$\langle Az, z \rangle_H \leq \omega |z|_H^2, \qquad \forall\, z \in D(A),$$
$$\langle A^* y, y \rangle_H \leq \omega |y|_H^2, \qquad \forall\, y \in D(A^*).$$

Let us recall that an operator $(A, D(A))$ on a Hilbert space V is *self-adjoint* if $D(A) = D(A^*)$ and $A = A^*$. If in addition $\langle Az, z \rangle_V \leq 0$ for all $z \in D(A)$ then A is called *negative-definite*.

Let us assume that A is a negative-definite operator on a Hilbert space V, and define

$$H := \begin{pmatrix} D\left((-A)^{1/2}\right) \\ V \end{pmatrix}.$$

Then H with the scalar product

$$\left\langle \begin{pmatrix} x \\ y \end{pmatrix}, \begin{pmatrix} u \\ v \end{pmatrix} \right\rangle_H := \langle (-A)^{1/2} x, \ (-A)^{1/2} u \rangle_V + \langle y, v \rangle_V$$

is a real separable Hilbert space. Let

$$\mathcal{A} := \begin{pmatrix} 0 & I \\ A & 0 \end{pmatrix}, \qquad D(\mathcal{A}) := \begin{pmatrix} D(A) \\ D\left((-A)^{1/2}\right) \end{pmatrix}.$$

Lemma B.3   *If $(A, D(A))$ is a negative-definite operator on V then $(A, D(A))$ and $(\mathcal{A}, D(\mathcal{A}))$ are generators of C_0-semigroups of contractions on V and H respectively.*

The lemma is a consequence of Theorem B.2. We will derive it from the following important *generation theorem*, due to G. Lumer and R. S. Phillips.

Theorem B.4 (Lumer–Phillips)   *Let $(T, D(T))$ be a linear operator on a Hilbert space U. Assume that T is densely defined, that is, $\overline{D(T)} = U$, and that it is dissipative, that is, $\langle Tx, x \rangle_U \leq 0$ for every $x \in D(T)$. Moreover, assume that for some $\lambda > 0$ the image $\mathrm{Im}\,(\lambda I - T)$ is equal to U. Then $(T, D(T))$ generates a C_0-semigroup of contractions on U.*

For the proof of the Lumer–Phillips theorem we refer the reader to e.g. Pazy (1983), p. 13.

Proof of Lemma B.3   We consider only the more difficult case of the operator $(\mathcal{A}, D(\mathcal{A}))$ and show that the assumptions of the Lumer–Phillips theorem are satisfied. The domain $D(\mathcal{A})$ is dense in H since $D((-A)^{1/2})$ is dense in V. In order to show that $\mathcal{A}$ is dissipative, we fix

$$z = \begin{pmatrix} x \\ y \end{pmatrix} \in \begin{pmatrix} D(A) \\ D\left((-A)^{1/2}\right) \end{pmatrix} = D(\mathcal{A}).$$

Then

$$\langle \mathcal{A}z, z\rangle_H = \left\langle \begin{pmatrix} y \\ Ax \end{pmatrix}, \begin{pmatrix} x \\ y \end{pmatrix} \right\rangle_H$$

$$= \langle (-A)^{1/2}\, y, \ (-A)^{1/2}\, x\rangle_V + \langle Ax, y\rangle_V = 0.$$

We now show that Im $(I - \mathcal{A}) = H$. To do this we need to show that for all $x \in D\big((-A)^{1/2}\big)$ and $y \in V$ there are $u \in D(A)$ and $v \in D\big((-A)^{1/2}\big)$ such that

$$(I - \mathcal{A}) \begin{pmatrix} u \\ v \end{pmatrix} = \begin{pmatrix} x \\ y \end{pmatrix},$$

that is, $u - v = x$ and $-Au + v = y$. Equivalently, $(I - A)u = x + y$ and $v = u - x$. Since A is negative-definite, $I - A$ is invertible. Thus the above equality holds for $u = (I - A)^{-1}\,(x + y)$ and $v = u - x = (I - A)^{-1}\,(x + y) - x$.

B.1.1 Bounded perturbation of a generator

Assume that $(A, D(A))$ generates a C_0-semigroup S on a Banach space B. Let G be a possibly unbounded linear operator on B. It is of great interest to find conditions under which $A + G$ generates a C_0-semigroup or, equivalently, the problem

$$\frac{dy}{dt} = (A + G)y(t), \qquad y(0) = y_0 \tag{B.2}$$

is well posed on B. Here we restrict our attention to the simplest case, where G is bounded. For more information on the theory of perturbations of C_0-generators we refer the reader to Davies (1980), Engel and Nagel (2000) or Pazy (1983).

Theorem B.5 *Assume that $(A, D(A))$ generates a C_0-semigroup S satisfying $\|S(t)\|_{L(B,B)} \leq Me^{\omega t}$ for all $t \geq 0$, for some $M \geq 1$ and $\omega \in \mathbb{R}$. If $G \in L(B, B)$ then the operator $C := A + G$ with domain $D(C) = D(A)$ generates a C_0-semigroup T satisfying*

$$\|T(t)\|_{L(B,B)} \leq Me^{(\omega + M\|G\|_{L(B,B)})t}, \qquad t \geq 0. \tag{B.3}$$

Proof   Below, we merely sketch the proof. For more details we refer the reader to Davies (1980). Write

$$T(t)z := \sum_{k=0}^{\infty} I_k(t)z, \qquad t \geq 0, \ z \in B, \tag{B.4}$$

where $I_0(t) := S(t)$ and

$$I_{k+1}(t) := \int_0^t S(t - s)GI_k(s)\,ds.$$

It is easy to verify that the series converges in the operator norm on $L(B, B)$ and that (B.3) holds. By a standard calculation, $(T(t), t \geq 0)$ forms a semigroup. It is strongly continuous since, for every $z \in B$,

$$\limsup_{t \downarrow 0} |T(t)z - z|_B$$

$$\leq \lim_{t \downarrow 0} \left\{ |S(t)z - z|_B + M e^{\omega t} |z|_B \sum_{k=1}^{\infty} \frac{\left(tM \|G\|_{L(B,B)}\right)^n}{n!} \right\} = 0.$$

We now show that C is the generator of T. To this end note that, for every $z \in B$,

$$\lim_{t \downarrow 0} \left| \frac{T(t)z - z}{t} - \frac{S(t)z - z}{t} - Gz \right|_B$$

$$\leq \lim_{t \downarrow 0} \left| \frac{1}{t} \int_0^t S(t-s)GS(s)z \, ds - Gz \right|_B$$

$$+ \lim_{t \downarrow 0} \frac{M e^{\omega t}}{t} |z|_B \sum_{k=2}^{\infty} \frac{\left(tM \|G\|_{L(B,B)}\right)^k}{k!} = 0.$$

Hence if Z denotes the generator of T then $z \in D(Z)$ if and only if $z \in D(A)$. Moreover, for $z \in D(A)$, $Zz = Az + Gz$. □

As a consequence of (B.4),

$$T(t)z = S(t)z + \int_0^t S(t-s)GT(s)z \, ds.$$

In other words, to solve (B.2) one needs to solve

$$y(t) = S(t)y(0) + \int_0^t S(t-s)Gy(s) \, ds.$$

The equation above is a version of the so-called *variation-of-constants formula*. This formula is introduced in Section 9.1.

B.1.2 Analytic semigroups and fractional powers

An important class of semigroups is formed by the so-called analytic semigroups; see the definition below. For an equivalent definition see e.g. Engel and Nagel (2000).

Definition B.6   A C_0-semigroup on a Banach space B is *analytic* if, for every $t > 0$, $S(t)(B) \subset D(A)$ and $\sup_{t \in (0,1]} \|t A S(t)\|_{L(B,B)} < \infty$.

To study the regularity properties of solutions to equations with a linear part A that is the generator of an analytic semigroup S on a Banach space B, it is convenient to introduce the concept of *fractional powers* of A. For simplicity we assume that $(0, \infty)$ is contained in the resolvent set of A and that there are $\epsilon > 0$ and $M < \infty$ such that

$$\|(\lambda - A)^{-1}\|_{L(B,B)} \leq \frac{M}{\epsilon + \lambda}, \qquad \forall \lambda > 0. \tag{B.5}$$

For $\gamma \in (0, 1)$ we define $(-A)^{-\gamma}$ by the formula

$$(-A)^{-\gamma} = \frac{1 - e^{-2\pi i \gamma}}{2\pi i} \int_0^\infty \lambda^{-\gamma}(\lambda - A)^{-1} \, d\lambda.$$

Then one can show that $(-A)^{-\gamma}$ is a bounded invertible operator and define $(-A)^{\gamma}$ as the inverse of $(-A)^{-\gamma}$. It is known (see e.g. Davies 1980, Engel and Nagel 2000, Pazy 1983 or Yosida 1965) that, for all $t > 0$ and $\gamma \in (0, 1)$, the range $S(t)(B)$ is contained in the domain $D((-A)^{\gamma})$ and that

$$\sup_{t \in (0,T]} t \|(-A)^{\gamma} S(t)\|_{L(B,B)} < \infty, \qquad \forall T < \infty.$$

To define fractional powers one can also use the concept of subordination.

B.1.3   Generators obtained by subordination

Let $(S(t), t \geq 0)$ be a C_0-semigroup generated on a Hilbert space H by an operator A, and let (μ_t) be a convolution semigroup of measures on $[0, +\infty)$ (see Definition 4.2). One may easily check that the formula

$$\widetilde{S}(t) = \int_0^{\infty} S(s)\,\mu_t(\mathrm{d}s), \qquad t \geq 0, \tag{B.6}$$

defines a C_0-semigroup on H as well. Assume that

$$\widetilde{\mu}_t(r) := \int_0^{\infty} \mathrm{e}^{-rs}\mu_t(\mathrm{d}s), \qquad r > 0,$$

is the Laplace transforms of μ_t. Then, by Theorem 4.31, $\widetilde{\mu}_t(r) = \mathrm{e}^{-t\widetilde{\psi}(r)}$ for some function $\widetilde{\psi}$. Since formally $S(t) = \mathrm{e}^{tA}$, $t \geq 0$, it follows that, also formally,

$$\widetilde{S}(t) = \int_0^{\infty} \mathrm{e}^{-s(-A)}\mu_t(\mathrm{d}s) = \mathrm{e}^{-t\widetilde{\psi}(-A)}, \qquad t > 0. \tag{B.7}$$

Consequently, $-\widetilde{\psi}(-A)$ is a candidate for the generator of $\big(\widetilde{S}(t), t \geq 0\big)$. In particular, if $\widetilde{\psi}(\lambda) = \lambda^{\beta}$ for some $\beta > 0$ then formula (B.6) defines a semigroup with generator $\widetilde{A} = -(-A)^{\beta}$. Taking into account that there exists an explicit formula for the density of the subordinator with $\beta = 1/2$, one obtains an explicit formula for the semigroup generated by $-(-A)^{1/2}$.

The procedure above can be made precise in many important cases; see Yosida (1965). For instance, if A is the negative definite generator of a semigroup $(S(t), t \geq 0)$, then its fractional power $\widetilde{A} = -(-A)^{\beta}$ generates a C_0-semigroup defined by (B.6).

B.2   Parabolic semigroups

We use the notation of Chapter 2. Namely, $\mathcal{A}$ is an elliptic differential operator of order $2m$ on $\mathbb{R}^d$ or on a bounded domain $\mathcal{O} \subset \mathbb{R}^d$. In the case of a bounded domain, $\mathcal{A}$ is considered together with a system of boundary operators B_j, $j = 1, \ldots, m$. We will assume that all conditions on $\mathcal{O}$, the coefficients of $\mathcal{A}$ and the set of boundary operators $\{B_j\}$ formulated in Section 2.5 are satisfied.

Given $p \in (1, \infty)$, define

$$D(A_p) = \{\psi \in W^{2m,p}(\mathcal{O})\colon B_j\psi = 0 \text{ on } \partial\mathcal{O} \text{ for } j = 1, \ldots, m\},$$
$$A_p\psi = \mathcal{A}\psi \qquad \text{for } \psi \in D(A_p)$$

in the case of a bounded domain, and $D(A_p) = W^{2m,p}(\mathbb{R}^d)$, $A_p\psi = \mathcal{A}\psi$ for $\psi \in D(A_p)$ in the case of $\mathbb{R}^d$. Then A_p is the infinitesimal generator of an analytic C_0-semigroup S_p on $L^p = L^p(\mathcal{O})$ or on $L^p = L^p(\mathbb{R}^d)$, respectively (see Tanabe 1979, Theorem 3.8.2). Let $\mathcal{G}$ be the Green function introduced in Section 2.5. Then (see (2.3) in Section 2.5)

$$S_p(t)\psi(\xi) = \int_{\mathcal{D}} \mathcal{G}(t, \xi, \eta)\psi(\eta)\,\mathrm{d}\eta, \qquad t \in (0, \infty), \tag{B.8}$$

where $\mathcal{D}$ is either $\mathcal{O}$ or $\mathbb{R}^d$.

Formula (B.8) also defines a C_0-semigroup S_1 on L^1. Clearly $S_q(t) = S_p(t)$ on $L^p \cap L^q$ for all q, p. Thus it will cause no confusion if we omit the subscript. Finally, by the Aronson estimates, for all $t > 0$ and $\psi \in L^p$, $S(t)\psi$ is a function with bounded and continuous derivatives up to order $2m$. In Theorem B.9 below we will show that S is a C_0-semigroup on the weighted spaces L_ρ^p, C_ρ, $\mathcal{L}_\rho^p$ and $\mathcal{C}_\rho$ introduced in Section 2.3. To this end we will need the following results, of independent interest.

Theorem B.7   *Let $\rho \in \mathbb{R}$, $1 \leq p \leq q < \infty$, and let $T < \infty$. Then:*

$$S(t)\colon L_\rho^p \mapsto L_\rho^q, \ t > 0, \qquad \sup_{t \in (0,T]} t^{(d/2m)(1/p-1/q)}\|S(t)\|_{L(L_\rho^p, L_\rho^q)} < \infty, \tag{B.9}$$

$$S(t)\colon C_\rho \mapsto C_\rho, \ t \geq 0, \qquad \sup_{t \in [0,T]} \|S(t)\|_{L(C_\rho, C_\rho)} < \infty, \tag{B.10}$$

$$S(t)\colon L_\rho^p \mapsto C_\rho, \ t > 0, \qquad \sup_{t \in (0,T]} t^{d/2mp}\|S(t)\|_{L(L_\rho^p, C_\rho)} < \infty. \tag{B.11}$$

Proof   Let $\mathfrak{g}_m$ be the function defined by (2.5) in Section 2.5. Let us observe first that for each $K_2 > 0$ there exists a constant K such that, for all $t \leq T$ and $\eta \in \mathbb{R}^d$,

$$\int_{\mathbb{R}^d} \mathfrak{g}_m(K_2 t, |\xi - \eta|)\vartheta_\rho(\xi)\,\mathrm{d}\xi \leq K\vartheta_\rho(\eta). \tag{B.12}$$

To this end let us fix $T \in (0, \infty)$ and $\rho \in \mathbb{R}$. Note that there is a constant $C_1 < \infty$ such that

$$\forall \xi, \eta \in \mathbb{R}^d, \qquad -\rho|\xi| + \rho|\eta| - \frac{1}{2}\left(\frac{|\xi - \eta|^{2m}}{K_2 T}\right)^{1/(2m-1)} \leq C_1. \tag{B.13}$$

Obviously there is a constant $C_2 \in (0, \infty)$, such that

$$\forall \eta \in \mathbb{R}^d, \qquad C_2^{-1}\mathrm{e}^{-\rho|\eta|} \leq \vartheta_\rho(\eta) \leq C_2\mathrm{e}^{-\rho|\eta|}.$$

Hence, for all $\eta \in \mathbb{R}^d$ and $t \in (0, T]$,

$$\int_{\mathbb{R}^d} \mathfrak{g}_m(K_2 t, |\xi - \eta|) \vartheta_\rho(\xi) \, d\xi$$

$$\leq C_2 K_2^{-d/2m} \int_{\mathbb{R}^d} t^{-d/2m} \exp\left\{ -\rho|\xi| - \left(\frac{|\xi - \eta|^{2m}}{K_2 t} \right)^{1/(2m-1)} \right\} d\xi$$

$$\leq C_2^2 K_2^{-d/2m} \vartheta_\rho(\eta) \int_{\mathbb{R}^d} t^{-d/2m} \exp\left\{ -\frac{1}{2} \left(\frac{|\eta - \xi|^{2m}}{K_2 t} \right)^{1/(2m-1)} \right.$$

$$\left. -\rho|\xi| + \rho|\eta| - \frac{1}{2} \left(\frac{|\xi - \eta|^{2m}}{K_2 T} \right)^{1/(2m-1)} \right\} d\xi$$

$$\leq C_2^2 K_2^{-d/2m} \vartheta_\rho(\eta) \int_{\mathbb{R}^d} t^{-d/2m} \exp\left\{ -\frac{1}{2} \left(\frac{|\eta - \xi|^{2m}}{K_2 t} \right)^{1/(2m-1)} \right\} d\xi \, e^{C_1}$$

$$\leq C_2^2 K_2^{-d/2m} e^{C_1} C_3 \vartheta_\rho(\eta),$$

where

$$C_3 = \int_{\mathbb{R}^d} t^{-d/2m} \exp\left\{ -\frac{1}{2} \left(\frac{|\eta - \xi|^{2m}}{K_2 t} \right)^{1/(2m-1)} \right\} d\xi$$

$$= \int_{\mathbb{R}^d} \exp\left\{ -\frac{1}{2} \left(\frac{|\xi|^{2m}}{K_2} \right)^{1/(m-1)} \right\} d\xi.$$

We can now proceed to prove (B.9). Assume that $1 \leq p \leq q < \infty$ and take $\psi \in L_\rho^p$ and $t > 0$. By Theorem 2.6 and the Jensen inequality (for the convex function $x \mapsto |x|^p$),

$$|S(t)\psi|_{L_\rho^q}^q = \int_{\mathbb{R}^d} \left| \int_{\mathbb{R}^d} \mathcal{G}(t, \xi, \eta) \psi(\eta) \, d\eta \right|^q \vartheta_\rho^q(\xi) \, d\xi$$

$$\leq K_1^q \int_{\mathbb{R}^d} \left| \int_{\mathbb{R}^d} \mathfrak{g}_m(K_2 t, |\xi - \eta|) |\psi(\eta)| \, d\eta \right|^q \vartheta_\rho^q(\xi) \, d\xi$$

$$\leq K_3 \int_{\mathbb{R}^d} \left| \int_{\mathbb{R}^d} \mathfrak{g}_m(K_2 t, |\xi - \eta|) |\psi(\eta)|^p \, d\eta \right|^{q/p} \vartheta_\rho^q(\xi) \, d\xi,$$

where

$$K_3 := K_1^q \left| \int_{\mathbb{R}^d} \mathfrak{g}_m(K_2 t, |\eta|) \, d\eta \right|^{q-1}.$$

Applying the Jensen inequality again we obtain

$$\left| \int_{\mathbb{R}^d} \mathfrak{g}_m(K_2 t, |\xi - \eta|) |\psi(\eta)|^p \, d\eta \right|^{q/p}$$

$$= \left| \int_{\mathbb{R}^d} \mathfrak{g}_m(K_2 t, |\xi - \eta|) \vartheta_\rho^{-p}(\eta) |\psi(\eta)|^p \vartheta_\rho^p(\eta) \, d\eta \right|^{q/p}$$

$$\leq |\psi|_{L_\rho^p}^{q-p} \int_{\mathbb{R}^d} \mathfrak{g}_m^{q/p}(K_2 t, |\xi - \eta|) \vartheta_\rho^{-q}(\eta) |\psi(\eta)|^p \vartheta_\rho^p(\eta) \, d\eta.$$

Note that there are constants K_4 and K_5 independent of t and ξ, η such that

$$\mathfrak{g}_m^{q/p}\left(K_2t, |\xi - \eta|\right) \le K_4 t^{-dq/2mp+d/2m} \mathfrak{g}_m\left(K_5t, |\xi - \eta|\right).$$

Next, using the equivalence of ϑ_ρ^q and $\vartheta_{q\rho}$ and estimate (B.12) we can find a constant K_6 such that

$$\int_{\mathbb{R}^d} \mathfrak{g}_m^{q/p}\left(K_5t, |\xi - \eta|\right)\vartheta_\rho^q(\xi)\,\mathrm{d}\xi \le K_6 t^{-dq/2mp+d/2m}\vartheta_\rho^q(\eta).$$

Summing up, we obtain

$$|S(t)\psi|_{L_\rho^q}^q \le K_7 t^{-dq/2mp+d/2m}|\psi|_{L_\rho^p}^q,$$

which gives (B.9).

Let $\psi \in C_\rho$. Then

$$
\begin{aligned}
|S(t)\psi(\xi)| &\le K_1 \int_{\mathbb{R}^d} \mathfrak{g}_m\left(K_2t, |\xi - \eta|\right)|\psi(\eta)|\,\mathrm{d}\eta \\
&\le C_1 \int_{\mathbb{R}^d} \mathfrak{g}_m\left(K_2t, |\xi - \eta|\right)|\psi(\eta)|\vartheta_\rho(\eta)\vartheta_{-\rho}(\eta)\,\mathrm{d}\eta \\
&\le C_1|\psi|_{C_\rho} \int_{\mathbb{R}^d} \mathfrak{g}_m\left(K_2t, |\xi - \eta|\right)\vartheta_{-\rho}(\eta)\,\mathrm{d}\eta \\
&\le C_2|\psi|_{C_\rho}\vartheta_{-\rho}(\xi).
\end{aligned}
$$

Consequently,

$$\exists C < \infty\colon \forall t \in (0, T], \ \forall \psi \in C_\rho, \qquad \sup_{\xi \in \mathbb{R}^d} |S(t)\psi(\xi)|\vartheta_\rho(\xi) \le C\,|\psi|_{C_\rho}. \qquad \text{(B.14)}$$

What is left to prove is the continuity of $S(t)\psi(\xi)$ with respect to ξ. It is easy to see that $S(t)\colon C_c^\infty(\mathbb{R}^d) \mapsto C_\rho$, where $C_c^\infty(\mathbb{R}^d)$ denotes the space of all infinitely differentiable functions with compact supports. Thus, since $C_c^\infty(\mathbb{R}^d)$ is dense in C_ρ, (B.10) follows from (B.14).

We now show (B.11). Again using the Aronson estimates, we obtain

$$
\begin{aligned}
|S(t)\psi(\xi)|^p &\le \left(K_1 \int_{\mathbb{R}^d} \mathfrak{g}_m\left(K_2t, |\xi - \eta|\right)|\psi(\eta)|\,\mathrm{d}\eta\right)^p \\
&\le K_1^p \left(\int_{\mathbb{R}^d} \mathfrak{g}_m\left(K_2t, |\xi - \eta|\right)|\psi(\eta)|^p\,\mathrm{d}\eta\right)\left(\int_{\mathbb{R}^d} \mathfrak{g}_m(K_2t, |\xi|)\,\mathrm{d}\xi\right)^{p-1} \\
&\le K_3 \int_{\mathbb{R}^d} \mathfrak{g}_m\left(K_2t, |\xi - \eta|\right)|\psi(\eta)|^p\,\mathrm{d}\eta.
\end{aligned}
$$

Thus

$$\sup_{\xi \in \mathbb{R}^d} \left|S(t)\psi(\xi)\right|\vartheta_\rho(\xi) \le C(t)\,|\psi|_{L_\rho^p},$$

where

$$C(t) := \sup_{\xi, \eta \in \mathbb{R}^d} K_3^{1/p}\,\mathfrak{g}_m^{1/p}\left(K_2t, |\xi - \eta|\right)\vartheta_\rho(\xi)\vartheta_{-\rho}(\eta).$$

Taking into account (B.13) we obtain $\sup_{t \in (0,T]} t^{d/2mp}\,C(t) < \infty$, and (B.11) follows. $\square$

Remark B.8   The proof of Theorem B.7 is based only on (B.12) and the Aronson estimates. It is easy to check that the polynomial weights $\{\theta_\rho\}$ satisfy the following analogue of (B.12):

$$\forall \rho \in \mathbb{R}^d, \ \forall T > 0, \exists C > 0 \colon \forall \eta \in \mathbb{R}^d,$$

$$\int_{\mathbb{R}^d} \mathfrak{g}_m\big(K_2 t, |\xi - \eta|\big)\theta_\rho(\xi)\, \mathrm{d}\xi \le C\,\theta_\rho(\eta).$$

Thus one can easily obtain a polynomial version of Theorem B.7 by replacing in the spaces L_ρ^p and C_ρ (B.9)–(B.11) by $\mathcal{L}_\rho^p$ and $\mathcal{C}_\rho$.

Now note that for any bounded domain $\mathcal{O}$ one has

$$\sup_{t>0,\xi\in\mathcal{O}} \int_{\mathcal{O}} \mathfrak{g}_m\big(K_2 t, |\xi - \eta|\big)\,\mathrm{d}\xi \le \int_{\mathbb{R}^d} \mathfrak{g}_m(K_2, |\eta|)\,\mathrm{d}\eta < \infty.$$

For a bounded domain $\mathcal{O}$ the estimate above can play the role of (B.12) in the proof of an analogue of Theorem B.7. In fact the family S corresponding to the Green function for the system $\mathcal{A}, \{B_j\}$ considered on a bounded domain $\mathcal{O}$ has the following properties: for all $1 \le p \le q < \infty$ and for all $T < \infty$ and $t > 0$,

$$S(t)\colon L^p \mapsto L^q, \qquad \sup_{t\in(0,T]} t^{(d/2m)(1/p-1/q)}\|S(t)\|_{L(L^p, L^q)} < \infty, \tag{B.15}$$

$$S(t)\colon C_b \mapsto C_b, \qquad \sup_{t\in[0,T]} \|S(t)\|_{L(C_b, C_b)} < \infty, \tag{B.16}$$

$$S(t)\colon L^p \mapsto C_b, \qquad \sup_{t\in(0,T]} t^{d/2mp}\|S(t)\|_{L(L^p, C_b)} < \infty, \tag{B.17}$$

where $L^p := L^p(\mathcal{O})$ and $C_b := C_b(\mathcal{O})$.

Theorem B.9

(i) *For all $p \in [1, \infty)$ and $\rho \in \mathbb{R}$, S is a C_0-semigroup on the spaces L_ρ^p and $\mathcal{L}_\rho^p$.*
(ii) *For every $\rho \in \mathbb{R}$, S is a C_0-semigroup on C_ρ and $\mathcal{C}_\rho$.*

Proof   We will restrict our attention to the exponential weights. The proof in the case of polynomial weights is similar. Let $H = L_\rho^p$ or $H = C_\rho$ for certain $\rho \in \mathbb{R}$ and $p \in [1, \infty)$. By Theorem B.7, $(S(t), \ t \ge 0)$ is a family of bounded linear operators on H and,

$$\forall T < \infty, \qquad \sup_{t\in[0,T]} \|S(t)\|_{L(H,H)} < \infty. \tag{B.18}$$

We need to show strong continuity, that is, for every $\psi \in H$,

$$\lim_{t\to 0} |S(t)\psi - \psi|_H = 0. \tag{B.19}$$

Taking into account (B.18) it is sufficient to verify (B.19) for ψ from an arbitrary dense subspace H_0 of H. For H_0 we can take $C_c(\mathbb{R}^d)$. Thus the proof of the theorem is complete if we can show that, for every $\psi \in C_c(\mathbb{R}^d)$ and $M > 0$,

$$\lim_{t\to 0} \sup_{\xi\in\mathcal{O}} |S(t)\psi(\xi) - \psi(\xi)| e^{M|\xi|} = 0.$$

To do this we fix a $\psi \in C_c(\mathbb{R}^d)$ and M. Since $\psi \in C_{2M}$, (B.10) yields

$$C := \sup_{t\in[0,1]} \sup_{\xi\in\mathbb{R}^d} e^{2M|\xi|} |S(t)\psi(\xi)| < \infty,$$

and consequently $|S(t)\psi(\xi)| e^{M|\xi|} \le C e^{-M|\xi|}$ for all $t \in (0, 1]$ and $\xi \in \mathbb{R}^d$.

Let $\varepsilon > 0$. Then there is an $R > 0$ such that $\psi(\xi) = 0$ for ξ satisfying $|\xi| \geq R$ and $e^{M|\xi|}|S(t)\psi(\xi)| \leq \varepsilon$ for all $\xi \in \mathcal{O}$: $|\xi| \geq R$. Thus, for all ξ: $|\xi| \geq R$ and $t \in (0, 1]$, $|S(t)\psi(\xi) - \psi(\xi)|e^{M|\xi|} \leq \varepsilon$. What is left is to show that, on every compact set $K \subset \mathbb{R}^d$,

$$\limsup_{t \to 0} \sup_{\xi \in K} |S(t)\psi(\xi) - \psi(\xi)| = 0.$$

For the proof see Arima (1964), p. 241. □

Remark B.10   One can show that a second-order operator A generates a C_0-semigroup on a weighted space $L^2(\mathbb{R}^d, \mathcal{B}(\mathbb{R}^d), \vartheta(\xi)\,d\xi)$ if and only if

$$\exists\, C < \infty: \forall\, \xi \in \mathbb{R}^d, \qquad A^*\vartheta(\xi) \leq C\vartheta(\xi).$$

For details see Da Prato and Zabczyk (1996). Thus weights of the types $\xi \mapsto e^{\pm|\xi|^2}$ are not admissible.

B.3 Semigroups for classical equations

Here are some examples of specific semigroups often used in applications.

Example B.11 (Transport equation)   Let $H = L^2(0, \infty)$, and let $A_0 = \partial/\partial\xi$. Thus (B.1) has the form

$$\frac{\partial y}{\partial t}(t, \xi) = \frac{\partial y}{\partial \xi}(t, \xi), \qquad y(0, \xi) = z(\xi), \ \xi \geq 0. \tag{B.20}$$

Denote by D_0 the set of all $z \in L^2(0, \infty)$ that have continuous first derivative at each $\xi \geq 0$. Then the solution y is given by the formula

$$y(t, \xi) = z(t + \xi), \qquad t, \xi \geq 0, \tag{B.21}$$

and the semigroup has the form $(S(t)z)(\xi) = z(t + \xi), t, \xi \geq 0, z \in L^2(0, \infty)$. Note that if $z \in L^2(0, \infty)$ is not differentiable at some point then y as given by (B.21) does not satisfy (B.20) for all $t \geq 0$ and $\xi \geq 0$. This is the reason that y is only a generalized solution. Sometimes for the transport semigroups one writes $S(t) = e^{t\partial/\partial\xi}, t \geq 0$.

Example B.12 (Heat equation)   Let $H = L^2(0, 1)$, and let $A_0 = \partial^2/\partial\xi^2$. More precisely we are concerned with the equation

$$\frac{\partial y}{\partial t}(t, \xi) = \frac{\partial^2 y}{\partial \xi^2}(t, \xi), \qquad y(0, \xi) = z(\xi), \ \xi \in (0, 1), \tag{B.22}$$

together with the Dirichlet boundary condition

$$y(t, 0) = y(t, 1) = 0, \qquad t \geq 0. \tag{B.23}$$

Note that we are using the general framework introduced at the beginning of Section B.2, with $\mathcal{A} = \partial^2/\partial\xi^2$, $B_1u(\xi) = u(0)$ and $B_2u(\xi) = u(1)$. Using the Fourier method we are able, however, to solve the problem (B.22), (B.23) explicitly. Namely, we look for a solution in the form

$$y(t, \xi) = \sum_{n=1}^{\infty} y_n(t)\sqrt{2}\,\sin(\pi n\xi), \qquad t \geq 0, \; \xi \in [0, 1].$$

Note that the functions $\{e_n\}$, where $e_n(\xi) = \sqrt{2}\,\sin(\pi n\xi)$, form an orthonormal basis in $L^2(0, 1)$; they are classical solutions of

$$\frac{\partial^2 e_n}{\partial \xi^2} = -\pi^2 n^2 e_n, \qquad e_n(0) = 0 = e_n(1).$$

Since

$$\frac{\partial y}{\partial t}(t, \xi) = \sum_{n=1}^{\infty} y_n'(t)\sqrt{2}\,\sin(\pi n\xi),$$

$$\frac{\partial^2 y}{\partial \xi^2}(t, \xi) = -\sum_{n=1}^{\infty} y_n(t)\pi^2 n^2 \sqrt{2}\,\sin(\pi n\xi),$$

we have

$$y(t, \xi) = S(t)z(\xi) = \sum_{n=1}^{\infty} e^{-\pi^2 n^2 t}\langle z, e_n\rangle_{L^2(0,1)}e_n(\xi). \tag{B.24}$$

This formula defines a C_0-semigroup $(S(t), \; t \geq 0)$ on $L^2(0, 1)$ and the generalized solution y. Its generator is the second-order differential operator $d^2/d\xi^2$ with domain $W^{2,2}(0, 1) \cap W_0^{1,2}(0, 1)$ (see Section 2.2 on Sobolev spaces). Clearly the generator of a heat semigroup is a self-adjoint negative-definite operator on $L^2(0, 1)$.

Example B.13 (Wave equation)   Consider now the problem

$$\frac{\partial^2 u}{\partial t^2}(t, \xi) = \frac{\partial^2 u}{\partial \xi^2}(t, \xi), \qquad t \geq 0, \; \xi \in (0, 1),$$

$$u(0, \xi) = u_0(\xi), \quad \frac{\partial u}{\partial \xi}(0, \xi) = v_0(\xi), \qquad \xi \in (0, 1), \tag{B.25}$$

$$u(t, 0) = u(t, 1) = 0, \qquad t \geq 0,$$

which is known as the *wave equation* or *vibrating string equation*. We write the first equation of (B.25) in the form

$$\frac{\partial u}{\partial t}(t, \xi) = v(t, \xi), \qquad \frac{\partial v}{\partial t}(t, \xi) = \frac{\partial^2 u}{\partial \xi^2}(t, \xi).$$

Let $e_n(\xi) = \sqrt{2}\,\sin(\pi n\xi), n \in \mathbb{N}$, be eigenvectors of the Laplace operator $\Delta = \partial^2/\partial\xi^2$ with Dirichlet boundary conditions. These functions have appeared already in our derivation of the solution of the heat equation on the interval $(0, 1)$.

As in the case of the heat equation, we are able find an explicit formula for the wave semigroup and hence a solution to (B.25). Namely, we look for a solution $y(t) = (u(t), v(t))^{\mathrm{T}}$, $t \geq 0$, of the form

$$u(t) = \sum_{n=1}^{\infty} u_n(t)e_n, \qquad v(t) = \sum_{n=1}^{\infty} v_n(t)e_n, \qquad t \geq 0.$$

In the same way as for the heat equation one arrives at the sequence of ordinary equations

$$u'_n(t) = v_n(t), \quad v'_n(t) = -\pi^2 n^2 u_n(t), \qquad t \geq 0, \ n = 1, 2, \dots,$$

which can be solved explicitly to give

$$u(t, \xi) = \sqrt{2} \sum_{n=1}^{\infty} \left(u_n(0) \cos(\pi nt) \sin(\pi nt) + \frac{v_n(0)}{n\pi} \sin(\pi nt) \sin(\pi n\xi) \right),$$

$$v(t, x) = \sqrt{2} \sum_{n=1}^{\infty} \left(-n\pi u_n(0) \sin(\pi nt) \sin(\pi nt) + v_n(0) \cos(\pi nt) \sin(\pi n\xi) \right),$$

and the semigroup $(S(t), \ t \geq 0)$ corresponding to the wave equation is given by

$$S(t) \begin{pmatrix} u \\ v \end{pmatrix} = \begin{pmatrix} u(t) \\ v(t) \end{pmatrix}. \tag{B.26}$$

Below, we derive from Lemma B.3 the existence of the semigroup S corresponding to the wave equation. First we have to specify the space on which S is defined. To this end, define H_0^1 and H^{-1} as the spaces of all sequences $x = \sum_{n=1}^{\infty} x_n e_n$ satisfying

$$|x|_{H_0^1}^2 := \pi^2 \sum_{n=1}^{\infty} n^2 x_n^2 < \infty \quad \text{and} \quad |x|_{H^{-1}}^2 := \frac{1}{\pi^2} \sum_{n=1}^{\infty} \frac{x_n^2}{n^2} < \infty,$$

respectively. Clearly, H_0^1 and H^{-1} are Hilbert spaces. In fact H_0^1 is equal to $W_0^{1,2}(0, 1)$. On H^{-1} we define the Laplace operator by

$$\Delta x = \sum_{n=1}^{\infty} \pi^2 n^2 x_n e_n.$$

Then Δ with domain H_0^1 is a self-adjoint negative definite operator. Moreover, $D((-\Delta)^{1/2}) = L^2(0, 1)$. By Lemma B.3 the operator

$$A := \begin{pmatrix} 0 & I \\ \Delta & 0 \end{pmatrix}, \qquad D(A) = \begin{pmatrix} H_0^1 \\ L^2(0, 1) \end{pmatrix}$$

generates a C_0-semigroup on the Hilbert space $H = \left(L^2(0, 1), \ H^{-1} \right)^T$. Note that this semigroup is given by (B.26). Using the same arguments we can show that S given by (B.26) is a C_0-semigroup on $\left(H_0^1, L^2(0, 1) \right)^T$.

B.4 Semigroups for non-local operators

Generators such as differential operators raised to fractional powers are non-local. The semigroups generated by them are, however, important examples of transition semigroups of Markovian processes.

Definition B.14   Let $P = (P_t)$ be a transition semigroup on a measurable space $(E, \mathcal{E})$. A σ-finite measure λ on $(E, \mathcal{E})$ is said to be *ω-excessive* for P if, for all $t \geq 0$ and $\Gamma \in \mathcal{E}$,

$$\int_E P_t \chi_\Gamma(\xi) \lambda(d\xi) \leq e^{\omega t} \lambda(\Gamma).$$

Let $L^p(d\lambda) := L^p(E, \mathcal{E}, \lambda)$. The following result is taken from Zabczyk (2001a).

Proposition B.15 *If λ is an excessive measure then, for all $p > 1$ and $t > 0$, P_t has an extension from $B_b(E) \cap L^p(d\lambda)$ to $L^p(d\lambda)$ and, for the extension,*

$$\|P_t\|_{L(L^p(d\lambda), L^p(d\lambda))} \leq e^{\omega t/p}, \qquad t \geq 0.$$

Let $\langle \cdot, \cdot \rangle$ be the scalar product on $\mathbb{R}^d$, and let $|\cdot|$ be the corresponding norm. Assume that (P_t) is the transition semigroup on $B_b(\mathbb{R}^d)$ corresponding to the Lévy process with characteristic (a, Q, ν). Thus

$$P_t \varphi(\xi) = \int_{\mathbb{R}^d} \varphi(\xi + \eta) \mu_t(d\eta), \tag{B.27}$$

where

$$\int_{\mathbb{R}^d} e^{i\langle \xi, \eta \rangle} \mu_t(d\eta) = e^{-t\psi(\xi)},$$

and

$$\psi(\xi) = -i\langle a, \xi \rangle_U + \tfrac{1}{2} \langle Q\xi, \xi \rangle_U$$
$$+ \int_U \left(1 - e^{i\langle \xi, \eta \rangle_U} + \chi_{\{|\eta|_U < 1\}}(\eta) i \langle \xi, \eta \rangle_U \right) \nu(d\eta).$$

The following theorems from Zabczyk (2001a) hold.

Theorem B.16 *If, for some $\gamma > 0$,*

$$\int_{\{|\xi|>1\}} e^{\gamma |\xi|} \nu(d\xi) < \infty \tag{B.28}$$

then P has an extension to $L^p(d\lambda^\gamma) := L^p(\mathbb{R}^d, \mathcal{B}(\mathbb{R}^d), d\lambda^\gamma)$, where $\lambda^\gamma(d\xi) := e^{-\gamma|\xi|} d\xi$. Moreover, for large enough $\omega > 0$,

$$\|P_t\|_{L(L^p(d\lambda^\gamma), L^p(d\lambda^\gamma))} \leq e^{\omega t/p}, \qquad \forall\, t \geq 0.$$

Theorem B.17 *Let $r > d$ and $p > 1$. If*

$$\int_{\{|\xi|>1\}} |\xi|^r \nu(d\xi) < \infty \tag{B.29}$$

then, for κ sufficiently large, (P_t) has an extension to the space $L^p(d\lambda_{r,\kappa}) := L^p(\mathbb{R}^d, \mathfrak{B}(\mathbb{R}^d), d\lambda_{r,\kappa})$, where

$$\lambda_{r,\kappa}(d\xi) := \frac{1}{1 + \kappa |\xi|^r} \, d\xi.$$

Moreover, for large enough $\omega > 0$,

$$\|P_t\|_{L(L^p(d\lambda_{r,\kappa}), L^p(d\lambda_{r,\kappa}))} \leq e^{\omega t/p}, \qquad \forall\, t \geq 0.$$

Conditions (B.28) and (B.29) are trivially satisfied if the jump measure ν is concentrated on bounded sets; thus they are satisfied for all processes X with only bounded jumps. Note that for a fixed $r > d$ the spaces $L^2(d\lambda_{r,\kappa})$, $\kappa > 0$, are isomorphic. The condition $r > d$ is equivalent to the requirement that the measure $\lambda_{r,\kappa}$ is finite.

For the proofs of the theorems above we need the following proposition, which is an easy corollary to Theorem 5.3.

Proposition B.18   *The semigroup (P_t) given by (B.27) is strongly continuous on $C_0(\mathbb{R}^d)$. Moreover, $C_0^2(\mathbb{R}^d)$ is contained in the domain $D(A)$ of the generator A of (P_t), and*

$$A\varphi(\xi) = \langle a, D\varphi(\xi)\rangle + \tfrac{1}{2}\operatorname{Tr} QD^2\varphi(\xi)$$

$$+ \int_{\{|\eta|\leq 1\}} \big(\varphi(\xi + \eta) - \varphi(\xi) - \langle D\varphi(\xi), \eta\rangle\big)\nu(d\eta)$$

$$+ \int_{\{|\eta|>1\}} \big(\varphi(\xi + \eta) - \varphi(\xi)\big)\nu(d\eta), \qquad \varphi \in C_0^2(\mathbb{R}^d),\ \xi \in \mathbb{R}^d.$$

Finally, $C_0^2(\mathbb{R}^d)$ is invariant for (P_t).

Given a measure λ and a function φ, write $(\lambda, \varphi) := \int_{\mathbb{R}^d} \varphi(\xi)\lambda(d\xi)$. Assume that $\omega \geq 0$. We need also the following characterization of excessive measures for (P_t).

Proposition B.19   *A finite measure λ is ω-excessive for (P_t) if and only if, for any nonnegative function $\varphi \in C_0^2(\mathbb{R}^d)$,*

$$(\lambda, A\varphi) \leq \omega\,(\lambda, \varphi). \tag{B.30}$$

Proof   If λ is ω-excessive and $\varphi \in D(A), \varphi \geq 0$, then $(\lambda, P_t\varphi) \leq e^{\omega t}(\lambda, \varphi)$ or, equivalently,

$$\left(\lambda, \ \frac{1}{t}\left(e^{-\omega t}P_t\varphi - \varphi\right)\right) \leq 0, \qquad \forall\, t > 0.$$

Therefore

$$(\lambda, A\varphi) = \left(\lambda, \lim_{t\downarrow 0}\frac{1}{t}\left(e^{-\omega t}P_t\varphi - \varphi\right)\right) \leq 0,$$

which proves (B.30). However, if $\varphi \geq 0$ and $\varphi \in C_0^2(\mathbb{R}^d)$ then $P_s\varphi \in C_0^2(\mathbb{R}^d)$, $P_s\varphi \geq 0$ for $s \geq 0$ and

$$e^{-\omega t}P_t\varphi = \varphi + \int_0^t e^{-\omega s}(A - \omega)P_s\varphi\, ds.$$

If (B.30) holds and $\varphi \in C_0^2(\mathbb{R}^d)$ then

$$\left(\lambda, \ e^{-\omega t}P_t\varphi - \varphi\right) = \int_0^t e^{-\omega s}\left(\lambda, \ P_s(A - \omega)\varphi\right) ds$$

$$= \int_0^t e^{-\omega s}\left(\lambda, \ (A - \omega)P_s\varphi\right) ds \leq 0,$$

because (B.30) holds also for $P_s\varphi$ and $(\lambda, \ P_s\varphi) \geq 0$. We have proved that if (B.30) holds then $(\lambda, P_t\varphi) \leq e^{\omega t}(\lambda, \varphi)$ for every non-negative $\varphi \in C_0^2(\mathbb{R}^d)$. By approximation, $(\lambda, P_t\chi_\Gamma) \leq e^{\omega t}(\lambda, \chi_\Gamma)$ for all $\Gamma \in \mathcal{B}(\mathbb{R}^d)$. $\square$

Clearly, in order to prove Theorems B.16 and B.17 it is sufficient to show that for ω large enough the measures λ^γ and $\lambda_{r,\kappa}$ are ω-excessive. To do this we give sufficient conditions for ω-excessivness in the following four cases.

(i)   Translation: $Q = 0$, $\nu = 0$.
(ii) Wiener process: $a = 0$, $\nu = 0$.
(iii) Large jumps: $a = 0$, $Q = 0$, ν finite.
(iv) Small jumps: $a = 0$, $Q = 0$, ν concentrated on $\{\eta : |\eta| \leq 1\}$.

Note that our infinite divisible family μ_t is of the form

$$\mu_t = \mu_t^1 * \mu_t^2 * \mu_t^3 * \mu_t^4, \qquad t \geq 0,$$

where μ_t^j, $j = 1, 2, 3, 4$, are the convolution semigroups of measures corresponding to cases (i), (ii), (iii) and (iv). Note that if a measure λ is ω_j-excessive for the semigroups (P_t^j) defined by (μ_t^j), $j = 1, 2, 3, 4$, then λ is $\omega(= \omega_1 + \omega_2 + \omega_3 + \omega_4)$-excessive for the semigroup determined by (μ_t). Therefore we can restrict our considerations to the specific cases listed above.

In the following formulations we denote by q the maximal eigenvalue of Q.

Theorem B.20   *The measure λ^γ is ω-excessive for (P_t) if, in cases (i), (ii), (iii) and (iv), respectively,*

$$\text{(i) } \gamma|a| \leq \omega, \quad \text{(ii) } \tfrac{1}{2}\gamma^2 q \leq \omega, \quad \text{(iii) } \int_{\mathbb{R}^d} (e^{\gamma|\eta|} - 1)\nu(d\eta) \leq \omega,$$

$$\text{(iv) } \tfrac{1}{2}\gamma^2 \int_{\{|\eta|\leq 1\}} e^{\gamma|\eta|}|\eta|^2 \nu(d\eta) \leq \omega.$$

Conditions for the ω-excessivness of the measures $\lambda_{r,\kappa}$ are more involved. To formulate them it is convenient to introduce the following family of functions:

$$\psi_r(z) = \sup_{\alpha \geq 0} \frac{(\alpha + z)^r - \alpha^r}{1 + \alpha^r}, \qquad r > 1, \; z \geq 0.$$

Note that

$$\sup_{\alpha \geq 0} \frac{(\alpha + z)^r - \alpha^r}{1 + \kappa\alpha^r} = \kappa^{-1/r}\psi_r\big(z\kappa^{1/r}\big), \qquad \forall \kappa > 0, \; \forall r > 1.$$

In addition, if $r > 1$ is a non-negative integer then

$$\psi_r(z) = \sup_{\alpha \geq 0} \frac{\sum_{j=0}^{r-1} \binom{r}{j}}{\alpha^j z^{r-j}} \, 1 + \alpha^r \leq \psi_r(1) \max_{k=1,\dots,r} z^k. \tag{B.31}$$

Theorem B.21   *The measure $\lambda_{r,\kappa}$ is ω-excessive for (P_t) if, in cases (i), (ii), (iii) and (iv), respectively,*

(i) $\kappa^{1-1/r}\psi_r(t|a|\kappa^{1/r}) \leq e^{\omega t} - 1, \quad t \geq 0,$
(ii) $\tfrac{1}{2}\kappa^{2/r}(r-1)^{2(1-1/r)}(r+2)q \leq \omega,$
(iii) $\kappa^{1-1/r} \int_{\mathbb{R}^d} \psi_r(|\eta|\kappa^{1/r})\nu(d\eta) \leq \omega,$
(iv) $\kappa^2 r^2 \left(\sup_{\alpha > 0} \frac{\alpha^{2(r-1)}}{(1+\kappa\alpha^r)^2} \right) \int_{\{|\eta|\leq 1\}} \big(\kappa^{1-1/r}\psi_r\big(|\eta|\kappa^{1/r}\big) + 1\big)|\eta|^2 \nu(d\eta) \leq \omega.$

If $r > 1$ is a natural number then, by (B.31), conditions (i), (iii) and (iv) may be simplified. In particular, (i) holds if

$$\kappa^{(k-1)/r+1}|a|(k!)^{1/k} \leq \omega, \qquad k = 1, 2, \dots, r.$$

Taking into account (B.31) we obtain

$$\psi_r\left(|\eta|\kappa^{1/r}\right) \le \psi_r(1) \max_{k=1,\dots,r}\left(|\eta|\kappa^{1/r}\right)$$

and therefore, instead of (iii) and (iv), simpler inequalities can be stated.

We proceed to the proofs of Theorems B.20 and B.21. We will prove the theorems simultaneously by considering the four cases separately.

Proof for case (i)   We use the definition of excessiveness. Since

$$-\gamma|\eta - at| + \gamma|\eta| \le \gamma|a|t \le \omega t, \qquad t \ge 0, \ \eta \in \mathbb{R}^d,$$

Theorem B.20(i) follows. In the case of Theorem B.21 it is sufficient to show that

$$\frac{1}{1 + \kappa|\eta - at|^r} \le \frac{e^{\omega t}}{1 + \kappa|\eta|^r}, \qquad \eta \in \mathbb{R}^d, \ t \ge 0,$$

or

$$\frac{\kappa\left(|\eta|^r - |\eta - at|^r\right)}{1 + \kappa|\eta - at|^r} \le e^{\omega t} - 1, \qquad \eta \in \mathbb{R}^d, \ t \ge 0.$$

But

$$\sup_{\eta \in \mathbb{R}^d} \frac{|\eta|^r - |\eta - at|^r}{1 + \kappa|\eta - at|^r} = \sup_{\eta \in \mathbb{R}^d} \frac{|\eta - at|^r - |\eta|^r}{1 + \kappa|\eta|^r}$$

$$= \sup_{\alpha > 0} \frac{(\alpha + |at|)^r - \alpha^r}{1 + \kappa\alpha^r}$$

and so the result holds. $\square$

In the proofs of the remaining cases we will use Propositions B.18 and B.19. We denote by A^* the formal adjoint of A, given by

$$A^*\varphi(\xi) = -\langle a, D\varphi(\xi)\rangle + \tfrac{1}{2}\operatorname{Tr} Q D^2\varphi(\xi)$$

$$+ \int_{\{|\eta| \le 1\}}\left(\varphi(\xi - \eta) - \varphi(\xi) + \langle D\varphi(\xi), \eta\rangle\right)\nu(d\eta)$$

$$+ \int_{\{|\eta| > 1\}}\left(\varphi(\xi - \eta) - \varphi(\xi)\right)\nu(d\eta).$$

Denote the weights by $G(\xi) = g(|\xi|)$. If we show that, away from a set of Lebesgue measure 0,

$$A^*G \le \omega G \tag{B.32}$$

then ω-excessiveness follows easily using integration by parts.

Proof for case (ii)   We have

$$\operatorname{Tr} D^2 G(\xi) = \left(\frac{g''(|\xi|)}{|\xi|^2} - \frac{g'(|\xi|)}{|\xi|^3}\right)\langle Q\xi, \xi\rangle + \frac{g'(|\xi|)}{|\xi|}\operatorname{Tr} Q \qquad \text{for } \xi \ne 0.$$

So (B.32) is equivalent to

$$\tfrac{1}{2}\left(g''(|\xi|) - \frac{g'(|\xi|)}{|\xi|}\right)\frac{\langle Q\xi, \xi\rangle}{|\xi|^2} + \tfrac{1}{2}\frac{g'(|\xi|)}{|\xi|}\operatorname{Tr} Q - \omega g(|\xi|) \le 0,$$

$$\forall \xi \in \mathbb{R}^d, \ \xi \ne 0. \tag{B.33}$$

In the case of Theorem B.20, $g'(u) = -\gamma e^{-\gamma u}$ and $g''(u) = \gamma^2 e^{-\gamma u}$ and so (B.33) becomes

$$\tfrac{1}{2}\left(\gamma^2 + \frac{\gamma}{u}\right)\frac{\langle Q\xi, \xi\rangle}{|\xi|^2} - \tfrac{1}{2}\frac{\gamma}{u}\,\mathrm{Tr}\,Q \le \omega \qquad \text{for } u = |\xi| > 0.$$

Since

$$q = \sup_{\xi \ne 0} \frac{\langle Q\xi, \xi\rangle}{|\xi|^2} \le \mathrm{Tr}\,Q,$$

it is therefore enough that $\tfrac{1}{2}\gamma^2 q \le \omega$, as required.

In the case of Theorem B.21 we assume that $d > 1$, as the case $d = 1$ can be treated in a similar way. Note that

$$g'(u) = -\frac{\kappa r u^{r-1}}{(1 + \kappa u^r)^2}, \qquad g''(u) = 2\frac{\kappa^2 r^2 u^{2r-2}}{(1 + \kappa u^r)^3} - \frac{\kappa r(r-1)u^{r-2}}{(1 + \kappa u^r)^2}, \qquad u > 0,$$

and that (B.33) holds if

$$\tfrac{1}{2}\left(2\frac{\kappa^2 r^2 u^{2r-2}}{(1 + \kappa u^r)^3} - \frac{\kappa r(r-2)u^{r-2}}{(1 + \kappa u^r)^2}\right)\frac{\langle Q\xi, \xi\rangle}{|\xi|^2} \le \frac{\omega_2}{1 + \kappa u^r}$$

for $u = |\xi| > 0$, that is, if

$$\tfrac{1}{2}\left(2\kappa^2 r^2 u^{2r-2} - \kappa^2 r(r-2)u^{2r-2}\right)q \le \omega_2(1 + \kappa u^r)^2, \qquad u > 0.$$

Rearranging, the condition for (B.33) to hold is that

$$\tfrac{1}{2}\kappa^2 r(r+2)q\frac{u^{2r-2}}{(1 + \kappa u^r)^2} \le \omega_2, \qquad u > 0. \tag{B.34}$$

Taking into account that the left-hand side of (B.34) attains its maximum at $u = (r-1)/\kappa$, the required inequality holds.   □

Proof for case (iii) We consider only the more complicated situation in Theorem B.21. Notice that the operator A^* is of the form

$$A^*G(\xi) = \int_{\mathbb{R}^d}\big(G(\xi - \eta) - G(\xi)\big)\nu(d\eta), \qquad \xi \in \mathbb{R}^d,$$

and the left-hand side of (B.32) is equal to

$$\int_{\mathbb{R}^d}\left(\frac{1}{1 + \kappa|\xi - \eta|^r} - \frac{1}{1 + \kappa|\xi|^r}\right)\nu(d\eta)$$

$$= \kappa\int_{\mathbb{R}^d}\frac{|\xi|^r - |\xi - \eta|^r}{(1 + \kappa|\xi - \eta|^r)(1 + \kappa|\xi|^r)}\nu(d\eta), \qquad \xi \in \mathbb{R}^d.$$

Since

$$\sup_{\xi \in \mathbb{R}^d}\frac{|\xi|^r - |\xi - \eta|^r}{1 + \kappa|\xi - \eta|^r} = \sup_{\xi \in \mathbb{R}^d}\frac{|\xi - \eta|^r - |\xi|^r}{1 + \kappa|\xi|^r}$$

$$= \psi_r(|\eta|) = \kappa^{-1/r}\psi_r\big(|\eta|\kappa^{1/r}\big),$$

the result follows.   □

Proof for case (iv) By Taylor's formula,

$$G(\xi - \eta) = G(\xi) - \langle DG(\xi), \eta \rangle + \tfrac{1}{2} \langle D^2 G(\xi - v_\eta \eta) \eta, \eta \rangle, \qquad v_\eta \in [0, 1],$$

and the inequality (B.32) becomes

$$\tfrac{1}{2} \int_{\{|\eta| \le 1\}} \langle D^2 G(\xi - v_\eta \eta) \eta, \ \eta \rangle v(d\eta) \le \omega G(\xi), \qquad \xi \in \mathbb{R}^d. \tag{B.35}$$

But

$$D^2 G(\xi) = \left(g''(|\xi|) - \frac{g'(|\xi|)}{|\xi|} \right) \frac{\xi \otimes \xi}{|\xi|^2} + \frac{g'(|\xi|)}{|\xi|}, \qquad \xi \in \mathbb{R}^d.$$

In the case of Theorem B.20, $g'(u) = -\gamma e^{-\gamma u}$ and $g''(u) = \gamma^2 e^{-\gamma u}$. Therefore

$$D^2 G(\xi) = e^{-\gamma |\xi|} \left(-\frac{\gamma}{|\xi|} + \frac{\gamma^2}{|\xi|^2} (\xi \otimes \xi) + \gamma \frac{\xi \otimes \xi}{|\xi|^3} \right), \qquad \xi \in \mathbb{R}^d,$$

and, for $z = \xi - v_\eta \eta$,

$$\left\langle \frac{D^2 G(z)}{G(\xi)} \eta, \ \eta \right\rangle$$

$$\le e^{\gamma (|\xi| - |z|)} \left(\gamma^2 \left\langle \frac{z \otimes z}{|z|^2} \eta, \ \eta \right\rangle + \gamma \left\langle \left(-\frac{1}{|z|} I + \frac{z \otimes z}{|z|^3} \right) \eta, \ \eta \right\rangle \right)$$

$$\le e^{\gamma |y|} \gamma^2 \left\langle \frac{z \otimes z}{|z|^2} \eta, \ \eta \right\rangle \le e^{\gamma |\eta|} \gamma^2 |\eta|^2,$$

so the result follows.

It remains to consider the case of Theorem B.21. If $d > 1$ then, for $u = |\xi|$,

$$g'(u) = -\frac{\kappa r u^{r-1}}{(1 + \kappa u^r)^2}, \qquad g''(u) = 2 \frac{\kappa^2 r^2 u^{2r-2}}{(1 + \kappa u^r)^3} - \frac{\kappa r (r - 1) u^{r-2}}{(1 + \kappa u^r)^2}$$

and

$$D^2 G(\xi) = \frac{\kappa r |\xi|^{r-2}}{(1 + \kappa |\xi|^r)^2} \left(-I - \frac{\xi \otimes \xi}{|\xi|^2} (r - 2) + \frac{\xi \otimes \xi}{|\xi|^2} \frac{2\kappa r |\xi|^r}{1 + \kappa |\xi|^r} \right).$$

However, for arbitrary $r > 1$ and $\xi \in \mathbb{R}^d$,

$$-I - \frac{\xi \otimes \xi}{|\xi|^2} (r - 2) \le 0$$

and (B.35) holds provided that

$$\tfrac{1}{2} \int_{\{|\eta| \le 1\}} \frac{2(\kappa r)^2 |z|^{2r-2}}{(1 + \kappa |z|^r)^3} \frac{\langle z, y \rangle^2}{|z|^2} v(d\eta) \le \omega. \tag{B.36}$$

In (B.36), $z = \xi - v_\eta \eta$ is from Taylor's formula and $v_\eta \in [0, 1]$. Elementary estimates lead to the required result. The case of $d = 1$ can be treated in a similar way. $\square$

Thus the proofs of Theorems B.20 and B.21 are complete.

B.5 Fractional powers of the heat semigroup

Let S be the heat semigroup generated by the Laplace operator Δ on $\mathcal{O}$ with Dirichlet or Neumann boundary conditions. Consider the semigroup S_γ generated by the fractional operator $-(-\Delta)^\gamma$. By (B.6),

$$S_\gamma(t) = \int_0^\infty S(s)\mu_t^\gamma(ds),$$

where (μ_t^γ) is the convolution semigroup of measures on $[0, +\infty)$ such that

$$\tilde{\mu}_t(r) = \int_0^{+\infty} e^{-r\xi} \mu_t(d\xi) = \exp\{-tr^\gamma\}, \qquad t, r > 0.$$

Since the Fourier transform of each μ_t^γ is integrable, μ_t^γ is absolutely continuous with respect to Lebesgue measure. Let us denote by ϕ_t^γ the density of μ_t^γ.

Lemma B.22   *The family* (ϕ_t^γ) *has the following scaling property:* $\phi_t^\gamma(s) = t^{-1/\gamma}\phi_1^\gamma$ $\times(st^{-1/\gamma})$ *for* $t, s > 0$.

Proof   We have

$$\int_0^{+\infty} e^{-rs} t^{-1/\gamma} \phi_1^\gamma(st^{-1/\gamma})\, ds = \int_0^{+\infty} \exp\left\{-rt^{1/\gamma}u\right\} \phi_1^\gamma(u)\, du = \exp\left\{-tr^\gamma\right\}.$$

$\square$

Let $\mathcal{G}$ be the Green function for Δ on $\mathcal{O}$. Then

$$S_\gamma(t)\psi(\xi) = \int_{\mathcal{O}} \mathcal{G}_\gamma(t - s, \xi, \eta)\psi(\eta)\, d\eta,$$

where

$$\mathcal{G}_\gamma(t, \xi, \eta) = \int_0^{+\infty} \mathcal{G}(s, \xi, \eta)\phi_t^\gamma(s)\, ds = \int_0^{+\infty} \mathcal{G}(t^{1/\gamma}u, \xi, \eta)\phi_1^\gamma(u)\, du.$$

The following lemma plays an important role in Section 12.5, which deals with applications to equations driven by a fractional Laplace operator.

Lemma B.23   *Let* $p > 1$. *Then there is a constant* C *such that, for all* $t > 0$, $\xi \in \mathcal{O}$,

$$\int_{\mathcal{O}} |\mathcal{G}_\gamma(t, \eta, \xi)|^p\, d\eta \le Ct^{(d/2\gamma)(1-p)}.$$

Proof   Taking into account Lemma B.22 and the fact that ϕ_1^γ is the density of a probability measure we obtain

$$\int_{\mathcal{O}} |\mathcal{G}_\gamma(s, \eta, \xi)|^p\, d\eta \le \int_{\mathcal{O}} \int_0^{+\infty} |\mathcal{G}(t^{1/\gamma}u, \eta, \xi)|^p\, \phi_1^\gamma(u)\, du\, d\eta.$$

Since (see Aronson 1967) the Aronson-type estimates hold for the Laplace operator uniformly on each time interval, one obtains

$$\int_{\mathcal{O}} |\mathcal{G}(t^{1/\gamma}u, \eta, \xi)|^p\, d\eta \le K_1 \int_{\mathbb{R}^d} (t^{1/\gamma}u)^{-dp/2} \exp\left\{-\frac{p|\xi - \eta|^2}{K_2 t^{1/\gamma}u}\right\}\, d\eta$$

$$\le K_3 (t^{1/\gamma}u)^{-d(p-1)/2}.$$

Thus the proof is complete if we can show that

$$\int_0^{+\infty} u^{-d(p-1)/2} \phi_1^\gamma(u)\, du < \infty.$$

In fact, for all $\alpha > 0$,

$$\int_0^{+\infty} \frac{\phi_1^\gamma(u)}{u^\alpha}\, du < \infty.$$

This follows from Feller (1971), Chapter XIII, Section 6. For, let X be a random variable with density $\phi := \phi_1^\gamma$. Then

$$\int_0^{+\infty} \frac{\phi(u)}{u^\alpha}\, du = \mathbb{E}\, X^{-\alpha} = \mathbb{E} \int_0^{X^{-\alpha}} ds = \mathbb{E} \int_0^{+\infty} \chi_{[0, X^{-\alpha}]}\, ds$$

$$= \int_0^{+\infty} \mathbb{P}\left(X^{-\alpha} \geq s\right) ds = \int_0^{+\infty} \mathbb{P}\left(X \leq s^{-\alpha}\right) ds$$

$$= \int_0^1 \mathbb{P}\left(X \leq s^{-\alpha}\right) ds + \int_1^\infty \mathbb{P}\left(X \leq s^{-\alpha}\right) ds =: I_1 + I_2.$$

It is clear that $I_1 < \infty$. In order to estimate I_2, note that

$$I_2 = \alpha \int_0^1 \mathbb{P}(X \leq r)\, \frac{dr}{r^{1+\alpha}}.$$

Since (see Feller 1971) $e^{r^{-\alpha}} \mathbb{P}(X \leq r) \to 0$ as $r \to 0$, we have $I_2 < \infty$. □

Remark B.24   The same estimates can be obtained (with the same arguments) for the fractional Laplace operator on $\mathbb{R}^d$.

B.6 Proof of Lemma 10.17

We have to show that there is a constant c_p such that

$$\langle A_p \psi, \psi^* \rangle \leq c_p |\psi|_{L^p}, \qquad \forall\, \psi \in D(A_p),\ \psi^* \in \partial |\psi|_{L^p}. \tag{B.37}$$

If $\psi \neq 0$ then $\partial |\psi|_{L^p}$ consists of one element,

$$\psi^*(\xi) = |\psi|_{L^p}^{p/q} \operatorname{sgn} \psi(\xi) |\psi(\xi)|^{p-1} = |\psi|_{L^p}^{p/q} \psi(\xi) |\psi(\xi)|^{p-2}, \qquad \xi \in \mathcal{O},$$

where q is such that $1/p + 1/q = 1$. If $\psi = 0$ then $\partial |\psi|_{L^p}$ is equal to the unit ball in $L^q(\mathcal{O})$.

Assume first that $p = 2k + 2$ for some $k \in \mathbb{N}$. Then

$$|\psi|^{p/q} \langle A_p \psi, \psi^* \rangle = I_1 + I_2 + I_3,$$

where

$$I_1 := \int_{\mathcal{O}} \sum_{i,j=1}^{d} (\psi(\xi))^{2k+1} a_{i,j}(\xi) \frac{\partial^2 \psi}{\partial \xi_i \partial \xi_j}(\xi) \, d\xi,$$

$$I_2 := \int_{\mathcal{O}} (\psi(\xi))^{2k+1} \sum_{i=1}^{d} b_i(\xi) \frac{\partial \psi}{\partial \xi_i}(\xi) \, d\xi,$$

$$I_3 := \int_{\mathcal{O}} (\psi(\xi))^{2k+1} c(\xi) \, d\xi.$$

Assume that $\psi \in C_c^2(\mathcal{O})$. Then, integrating by parts, we obtain $I_1 = I_{1,1} + I_{1,2}$, where

$$I_{1,1} := -\int_{\mathcal{O}} \sum_{i,j=1}^{d} (\psi(\xi))^{2k+1} a_{i,j}(\xi) \frac{\partial \psi}{\partial \xi_i}(\xi) \frac{\partial \psi}{\partial \xi_j}(\xi) \, d\xi,$$

$$I_{1,2} := -(2k+1) \int_{\mathcal{O}} (\psi(\xi))^{2k} \sum_{i,j=1}^{d} a_{i,j}(\xi) \frac{\partial \psi}{\partial \xi_i}(\xi) \frac{\partial \psi}{\partial \xi_j}(\xi) \, d\xi$$

$$\leq -(2k+1)\delta \int_{\mathcal{O}} \sum_{j=1}^{d} \left(\frac{\partial \psi}{\partial \xi_j}(\xi) \psi^k(\xi) \right)^2 \, d\xi.$$

Note that

$$I_{1,1} + I_2 = \int_{\mathcal{O}} \sum_{j=1}^{d} \tilde{b}_j(\xi) \frac{\partial \psi}{\partial \xi_j}(\xi) \psi^{2k+1}(\xi) \, d\xi,$$

where

$$\tilde{b}_j(\xi) := b_j(\xi) - \sum_{i=1}^{d} \frac{\partial a_{i,j}}{\partial \xi_i}(\xi).$$

Hence, for any $\varepsilon > 0$,

$$I_1 + I_2 + I_3 \leq -(2k+1) \int_{\mathcal{O}} (\psi(\xi))^{2k} \sum_{i,j=1}^{d} a_{i,j}(\xi) \frac{\partial \psi}{\partial \xi_i}(\xi) \frac{\partial \psi}{\partial \xi_j}(\xi) \, d\xi$$

$$+ \frac{1}{2\varepsilon^2} \max_j \left\| \tilde{b}_j \right\|_{\infty} \int_{\mathcal{O}} \sum_{j=1}^{d} \frac{\partial \psi}{\partial \xi_j}(\xi) \psi^{2k+1}(\xi) \, d\xi$$

$$+ \frac{\varepsilon^2 d}{2} \max_j \left\| \tilde{b} \right\|_{\infty} \int_{\mathcal{O}} |\psi(\xi)|^p \, d\xi + ||c||_{\infty} \int_{\mathcal{O}} |\psi(\xi)|^p \, d\xi$$

$$\leq \left(-(2k+1)\delta + \frac{1}{2\varepsilon^2} \max_j \left\| \tilde{b}_j \right\|_{\infty} \right) \int_{\mathcal{O}} |\psi(\xi)|^{2k} \sum_{j=1}^{d} \left| \frac{\partial \psi}{\partial \xi_j}(\xi) \right|^2 \, d\xi$$

$$+ \left(\frac{d\varepsilon^2}{2} \max_j \left\| \tilde{b}_j \right\|_{\infty} + ||c||_{\infty} \right) \int_{\mathcal{O}} |\psi(\xi)|^p \, d\xi.$$

Choosing ε such that

$$-(2k+1)\delta + \frac{1}{2\varepsilon^2}\max_j \|\tilde{b}_j\|_\infty < 0$$

we find that, for some $c_p > 0$,

$$\langle A\psi, \psi^* \rangle \le c_p |\psi|^{-p/q} L^p |\psi|^p_{L^p} = c_p |\psi|_{L^p}, \tag{B.38}$$

as required. If $\psi \in W^{2,p}(\mathcal{O})$ vanishes on $\partial\mathcal{O}$ then there is a sequence $(\psi_n) \subset D(A_p)$ of $C_0^2(\mathcal{O})$ functions converging to ψ in the graph norm of A. Taking into account that (B.38) holds for all ψ_n, (B.38) holds for ψ.

To obtain (B.37) for any p, note that for $p = 2k + 2$ there is an ω_p such that

$$\|S(t)\|_{L(L^p, L^p)} \le e^{\omega_p t}, \qquad t \ge 0. \tag{B.39}$$

By the Riesz–Thorin theorem, (B.39) holds for all $p \ge 2$ (with appropriate ω_p). But (B.39) implies that the operator $A_p - \omega_p I$ is dissipative, and so the proof is complete. $\qquad\square$

Remark B.25   Under some assumptions, we have that $\|S(t)\|_{L(L^2, L^2)} \le e^{\omega t}$, $t \ge 0$, with a negative ω. Moreover, given A and $\omega \in \mathbb{R}$ we can find an α such that the semigroup $T = (T(t))$ generated by $A - \lambda I$ satisfies $\|T(t)\| \le e^{\omega t}$.

Indeed, we have

$$\langle A\psi, \psi \rangle \le -\delta \int_\mathcal{O} \left|\frac{\partial\psi}{\partial\xi_i}(\xi)\right|^2 d\xi + \int_\mathcal{O} \psi(\xi) \sum_{j=1}^d \tilde{b}_j(\xi)\frac{\partial\psi}{\partial\xi_j}(\xi)\,d\xi$$

$$+ \int_\mathcal{O} c(\xi)\psi^2(\xi)\,d\xi,$$

where

$$\tilde{b}_j(\xi) := b_j(\xi) - \sum_{i=1}^d \frac{\partial a_{i,j}}{\partial\xi_i}(\xi).$$

By the Poincaré inequality,

$$\int_\mathcal{O} |\psi(\xi)|^2\,d\xi \le C_\mathcal{O} \int_\mathcal{O} \sum_{j=1}^d \left|\frac{\partial\psi}{\partial\xi_j}(\xi)\right|^2 d\xi, \qquad \forall \psi \in W_0^{1,2}(\mathcal{O}),$$

where $C_\mathcal{O}$ is a constant depending on the volume of $\mathcal{O}$ and d.

If the partial derivatives $\partial a_{i,j}/\partial\xi_i$, $i, j = 1, \ldots, d$, are bounded then

$$\int_\mathcal{O} \psi(\xi) \sum_{j=1}^d \tilde{b}_j(\xi)\frac{\partial\psi}{\partial\xi_j}(\xi)\,d\xi \le \max_j \|\tilde{b}_j\|_\infty \int_\mathcal{O} \sum_{j=1}^d \left|\frac{\partial\psi}{\partial\xi_j}(\xi)\right| |\psi(\xi)|\,d\xi$$

$$\le \frac{1}{2}\max_j \|\tilde{b}_j\|_\infty \left(\sum_{j=1}^d \int_\mathcal{O} \left(\frac{\partial\psi}{\partial\xi_j}(\xi)\right)^2 d\xi + d\int_\mathcal{O} |\psi(\xi)|^2 d\xi\right).$$

Consequently, provided that $-\delta + \frac{1}{2}\max_j \|\tilde{b}_j\|_\infty = -\tilde{\delta} < 0$,

$$\langle A\psi, \psi \rangle \leq -\tilde{\delta}\,|D\psi|_{L^2}^2 + \left(\sup_{\xi \in \mathcal{O}} c(\xi) + \tfrac{1}{2}d \max_j \|\tilde{b}_j\|_\infty\right)|\psi|_{L^2}^2$$

$$\leq \left(\frac{1}{C_{\mathcal{O}}}\left(-\delta + \tfrac{1}{2}\|\tilde{b}\|_\infty\right) + \sup_{\xi \in \mathcal{O}} c(\xi) + \tfrac{1}{2}d \max_j \|\tilde{b}_j\|_\infty\right)|\psi|_{L^2}^2.$$

Example B.26   Assume that $c \equiv 0 \equiv b_j$ and that $a_{i,j}$ are constants. Then

$$\langle A\psi, \psi \rangle \leq -\frac{\delta}{C_{\mathcal{O}}}|\psi|_{L^2}^2.$$

Appendix C   Regularization of Markov processes

Let (P_t) be the transition semigroup of a Markov process $(X(t), \, t \in [0, T])$ on a complete separable metric space (E, ρ). We denote by $B(x, r)$ the ball in E with center at x and radius r. We now prove Theorem 3.23. The proof follows Gikhman and Skorokhod (1974), Vol. I, and is based on some lemmas. Define

$$\alpha_t(r) := \sup_{s \leq t} \sup_{x \in E} P_s(x, B^c(x, r)).$$

Lemma C.1   *Condition (3.3) in Theorem 3.23 holds if and only if*

$$\lim_{t \to 0} \alpha_t(r) = 0, \qquad \forall r > 0. \tag{C.1}$$

Proof   If, for all $t \leq \delta$, $\sup_{x \in E} P_t(x, B^c(x, r)) \leq \varepsilon$ then also $\alpha_t(r) \leq \varepsilon$; so the equivalence follows.   □

Lemma C.2   *If, for all $x \in E$ and $r > 0$, $\lim_{h \to 0} P_h(x, B^c(x, r)) = 0$ then for all $s \in [0, T)$ we have*

$$\lim_{t \downarrow s} \mathbb{P}\big(\rho(X(s), X(t)) \geq r\big) = 0.$$

Proof   By the Markov property for $s \leq t \leq T$, one has

$$\mathbb{P}\big(\rho(X(s), X(t)) > r\big) = \mathbb{E} \, P_{t-s}\big(X(s), B^c(X(s), r)\big),$$

and the result follows from the Lebesgue dominated convergence theorem.   □

For every sequence (x_n) in a metric space E and any $r > 0$ define the *number $o_r(x_n)$ of r-oscillations of* (x_n) as the supremum of those natural numbers k for which there exist $n_0 < n_1 < n_2 < \cdots < n_k$ such that $\rho\big(x_{n_j}, x_{n_{j-1}}\big) \geq r$ for $j = 1, 2, \ldots, k$.

Lemma C.3   *A sequence (x_n) satisfies the Cauchy condition if and only if, for each $r > 0$, $o_r(x_n) < \infty$.*

Proof   Assume that, for each $r > 0$, $o_r(x_n) < \infty$ but (x_n) is not a Cauchy sequence. Then there exist an $r > 0$ and an increasing sequence (m_k) of natural numbers such that $\rho\big(x_{m_{2k}}, x_{m_{2k-1}}\big) \geq r$. Now construct an infinite subsequence $(x_{\bar{m}_k})$ such that $\rho\big(x_{\bar{m}_k}, x_{\bar{m}_{k-1}}\big) \geq r/2$. The new subsequence consists of the two first elements x_{m_1}, x_{m_2} of the sequence $\big(x_{m_k}\big)$ and one element from each pair $x_{m_{2k}}, x_{m_{2k-1}}, k = 2, 3, \ldots$

Since $\rho(x_{m_{2k}}, x_{m_{2k-1}}) \geq r$, for any element $a \in E$ either $\rho(x_{m_{2k}}, a) \geq r/2$ or $\rho(x_{m_{2k-1}}, a) \geq r/2$. Thus if the elements $x_{\bar{m}_1}, \ldots, x_{\bar{m}_k}$ have already been selected and each $\bar{m}_j$ is either m_{2j} or m_{2j-1}, $j = 1, 2, \ldots, k$, then we select from the pair $x_{m_{2k+2}}, x_{m_{2k+1}}$ that element which is at a distance at least $r/2$ from $x_{\bar{m}_k}$. Consequently, $o_{r/2}(x_n) = \infty$, a contradiction.

If (x_n) is Cauchy then for each $r > 0$ there is an N such that $\rho(x_n, x_m) < r$ for all $n, m \geq N$. So, if $\rho(x_{n_j}, x_{n_{j-1}}) \geq r$, $n_{j-1} < n_j$, $j = 1, \ldots, k$, then $n_{k-1} < N$; therefore $o_r(x_n) < N$. $\qquad\square$

Lemma C.4 *Let $0 \leq s \leq t_1 < t_2 < \cdots < t_n \leq t < T$, and let A_m, $m = 1, 2, \ldots$, be the event that the number of r-oscillations in the sequence $X(t_1), \ldots, X(t_n)$ is at least m. Then*

$$\mathbb{P}(A_m) \leq (2\alpha_{t-s}(r/4))^m, \qquad m = 1, 2, \ldots \tag{C.2}$$

Proof
    Step 1   We show first that the result is true for $m = 1$. We can assume that $s = 0$. It is clear that

$$A_1 = \left(A_1 \cap \{\rho(X(0), X(t)) < r/4\}\right) \cup \left(A_1 \cap \{\rho(X(0), X(t)) \geq r/4\}\right)$$

and that

$$\mathbb{P}\left(A_1 \cap \{\rho(X(0), X(t)) \geq r/4\}\right) \leq \mathbb{P}\left(\rho(X(0), X(t)) \geq r/4\right) \leq \alpha_t(r/4).$$

Note that

$$A_1 \cap \{\rho(X(0), X(t)) < r/4\}$$

$$\subset \bigcup_{l=1}^{n} \{\rho(X(0), X(t_j)) < r/2, \ j = 1, 2, \ldots, l-1\}$$

$$\cap \{\rho(X(0), X(t)) < r/4\}$$

$$\subset \bigcup_{l=1}^{n} \{\rho(X(0), X(t_j)) < r/2, \ j = 1, 2, \ldots, l-1,$$

$$\text{and } \rho(X(0), X(t_l)) \geq r/2\} \cap \{\rho(X(0), X(t)) < r/4\}.$$

Therefore

$$J = \mathbb{P}\left(A_1 \cap \{\rho(X(0), X(t)) < r/4\}\right)$$

$$\leq \sum_{l=1}^{n} \mathbb{E} \prod_{j=1}^{l-1} \chi_{B(X(0), r/2)}(X(t_j)) \chi_{B^c(X(0), r/2)}(X(t_l)) \chi_{B(X(0), r/4)}(X(t)).$$

If $\rho(X(0), X(t_l)) \geq r/2$ and $\rho(X(0), X(t)) < r/4$ then $\rho(X(t_l), X(t)) \geq r/4$. Consequently,

$$J \leq \sum_{l=1}^{n} \mathbb{E} \prod_{j=1}^{l-1} \chi_{B(X(0), r/2)}(X(t_j)) \chi_{B^c(X(0), r/2)}(X(t_l)) \chi_{B^c(X(t_l), r/4)}(X(t))$$

and, by the Markov property, J is less than or equal to

$$\sum_{l=1}^{n} \mathbb{E} \prod_{j=1}^{l-1} \chi_{B(X(0), r/2)}(X(t_j)) \chi_{B^c(X(0), r/2)}(X(t_l)) P_{t-t_l}(X(t_l), B^c(X(t_l), r/4))$$

$$\leq \alpha_t(r/4) \, \mathbb{P}\left(\rho(X(0), X(t_l)) \geq r/2 \text{ for some } l = 1, 2, \ldots, n\right) \leq \alpha_t(r/4).$$

Step 2   We can now show that the result is true for $m \geq 1$. Let M_k be the number of r-oscillations in the sequence $X(t_1), \ldots, X(t_k)$, and let $\tau_m := \inf\{k \colon M_k = m\}$. Then

$$A_m = \bigcup_{k=1}^{n} \{\tau_m = k\} \qquad \text{and} \qquad A_{m+1} = \bigcup_{k=1}^{n} \{\tau_m = k\} \cap {}^{\circ}A_{m+1}.$$

But

$$\{\tau_m = k\} \cap A_{m+1} \subseteq \{\tau_m = k\} \cap A_1^{k,n},$$

where $A_1^{k,n}$ is the event that in the sequence $X(t_k), X(t_{k+1}), \ldots, X(t_n)$ there is at least one r-oscillation. By the Markov property,

$$\mathbb{P}\left(\{\tau_m = k\} \cap A_1^{k,n}\right) = \mathbb{E}\, \chi_{\{\tau_m = k\}} \mathbb{P}\left(A_1^{k,n} \mid \mathcal{F}_{t_k}\right)$$
$$\leq \mathbb{P}\left(\tau_m = k\right) 2\alpha_{t_n - t_k}(r/4).$$

Consequently, $\mathbb{P}(A_{m+1}) \leq 2\alpha_{t-s}(r/4)\,\mathbb{P}(A_m)$ and, by an obvious induction argument, (C.2) follows. □

Proof of Theorem 3.23   Let Q be a countable dense subset of $[0, T]$ containing 0 and T. We show that, for each $r > 0$, the number $o_r(X(t), t \in Q)$ of r-oscillations in $\{X(t), t \in Q\}$ is finite. Let (Q_n) be an increasing sequence of finite subsets of Q such that $\bigcup_n Q_n = Q$. Let m be a natural number such that $2\alpha_{T/m}(r/4) < 1$. Then, for each n, the probability that the number of r-oscillations in the sequence

$$Q_n \cap \left[\frac{j-1}{m}T, \ \frac{j}{m}T\right], \qquad j = 1, \ldots, m,$$

is at least k can be estimated as $\left(2\alpha_{T/m}(r/4)\right)^k$. This easily implies that, for each $r > 0$, $o_r(X(t), t \in Q) < \infty$ with probability 1. Let Ω_0 consist of all $\omega \in \Omega$ such that, for all $r > 0$, $o_r(X(t, \omega), t \in Q)$ is finite. We define a modification $\widehat{X}$ of X by putting

$$\widehat{X}(t, \omega) = X(t, \omega) \qquad \text{for } t \in Q \text{ and } \omega \in \Omega$$

and

$$\widehat{X}(t, \omega) = \lim_{\substack{s \downarrow t \\ s \in Q}} \widehat{X}(s, \omega) \qquad \text{for } t \in [0, T] \cap Q^c.$$

It follows from Lemma C.3 that $\widehat{X}(t, \omega), t \in (0, T]$, is right-continuous with left limits. By Lemma C.2 the process $\widehat{X}$ is a modification of X. □

Appendix D   Itô formulae

In this appendix, Itô formulae for finite-dimensional and infinite-dimensional semimartingales are presented.

D.1 Main results

For the proof of the following result see Métivier (1982), Theorem 27.1. Below, $\Delta Y(t) := Y(t) - Y(t-)$.

Theorem D.1   *Assume that $X = M + A$ is an $\mathbb{R}^d$-valued semimartingale with martingale and bounded-variation parts M and A, respectively. Then, for every $\psi \in C^2(\mathbb{R}^d)$, $\psi(X)$ is a local semimartingale and, for all $t \geq 0$, $\mathbb{P}$-a.s.,*

$$\psi(X(t)) = \psi(X(0)) + \sum_{i=1}^{d} \int_0^t \frac{\partial \psi}{\partial \xi_i}(X(s-)) \, dX^i(s)$$

$$+ \frac{1}{2} \sum_{i,j=1}^{d} \int_0^t \frac{\partial^2 \psi}{\partial \xi_i \partial \xi_j}(X(s-)) \, d[M^i, M^j]_s + \sum_{s \leq t} (\Delta \psi(X)(s) - Y(s)),$$

where

$$Y(s) := \sum_{i=1}^{d} \frac{\partial \psi}{\partial \xi_i}(X(s-)) \, \Delta X^i(s) - \frac{1}{2} \sum_{i,j=1}^{d} \frac{\partial^2 \psi}{\partial \xi_i \partial \xi_j}(X(s-)) \, \Delta X^i(s) \, \Delta X_j(s).$$

The formula can be rewritten in terms of the so-called *continuous part* $[M_i, M_i]^c$ of the square bracket $[M_i, M_i]$:

$$[M_i, M_i]_t^c := [M_i, M_i]_t - \sum_{s \leq t} \Delta[M_i, M_i]_s.$$

391

Namely,

$$\psi(X(t)) = \psi(X(0)) + \sum_{i=1}^{d} \int_0^t \frac{\partial \psi}{\partial \xi_i}(X(s-)) \, dX^i(s)$$

$$+ \frac{1}{2} \sum_{i,j=1}^{d} \int_0^t \frac{\partial^2 \psi}{\partial \xi_i \partial \xi_j}(X(s-)) \, d[M^i, M^j]_s^c$$

$$+ \sum_{s \le t} \left(\Delta \psi(X)(s) - \sum_{i=1}^{d} \frac{\partial \psi}{\partial \xi_i}(X(s-)) \, \Delta X^i(s) \right).$$

If

$$\pi((0, t], \Gamma) := \sum_{s \le t} \chi_\Gamma(\Delta X(s))$$

is the jump measure of X then

$$\psi(X(t)) = \psi(X(0)) + \sum_{i=1}^{d} \int_0^t \frac{\partial \psi}{\partial \xi_i}(X(s-)) \, dX^i(s)$$

$$+ \frac{1}{2} \sum_{i,j=1}^{d} \int_0^t \frac{\partial \psi}{\partial \xi_i \partial \xi_j}(X(s-)) \, d[M^i, M^j]_s^c$$

$$+ \int_0^t \int_{\mathbb{R}^d} \left(\psi(X(s-) + y) - \psi(X(s-)) - \sum_{i=1}^{d} y_i \frac{\partial \psi}{\partial \xi_i}(X(s-)) \right) \pi(ds, dy).$$

The next results deal with the Itô formula for Hilbert-valued semimartingales (see Métivier 1982, Theorem 27.2) and require some notation to be defined. We denote by $H \hat{\otimes} H$ the space $H \otimes H$ completed with respect to the Hilbert–Schmidt norm. Clearly, $L(H \hat{\otimes} H, \mathbb{R}) \equiv L_{(HS)}(H, H)$. Let $\{e_n\}$ be an orthonormal basis of H. Then $M^k = \langle M, e_k \rangle_H$ and (see Section 8.1),

$$[M, M]_s := \sum_{i,j} e_k \otimes e_j [M^k, M^j]_s,$$

where on the left-hand side the square brackets of Theorem 3.48 have been doubled (cf. (8.2)). Next,

$$[M, M]_t^c := \sum_{i,j} e_k \otimes e_j \left([M^k, M^j]_t - \sum_{s \le t} \Delta M^k(s) M^j(s) \right).$$

Theorem D.2   *Assume that $X = M + A$ is a semimartingale taking values in a Hilbert space H. Let $\psi \colon H \mapsto \mathbb{R}$ be of class C^2. Assume that, for each $x \in H$, $D^2\psi(x) \in L_{(HS)}(H, H)$ and the mapping*

$$H \ni x \mapsto D^2\psi(x) \in L_{(HS)}(H, H) \cong L(H \hat{\otimes} H, \mathbb{R})$$

is uniformly continuous on any bounded subset of H. Then $\psi(X)$ is a local semimartingale and, for all $t \geq 0$, $\mathbb{P}$-a.s.,

$$\psi(X(t)) = \psi(X(0)) + \int_0^t \left\langle D\psi(X(s-)), \mathrm{d}X(s) \right\rangle_H + \frac{1}{2} \int_0^t D^2\psi(X(s-)) \, \mathrm{d}[M, M]_s$$
$$+ \sum_{s \leq t} \left(\Delta(\psi X)(s) - \langle D\psi(X(s-)), \ \Delta X(s) \rangle_H - Y(s) \right)$$

where

$$Y(s) := \tfrac{1}{2} D^2\psi(X(s-)) \, \Delta X(s) \otimes \Delta X(s).$$

We also have

$$\psi(X(t)) = \psi(X(0)) + \int_0^t \left\langle D\psi(X(s-)), \mathrm{d}X(s) \right\rangle_H + \frac{1}{2} \int_0^t D^2\psi(X(s-)) \, \mathrm{d}[M, M]_s^c$$
$$+ \sum_{s \leq t} \left(\Delta(\psi X)(s) - \langle D\psi(X(s-)), \ \Delta X(s) \rangle_H \right).$$

D.2 An application

Let Z be a square integrable Lévy martingale in $\mathbb{R}^d$ with covariance equal to the identity matrix.

Lemma D.3   *Let a and b be predictable $\mathbb{R}$- and $\mathbb{R}^d$-valued processes satisfying*

$$\mathbb{E} \int_0^T \left(|a(t)|^2 + |b(t)|^2 \right) \mathrm{d}t < \infty, \qquad \forall \, T > 0.$$

Then, for any $x \in \mathbb{R}$, the process

$$X(t) = x + \int_0^t a(s) \, \mathrm{d}s + \int_0^t \langle b(s), \mathrm{d}Z(s) \rangle$$

is square integrable and

$$\mathbb{E} \, X^2(t) = x^2 + \mathbb{E} \int_0^t X(s) a(s) \, \mathrm{d}s + \mathbb{E} \int_0^t |b(s)|^2 \, \mathrm{d}s.$$

Proof   The lemma follows from the Itô formula for $\psi(x) = |x|^2$. Indeed, we have

$$\psi(X(s-) + y) - \psi(X(s-)) - \langle D\psi(X(s-)), y \rangle = |y|^2.$$

$\square$

Appendix E Lévy–Khinchin formula
on $[0, +\infty)$

This appendix is devoted to the proof of Theorem 4.31. Let B be the Banach space of continuous functions on $[0, +\infty)$ having finite limits at $+\infty$. If (λ_t) is a convolution semigroup of measures on $[0, +\infty)$ then the formula

$$P_t f(\xi) = \int_0^{+\infty} f(\xi + \eta) \lambda_t(d\eta), \qquad f \in B, \ \xi \geq 0,$$

defines a C_0-semigroup of linear operators on B. By the Hille–Yosida theorem, $P_t f = \lim_{\alpha \to +\infty} P_t^\alpha f$, $f \in B$, where

$$P_t^\alpha f(\xi) := \int_0^{+\infty} f(\xi + \eta) \, \lambda_t^\alpha(d\eta), \qquad f \in B, \ \xi \geq 0,$$

$$\lambda_t^\alpha := e^{-\alpha t} \sum_{n=0}^{+\infty} \frac{(\alpha t)^n}{n!} (\gamma_\alpha)^{*n},$$

$$\gamma_\alpha := \alpha \int_0^{+\infty} e^{-\alpha t} \lambda_t \, dt, \qquad \alpha > 0.$$

Note that (λ_t^α) is the convolution semigroup of measures corresponding to the compound Poisson family with intensity α and jump measure γ_α. Since convergence holds for all $f \in B$, it holds in particular for $f(\xi) = e^{-r\xi}$, $\xi \geq 0$, $r \geq 0$. Consequently, taking $\xi = 0$ we obtain

$$\lim_{\alpha \uparrow +\infty} \int_0^{+\infty} e^{-r\eta} \lambda_t^\alpha(d\eta) = \int_0^{+\infty} e^{-r\eta} \lambda_t(d\eta), \qquad \forall \, r \geq 0.$$

However,

$$\int_0^{+\infty} e^{-r\eta} \lambda_t^\alpha(d\eta) = \exp \left\{ -t\alpha \int_0^{+\infty} (1 - e^{-r\eta}) \gamma_\alpha(d\eta) \right\}.$$

Thus for the Laplace transform we have $\widetilde{\lambda}_t(r) = e^{-t\widetilde{\psi}(r)}$, where

$$\widetilde{\psi}(r) = \lim_{\alpha \to +\infty} \widetilde{\psi}_\alpha(r),$$

$$\widetilde{\psi}_\alpha(r) := \int_0^{+\infty} (1 - e^{-r\eta}) \nu_\alpha(d\eta) \quad \text{and} \quad \nu_\alpha := \alpha \gamma_\alpha, \qquad \alpha > 0.$$

Let

$$\varphi(\eta) := \begin{cases} \eta & \text{if } 0 \le \eta \le 1, \\ 1 & \text{if } \eta \ge 1. \end{cases}$$

Then

$$\widetilde{\psi}_\alpha(r) = \int_0^{+\infty} \frac{1 - e^{-r\eta}}{\varphi(\eta)} \widetilde{\nu}_\alpha(d\eta),$$

where $\widetilde{\nu}_\alpha(d\eta) = \varphi(\eta)\nu_\alpha(d\eta)$, $\alpha > 0$, is a family of measures on $(0, +\infty)$ with uniformly bounded masses. Consequently, for a sequence $\alpha_k \uparrow +\infty$, $(\widetilde{\nu}_{\alpha_k})$ converges weakly to a measure $\widetilde{\nu}$ concentrated on $[0, +\infty]$. But

$$\lim_{\eta \to 0} \frac{1 - e^{-r\eta}}{\varphi(\eta)} = r, \qquad \lim_{\eta \to +\infty} \frac{1 - e^{-r\eta}}{\varphi(\eta)} = 1$$

and

$$\lim_{\alpha \to +\infty} \widetilde{\psi}_\alpha(r) = \int_{[0, +\infty]} \frac{1 - e^{-r\eta}}{\varphi(\eta)} \widetilde{\nu}(d\eta)$$

$$= r\widetilde{\nu}(\{0\}) + \int_{(0, +\infty)} \frac{1 - e^{-r\eta}}{\varphi(\eta)} \widetilde{\nu}(d\eta) + \widetilde{\nu}(\{+\infty\}).$$

Since the λ_t are probability measures, $\widetilde{\nu}(\{+\infty\}) = 0$. Thus the measure

$$\nu(d\eta) := \frac{1}{\varphi(\eta)} \widetilde{\nu}(d\eta)$$

has the required properties and the result has been proved in one direction. To prove sufficiency is easy and is left to the reader. For more details, refer to Zabczyk (2003).

Appendix F   Proof of Lemma 4.24

Let A be a Borel set separated from 0. Following Gikhman and Skorokhod (1974), Vol. II, we will show that $(L_A(t),\ t \geq 0)$ and $(L(t) - L_A(t),\ t \geq 0)$ are independent Lévy processes.

Assume that A is an open set. Let $0 = t_0 < t_1 < \cdots < t_n = t$ be a partition, and let

$$\delta := \max\{(t_{k+1} - t_k),\ k = 0, 1, \ldots, n-1\}.$$

Given $k = 0, 1, \ldots, n$, define

$$Z_k := \chi_A\big(L(t_{k+1}) - L(t_k)\big)\big(L(t_{k+1}) - L(t_k)\big),$$
$$Y_k := \big(L(t_{k+1}) - L(t_k)\big) - Z_k.$$

Since A is open, we have, $\mathbb{P}$-a.s., as $n \to \infty$ and $\delta \to 0$,

$$L_A(t) = \lim_n \sum_{k=0}^{n-1} Z_k, \qquad L(t) - L_A(t) = \lim_n \sum_{k=0}^{n-1} Y_k.$$

Consequently, for all $x, y \in U$,

$$\mathbb{E} \exp\big\{i\langle x, L_A(t)\rangle_U + i\langle y, L(t) - L_A(t)\rangle_U\big\}$$

$$= \lim_n \mathbb{E} \exp\left\{i \sum_{k=0}^{n-1} \langle x, Z_k\rangle_U + i \sum_{k=0}^{n-1} \langle y, Y_k\rangle_U\right\}$$

$$= \lim_n \mathbb{E} \exp\left\{i \sum_{k=0}^{n-1} \big(\langle x, Z_k\rangle_U + \langle y, Y_k\rangle_U\big)\right\}$$

$$= \lim_n \prod_{k=0}^{n-1} \mathbb{E} \exp\big\{i\langle x, Z_k\rangle_U + i\langle y, Y_k\rangle_U\big\}.$$

Define

$$\gamma := \prod_{k=0}^{n-1} \mathbb{E}\, e^{i\langle x, Z_k\rangle_U + i\langle y, Y_k\rangle_U} - \prod_{k=0}^{n-1} \mathbb{E}\, e^{i\langle x, Z_k\rangle_U}\, \mathbb{E}\, e^{i\langle y, Y_k\rangle_U}.$$

Using $|ab - cd| \leq |a - c| + |b - d|$ for $a, b, c, d \in \mathbb{C}$ such that $|a| \leq 1, |b| \leq 1, |c| \leq 1$,

$|d| \leq 1$, we obtain

$$|\gamma| \leq \sum_{k=0}^{n-1} \left| \mathbb{E}\, e^{i\langle x, Z_k \rangle_U + i\langle y, Y_k \rangle_U} - \mathbb{E}\, e^{i\langle x, Z_k \rangle_U}\, \mathbb{E}\, e^{i\langle y, Y_k \rangle_U} \right|.$$

Since $\langle x, Z_k \rangle_U \langle y, Y_k \rangle_U = 0$,

$$e^{i\langle x, Z_k \rangle_U + i\langle y, Y_k \rangle_U} = e^{i\langle x, Z_k \rangle_U} + e^{i\langle y, Y_k \rangle_U} - 1,$$

and therefore

$$|\gamma| \leq \sum_{k=0}^{n-1} \left| \mathbb{E}\, e^{i\langle x, Z_k \rangle_U} + \mathbb{E}\, e^{i\langle y, Y_k \rangle_U} - 1 - \mathbb{E}\, e^{i\langle x, Z_k \rangle_U}\, \mathbb{E}\, e^{i\langle y, Y_k \rangle_U} \right|$$

$$\leq \sum_{k=0}^{n-1} \left| \mathbb{E}\, e^{i\langle x, Z_k \rangle_U} - 1 \right| \left| \mathbb{E}\, e^{i\langle y, Y_k \rangle_U} - 1 \right|$$

$$\leq 2 \sup_k \left| \mathbb{E}\, e^{i\langle y, Y_k \rangle_U} - 1 \right| \sum_{j=0}^{\infty} \mathbb{P}\left(|Z_j|_U > 0 \right).$$

The stochastic continuity of L implies stochastic uniform continuity on $[0, t]$ and, consequently,

$$\sup_k \left| \mathbb{E}\, e^{i\langle y, Y_k \rangle_U} - 1 \right| \to 0 \qquad \text{as } n \to \infty.$$

But $\mathbb{P}\left(|Z_j|_U > 0 \right) = \mathbb{P}\left(\pi_A(t_{j+1}) - \pi_A(t_j) \geq 1 \right)$. Since π_A is a Poisson process with intensity $\nu(A)$, it follows easily that

$$\lim_n \sum_{j=0}^{n-1} \mathbb{P}\left(\pi_A(t_{j+1}) - \pi_A(t_j) \geq 1 \right) = t\nu(A).$$

Therefore $|\gamma| \to 0$ as $n \to \infty$ and

$$\mathbb{E} \exp\left\{ i\langle x, L_A(t) \rangle_U + i\langle y, L(t) - L_A(t) \rangle_U \right\} = \mathbb{E} \exp\left\{ i\langle x, L_A(t) \rangle_U \right\}$$
$$\times \mathbb{E} \exp\left\{ i\langle y, L(t) - L_A(t) \rangle_U \right\}. \qquad \text{(F.1)}$$

Now assume that A is an arbitrary Borel set separated from 0. If $A = \bigcup_{n=1}^{\infty} A_n$, where $A_1, A_2, \ldots$ are disjoint Borel sets, then

$$L_A(t) = \sum_{n=1}^{\infty} L_{A_n}(t).$$

Consequently if (F.1) holds for disjoint sets A_n, then it holds also for their union. In addition, if (F.1) holds for A and $A \cap B(0, r) = \emptyset$ then it holds also for $A^c \cap B(0, r)$. By the Dynkin theorem on λ-π systems, (F.1) holds for every Borel A such that $A \cap B(0, r) = \emptyset$ for some positive r. Now let $0 = t_0 \leq t_1 < \cdots < t_m = t$. To complete the proof we have to show that the two vector random variables

$$\left(L_A(t_1), \ldots, L_A(t_m) \right), \quad \left(L(t_1) - L_A(t_1), \ldots, L(t_m) - L_A(t_m) \right)$$

are independent. Since the process with components $L_A(t)$, $L(t) - L_A(t)$, $t \geq 0$, has independent increments, we have

$$
\mathbb{E} \exp \left\{ i \sum_{k=0}^{m-1} \langle x_k, \ L_A(t_{k+1}) - L_A(t_k) \rangle_U + i \sum_{k=0}^{m-1} \langle y_k, \ L(t_{k+1}) - L(t_k) - \left(L_A(t_{k+1}) - L_A(t_k) \right) \rangle_U \right\}
$$

$$
= \prod_{k=0}^{m-1} \mathbb{E} \exp \left\{ i \langle x_k, \ L_A(t_{k+1}) - L_A(t_k) \rangle_U \right\}
$$

$$
\times \prod_{k=0}^{m-1} \mathbb{E} \exp \left\{ i \langle y_k, \ L(t_{k+1}) - L(t_k) - \left(L_A(t_{k+1}) - L_A(t_k) \right) \rangle_U \right\}.
$$

Consequently, the random variables

$$
L_A(t_1), \ L_A(t_2) - L_A(t_1), \ \ldots, \ L_A(t_m) - L_A(t_{m-1}),
$$
$$
\left(L(t_1) - L_A(t_1), \ldots, L(t_m) - L(t_{m-1}) - (L_A(t_m) - L_A(t_{m-1})) \right)
$$

are independent, and by an easy inductive argument the required independence follows.

List of symbols

General

$\mathbb{Z}, \mathbb{N}, \mathbb{Z}_+$	all integers, positive integers, non-negative integers
$\mathbb{Q}, \mathbb{Q}_+$	rational, non-negative-rational numbers
$\mathbb{C}, \mathbb{R}, \mathbb{R}_+$	complex, real, non-negative-real numbers
δ_x	Dirac δ-measure at a point x
$\delta_{n,m}$	Kronecker symbol
$\Rightarrow$	weak convergence or implication
χ_A	indicator function of a set A
ℓ_d	Lebesgue measure on $\mathbb{R}^d$
σ_t^d	surface measure on the sphere in $\mathbb{R}^d$ with center at 0 and radius t
$\#A$	cardinality of a (usually finite) set A
$\dim V$	dimension of V
$[x]$	integer part of a real number x
$a \wedge b, a \vee b$	$\min\{a, b\}, \max\{a, b\}$
a^+, a^-	$a \vee 0, (-a) \vee 0$
$X(t-)$	left limit of a function (process) X at t
$\mu * \lambda$	convolution of measures or functions
$\otimes$	product of measures, or (see p. 107) tensor product on a Hilbert space
$\| \cdot \|_\infty$	supremum norm; see p. 13
$\| \cdot \|_{\text{Lip}}$	smallest Lipschitz constant; see p. 16
$\mathcal{B}(E)$	σ-field of Borel subsets of E
$\mathcal{P}_I, \mathcal{P}$	σ-fields of predictable sets; see p. 21
$(\mathcal{F}_t)$	filtration; see p. 21
$(\mathcal{F}_{t+}), (\mathcal{F}_t^X), \overline{\mathcal{F}}_t^X, \overline{\mathcal{F}}_{t+}^X$	see pp. 21–22

Spaces of functions or processes

$B_b(E)$	space of bounded measurable functions
$C_b(E)$	space of bounded continuous functions
$UC_b(U)$	space of uniformly continuous bounded functions
$C_0(\mathbb{R}^d)$	space of continuous functions vanishing (i.e. having limit 0) at infinity
$C_0^\infty(\mathbb{R}^d)$	space of infinitely differentiable functions vanishing at infinity with all derivatives
$C_c^n(\mathcal{O})$	space of compactly supported functions having continuous derivatives up to order n
$C^\gamma(\mathcal{O})$	space of Hölder continuous functions; see p. 14
$C^{n+\gamma}(\mathcal{O})$	see p. 14
$C_b^\infty(\mathcal{O})$	space of functions with bounded derivatives
$W^{n,p}(\mathcal{O})$	Sobolev space; see p. 13
$W_0^{n,p}(\mathcal{O})$	see p. 14
$S(\mathbb{R}^d), S(\mathbb{R}^d;\mathbb{C})$	space of tempered distributions; see p. 240
$L^2_{(s)}(\mathbb{R}^d, \mu;\mathbb{C})$	see p. 253
$D_D(\theta, p), D_N(\theta, p)$	real interpolation spaces; see p. 277
$L^p(S, \mathcal{S}, \lambda)$	space of p-integrable functions
$L^p_\rho, \mathcal{L}^p_\rho$	weighted L^p-spaces; see p. 15
$C_\rho, \mathcal{C}_\rho$	space of weighted continuous functions; see p. 15
$\text{Lip}(E), \text{Lip}(p, S, \lambda)$	space of Lipschitz functions; see p. 16
$\mathcal{M}^2(U)$	space of càdlàg square integrable martingales taking values in a Hilbert space U
$\mathcal{M}^2$	$\mathcal{M}^2(\mathbb{R})$
$\mathcal{M}_{\text{loc}}(B), \mathcal{M}^2_{\text{loc}}(B)$	space of local, local square integrable, martingales; see p. 36
$\mathcal{BV}$	see p. 36
$\mathcal{L}^2_{M,T}(H)$	space of admissible integrands; see p. 114
$\mathbf{L}^2_{\mathcal{H},T}$	see p. 122
$\mathcal{L}^p_{\mu,T}$	see p. 128
$\mathcal{L}^p_{\hat{\pi},T}(L^p)$	see subsection 8.8.1
$\mathcal{L}^p_{W,T}(L^p)$	see subsection 8.8.2
$R_{U,0}(\mathcal{H}, L^p)$	see p. 130
$R(\mathcal{H}, L^p), \mathfrak{R}(\mathcal{H}, L^p)$	see subsection 8.8.2
$\mathcal{H}$	usually reproducing kernel Hilbert space

Functions, processes and distributions

$\mathcal{L}(X)$	law (distribution) of a random element X; see p. 21
$\mathcal{N}(m, Q)$	normal distribution; see p. 28
$\mathcal{P}(a)$	Poisson distribution; see p. 41
$\mathcal{G}$	Green function or fundamental solution
$\mathfrak{g}_m$	see p. 19
$\mathfrak{g}$	see p. 181
$\vartheta_\rho, \theta_\rho$	see p. 15
Π	Poisson process, see p. 42
π	π-number or Poisson random measure
$\widehat{\pi}$	compensated Poisson random measure
W	Wiener process
$\mathcal{W}$	usually a Brownian sheet; see p. 99
$\widehat{Z}$	compensated process; see e.g. p. 48
P_t	transition semigroup; see p. 7
$P(t, \Gamma)$	transition function; see p. 6
$\langle \cdot, \cdot \rangle$	scalar product on $\mathbb{R}^d$
Γ	Euler Γ-function or (see p. 248) covariance
λ, μ, ν	usually measures
M	usually a martingale
L, Z	usually Lévy processes
$\mathcal{Z}$	usually a Lévy sheet; see p. 103

Spaces of operators

$M(n \times m)$	space of matriices of dimension $n \times m$
$M_s^+(n \times n)$	space of symmetric non-negative-definite matrices
V^*	adjoint space
$L(U, H)$	space of bounded (i.e. continuous) linear operators; see p. 355
$L(U)$	$L(U, U)$
$K(U, H)$	space of compact linear operators; see p. 355
$L_1(U, H)$	space of nuclear operators; see p. 356
$L_1(U)$	$L_1(U, U)$
$L_1^+(U)$	space of nuclear symmetric positive-definite operators
$L_{(HS)}(U, H)$	space of Hilbert–Schmidt operators; see p. 356
$L_{(HS)}(U)$	$L_{(HS)}(U, U)$

Operators

N_f	composition operator; see p. 16
D	Fréchet derivative (i.e. gradient) operator
Δ	Laplace operator, or $\Delta X(t) = X(t) - X(t-)$
Δ_D	Laplace operator with Dirichlet boundary conditions; see p. 279
Δ_N	Laplace operator with Neumann boundary conditions; see p. 279
$D_{B,\gamma}$	boundary operator; see p. 273
$\mathfrak{F}$	Fourier transform; see p. 231
$\tilde{\lambda}$	Laplace transform of λ; see p. 59
Tr	trace operator; see p. 357
A^*	adjoint operator; see p. 355
A^{-1}	inverse or pseudo-inverse of A; see p. 92
$D(A)$	domain of an operator A
$\rho(A)$	resolvent set of an operator A; see p. 365
$R(\alpha)$	resolvent of A; see p. 365
$S(t), t \geq 0$	C_0-semigroup; see Appendix B
$\partial\lvert \cdot \rvert_B$	subdifferential; see p. 179
$\mathbb{E}\,X$	mathematical expectation of a random element X; see e.g. p. 24
$\mathbb{E}(X\lvert\mathcal{F}_t)$	conditional expectation; see p. 24
$\langle M, M\rangle$	angle bracket (predictable variation process) of a martingale M; see p. 35
$[M, M]$	quadratic variation of a martingale M; see p. 36
$\langle\langle M, M\rangle\rangle$	operator angle bracket of a martingale; see p. 109

References

Albeverio, S., Lytvynov, E. and Mahnig, A. (2004). A model of the term structure of interest rates based on Lévy fields, *Stochastic Process. Appl.* **114**, 251–263.

Albeverio, S. and Rüdiger, B. (2005). Stochastic integrals and the Lévy–Itô decomposition on separable Banach spaces, *Stoch. Anal. Appl.* **23**, 217–253.

Albeverio, S., Wu, J. L. and Zhang, T. S. (1998). Parabolic SPDEs driven by Poissonian noise, *Stochastic Process. Appl.* **74**, 21–36.

Alòs, E and Bonaccorsi, S. (2002a). Stochastic partial differential equations with Dirichlet white-noise boundary conditions, *Ann. Inst. H. Poincaré Probab. Statist.* **38**, 125–154.

Alòs, E. and Bonaccorsi, S. (2002b), Stability for stochastic partial differential equations with Dirichlet white-noise boundary conditions, *Infin. Dimens. Anal. Quantum Probab. Relat. Top.* **5**, 465–481.

Ambrosetti, A. and Prodi, G. (1995). *A Primer of Nonlinear Analysis* (Cambridge, Cambridge University Press).

Applebaum, D. (2004). *Lévy Processes and Stochastic Calculus* (Cambridge, Cambridge University Press).

Applebaum, D. (2005). Martingale-valued measures, Ornstein–Uhlenbeck processes with jumps and operator self-decomposability in Hilbert space, in *Séminaire de Probabilité* **39**, Lecture Notes in Mathematics, vol. 1847, 173–198.

Applebaum, D. and Wu, J. L. (2000). Stochastic partial differential equations driven by Lévy space–time white noise, *Random Oper. Stochastic Equations* **3**, 245–261.

Arima, R. (1964). On general boundary value problem for parabolic equations, *J. Math. Kyoto Univ.* **4**, 207–243.

Aronson, D. G. (1967). Bounds for the fundamental solution of a parabolic equation, *Bull. Amer. Math. Soc.* **73**, 890–896.

Aubin, J. P. and Da Prato. G. (1990). Stochastic viability and invariance, *Ann. Scuola Norm. Sup. Pisa* **17**, 595–613.

Bachelier, L. (1900). Théorie de la spéculation, *Ann. Sci École Norm. Sup.* **17**, 21–86.

Bally, V., Gyöngy, I. and Pardoux, E. (1994). White noise driven parabolic SPDEs with measurable drift, *J. Funct. Anal.* **120**, 484–510.

Bertoin, J. (1996). *Lévy Processes* (Cambridge, Cambridge University Press).

Bichteler, K. (2002). *Stochastic Integration with Jumps* (Cambridge, Cambridge University Press).

Billingsley, P. (1968). *Convergence of Probability Measures* (New York, Wiley).

Billingsley, P. (1986). *Probability and Measure* (New York, Wiley).

Björk, T. and Christensen, B. J. (1999). Interest rate dynamics and consistent forward rate curves, *Math. Finance* **9**, 323–348.

Björk, T., DiMassi, G., Kabanov, Y. and Runggaldier, W. (1997). Towards a general theory of bound markets, *Finance Stoch.* **1**, 141–174.

Björk, T., Kabanov, Y. and Runggaldier, W. (1997). Bond market structure in the presence of marked point process, *Math. Finance* **7**, 211–239.

Blumenthal, R. M. and Getoor, R. K. (1968). *Markov Processes and Potential Theory* (Academic Press, New York).

Bo, L. and Wang, Y. (2006). Stochastic Cahn–Hilliard partial differential equations with Lévy spacetime white noises, *Stoch. Dyn.* **6**, 229–244.

Bojdecki, T. and Jakubowski, J. (1989). Ito stochastic integral in the dual of a nuclear space, *J. Multivariate Anal.* **31**, 40–58.

Brace, A., Gątarek, D. and Musiela, M. (1997). The market model of interest rate dynamics, *Math. Finance* **7**, 127–147.

Brzeźniak, Z. (1997). On stochastic convolutions in Banach spaces and applications, *Stochastics Stochastics Rep.* **61**, 245–295.

Brzeźniak, Z. (2006). Asymptotic compactness and absorbing sets for $2D$ stochastic Navier–Stokes equations on some unbounded domains, in *Stochastic Partial Differential Equations and Applications VII*, eds. G. Da Prato and L. Tubaro, Lecture Notes in Pure and Applied Mathematics, Vol. 245, (Boca Raton, Chapman & Hall/CRC), 35–52.

Brzeźniak, Z., Maslowski, B. and Seidler, J. (2005). Stochastic nonlinear beam equations, *Probab. Theory Related Fields* **132**, 119–149.

Brzeźniak, Z. and Peszat, S. (1999). Space–time continuous solutions to SPDEs driven by a homogeneous Wiener process, *Studia Math.* **137**, 261–299.

Burgers, J. M. (1939). Mathematical examples illustrating relations occurring in the theory of turbulent fluid motion, *Verh. Nederl. Akad. Wetensch. Afd. Natuurk.* **17**.

Cabana, E. (1970). The vibrating string forced by white noise, *Z. Wahrscheinlichkeitstheorie und Verw. Gebiete* **15**, 111–130.

Capiński, M. and Peszat, S. (2001). On the existence of a solution to stochastic Navier–Stokes equations, *Nonlinear Anal.* **44**, 141–177.

Carathéodory, C. (1918). *Vorlesung über Reelle Funktionen*, Leipzig, Berlin.

Carmona, R. and Molchanov, S. A. (1994). Parabolic Anderson problem and intermittency, *Mem. Amer. Math. Soc.* **108**, 1–125.

Carmona, R. and Nualart, D. (1988). Random nonlinear wave equations: smoothness of the solutions, *Probab. Theory Related Fields* **79**, 469–508.

Cerrai, S. (2001). *Second Order PDEs in Finite and Infinite Dimensions. A Probabilistic Approach*, Lecture Notes in Mathematics, Vol. 1762 (Berlin, Springer).

Chojnowska-Michalik, A. (1978). Representation theorem for general stochastic delay equations, *Bull. Acad. Pol. Sci., Ser. Sci. Math.* **26**, 634–641.

Chojnowska-Michalik, A. (1987). On processes of Ornstein–Uhlenbeck type in Hilbert spaces, *Stochastics* **21**, 251–286.

Ciesielski, Z. (1961). Hölder condition for realizations of Gaussian processes, *Trans. Amer. Math. Soc.* **99**, 403–413.

Cont, R. and Tankov, P. (2004). *Financial Modelling with Jump Processes* (Boca Raton, Chapman and Hall/CRC).

Courrège, Ph. (1965/66). Sur la forme integro-différentielle des opérateurs de C_K^∞ dans C_0 satisfaisant au principe du maximum, in *Sém. Théorie du Potentiel*, Exposé 2.

Dalang, R. C. (1999). Extending the martingale measure stochastic integral with applications to spatially homogeneous s.p.d.e.s, *Electron. J. Probab.* **4**, 1–29.

Dalang, R. C. and Frangos, N. (1998). The stochastic wave equation in two spatial dimensions, *Ann. Probab.* **26**, 187–212.

Dalang, R. C. and Mueller, C. (2003), Some non-linear SPDEs that are second order in time, *Electron. J. Probab.* **8**, 304–337.

Dalang, R. C. and Sanz-Solé, M. (2005). Regularity of the sample paths of a class of second-order spde's, *J. Funct. Anal.* **227**, 304–337.

Dalang, R. C. and Sanz-Solé, M. (2006). Hölder–Sobolev regularity of the solution to the stochastic wave equation in dimension 3, preprint.

Daniell, P. J. (1918). Integrals in an infinite number of dimensions, *Ann. Math.* **20**, 281–288.

Da Prato, G. and Gatarek, D. (1995). Stochastic Burgers equation with correlated noise, *Stochastics Stochastics Rep.* **52**, 29–41.

Da Prato, G. and Giusti, E. (1967). Equazioni di Schrödinger e delle onde par l'operatore di Laplace iterato in $L^p(\mathbb{R}^n)$, *Ann. Mat. Pura Appl.* **76**, 378–397.

Da Prato, G., Kwapień, S. and Zabczyk, J. (1987). Regularity of solutions of linear stochastic equations in Hilbert spaces, *Stochastics* **23**, 1–23.

Da Prato, G. and Zabczyk, J. (1992a). *Stochastic Equations in Infinite Dimensions* (Cambridge, Cambridge University Press).

Da Prato, G. and Zabczyk, J. (1992b). Non-explosion, boundedness and ergodicity for stochastic semilinear equations, *J. Differential Equations* **98**, 181–195.

Da Prato, G. and Zabczyk, J. (1993). Evolution equations with white-noise boundary conditions, *Stochastics Stochastics Rep.* **42**, 167–182.

Da Prato, G. and Zabczyk, J. (1996). *Ergodicity for Infinite Dimensional Systems* (Cambridge, Cambridge University Press).

Da Prato, G. and Zabczyk, J. (2002). *Second Order Partial Differential Equations in Hilbert Spaces*, London Mathematical Society Lecture Note Series, Vol. 293 (Cambridge, Cambridge University Press).

Davies, E. B. (1980). *One-parameter Semigroups* (New York, Academic Press).

Dawson, D. A. and Salehi, H. (1980). Spatially homogeneous random evolutions, *J. Multivariate Anal.* **10**, 141–180.

De Acosta, A. (1980). Exponential moments of vector valued random series and triangular arrays, *Ann. Probab.* **8**, 381–389.

Delbaen, F. and Schachermayer, W. (1994). A general version of the fundamental theorem of asset pricing, *Math. Ann.* **30**, 463–520.

Dellacherie, C. (1980). Un survol de la theorie de l'integral stochastique, *Stochastic Process. Appl.* **10**, 115–144.

Dettweiler, E. (1991). Stochastic integration relative to Brownian motion on a general Banach space, *Doğa Mat.* **15**, 6–44.

Doob, J. L. (1940). The law of large numbers for continuous stochastic processes, *Duke Math. J.* **6**, 290–306.

Doob, J. L. (1953). *Stochastic Processes* (New York, Wiley).

Douglas, R. (1966). On majorization, factorization and range inclusion of operators in Hilbert spaces, *Proc. Amer. Math. Soc.* **17**, 413–45.

Dynkin, E. B. (1965). *Markov Processes*, Vols. I, II (Berlin, Springer).

Dynkin, E. B. and Yushkevich, A. A. (1978). *Controlled Markov Processes* (Berlin, Springer).

Eberlein, E. and Raible, S. (1999). Term structure models driven by general Lévy processes, *Math. Finance* **9**, 31–53.

Eidelman, S. D. (1969). *Parabolic Systems* (Amsterdam, North-Holland).

Eidelman, S. D. and Zhitarashu, N.V. (1998). *Parabolic Boundary Value Problems* (Basel, Birkhäuser).

Encyclopedia of Mathematics (1987). MIT Press.

Engel, K. and Nagel, R. (2000). *One-parameter Semigroups for Linear Evolution Equations*, Springer Graduate Texts in Mathematics, Vol. 194 (Berlin, Springer).

Ethier, S. N. and Kurtz, T. G. (1986). *Markov Processes. Characterization and Convergence*, Wiley Series in Probability and Mathematical Statistics: Probability and Mathematical Statistics (New York, Wiley).

Evans, L. C. (1998). *Partial Differential Equations* (Providence R.I., American Mathematical Society).

Fernique, M. X. (1970). Intégrabilité de vecteurs Gaussiens, *C.R. Acad. Sci. Paris Sér. I Math.* **270**, 1698–1699.

Feller, W. (1971). *An Introduction to Probability Theory and Its Applications*, Vol. II (New York, Wiley).

Filipović, D. (2001). *Consistency problems for Heath–Jarrow–Morton interest rate models*, Lecture Notes in Mathematics, Vol. 1760 (Berlin, Springer).

Filipović, D. and Tappe, S. (2006). Existence of Lévy term structure models, submitted.

Freidlin, M. and Sowers, R. (1992). Central limit results for a reaction–diffusion equation with fast-oscillating boundary perturbations, in *Proc. conf. on Stochastic Partial Differential Equations and Their Applications (Charlotte NC, 1991)*, Lecture Notes in Control and Inform. Sci., Vol. 176 (Berlin, Springer), pp. 101–112.

Freidlin, M. and Wentzell, A. (1992). Reaction–diffusion equations with randomly perturbed boundary conditions, *Ann. Probab.* **20**, 963–986.

Funaki, T. (1991). Regularity properties for stochastic partial differential equations of parabolic type, *Osaka J. Math.* **28**, 495–516.

Fournier, N. (2000). Malliavin calculus for parabolic SPDEs with jumps, *Stochastic Processes Appl.* **87**, 115–147.

Fournier, N. (2001). Support theorem for solutions of a white-noise driven parabolic stochastic partial differential equations with temporal Poissonian jumps, *Bernoulli* **7**, 165–190.

Fuhrman, M. and Röckner. M. (2000). Generalized Mahler semigroups: the non-Gaussian case, *Potential Anal.* **12**, 1–47.

van Gaans, O. (2007). Invariant measures for stochastic evolution equations with Hilbert space valued Lévy noise, preprint.

García-Cuerva, J. and Kazarian, K. S. (1985). Spline wavelet bases of weighted spaces, in *Proc. Int. Conf. on Fourier Analysis and Partial Differential Equations*, eds. J. Garcia-Cuerva, E. Hernández, F. Soria and J.-L. Torrea (Boca Raton, CRC Press).

Gilbarg, D. and Trudinger, N. T. (1998). *Elliptic Partial Differential Equations of Second Order* (Berlin, Springer).

Gelfand, I. M. and Vilenkin, N. Ya. (1964). *Generalized Functions*, Vol. IV, *Applications of Harmonic Analysis* (New York, Academic Press).

Gikhman, I. I. and Skorokhod, A. V. (1974). *The Theory of Stochastic Processes*, Vols. I, II (Berlin, Springer).

Goncharuk, N. and Kotelenez, P. (1998). Fractional step method for stochastic evolution equations, *Stochastic Process. Appl.* **73**, 1–45.

Grisvard, P. (1966). Commutativité de deux foncteurs d'interpolation at applications, *J. Math. Pures Appl.* **45**, 143–290.

Gyöngy, I. and Krylov, N. V. (1980). On stochastic equations with respect to semimartingales I, *Stochastics* **4**, 1–21.

Hamedani, H. D. and Zangeneh, B. H. (2001). Stopped Doob inequality for pth moment, $0 < p < \infty$, stochastic convolution integrals, *Stochastic Anal. Appl.* **19**, 771–798.

Hartman, P. (1964). *Ordinary Differential Equations* (New York, Wiley).

Hausenblas, E. (2005). SPDEs driven by Poisson Random Measure: existence and uniqueness, *Electron. J. Probab.* **11**, 1496–1546.

Hausenblas, E. and Seidler, J. (2001). A note on maximal inequality for stochastic convolutions, *Czechoslovak Math. J.* **51**, 785–790.

Hausenblas, E. and Seidler, J. (2006). Maximal inequalities and exponential integrability for stochastic convolutions driven by martingales, preprint.

Heath, D., Jarrow, R. and Morton, A. (1992). Bond pricing and the term structure of interest rates: a new methodology, *Econometrica* **61**, 77–105.

Hunt, G. A. (1957, 1958). Markoff processes and potentials, I, II, III, *Illinois J. Math.* **1**, 44–93, 316–369, **2**, 151–213.

Ichikawa, A. (1986). Some inequalities for martingales and stochastic convolutions, *Stochastic Anal. Appl.* **4**, 329–339.

Ikeda, N. and Watanabe, S. (1981). *Stochastic Differential Equations and Diffusion Processes* (Amsterdam, North-Holland).

Itô, K. (1951). On stochastic differential equations, *Mem. Amer. Math. Soc.* **4**, 1–51.

Itô, K. (1984). *Foundations of Stochastic Differential Equations in Infinite Dimensional Spaces* (Philadelphia, SIAM).

Itô, K. and Nisio, M. (1968). On the convergence of sums of independent Banach space valued random variables, *Osaka Math. J.* **5**, 35–48.

Jachimiak, W. (1996). A note on invariance for semilinear differential equations, *Bull. Acad. Sci. Math.* **44**, 179–183.

Jacod, J. and Shiryaev, A. N. (1987). *Limit Theorems for Stochastic Processes* (Berlin, Springer).

Jacob, N. (2001). *Pseudodifferential Operators and Markov Processes*, Vol. I, *Fourier Analysis and Semigroups* (London, Imperial College Press).

Jacob, N. (2002). *Pseudodifferential Operators and Markov Processes*, Vol. II, *Generators and their Potential Theory* (London, Imperial College Press).

Jacob, N. (2005). *Pseudodifferential Operators and Markov Processes*, Vol. III, *Markov Processes and Applications* (London, Imperial College Press).

Jakubowski, A. (2006). Towards a general Doob–Meyer decomposition theorem, Probab. Math. Statist. **26**, 143–153.

Jakubowski, J., Niewęgłowski, M. and Zabczyk, J. (2006). BGM equation for bonds with Lévy noise, private communication.

Jakubowski, J. and Zabczyk, J. (2004). HJM condition for models with Lévy noise, *IM PAN Preprint 651*, (Warsaw).

Jakubowski, J. and Zabczyk, J. (2007). Exponential moments for HJM models with jumps, *Finance and Stochastics*, to appear.

Kallenberg, O. (2002). *Foundations of Modern Probability* (Berlin, Springer).

Kallianpur, G. and Pérez-Abreu, V. (1988). Stochastic evolution equations driven by nuclear space valued martingales, *Appl. Math. Optimiz.* **17**, 237–272.

Kallianpur, G. and Xiong, J. (1995). *Stochastic Differential Equations in Infinite Dimensional Spaces*, Institute of Mathematical Statistics, Lecture Notes, Monograph Series, Vol. 26 (Philadelphia, SIAM).

Karczewska, A. and Zabczyk, J. (2000a). Stochastic PDEs with function-valued solutions, in *Proc. Conf. on Infinite Dimensional Stochastic Analysis (Amsterdam 1999)*, eds. Ph. Clément, F. den Hollander, J. van Neerven and B. de Pagter, Verh. Afd. Natuurkd. 1. Reeks. K. Ned. Akad. Wet., Vol. 52 (Amsterdam, R. Neth. Acad. Arts Sci.), pp. 197–216.

Karczewska, A. and Zabczyk, J. (2000b). Regularity of solutions to stochastic Volterra equations, *Rend. Mat. Acc. Lincei* **11**, 117–213.

Karczewska, A. and Zabczyk, J. (2001). A note on stochastic wave equations, in *Proc. Conf. on Evolution Equations and their Applications in Physical and Life Sciences (Bad Herrenhalb 1998)*, eds. G. Lumre and L. Weis (Marcel Dekker), pp. 501–511.

Kifer, Yu. (1986). *Ergodic Theory of Random Transformations* (Bassel, Birkhauser).

Kifer, Yu. (1997). The Burgers equation with a random force and a general model for directed polymers in random environments, *Probab. Theory Related Fields* **108**, 29–65.

Kingman, J. F. C. (1993). *Poisson Processes* (Oxford, Clarendon Press).

Kinney, J. H. (1953). Continuity properties of sample functions of Markov processes, *Trans. Amer. Math. Soc.* **74**, 280–302.

Knoche, C. (2004). SPDEs in infinite dimension with Poisson noise, *C. R. Acad. Sci. Paris, Ser. I* **339**, 647–652.

Kolmogorov, A. N. (1931). Über die analytischen Methoden in der Wahrscheinlichkeitsrechnung, *Math. Ann.* **104**, 415–458.

Kolmogorov, A. N. (1933). *Grundbegriffe der Wahrscheinlichkeitsrechnung*, Erg. der Math., Vol. 2 (Berlin, Springer).

Kolmogorov, A. N. (1937). Zur Umkehrbarkeit der statischen Naturgestze, *Math. Ann.* **113**, 766–772.

Kotelenez, P. (1982). A submartingale type inequality with applications to stochastic evolution equations, *Stochastics* **8**, 139–151.

Kotelenez, P. (1984). A stopped Doob inequality for stochastic convolution integrals and stochastic evolution equations, *Stochastic Anal. Appl.* **2**, 245–265.

Kotelenez, P. (1987). A maximal inequality for stochastic convolution integrals on Hilbert space and space–time regularity of linear stochastic partial differential equations, *Stochastics* **21**, 345–458.

Kotelenez, P. (1992a). Existence, uniqueness and smoothness for a class of function valued stochastic partial differential equations, *Stochastics Stochastics Rep.* **41**, 177–199.

Kotelenez, P. (1992b). Comparison methods for a class of function valued stochastic partial differential equations, *Probab. Theory Related Fields*, **93**, 1–19.

Kruglov, V. M. (1972). Integrals with respect to infinitely divisible distributions in Hilbert spaces, *Matematičeskie Zamietki* **11**, 669–676.

Kruglov, V. M. (1984). *Additional Chapters of Probability Theory* (Moscow), in Russian.

Krylov, N. V. (1995). *Introduction to the Theory of Diffusion Processes* (Providence R.I., American Mathematical Society).

Krylov, N. V. (1999). An analytic approach to spde's, in *Stochastic Partial Differential Equations: Six Perspectives*, eds. R. A. Carmona and B. Rozovskii, Mathematical Surveys and Monographs, Vol. 64 (Providence R.I., American Mathematical Society), 185–242.

Krylov, N. V. (2005). On the foundation of the L_p-theory of stochastic partial differential equations, in *Stochastic Partial Differential Equations and Applications VII*, eds. G. Da Prato and L. Tubaro (Taylor and Francis Group), pp. 179–192.

Kunita, H. (1970). Stochastic integrals based on martingales taking values in Hilbert spaces, *Nagoya Math. J.* **38**, 41–52.

Kuo, H. H. (1975). *Gaussian Measures in Banach Spaces*, Lecture Notes in Mathematics, Vol. 463 (Berlin, Springer).

Kuratowski, K. (1934). Sur une généralisation de la notion d'homéomorphie, *Fund. Math.* **22**, 206–220.

Kuratowski, K. and Ryll-Nardzewski, C. (1965). A general theorem on selectors, *Bull. Acad. Polon. Sér. Sci. Math. Astr. Phys.* **13**, 397–403.

Kwapień, S. and Woyczyński, W. A. (1992). *Random Series and Stochastic Integrals: Single and Multiple* (Boston, Basel, Berlin, Birkhauser).

Lasota, A. and Szarek, T. (2006). Lower bound technique in the theory of a stochastic differential equation, *J. Diff. Eq.* **231**, 513–533.

Lasota, A. and Yorke, J. A. (1994). Lower bound technique for Markov operators and iterated functions systems, *Random Computat. Dynamics* **2**, 41–77.

Lebesgue, H. (1902). Intégrale, longeur, air, *Ann. Math.* **3**, 231–359.

Lévy, P. (1948). *Processus Stochastiques et Mouvement Brownien* (Paris, Gauthier–Villars).

Linde, W. (1986). *Probability in Banach Spaces – Stable and Infinitely Divisible Distributions* (New York, Wiley).

Lions, J. L. and Magenes, E. (1972). *Non-Homogeneous Boundary Value Problems and Applications I* (Berlin, Springer).

Littman, W. (1963). The wave operator and L_p norms, *J. Math. Mech.* **12**, 55–68.

Løkka, A., Øksendal, B. and Proske, F. (2004). Stochastic partial differential equations driven by Lévy space–time white noise, *Ann. Appl. Probab.* **14**, 1506–1528.

Lunardi, A. (1995). *Analytic Semigroups and Optimal Regularity in Parabolic Problems* (Basel, Birkhauser).

Lundberg, F. (1909). *Zur Theorie def Rückversicherung*, (Vienna, Verhandl. Kongr. Versicherungsmath.).

Manthey, R. and Zausinger, T. (1999). Stochastic evolution equations in $L_\rho^{2\nu}$, *Stochastic Stochastic Rep.* **66**, 37–85.

Márquez, D., Mellouk, M. and Sarrá, M. (2001). On stochastic partial differential equations with spatially correlated noise: smoothness of the law, *Stochastic Process. Appl.* **93**, 269–284.

Maslowski, B. (1995). Stability of semilinear equations with boundary and pointwise noise, *Ann. Scuola Norm. Sup. Pisa* **22**, 55–93.

Métivier, M. (1982). *Semimartingales: A Course on Stochastic Processes* (Berlin, De Gruyer).

Métivier, M. (1988). *Stochastic Partial Differential Equations in Infinite Dimensional Spaces* (Pisa, Scuola Normale Superiore).

Métivier, M. and Pellaumail, J. (1980). *Stochastic Integration* (New York, Academic Press).

Métivier, M. and Pistone, G. (1975). Une formule d'isométrie pour l'intégrale stochastique d'évolution linéaire stochastique, *Z. Wahrscheinlichkeitstheorie verw. Gebiete* **33**, 1–18.

Meyer, P. A. (1976). *Un cours sur les intégrales stochastiques*, Lecture Notes in Mathematics, Vol. 511, pp. 245–400.

Mikulevicius, R. and Rozovskii, B. L. (1998). Normalized stochastic integrals in topological vector spaces, in *Proc. Séminaire de Probabilité XXXII*, Lecture Notes in Mathematics, Vol. 1686 (Berlin, Springer), 137–165.

Milian, A. (2002). Comparison theorems for stochastic evolution equations, *Stochastic Stochastic Rep.* **72**, 79–108.

Millet, A. and Morien, P. (2001). On a nonlinear stochastic wave equation in the plane: existence and uniqueness of the solution, *Ann. Appl. Probab.* **11**, 922–951.

Millet, A. and Sanz-Solé, M. (1999). A stochastic wave equation in two space dimensions: smoothness of the law, *Ann. Probab.* **27**, 803–844.

Millet, A. and Sanz-Solé, M. (2000). Approximation and support theorem for a wave equation in two space dimensions, *Bernoulli* **6**, 887–915.

Mizohata, S. (1973). *The Theory of Partial Differential Equations* (Cambridge, Cambridge University Press).

Mlak, W. (1991). *Hilbert Spaces and Operator Theory* (Dordrecht, Kluwer).

Mueller, C. (1997). Long time existence for the wave equation with a noise term, *Ann. Probab.* **25**, 133–151.

Mueller, C. (1998). The heat equation with Lévy noise, *Stochastic Process. Appl.* **74**, 67–82.

Mytnik, L. (2002). Stochastic partial differential equation driven by stable noise, *Probab. Theory Related Fields* **123**, 157–201.

Nagy, S. and Foiaş, C. (1970). *Harmonic Analysis of Operators on Hilbert Space* (Amsterdam, North-Holland).

Nakayama, T. (2004a). Support theorem for mild solutions of SDEs in Hilbert spaces, *J. Math. Sci. Univ. Tokyo* **11**, 245–311.

Nakayama, T. (2004b). Viability theorem for SPDEs including HJM framework, *J. Math. Sci. Univ. Tokyo* **11**, 313–324.

Neidhardt, A. L. (1978). Stochastic integrals in 2-uniformly smooth Banach spaces, University of Wisconsin, D. Phil. thesis.

Nemirovskii, A. S. and Semenov, S. M. (1973). On polynomial approximation of functions on Hilbert space, *Mat. Sb.* **21**, 251–277.

Nikodym, O. (1930). Sur une généralisation des intégrales de M. J. Radon, *Fund. Math.* **15**, 131–179.

Nualart, D. (1995). *The Malliavin Calculus and Related Topics* (Berlin, Springer).

Oldham, K. B. and Spanier, J. (1974). *The Fractional Calculus* (New York, Academic Press).

Ondreját, M. (2004a). Existence of a global mild and strong solutions to stochastic hyperbolic equations driven by a spatially homogeneous Wiener process, *J. Evol. Equ.* **4**, 169–191.

Ondreját, M. (2004b). Uniqueness for stochastic evolution equations in Banach spaces, *Dissertationes Math.* **426**, 1–63.

Ondreját, M. (2005). Brownian representations of cylindrical martingales, martingale problem and strong Markov property of weak solutions of SPDEs in Banach spaces, *Czechoslovak Math. J.* **55**, 1003–1039.

Ondreját, M. (2006). Uniqueness for stochastic nonlinear wave equations, preprint.

Oxtoby, J. C. and Ulam, S. (1939). On the existence of a measure invariant under a transformation, *Ann. Math.* **2**, 560–566.

Paley, R. E. A. C. and Wiener, N. (1987). Fourier transforms in the complex domain, *Reprint of the 1934 original American Mathematical Society Colloquium Publications*, Vol. 19 (Providence R.I., American Mathematical Society).

Parthasarathy, K. R. (1967). *Probability Measures on Metric Spaces* (New York, Academic Press).

Pazy, A. (1983). *Semigroups of Linear Operators and Applications to Partial Differential Equations* (Berlin, Springer).

Peszat, S. (1995). Existence and uniqueness of the solution for stochastic equations on Banach spaces, *Stochastics Stochastics Rep.* **55**, 167–193.

Peszat, S. (2001). SPDEs driven by a homogeneous Wiener process, in *Proc. Conf. on SPDEs and Applications (Levico 2000)*, eds. G. Da Prato and L. Tubaro (New York, Marcel Dekker), pp. 417–427.

Peszat, S. (2002). The Cauchy problem for a nonlinear stochastic wave equation in any dimension, *J. Evol. Equ.* **2**, 383–394.

Peszat, S. Rusinek, A. and Zabczyk, J. (2007). On a stochastic partial differential equation of the bond market, submitted.

Peszat. S. and Russo, F. (2007). PDEs with the noise on the boundary, preprint.

Peszat, S. and Seidler, J. (1998). Maximal inequalities and space–time regularity of stochastic convolutions, *Math. Bohem.* **123**, 7–32.

Peszat, S. and Tindel, S. (2007). Stochastic heat and wave equations on a Lie group, submitted.

Peszat, S. and Zabczyk, J. (1997). Stochastic evolution equations with a spatially homogeneous Wiener process, *Stochastic Process. Appl.* **72**, 187–204.

Peszat, S. and Zabczyk, J. (2000). Nonlinear stochastic wave and heat equations, *Probab. Theory Relat. Fields* **116**, 421–443.

Peszat, S. and Zabczyk, J. (2004). *Stochastic Evolution Equations* (Warsaw, ICM), in Polish.

Peszat, S. and Zabczyk, J. (2006). Stochastic heat and wave equations driven by an impulsive noise, in *Stochastic Partial Differential Equations and Applications VII*, eds. G. Da Prato and L. Tubaro, Lecture Notes in Pure and Applied Mathematics, Vol. 245 (Boca Raton, Chapman & Hall/CRC), 229–242.

Priola, E. and Zabczyk, J. (2004). Liouville theorems for non-local operators, *J. Funct. Anal.* **216**, 455–490.

Priola, E. and Zabczyk, J. (2005). Harmonic Functions for Mehler Semigroups, in *Stochastic Partial Differential Equations and Applications VII*, eds. G. Da Prato and L. Tubaro, Lecture Notes in Pure and Applied Mathematics, Vol. 245 (Boca Raton, Chapman & Hall/CRC), pp. 243–256.

Priola, E. and Zabczyk, J. (2006). On bounded solutions to convolution equations, *Proc. Amer. Math. Soc.* **134**, 3275–3286.

Prohorov, Yu. V. (1956). Convergence of random processes and limit theorems in probability, *Theory Probab. Appl.* **1**, 157–214.

Protter, P. (2005). *Stochastic Integration and Differential Equations*, 2nd edition (Berlin, Springer).

Quer-Sardanyons, Ll. and Sanz-Solé, M. (2004a). Absolute continuity of the law of the solution to the 3-dimensional stochastic wave equation, *J. Funct. Anal.* **206**, 1–32.

Quer-Sardanyons, Ll. and Sanz-Solé, M. (2004b). A stochastic wave equation in dimension 3: smoothness of the law, *Bernoulli* **10**, 165–186.

Rogers, L. C. G. and Williams, D. (2000). *Diffusions, Markov Processes and Martingales*, Vols. I, II (Cambridge, Cambridge University Press).

Rosinski, J. (1995). Remarks on strong exponential integrability of vector-valued random series and triangular arrays, *Ann. Probab.* **23**, 464–473.

Rusinek, A. (2006a). Invariant measures for a class of stochastic evolution equations, Preprint IMPAN 667, Warszawa.

Rusinek, A. (2006b). Invariant measures for forward rate HJM model with Lévy noise, Preprint IMPAN 669.

Saint Loubert Bié, E. (1998). Étude d'une EDPS conduite par un bruit poissonien, *Probab. Theory Related Fields* **111**, 287–321.

Saks, S. (1937). *Theory of the Integral*, 2nd edition, Monografie Matematyczne No. 7.

Sanz-Solé, M. (2005). *Malliavin Calculus with Applications to Stochastic Partial Differential Equations*, Fundamental Sciences: Mathematics (EPFL Press, distributed by CRC Press).

Sanz-Solé, M. and Sarrà, M. (2002). Hölder continuity for the stochastic heat equation with spatially correlated noise, in *Stochastic Analysis, Random Fields and Applications*, eds. R. C. Dalang, M. Dozzi and F. Russo, Progress in Probability, Vol. 52 (Basel, Birkhäuser), pp. 259–268.

Sato, K. I. (1999). *Lévy Processes and Infinite Divisible Distributions* (Cambridge, Cambridge University Press).

Seidler, J. and Sobukawa, T. (2003). Exponential integrability of stochastic convolutions, *J. London Math. Soc.* **67**, 245–258.

Sharpe, M. (1988). *General Theory of Markov Processes*, Pure and Applied Mathematics Series (Boston MA, Academic Press).

Sinestrari, E. (1976). Accretive differential operators, *Boll. Un. Mat. It. B* **13**, 19–31.

Slucky, E. E. (1937). Alcuni proposizioni sulla teoria delle funzioni aleatorie, *Giorn. Ist. Ital. Attuari* **8**, 183–199.

Solonnikov, V. A. (1965). On boundary value problems for linear parabolic systems of differential equations of general form, *Trudy Mat. Inst. Steklov* **83**, 3–162, in Russian.

Solonnikov, V. A. (1969). On the Green matrices for parabolic boundary value problems, *Zap. Nauchn. Sem. Leningrad. Otdel. Mat. Inst. Steklov (LOMI)* **14**, 256–287, in Russian.

Sowers, R. (1994). Multi-dimensional reaction–diffusion equations with white noise boundary perturbations, *Ann. Probab.* **22**, 2071–2121.

Srivastava, S. M. (1998). *A Course on Borel Sets* (New York, Springer).

Steinhaus, H. (1922). Les probabilités dénombrables et leur rapport à la théorie de la mesure, *Fund. Math.* **4**, 286–310.

Stolze, S. (2005). Stochastic equations in Hilbert space with Lévy noise and their applications in finance, Diplomarbeit Universität Bielefeld.

Stroock, D. W. (2003). *Markov Processes from K. Itô's Perspective* (Princeton, Princeton University Press).

Stroock, D. V. and Varadhan, S. R. S. (1979). *Multidimensional Diffusion Processes* (Berlin, Springer).

Tanabe, H. (1979). *Equations of Evolution* (London, Pitman).

Tehranchi, M. (2005). A note on invariant measures for HJM models, *Finance Stochastics* **9**, 387–398.

Teichmann, J. (2005). Stochastic evolution equations in infinite dimension with applications to term structure problem, *Notes from Lectures at CREST (Paris 2003), RTN Workshop (Roscoff 2003), MPI Leipzig (Leipzig 2005) and RICAM (Linz 2005)*.

Tessitore, G. and Zabczyk, J. (1998a). Invariant measures for stochastic heat equations, *Probab. Math. Statist.* **18**, 271–287.

Tessitore, G. and Zabczyk, J. (1998b). Strict positivity for stochastic heat equations, *Stochastic Process. Appl.* **77**, 83–98.

Tessitore, G. and Zabczyk, J. (2001a). Wong–Zakai approximations of stochastic evolution equations, Warwick Preprint 9, University of Warwick.

Tessitore, G. and Zabczyk, J. (2001b). Trotter's formula for transition semigroups, *Semigroup Forum* **63**, 114–126.

Tessitore, G. and Zabczyk, J. (2006). Wong–Zakai approximations of stochastic evolution equations, *J. Evol. Equ.* **6**, 621–655.

Tortrat, A. (1967). Structure des lois indéfiniment divisibles ($\mu \in I(X)$) dans un espace vectoriel topologique (séparé) X, in *Proc. Symposium on Probability Methods in Analysis*, Lecture Notes in Mathematics, Vol. 31, pp. 299–328.

Tortrat, A. (1969). Sur la structure des lois indéfiniment divisibles (classe $I(X)$) dans les espaces vectoriels X (sur le corps réel), *Z. Wahrscheinlichkeitsteorie und verw. Gebiete* **11**, 311–326.

Triebel, H. (1978). *Interpolation Theory, Function Spaces, Differential Operators* (Berlin, Deutscher Verlag der Wissenschaften).

Truman, A. and Wu, J. L. (2005). Fractional Burgers' equations driven by Lévy noise, in *Stochastic Partial Differential Equations and Applications VII*, eds. G. Da Prato and L. Tubaro, Lecture Notes in Pure and Applied Mathematics, Vol. 245 (Boca Raton, Chapman & Hall/CRC), pp. 295–310.

Twardowska, K. and Zabczyk, J. (2004). A note on stochastic Burgers' system of equations, *Stoch. Anal. Appl.* **22**, 1641–1670.

Twardowska, K. and Zabczyk, J. (2006). Qualitative properties of solutions to stochastic Burgers's system of equations, in *Stochastic Partial Differential Equations and Applications VII*, eds. G. Da Prato and L. Tubaro, Lecture Notes in Pure and Applied Mathematics, Vol. 245 (Boca Raton, Chapman & Hall/CRC), pp. 311–322.

Ustunel, A. S. (1984). Additive processes on nuclear spaces, *Ann. Probab.* **12**, 858–868.

Wagner, D. H. (1977). Survey of measurable selection theorems, *SIAM J. Control Optim.* **15**, 859–903.

Walsh, J. B. (1986). An introduction to stochastic partial differential equations, in *École d'été de probabilités de Saint-Flour XIV (1984)*, Lecture Notes in Mathematics, Vol. 1180 (Berlin, New York, Springer), pp. 265–439.

Wiener, N. (1923). Differential space, *J. Math. Phys., Math. Inst. Tech.* **2**, 131–174.

Yosida, K. (1965). *Functional Analysis*, Grundlehren Math. Wiss., Vol. 123 (Berlin, Springer).

Zabczyk, J. (1976). An introduction to probability theory, in *Control Theory and Topics in Functional Analysis*, Vol. 1 (Vienna, International Atomic Energy Agency), pp. 419–462,

Zabczyk, J. (1983). Stationary distributions for linear equations driven by general noise, *Bull. Polish Acad. Sci.* **31**, 197–209.

Zabczyk, J. (1889). Symmetric solutions of semi-linear stochastic equations, in *Stochastic Partial Differential Equations and Applications II (Trento, 1988)*, Lecture Notes in Mathematics, Vol. 1390 (Berlin, Springer), pp. 237–256.

Zabczyk, J. (1993). The fractional calculus and stochastic evolution equations, in *Proc. Barcelona Seminar on Stochastic Analysis (St Feliu de Guixol 1991)*, (Basel, Boston, Berlin, Birkhäuser), pp. 222–234.

Zabczyk, J. (1995). *Mathematical Control Thory: An Introduction* (Boston, Birkhäuser).

Zabczyk, J. (1996). *Chance and Decision*, Quaderni (Pisa, Scuola Normale Superiore).

Zabczyk, J. (1999). Parabolic equations on Hilbert spaces, Lecture Notes in Mathematics, Vol. 1715, pp. 117–213.

Zabczyk, J. (2000). Stochastic invariance and consistency of financial models, *Atti Accad. Naz. Cl. Sci. Fis. Mat. Natur. Rend. Lincei (9) Mat. Appl.* **11**, 67–80.

Zabczyk, J. (2001a). Bellman's inclusions and excessive measures, *Probab. Math. Statist.* **21**, 101–122.

Zabczyk, J. (2001b). Leverhulme lectures, Warwick Preprint 8, University of Warwick.

Zabczyk, J. (2003). *Topics in Stochastic Processes*, Quaderni (Pisa, Scuola Normale Superiore).

Index

adapted process, 21
adjoint operator, 355
α-stable families, 61
angle bracket, 35, 107
arbitrage opportunity, 327
Aronson
    estimate, 18
    theorem, 18

Bochner
    integral, 24
    theorem, 72
Borel
    isomorphism theorem, 4
    space, 5
boundary conditions
    Dirichlet, 220
    Neumann, 221
bounded range, 301
Brace–Gątarek–Musiela (BGM) equation,
    347, 349
Brownian sheet, 72, 100
Burgers
    equation, 317
    system, 312
Burkholder–Davies–Gundy inequality,
    37

C_0-semigroup, **75**, **140**, 365–367
C_0-generator, **75**, **140**
càdlàg, definition, 26
Carathéodory theorem, 23
Cauchy family, symmetric, 61
Chapman–Kolmogorov equation, 7
characteristics, 56, 299
compact operator, 355

composition operator, 16
compound Poisson process, 9
configuration, 308
consistency, 350
consistency conditions, 22
consistency problem, 350
continuous part, 391
contraction, 6
convolution semigroup of
    measures, 39
core, 149
Courrège's theorem, 8
covariance, 29, 248
    form, 31
    operator, 30, 67
cylindrical process
    impulsive, 103
    Wiener, 97

De Acosta theorem, 39
dilation theorem, 160
Dirichlet boundary conditions,
    220
discounted
    capital, 326
    price, 326
dissipative mapping, 180
distribution, 21
    exponential, 41
    finite-dimensional, 22
    gamma, 191
    geometric, 41
    Poisson, 41
Dobrushin–Lanford–Ruelle (DLR) equation,
    308
Doléans measure, 115

415

Doob
  optional sampling theorem, 32
  regularity theorem, 33
  submartingale inequality, 32–34
Doob–Meyer decomposition, 35
Doob–Meyer theorem, 35
Douglas theorem, 364
dynamical system
  deterministic
    continuous-time, 5
    discrete-time, 3
  stochastic
    continuous-time, 6
    discrete-time, 3
Dynkin $\pi-\lambda$ theorem, 20

equation
  BGM, 347, 349
    classical, 349
  Burgers, 317
  Chapman–Kolmogorov, 7
  DLR, 308
  Focker–Planck, 306
  heat, 374
  HJMM, 333
  stochastic
    delay, 238
    difference, 4
    heat, 220, 221, 255, 261, 279, 283, 284
    wave, 226, 228, 255
  transport, 374
  wave, 375
Euler Γ-function, 190
evolution operator, 156
excessive measure, 376
expectation, 24
  conditional, 25
exponential distribution, 41
exponential mixing, 288
exponentially tempered one-sided stable
    coordinates, 64

factorization, 190
family, α-stable, 61
Feller property, family, 170
Fernique theorem, 30
filtered probability space, 21
filtration, 21
  generated by X, 22
  right-continuous, 21
  satisfying usual assumption, 22
Fortet–Mourier norm, 288

forward curve, 333
forward rate function, 324
Fourier transform, 231
Fubini stochastic theorem, 119
function
  non-negative-definite, 71, 360
  symmetric, 71

Gaussian
  measure, 28–31
  random process, 28–31
  random variable, 28–31
    centered (zero-mean), 29
generator
  of arbitrary Lévy process, 80
  of compound Poisson, 78
  of Wiener process, 79
geometric distribution, 41
Gibbs
  density, 308
  measure, 308, 309
gradient equation, 307
Green function, 17–19

Haar system, 45
Hamiltonian, 308
Heath–Jarrow–Morton–Musiela (HJMM)
    equation, 333
Hilbert–Schmidt
  norm, 356
  operator, 356
Hille–Yosida theorem, 365
HJM
  condition, 329
    classical, 329
  postulate, 327
Hölder spaces, 13–14

independent identically distributed (i.i.d.)
    random variables, 4
infinite divisible family, *see* convolution
    semigroup of measures
integrated volatility, 348
invariance problem, 351
invariant measure, 287
Ising model, 308
Itô formulae, 391–393
Itô–Nisio theorem, 23

jump intensity measure, *see* Lévy measure
jump position intensity, 103
jump size intensity, 103

Kinney theorem, 27
Kolmogorov
  existence theorem, 22
  formula, 306–307
  test, 26
Kolmogorov–Loeve–Chentsov theorem, 26
Kotelenez inequality, 156
Kruglov theorem, 39
Kuratowski–Ryll-Nardzewski theorem,
    114, 115

Lévy
  measure, 11, 45, 56
  semigroup, 8, 75–82
  sheet, 100, 103
Lévy process
  double-sided, 290
Lévy–Ciesielski construction, 45
Lévy–Khinchin
  decomposition, 53
  formula, 56
Lévy–Ottaviani inequality, 23
Laplace operator, 220, 221, 230
  discrete, 302
  fractional, 223
Laplace transform, 59
Laplacian, *see* Laplace operator,
    302
law, *see* distribution
LIBOR rates, 348
local martingale, 28
Lumer–Phillips theorem, 366

m-dissipative mapping, 180
mapping
  almost m-dissipative, 180
  dissipative, 180
  m-dissipative, 180
  maximal dissipative, *see* m-dissipative
Markov
  chain, 4
  property, 167
Markov time, *see* stopping time
martingale, 28
martingale covariance, 107, **111**
maximal dissipative mapping, 180
mean, 67
mean-reversion property, 350
measure
  equivalent martingale, 327
  Gibbs, 308, 309
  ω-excessive, 376

modification, 21
Musiela parametrization, 332

Nagy theorem, *see* dilation theorem
Nelson–Siegel family, 350
  augmented, 350
  degenerate, 350
Nemytskii operator, *see* composition
    operator
Neumann boundary conditions, 221
norm
  Fortet–Mourier, 288
  graph, 149, 196
  Hilbert–Schmidt, 356
  nuclear, 356
  operator, 355
normalizing constant, 308
nuclear space, 241

ω-excessive measure, 376
operator
  adjoint, 355, 366
  compact, 355
  convolution, 191, 248
  fractional power of, 368
  Hilbert–Schmidt, 356
  Laplace, 230
    Dirichlet boundary conditions, 220
    Neumann boundary conditions, 221
  Liouville–Riemann, 190, 193
  negative-definite, 366
  non-negative-definite, 355
  norm, 355
  nuclear, 356
  self-adjoint, 355, 366
orthonormal basis, 356

Parseval identity, 355
path, 3
π and λ systems, 20
Poincaré inequality, 14
Poisson
  process, 40–44
  random measure, 47, **83**
    compensated, 47, **83**
    stationary, 84
    stochastic integral, 85–87
Poisson distribution, 41
polarization, 37
Polish space, 4
portfolio, 325
positive-definite mapping, 160

potential, 308
  of local interactions, 299, 310
predictable
  modification, 22
  process, 21
predictable variation process, *see* angle bracket
process
  adapted, 21
  càdlàg, 26
  Cauchy, 11
  compensated, 48
  compound Poisson, 9, 45
  continuous, 26
  continuous in probability, 26
  E-valued, 21
  general Markov, 27
  Lévy, 38
    in $S'(\mathbb{R}^d)$, 241
    Ornstein–Uhlenbeck, 183
    spatially homogeneous, stationary, 247
  Markov, 27
  mean-square continuous, 26
  measurable, 21
  of finite variation, 28
  of jumps, 39
  of mean zero, 25
  predictable, 21
  right-continuous, 26
  square integrable, 25
  stochastically continuous, 26
  Wiener
    in $S'(\mathbb{R}^d)$, 242
    in Hilbert space, 50
    with independent increments, 38

quadratic variation, 36

Radon measure, 105
radonifying
  norm, 135
  operator, 135
random element, *see* random variable
random variable, 21
  integrable, 23
  of mean zero, 24
  square integrable, 24
reaction–diffusion equation, 187
renormalization, 307
reproducing kernel Hilbert space (RKHS),
  91–95
resolvent, 365

self-financing strategy, 326
semigroup
  analytic, 368
  Feller, 170
  of contractions, 156, 181
  property, 5
  spatially homogeneous, 76
  translation-invariant, 76
semimartingale, 28, 36
shift semigroup, 333
Skorokhod embedding theorem, 5
Sobolev spaces, 13–14
solution
  fundamental, 232
  generalized, 139
  mild, 140, **142**, 183, 314
  weak, 141, **149**, 313
space, real-interpolation, 277
spatial correlation, 70
spectral measure, 248
stable families
  of order $\beta \in (0, 1)$, 60
  of order $\beta \in (1, 2)$, 61
state space, 3
stochastic process, 21
stochastic convolution, 155
stopping time, 22
subdifferential, 179
submartingale, 31
subordinators, 59
supermartingale, 31
supremum norm, 13
symmetric Cauchy family, 61
system of Dirichlet conditions, 17

tempered distributions, 240
theorem
  Aronson, 18
  Bochner, 72
  Borel isomorphism, 4
  Carathéodory, 23
  Courrège, 8
  De Acosta, 39
  Delbaen–Schachermayer, 327
  dilation, 160
  Doob regularity, 33
  Doob–Meyer, 35
  Douglas, 364
  Dynkin π–λ, 20
  Fernique, 30
  Frobenius–Perron, 302

Hille–Yosida, 365
Itô–Nisio, 23
Kinney, 27
Kolmogorov, 22
Kruglov, 39
Krylov–Bogolyubov, 288
Kuratowski–Ryll-Nardzewski, 114
Lumer–Phillips, 366
Skorokhod embedding, 5
Tauberian, 315
Tortrat, 74
Ulam, 23
Tortrat theorem, 74
trace, 357
trace operator, 8
trajectory, 3, 5
transition
  function, 3
  probability, 3, 6, 10
  semigroup, 7
translation operator, 84
translations on $S'(\mathbb{R}^d)$, 246
two-sided exponentially tempered stable
  coordinates, 64

Ulam theorem, 23
uniform integrable sequence, 25
uniform motion, 78
usual conditions, 22

variation-of-constants formula, 140
volatility, 333

Wong–Zakai correction term, 351
weighted spaces, 15
weights
  exponential, 15
  polynomial, 15
white noise
  discrete-time, 4
  Gaussian space–time, 98
  impulsive, 100–105
Wiener
  process, 44
    cylindrical, 97
    standard, 44
    with respect to filtration, 44

Yosida approximation, 180, 184, 365